SEA–LEVEL CHANGES: AN INTEGRATED APPROACH

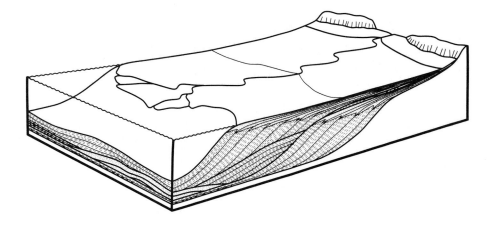

Edited by

Cheryl K. Wilgus
Bruce S. Hastings
Everest Geotech
Houston, Texas 77074

Henry Posamentier
John Van Wagoner
Exxon Production Research
Houston, Texas 77252-2189

Charles A. Ross
Chevron U.S.A., Inc.
Houston, Texas 77010

Christopher G. St. C. Kendall
University of South Carolina
Columbia, South Carolina 29208

SOCIETY OF ECONOMIC PALEONTOLOGISTS AND MINERALOGISTS

Special Publication No. 42
Barbara H. Lidz, Editor of Special Publications

Tulsa, Oklahoma, U.S.A.   *September 1988*

This book is published with funds from the SEPM Foundation Inc., including the Bruce H. Harlton Publications Fund made possible by Allan P. Bennison. Illustrations were supported by a major contribution from Exxon Production Research Company, and additional contributions from Canadian Hunter Exploration Ltd. and ARCO Oil & Gas Company.

ISBN # 0-918985-74-9

Printed in the United States of America

CONTENTS

SUBJECT INDEX

PREFACE

In October 1985, SEPM sponsored a four-day conference entitled "Sea-Level Changes—An Integrated Approach." The conference was organized by Everest Geotech, Inc., and hosted by Transco Exploration Company in Houston, Texas. Co-conveners of the conference were Dr. Cheryl Wilgus of Everest Geotech, Inc., Dr. Walter C. Pitman of Lamont-Dougherty Oceanographic Institute, and Dr. Christopher G. St. C. Kendall of the University of South Carolina. The purpose of the conference was to provide a forum for an interdisciplinary exchange of ideas on sea-level changes and to provide an opportunity for integrating various types of evidence in approaching unresolved issues.

The conference was successful in bringing together scientists from industry, academia, and government, representing all of the major geoscience disciplines. Presentations of many new papers, plus significant releases of data that were previously held proprietary, provided fertile ground for discussion in the workshop environment of the conference. This volume represents the best of the material presented at the conference, plus some additional papers on sea-level changes that were subsequently released.

The editors appreciate the patience of the authors in awaiting the publication of this volume, but we believe that the wait was more than justified by the quality and significance of the final product. If this publication even begins to generate the kinds of ideas, interdisciplinary discussions, and inquiries that arose during the SEPM conference, we will consider its purpose to be accomplished.

BRUCE H. HARLTON (1890–1983)

Bruce Harlton was of noble European birth, and he fought heroically in the skies over Europe during World War I. He was gravely wounded in an airplane crash, which affected his speech and arm muscles for the rest of his long, productive life. In spite of this handicap, he persevered in acquiring the best geological education possible in the leading universities of Europe and America. Bruce was graduated with high honors from Columbia University and, years later, he was awarded an honorary doctorate from that institution. He collaborated with Dr. J. J. Galloway, the famous micropaleontologist, on two outstanding papers on foraminifera, published in the Journal of Paleontology. One of his classmates at Columbia was the late Marshall Kay, well known for his innovative studies of geosynclines.

During his college days in New York, Bruce became devoted to the Ziegfield Follies and was proud of the fact that he dated many of the beautiful performers; however, his great romantic attachment was to Lucille Hardy, who consented to become his wife and whom he dearly loved all the rest of his life. Her death was a shock to him from which he never fully recovered.

Bruce's first professional employment was as a geologist for the Aguila Oil Company in the jungles of eastern Mexico in 1922. Later, he joined the geological staff of the Amerada Petroleum Corporation in Tulsa and was instrumental in that company's growth and its acquisition of large reserves of oil and gas. Southern Oklahoma became his Mecca, and Bruce contributed much to our understanding of that area by publishing numerous descriptions of its minerals and new fossils, naming new mappable formations and members, exploring its complex tectonic history, drawing numerous geologic maps and cross sections, and leading field trips into the area. Up to the time of his last illness, he was laboriously preparing a comprehensive manuscript on the tectonics of the Arkoma Basin. This would have been his fifty-seventh published article.

Bruce also had time for less scientific pursuits, such as polo, flying, gardening, and stamp collecting. In all of these avocations, he exhibited the proud stamp of excellence that characterized all of his endeavors. He was a long-time, active member of Trinity Episcopal Church in Tulsa.

After his retirement from Amerada Petroleum Corporation, he formed the Harlton Exploration Company and established an office in the Beacon Building in downtown Tulsa. It was at that time that I formed a friendship of many years with Bruce, owing to our many common interests.

Bruce became a charter member of the Society of Economic Paleontologists and Mineralogists in 1927. He contributed the first article in the first issue of the fledgling *Journal of Paleontology*. Inasmuch as he was proud of having 56 articles published on many phases of the geosciences, nothing would have pleased him more than a publication fund established in his name. He set a high standard for all of us to emulate.

Allan P. Bennison

PART I
ANALYSIS OF SEA LEVEL CHANGES

THE RISE AND FALL OF EUSTASY

CHRISTOPHER G. ST. C. KENDALL AND IAN LERCHE

Department of Geology, University of South Carolina, Columbia, South Carolina 29208

ABSTRACT: Techniques that can be used to determine the relative magnitude of eustatic excursions include the measurement of: (a) the amount of sedimentary onlap onto the continental margins; (b) the thickness of marine sedimentary cycles and the elevation and distance between indicators of old strandlines; (c) the perturbations on individual thermo-tectonic subsidence curves and stacked crustal subsidence curves; (d) the variations in deep-ocean oxygen isotopes found in sediments; and (e) the size of variables, such as rates of tectonic movement, sediment accumulation, and eustatic changes, used in graphical and numerical simulations of basin fill that "invert" the problem. To date, a combination of some or all of these methods can be used to construct relative (tectono/eustatic) sea-level curves; however, these are not unique solutions to absolute eustatic variations. Each method *assumes* some behavior for two of the three underlying processes (tectonic movement of the basement, sedimentary accumulation, and eustasy), and then determines the third process relative to the assumed model behavior of the other two. The sense of this result is confirmed by mathematical models which suggest that only the *sum* of tectonic basement subsidence and sea-level variations can be obtained.

INTRODUCTION

Although several indirect methods exist for determining the apparent vertical amplitude of eustatic change, no direct method has been derived, because there is no stationary datum from which these changes can be measured. This datum cannot be established (Burton and others, 1987), because the earth's surface has a history of constantly moving in response to: (1) sediment compaction; (2) the isostatic response of the crust to varying loads of the sedimentary and water columns that rest upon it; and (3) thermo-tectonic movement. Such a stationary datum might be the center of the earth, but since its position with respect to ancient sediments cannot be determined, only "relative" sea level can be observed. As Vella (1961) pointed out, apparent changes in the height of sea level can be expressed only as the height between pairs of four kinds of vertical reference point, none of which is fixed and all of which are subject to vertical movement. These are: (1) any point at mean sea level; (2) any point fixed relative to the lithosphere; (3) any point on the changing surface of the lithosphere; and (4) any point on the near compaction surface slightly below any depositional surface. Because no direct measurement can be made, geologists are forced to construct models that use measurements of the physical changes produced by movements in sea level. These indirect measurements, which are related to the magnitude of eustatic change, are the subject of this paper. The reader should note that the definition we use in this paper for eustasy is "a change in elevation in sea level on a world-wide basis relative to a stationary datum like the center of the earth." This definition may differ from the definitions of other geologists cited in this paper.

Our paper considers four methods that are used to describe the earth's physical response to eustasy: (1) the changing area of the continents onlapped by the sea; (2) the marine-sediment depositional record; (3) the crustal response to the weight of the onlapping sea water; and (4) the changing volume of ocean water. We now review these different methods.

The Use of the Changing Area of Continent Covered by Marine Sediments to Determine the Amplitude of Sea-Level Excursion

Hypsometric curves.—

Kossinna (1921, 1933) compiled hypsometric curves for the present topography of the continents. His curves, and those of scientists who use his modified concepts, have been the basis for many estimates of the magnitude of eustatic excursions. The hypsometric curve for today's topography is one tool for estimating how much the sea level rose in proportion to the area of the continent covered by the sea. The amount by which the sea advanced across the continent during any particular time interval can be derived from paleogcographic maps of the marine sedimentary sequences for that time period by using a planimeter and an equal area projection (Fig. 1). A number of recent papers (Eyged, 1956; Hallam, 1963, 1984; Forney, 1975; Bond, 1976, 1978a,b; Cogley, 1981, 1984; Harrison and others, 1981; and Wyatt, 1984) describe the use of the hypsometric curve (or the hypsographic curve) in the determination of the size of sea-level excursions.

The hypsometric curve expresses the area of land between pairs of contour lines as a percentage of the total land area. For instance, Harrison and others (1981) made their plots using fractional area versus fractional height and included the continental shelf area. They normalized the curves by dividing the observed height by the observed average height, which is expressed as an ordinate on their plot: the average height is therefore unity, and the area is a percentage of the total area of the continent (Fig. 1).

The problem with this method of determining the magnitude of the sea-level rise or fall is the assumption that the hypsometric curve describing a continent today is the same as that which existed in the past. The way this simple concept is applied, however, varies from scientist to scientist. Curves for present topography (e.g., Kossinna, 1921, 1933) may be too steep because of epeirogenic uplift in the Tertiary (Bond, 1976). According to Harrison and others (1981),

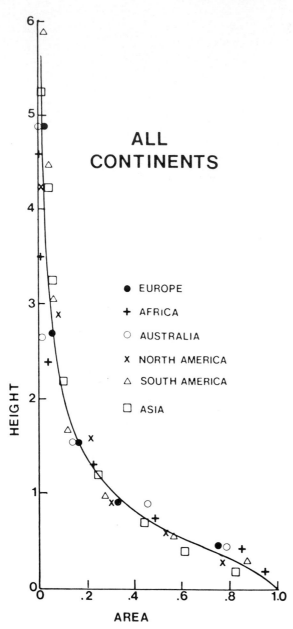

FIG. 1.—Normalized hypsometric curve, after Harrison and others (1981, p. 6): "The continental heights are normalized by dividing observed height by average height. The average height of each continent would therefore be plotted as unity on the ordinate scale. Area is unity on the abscissa scale. The best fitting hypsographic curve through these points is also shown."

a consequent underestimation of the areas of continent flooded back through time (Cogley, 1981).

"Modern area–altitude distribution provides important clues to that of ancient continent" (Cogley, 1984, p. 116) and can be used in modeling to determine the magnitude of sea-level excursions. There is no unequivocal way of deriving the "correct" hypsometric curve, however. One can use those of Kossinna (1921, 1933), Bond (1976), Harrison and others (1981), Southham and Whitman (1981), or Cogley (1984), or one's own system. Loss of section by erosion also complicates matters and results in underestimations of flooded areas. Thus, it is clear that the inaccuracies of paleogeographic mapping alone will result in inexact sea-level relationships, although some useful relative curves may result.

Sediment aggradation and onlapping geometries from seismic reflection data.—

Vail and others (1977), Hardenbol and others (1981), and Vail and others (1984) have developed a technique that is an extension of work by Wheeler (1958), Sloss (1963, 1972) and Sloss and Speed (1974). Seismic sequences on seismic cross sections are identified following the assumption that continuous seismic reflectors are close matches to chronostratigraphic surfaces, or to time boundaries such as bedding planes and unconformities. The unconformities that bound the sequences are marked by seismic reflectors onlapping and terminating against either the lower unconformity surface or each other (Fig. 2). The argument is that the position of the onlapping seismic reflectors is controlled by the base level of the mean high-water mark. Thus, a sediment encroachment chart will show how far the wedge of submarine, coastal, and alluvial sediment has onlapped the basin margin (Vail and others, 1984). A sediment aggradation chart will show the vertical amount that onlapped seismic reflectors have climbed or fallen (Fig. 2c; Vail and others, 1977, p. 66–67). Vail and others (1977, p. 77) pointed out that in using this method, "the measurements of coastal aggradation are made as closely as possible to the underlying unconformity to minimize the effect of differential basinal subsidence." They correlated the cycles of relative changes of sea level measured at many locations and constructed charts that incorporate the occurrence of global seismic-onlap cycles (which they called coastal onlap). Using the aggradational measurements from seismic data, Vail and others (1977) estimated the magnitude of the relative sea-level excursions. As Hardenbol and others (1981, p. 34) pointed out, however, "Quantifying eustatic sea-level changes from measured changes in coastal onlap does not provide an accurate measure, because of variations in subsidence in different basins." Presumably, they assume that once subsidence is factored out through geohistory analysis, only eustasy remains (in theory). It should be remembered that this is an assumption rather than fact.

Another problem with the sediment onlap curves of Vail and others (1977) is that, while they convincingly demonstrated the existence of eustatic events, the position that a eustatic event had on the continent is complicated by the local effects of tectonic subsidence (Bally, 1981; Watts, 1982;

the hypsometry of a continent is unlikely to be constant through time, because of: (1) changes related to the sediment fill of the continental margins; (2) the effects of fold-mountain generation; and (3) problems in constructing hypsometric curves, such as not including the continental shelves in the continental area covered by the hypsometric curve. In addition, the paleogeographic maps that are used in conjunction with the hypsometric curve are often in error, particularly when an older geologic system is mapped. There is a bias toward drawing shorter, simpler shorelines, with

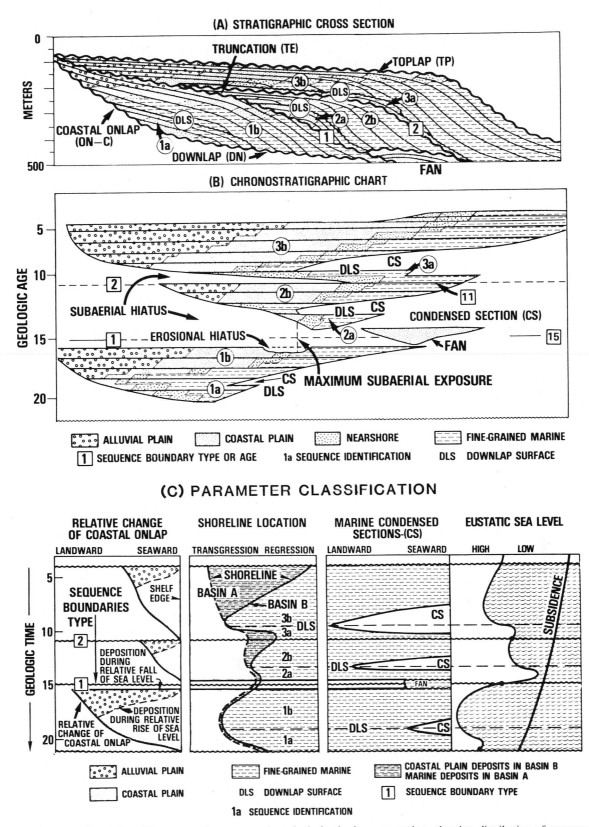

(A) STRATIGRAPHIC CROSS SECTION

TRUNCATION (TE)

TOPLAP (TP)

COASTAL ONLAP (ON—C)

DOWNLAP (DN)

FAN

METERS

(B) CHRONOSTRATIGRAPHIC CHART

DLS   CS

DOWNLAP DLS

CONDENSED SECTION (CS)

SUBAERIAL HIATUS

EROSIONAL HIATUS

DLS   CS

FAN

GEOLOGIC AGE

MAXIMUM SUBAERIAL EXPOSURE

CS

DLS

ALLUVIAL PLAIN   COASTAL PLAIN   NEARSHORE   FINE-GRAINED MARINE

1 SEQUENCE BOUNDARY TYPE OR AGE   1a SEQUENCE IDENTIFICATION   DLS DOWNLAP SURFACE

(C) PARAMETER CLASSIFICATION

RELATIVE CHANGE OF COASTAL ONLAP

SHORELINE LOCATION

MARINE CONDENSED SECTIONS-(CS)

EUSTATIC SEA LEVEL

LANDWARD   SEAWARD

TRANSGRESSION REGRESSION

LANDWARD   SEAWARD

HIGH   LOW

SEQUENCE BOUNDARIES TYPE

SHELF EDGE

SHORELINE

BASIN A

BASIN B

DLS

CS

SUBSIDENCE

GEOLOGIC TIME

DEPOSITION DURING RELATIVE FALL OF SEA LEVEL

DLS   CS

FAN

RELATIVE CHANGE OF COASTAL ONLAP

DEPOSITION DURING RELATIVE RISE OF SEA LEVEL

DLS   CS

ALLUVIAL PLAIN   FINE-GRAINED MARINE   COASTAL PLAIN DEPOSITS IN BASIN B MARINE DEPOSITS IN BASIN A

COASTAL PLAIN   DLS DOWNLAP SURFACE   1 SEQUENCE BOUNDARY TYPE

1a SEQUENCE IDENTIFICATION

FIG. 2.—(A) Diagrammatic stratigraphic cross section across a hypothetical seismic cross section, showing distribution of sequence boundary types, downlap surfaces (condensed sections), and facies of three idealized sequences in depth. (B) Chronostratigraphic chart or Wheeler diagram (Wheeler, 1958; Sloss, 1984) identified in A, showing nature of sequence boundaries, downlap surfaces, condensed sections, and facies. (C) Parameter classification chart for seismic sequences identified above, showing relation of coastal onlap, relative changes of sea level, transgression-regression, marine condensed sections, and eustatic sea-level changes and subsidence (Vail and others, 1984).

Thorne and Watts, 1984; Parkinson and Summerhayes, 1985; and Miall, 1986). This may explain why sea-level curves for the Jurassic compiled by Hallam (1981) and Vail and Todd (1981) from different data sources record different positions for the same sea-level stands. Similarly, sea-level curves for the Cretaceous compiled by Vail and others (1977), Kauffman (1977), Hancock and Kauffman (1979), Harris and others (1984), and Seiglie and Baker (1984) from different data sources (e.g., seismic, lithostratigraphic, and biostratigraphic data) are also different. Sea-level curves for the Late Tertiary compiled by Vail and others (1977) differ from those of Seiglie and Moussa (1984). Thus, these methods (which are used differently by different authors) recognize the occurrence of the same eustatic events but cannot be used to determine either their magnitude or their position with respect to other sea-level events.

The Sedimentary Record as a Means of Determining the Magnitude of Variations in Eustasy

Paleobathymetric markers tied to old strandline positions are used to estimate the magnitude of excursions in eustasy. These markers include sedimentary structures that indicate the position of the high-water mark in shoaling cycles; old beach lines; notches in old cliffs; and fossil indicators of paleobathymetry, such as benthic organisms, algal stromatolites, burrows, coral reef terraces, and peats. For example, Busch (1983) correlated shoaling-upward cycles or PACs (Punctuated Aggradation Cycles) in the Manlius Formation of central New York and related them to sea-level events. He correlated high sea-level stands in cycles, identifying the highstands by using the upper limit of vertical burrows or the lower limit of algal laminites within the cycles. Busch identified three cycles bound by these sea-level surfaces that have approximately the same thickness (1.1, 1.0, and 0.8 m or 3.7, 3.3, and 2.5 ft) at several localities, suggesting that relative sea level had changed by this amount at these locations (Fig. 3). Busch ignored the effect of compaction, so these values may in fact record less than the actual magnitude of relative sea-level change. He noted that the transgressive surface climbs stratigraphically in a seaward direction. This system of estimating the magnitude of sea-level excursions is dependent on the *assumptions* that the cycles are a result of eustatic cause, that each of these localities has the same tectonic history, and that this tectonic effect can be accurately modeled.

Other papers by Beukes (1977), Kauffman (1977), McKerrow (1979), Harris and others (1984), Seiglie and Baker (1984), Sieglie and Moussa (1984), and Weimer (1984), take similar approaches and, while recognizing a eustatic signal, do not neglect the importance of local tectonism in producing the accommodation for the sedimentary section.

The general philosophy of the use of reef terraces, peats, cliff notches, and paleobathymetric markers is the same as in the papers just cited. They all require assumptions either about the tectonic behavior of the depositional setting or about eustasy or sedimentation. Once again, the size of the sea-level excursion is dependent on an assumed model that cannot be proven independently.

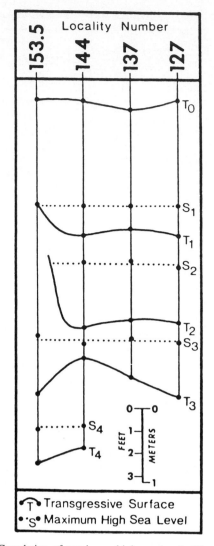

FIG. 3.—Correlation of maximum high sea levels (S) and transgressive surfaces (T) among four study locations from the Manlius Formation of central New York (Busch, 1983).

The Use of Backstripped Subsidence to Determine the Magnitude of Excursions of Eustasy

Differences between crustal subsidence and thermotectonic curves.—

Hardenbol and others (1981, p. 35) noted that: (1) the eustatic curves of Vail and Hardenbol (1979) and Vail and Todd (1981) are based on estimates of "changes in coastal onlap and from paleontologic studies"; and (2) the magnitude of the low-frequency eustatic events can be assessed by measuring the difference between crustal subsidence curves calculated from wells and the theoretical thermotectonic subsidence curves postulated for that location using the approach outlined by Royden and others (1980). Similarly, Hallam (1963) proposed the possibility of measuring sea-level excursions on the basis of assumed rates of subsidence of Pacific guyots.

Crustal-subsidence curves are obtained for a well first by

determining its burial path from the datum of present sea level (Fig. 4, the geohistory plot of Van Hinte, 1978). Next, using paleobathymetry as a datum, the effects of compaction and the isostatic effect of sediment and water weight on the crust are removed following Watts and Steckler (1979), who used Airy's (1855) concept of isostacy as modified by Bomford (1971). Thus, basement subsidence is estimated by compensating for the influence of sedimentation but ignoring the effect of sea-level fluctuations on this subsidence (Fig. 5).

Results obtained from calculation of the burial history are illustrated in Figure 6, which shows the reconstructed depositional history and basement subsidence curve of a sample well from the North Sea. The basement subsidence curves are then compared to the thermo-tectonic curves, and the difference between them is used to determine the size of the sea-level excursion. The hypothetical thermo-tectonic curves differ from author to author, however. Models have been developed by many geologists, including Falvey (1974), McKenzie (1978), Royden and others (1980), Beaumont (1981), Hellinger and Sclater (1983), and Nunn and others (1984). Each model is different and was developed to find ways around weaknesses in other models. If we are to measure the difference between crustal subsidence derived from a well and thermo-tectonic subsidence, we must choose one model from among many widely different models. Once we make our choice, the difference between the crustal-subsidence curve derived from a well and the arbitrarily chosen thermo-tectonic subsidence curve is given by a least-square

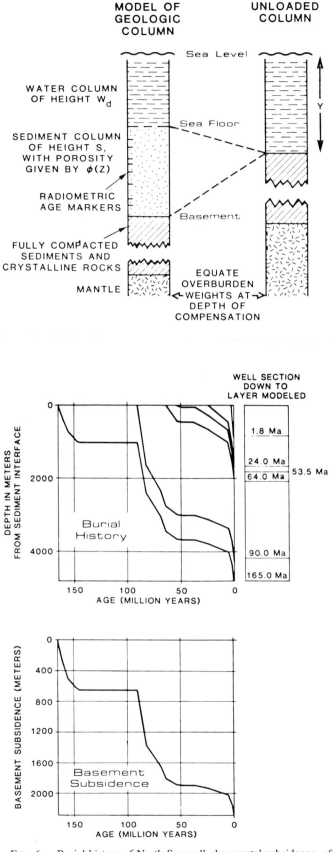

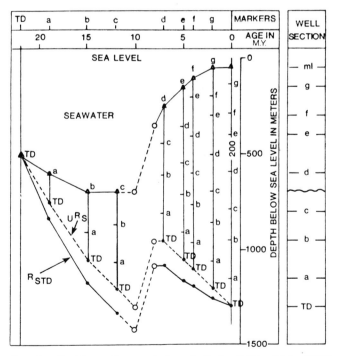

Fig. 4.—Geohistory diagram for hypothetical well (Van Hinte, 1978). Upper curve shows paleo-water depth at this locality. The dashed lower curve $_uR_s$ is the uncorrected burial path of TD through time, measured cumulatively from the water/sediment interface; the solid line lower curve R_{STD} is the corrected burial path of TD, which incorporates the progressive compaction of the overlying sediment as TD is buried.

Fig. 6.—Burial history of North Sea well plus crustal subsidence, after Guidish and others (1984).

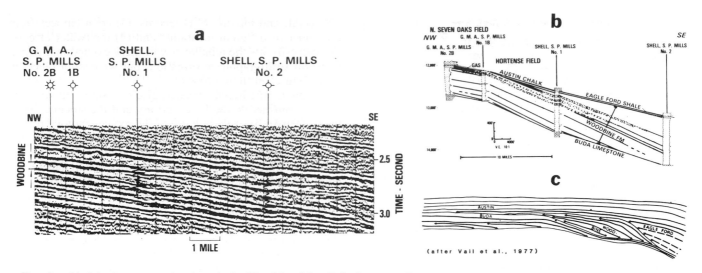

FIG. 9.—(a) Seismic cross section through the Woodbine delta, Polk Company, Texas: (b) Electric log cross section showing distribution and geometry of sandstone beds in Woodbine delta, Polk Company, Texas. (c) Diagrammatic interpretation of a and b, after Vail and others (1977).

the physical processes: rather, we attempt to reproduce the sedimentary geometries, which are the results of such processes averaged over long periods of time.

We simulate deposition over fixed time intervals that we define as sediment triangles of specified length and thickness (Fig. 10b). The distribution of these packages for each time interval is controlled by: (1) the configuration of the original depositional surface (Fig. 10a); (2) the position of the sea relative to that surface (Fig. 10d); (3) the quantity of sediment (shale and/or sand) deposited during that time interval (Fig. 10b); and (4) the subsidence behavior of the depositional surface, including hinged tectonic or thermal subsidence, isostatic loading, and sediment compaction in response to dewatering (Fig. 10c).

The simulation includes functions that allow the sediment to build to sea level, after which erosion and bypass take place. We simulated only marine clastic deposition, ignoring the alluvial wedge landward, in order to reproduce the gross sedimentary geometries of the marine sediments and to show the location of sand-prone and shale-prone portions of the section (Fig. 11). We acknowledge the fact that without the alluvial plain, our simulation was flawed, and we are working on this defect (Helland-Hansen and others, 1988). We think, however, that our simulations are quite informative.

In Figure 11, the area between dark lines represents the sediment deposited during a designated time interval. The proportion of shale and sand within the time step is displayed in the following way. The area immediately below a dark line represents the shale portion of the sediment, and the area below a light line represents the sand portion. Individual sand and shale bodies (such as point bars, filled channels, or deltas) are not specifically identified, although areas where sand or shale bodies are likely to occur are shown. This output is designed to aid in the interpretation of seismic data by simulating the general, rather than specific, depositional history of the rock sequence. In particular, the simulation can be utilized to help identify probable

zones of good source rocks and seals (shale-prone areas) and the general region in which potential stratigraphic traps (pinchouts of sand-prone areas within and/or beneath shale-prone areas) may occur.

Figures 11 and 12 show the progressive evolution of the shelf and basin margin geometry with a small element of hinged subsidence. Stage 1 shows the progressive onlap of shelf in response to a relative rise in sea level, which is a function of the interplay between rates of eustatic change, subsidence, and sediment supply. Stages 2 and 3 show toplapping progradation in which the fall in sea level is not rapid enough to drop the coastal sedimentary wedge below the shelf margin. Stage 4 shows the effect of a relative rise in sea level and the onlapping sedimentary wedges. The graphic output for each stage is the product of the interplay between eustasy, subsidence, and sediment supply rather than eustasy alone. One of the interesting responses visible in the simulation is the effect of the isostatic response of the crust to loading by sediment and water. In the first stage, the crust is bowed down close to the shore, while at the basin margin, the crust has a positive area which is ridged upward. In the third and fourth stages, the crust is bowed down seaward of the positive ridge. This happens because sediment accommodation is reduced on the shelf, and the sediment now occupies space seaward, causing the downwarp there. The resulting "basement" geometry is a product of sediment weight alone and assumes a perfect elastic Airy response of the crust.

In Figure 13, changes in sea-level position are plotted per time step after compaction and subsidence, with a maximum elevation in the onlap of 46 m (150 ft; the lower heavy line) plus a modified Vail aggradational curve with a maximum excursion of 85 m (280 ft; without the alluvial plain, which is not modeled). The latter curve plots the difference in height of onlapped sediment per time step and is measured from the final geometry of the simulation after deposition of the whole sedimentary package, using the criteria outlined by Vail and others (1977). The result is both

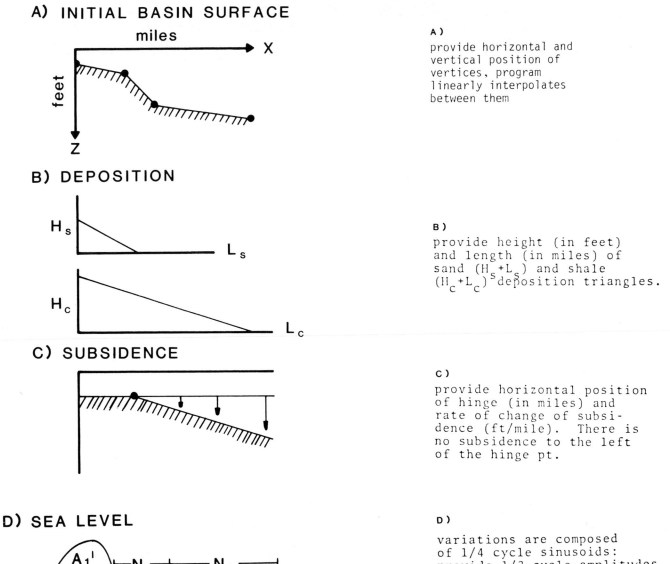

A) INITIAL BASIN SURFACE

A)

provide horizontal and vertical position of vertices, program linearly interpolates between them

B) DEPOSITION

B)

provide height (in feet) and length (in miles) of sand (H_s+L_s) and shale (H_c+L_c) deposition triangles.

C) SUBSIDENCE

C)

provide horizontal position of hinge (in miles) and rate of change of subsidence (ft/mile). There is no subsidence to the left of the hinge pt.

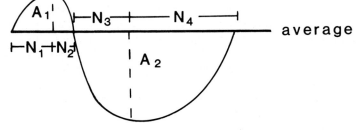

D) SEA LEVEL

D)

variations are composed of 1/4 cycle sinusoids: provide 1/2 cycle amplitudes + the number of time steps/1/4 cycle.

provide also the average value of sea level and whether the initial direction is up (as shown), or down.

FIG. 10.—Some of the input parameters for the simulation.

a two-fold exaggeration and a distortion of the sea-level excursion (or coastal-onlap curve in Vail and others, 1977) as determined by the simulation. In the aggradation of Vail and others (1977), the relative rise in sea level appears much larger than the fall. We expect the magnification and distortion to be even larger when we add the alluvial-plain sediments. This magnification effect may explain why the sea-level excursions of Vail and others (1977) are so large, whereas the crustal response seen by Guidish and others (1984) is so small. According to the simulation results, when small excursions in sea level are input, they have a marked effect on the sedimentary response. In addition, compaction and tectonic movement exaggerate the apparent size of the excursion when it is measured from the final geometry of the simulation and presumably when it is measured on seismic cross sections.

Figure 14 shows the simulated positions of POGO (point of maximum landward encroachment of coastal onlap), the

SEDIMENTARY GEOMETRY

DARK-SHALE TOP, LIGHT-SAND TOP

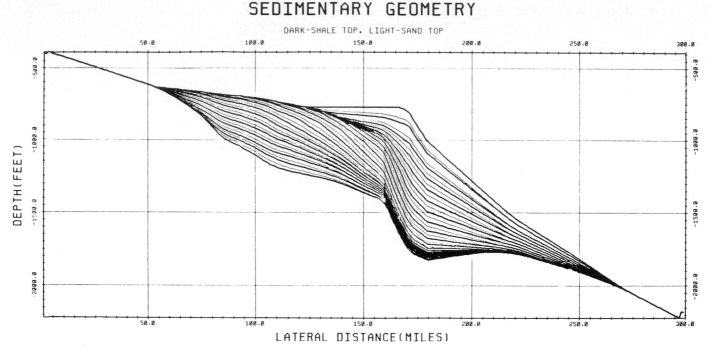

FIG. 11.—Cross section of sedimentary geometry produced by simulation of eustasy, compaction, tectonic movement, isostacy, and sediment fill, assuming no alluvial wedge landward and sediment fill to sea level.

shelf break, and the most basinward position of the sediment wedge through time. The sinusoidal shape of these curves and of the coastal-onlap curve is due to the variation in sea level, as well as to the fact that we are not modeling the alluvial plain sediments, which are responsible for the sawtooth appearance of the "coastal" or sediment onlap curves in Hardenbol and others (1981).

Despite the absence in our simulation of the alluvial wedge that onlaps landward of the coastline, the simulation gives a general sense of the response to sedimentation, sea level, and tectonism in the offshore marine and coastal sediments. We have yet to make a systematic test of the simulation's reliability and its effectiveness as an aid to interpretation. Once we have added the alluvial wedge, we plan to test the simulation in a well-studied area that provides good geologic control and to make reasonable estimates of the program inputs. Obviously, such an area should not have geologic phenomena that are not included in the program, such as faulting, carbonate buildups, and salt movement. Until that testing is done, we cannot apply our model with confidence to less well known areas.

GENERAL DISCUSSION OF METHODS FOR MEASURING SEA LEVEL

Because the magnitude of eustatic change cannot be measured directly, geologists are forced to use the variety of methods described in this paper. The problem is that these methods depend on some assumption about the size of one or more of the variables in the model.

When geologists use hypsometric curves and paleogeographic maps that show the area onlapped by marine sediments across the continent, they are forced to assume that the configuration of the hypsometric curve is the same for the past or to construct an assumed curve of their own. The result is a relative sea-level curve that helps unravel stratigraphic history but does not provide measurements of the size of excursions.

We agree with Hardenbol and others (1981) that the sedimentary aggradational curves are also not an accurate means of determining the size of a eustatic event. The measurement of sedimentary aggradation is also influenced by compaction, rate and magnitude of crustal subsidence, and, importantly, the extent of alluvial plain onlap. All of these variables are particularly difficult to determine accurately. For instance, the commonly used compaction algorithms are based on the assumptions that fluid escapes uniformly during burial, and the fluid pressure is always in equilibrium with the compacted sedimentary section (Guidish and others, 1984). In fact, fluid escape is much more complex, so that these algorithms do not handle compaction accurately. In schemes that use thermo-tectonic models to predict crustal-subsidence curves and to determine the size of eustatic excursions, the results are dependent on the choice of model and the parameters used. These choices of thermo-tectonic model are not grounded in fact. A tectonic model is also assumed when geologists use the height between strandline markers, such as coral reef terraces, peats, and sedimentary structures, to measure changes in sea level. Attempts at measuring high-frequency changes in sea level from stacked subsidence curves fail, because the crust does not respond noticeably to such small sea-level events. The use of oxygen isotopes does not give a clear-cut answer because the oxygen isotope ratio in the sea is controlled only in part by the ocean volume. Even if that portion of

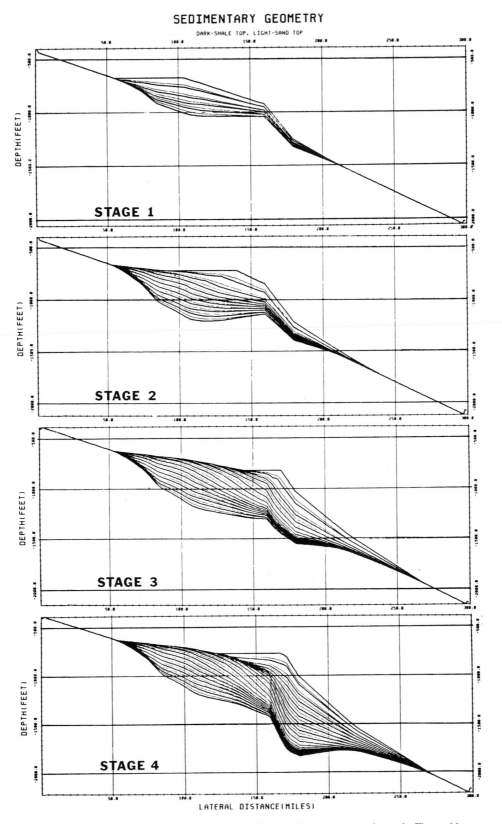

Fɪɢ. 12.—Four steps involved in sediment fill to produce geometry shown in Figure 11.

VAIL AGGRADATION

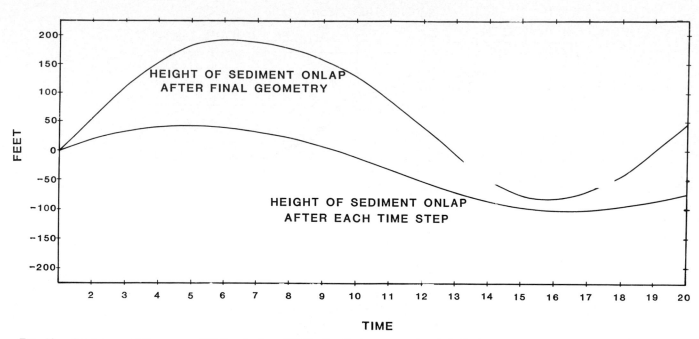

FIG. 13.—Coastal aggradation curve of Vail and others (1977) after final geometry (top light line) versus eustatic sea-level variation after each time step (bottom dark line).

the isotope ratio which is correlatable to water volume could be established, we cannot determine the size of eustatic excursions, since we cannot accurately model the crust's flexural response to changes in ocean volume except in direction and very general magnitude. Finally, our attempts to determine the variables through numerical and graphical simulation are equally unsatisfactory, because of the apparent lack of a unique solution for specific input variables.

All of these methods require making basic assumptions about the sizes of a number of variables that cannot be meaured. The exciting and imaginative concepts in these methods are not yet supported by unequivocal evidence.

SEDIMENTARY LIMITS VS TIME

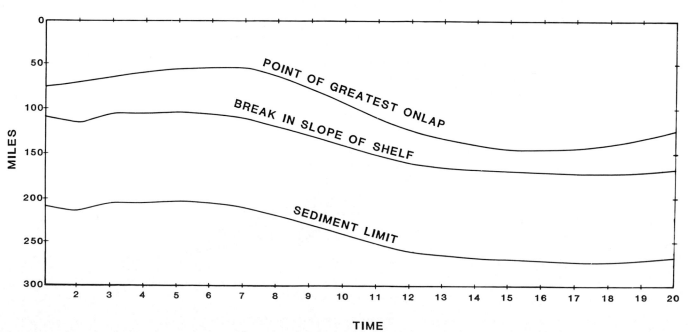

FIG. 14.—Position of POGO (top dark line), shelf break (middle line), and basinward position of sediment (bottom line).

AN OUTLINE OF A SYSTEM FOR DETERMINING SEA-LEVEL
VARIATIONS FROM BASIN FILL UNDER ISOSTATIC BURIAL HISTORY

After examining different schemes for estimating sea-level excursions, we still were not sure that we could determine the correct size for the variables used in each scheme. As a result, we tried to see if there were a mathematically unique answer. Our efforts were educational and explain why we could not determine the size of the variables even with forward modeling (Burton and others, 1987). The main conclusion of Burton and others (1987) is that the whole process is nonlinear. To determine change in sea-level from measurements of present sedimentary strata and their onlap points by inverting a forward model seems difficult if not impossible, because the forward model uses prisms of sediment that likely cannot be inverted.

Even if a forward model could be inverted, two major problems remain: (1) all we can deduce is *the sum of sea-level and basin motion;* and (2) we cannot know where the points of greatest onlap are in an onlap sequence without knowing the complete basin history. That requires knowledge of (a) sedimentary fill laterally in a basin (including erosion) together with the basement response; and (b) how overpressuring with time influences the sequence and onlap effects, since basins are not isostatically controlled for all time.

CONCLUSIONS

Sedimentary geometries that are commonly seen on seismic cross sections and in outcrop are the products of variations in rates of sediment accumulation, eustasy, and tectonic movement. To date, neither the use of hypsometric curves, the innovative concepts of Vail and his co-workers, the use of paleomarkers of strandline position, the ties to stacked crustal subsidence, the systems for measuring Pleistocene sea-level movement using either reef terrace positions or deeper ocean oxygen isotope ratios, nor our own forward modeling seems to provide a unique and unequivocal key for determining the size of tectonic movement, sea-level excursions, or rate of sedimentation (Table 1). The major conclusion from Vail's work, from our crustal subsidence work, from the forward modeling algorithm that we used, and from formal mathematical development is that unless one variable is dependent on the other two (which is not the case), then, at the very least, two of the three processes (sediment accumulation, eustasy, and tectonic subsidence) *must* be specified in order to determine

TABLE 1.—METHODS FOR ESTIMATING THE SIZE OF EUSTATIC SEA LEVEL EXCURSIONS

Method	Measured Variable	Assumptions	Problems
Hypsometric.	Area of continent covered by marine sediments for time interval on equal area projection, measured with planimeter.	The relationship between continental relief and area of continent at that elevation today was the same in the past.	1. Time interval may be too long. 2. Paleogeographic maps are inaccurate. 3. Tectonic behavior unknown. 4. Continental thickness unknown.
Vail sediment onlap.	1. Distance of onlap of seismic reflectors on unconformities, perpendicular to shore. 2. Height of onlap.	Onlap not a product of: (1) tectonic subsidence; (2) compaction; or (3) isostatic response.	Cannot put dimensions on tectonic subsidence, compaction, or isostatic response.
Paleo-bathymetric markers 1. Shoaling cycles.	Thickness of cycle between high-water-mark indicators.	1. Thickness a result of eustasy. 2. Tectonic subsidence, compaction, and isostacy negligible.	Effects of tectonic subsidence, compaction, and isostasy are unknown.
2. Strand-line markers (beaches, reefs, notches, peats, etc.).	Elevation above present sea level and between markers.	1. Result of eustasy and constant rate of tectonic uplift.	Tectonic uplift rate unknown. Constant behavior unknown.
Crustal subsidence curves 1. Divergence from thermo-tectonic curves.	Difference between crustal subsidence curve for a well and predicted thermotectonic curve for same location.	1. Depth of average 1% porosity and/or basement. 2. Compaction history, response to sediment load on crust. 3. Thermotectonic model.	Depth to 1% porosity and basement may not be known. Compaction history, isostatic response of crust unknown. Thermo-tectonic behavior unknown. Cannot determine the assumptions.
2. Perturbations on stacked crustal subsidence curves.	Size of perturbations from integrated stacked subsidence curves.	1. Depth of average 1% porosity and/or basement. 2. Compaction history. 3. Isostatic response to sediment and water load on crust. 4. Lithospheric rigidity and thermo-tectonic model.	
Oxygen isotopes	$\delta^{18}O$ values	1. Variation in $\delta^{18}O$ value is a result of ocean volume. 2. Isostatic response of crust to weight of water the same everywhere. 3. Can estimate volume of continental ice and volume of ocean ice as function of time. 4. No diagenetic effect.	Cannot prove any of the assumptions.

the third. The relative sizes of these processes can be estimated by modeling them in different basins, but absolute values remain elusive. Presumably, this lack of knowledge underlies the variety of good hypotheses often invoked to account for observed sedimentary geometries. The necessary information is lacking to decide which hypothesis represents the truth.

Thus, although an *accurate* chart of variations in eustatic sea level would be a boon to frontier basin exploration in the oil industry, particularly where seismic cross sections and occasional stratigraphic wells and/or wildcats are the only data sources, such a chart canot be made. When *relative* (combining tectonic and eustatic effects) sea-level charts are tied to wells, however, it is still possible to project sedimentary sequences related to the relative sea-level events across a basin on seismic cross sections, after the manner of Vail and others (1977) and Hallam (1981). Possibly the best *relative* (tectono/eustatic) sea-level chart would use sediment onlap of the continental margin in combination with oxygen isotope signals responding to continental and alpine glacial events back to at least 100 ma, if glacial events existed then. With such a chart and a computer simulation, geologists would produce a family of solutions. Then geologists could assess the probability of finding, say, sand-prone rather than shale-prone or carbonate-prone sections—an objective of so many hydrocarbon exploration companies. Models involving *relative* (tectono/eustatic) sea level will test novel play concepts and initiate exploration in frontier areas.

ACKNOWLEDGMENTS

This paper is an outgrowth of work begun at Gulf Research Laboratory. The simulation results were previously presented at Lamont-Doherty Geological Observatory in November 1979; and the stacked subsidence research results, in Guidish and others (1984). We express our appreciation to Willard Moore and Doug Williams at the University of South Carolina, who have helped us with our understanding of Pleistocene sea-level determination; to Peter Vail, who through numerous conversations has tried to straighten out our thinking on eustasy; to Ric Busch, Timothy Cross, Jim Sadd, and Doug Williams, who read and offered suggestions to improve our paper; and to Robert Ehrlich, whose geologic ideas, advice, and command of the English language have constantly delayed our submitting this paper. Funds used to carry out this research at the University of South Carolina were kindly provided by Chevron, Cities Service, Lytton Industries, Marathon, Norsk Hydro, Saga Petroleum, Statoil, Sun, Texaco, and Union of California.

REFERENCES

AHARON, P., 1983, 140,000-yr isotope climatic record from raised coral reefs in New Guinea: Nature, v. 304, p. 720–723.
AIRY, G. B., 1855, On the computation of the effect of the attraction of the mountain-masses as disturbing the apparent astronomical latitude of stations at geodetic surveys: Philosophical Transactions of the Royal Society of London, v. 145, p. 101–104.
BALLY, A. W., 1981, Basins and subsidence: American Geophysical Union Geodynamic Series, v. 1, p. 5–20.
BEAUMONT, C., 1981, Foreland basins: Geophysical Journal of Royal Astornomical Society, v. 65, p. 291–329.
BEUKES, N. J., 1977, Transition from siliciclastic to carbonate sedimentation near base of the Transvaal Superegroup, northern Cape Province, South Africa: Sedimentary Geology, v. 18, p. 201–221.
BLOOM, A. L., BROECKER, W. S., CHAPPELL, J. M. A., MATTHEWS, R. K., AND MESOLELLA, 1974, Quarternary sea level fluctuations on a tectonic coast: new ^{230}Th/^{234}U dates from the Huon Peninsula, New Guinea: Quaternary Research, v. 4, p. 185–205.
BOMFORD, G., 1971, Geodesy, second ed.: Oxford University Press, London, p. 441–443.
BOND, G., 1976, Evidence for continental subsidence in North America during the Late Cretaceous global submergence: Geology, v. 4, p. 557–560.
——, 1978a, Evidence for Late Tertiary uplift of Africa relative to North America, South America, Australia and Europe: Journal of Geology, v. 86, p. 47–65.
——, 1978b, Speculations on real sea level changes and vertical motions of continents at selected times in the Cretaceous and Tertiary Periods: Geology, v. 6, p. 247–250.
BROECKER, W. S., AND VAN DONK, J., 1970, Insolation changes, ice volumes, and the O^{18} record in deep-sea cores: Review of Geophysics and Space Physics, v. 8, p. 169–198.
BURTON, RANDE, KENDALL, C. G. ST. C. AND LERCHE, IAN, 1987, Out of our depth: on the impossibility of fathoming eustatic sea level from the stratigraphic record: Earth Science Reviews, v. 24, p. 237–277.
BUSCH, R. M., 1983, Sea level correlation of punctuated aggradational cycles (PACs) of the Manlius Formation, central New York: Northeastern Geology, v. 5, p. 82–91.
CHAPPELL, J. 1974, Geology of coral terraces, Huon Peninsula, New Guinea: a study of Quaternary tectonic movements and sea level changes: Geological Society of America Bulletin, v. 85, p. 553–570.
——, AND VEEH, H. H., 1978, Late Quaternary tectonic movements and sea level changes at Timor and at Atauro Island: Geological Society of America Bulletin, v. 89, p. 356–368.
COGLEY, J. G., 1981, Late Phanerozoic extent of dry land: Nature, v. 291, p. 56–58.
——, 1984, Continental margins and extent and number of the continents: Review of Geophysics and Space Physics, v. 22, p. 101–122.
DE PRATTER, C. B., AND HOWARD, J. D., 1981, Evidence for a sea level lowstand between 4500 and 2400 BP on the southeast coast of the United States: Journal of Sedimentary Petrology, v. 51, p. 1287–1295.
EYGED, L., 1956, Determination of changes in the dimensions of the earth from paleogeographic data: Nature, v. 173, p. 534.
FAIRBANKS, R. G., AND MATTHEWS, R. K., 1978, The marine oxygen isotope record in Pleistocene coral, Barbados, West Indies: Quaternary Research, v. 10, p. 181–196.
FALVEY, D. A., 1974, The development of continental margins in plate tectonic theory: Australian Petroleum Exploration Association Journal, v. 14, p. 95–106.
FILLON, R. H. AND WILLIAMS, D. F., 1983, Glacial evolution of the Plio-Pleistocene: role of continental and Arctic ice sheets: Palaeogeography, Palaeoclimatology, Palaeoecology, v. 43, p. 7–33.
——, 1984, Dynamics of meltwater discharge from Northern Hemisphere ice sheets during the last deglaciation: Nature, v. 310, p. 674–677.
FORNEY, G. G., 1975, Permo-Triassic sea level change: Journal of Geology, v. 83, p. 773–779.
GUIDISH, T. M., LERCHE, I., KENDALL, C. G. St. C., AND O'BRIEN, J. J., 1984, Relationship between eustatic sea level changes and basement subsidence: American Association of Petroleum Geologists Bulletin, v. 68, p. 164–177.
HALLAM, A., 1963, Major epeirogenic and eustatic changes since the Cretaceous, and their possible relationship to crustal structure: American Journal of Science, v. 261, p. 397–423.
——, A revised sea-level curve from the Early Jurassic: Quarterly Journal of Geology, Society of London, v. 138, p. 735–743.
——, 1984, Pre-Quaternary sea-level changes: Annual Review of Earth Planetary Science, v. 12, p. 205–243.
HANCOCK, J. M., AND KAUFFMAN, E. G., 1979, The great transgressions of the Late Cretaceous: Quarterly Journal of Geology, Society of London, v. 136, p. 175–186.
HARDENBOL, J., VAIL, P. R., AND FERRER, J., 1981, Interpreting paleoen-

vironments, subsidence history, and sea level changes of passive margins from seismic and biostratigraphy: Oceanology Acta, Proceedings, 26th International. Geologic Congress, Geology of Continental Margins Symposium, p. 33–44.

HARRIS, P. M., FROST, S. H., SEIGLIE, G. A., AND SCHNEIDERMANN, N., 1984, Regional unconformities and depositional cycles, Cretaceous of Arabian Peninsula, *in* Schlee, J. S., ed., Interregional Unconformities and Hydrocarbon Accumulations: American Association of Petroleum Geologists Memoir 36, p. 67–80.

HARRISON, C. G. A., BRASS, G. W., SALTZMAN, E., SLOAN, J., II., SOUTHAM, J., AND WHITMAN, J. M., 1981, Sea level variations, global sedimentation rates and the hypsographic curve: Earth and Planetary Science Letters, v. 54, p. 1–16.

HELLAND-HANSEN, W., KENDALL, C. G. ST. C., LERCHE, I., AND NAKAYAMAK, 1988, A simulation of continental basin margin sedimentation in response to crustal movements, eustatic sea level change and sediment accumulation rates: Mathematical Geology, v. 20, p. 777–802.

HELLINGER, S. J., AND SCLATER, J. G., 1983, Some comments on two-layer extensional models for the evolution of sedimentary basins: Journal of Geophysical Research, v. 88, p. 8251–8269.

KAUFFMAN, E. G., 1977, Geological and biological overview: Western Interior Cretaceous basin: Mountain Geologist, v. 14, p. 75–99.

KOSSINNA, E., 1921, Die Tiefen des Weltmeeres: Berlin Univ. Inst. Meereskunde, Veroff., Geogr. Naturwiss, n.s., n. 9, 70 pp.

———, 1933, Die Erdoberflache, *in* Gutenberg, B., ed., Handbuch der Geophysik, v. 2, Aufbau der Erde: Gebruder Borntraeger, Abschuitt VI, Berlin, p. 809–954.

MATTHEWS, R. K., 1984a, Dynamic Stratigraphy, An Introduction to Sedimentation and Stratigraphy: Prentice-Hall, New Jersey, 489 p.

———, 1984b, Oxygen-isotope record ocean-volume history: 100 million years of glacio-eustatic sea level fluctuation, *in* Schlee, J. S., ed., Interregional Unconformities and Hydrocarbon Accumulation: American Association of Petroleum Geologists Memoir 36, p. 97–107.

McKENZIE, D., 1978, Some remarks on the development of sedimentary basins: Earth and Planetary Science Letters, v. 40, p. 25–32.

McKERROW, W. S., 1979, Ordovician and Silurian changes in sea level: Quarterly Journal of Geology, Society of London, v. 136, p. 137–145.

MESOLELLA, K. J., MATTHEWS, R. K., BROECKER, W. S., AND THURBER, D. L., 1969, The astronomical theory of climatic change: Barbados data: Journal of Geology, v. 77, p. 250–274.

MIALL, A. D., 1986, Eustatic sea level changes interpreted from siesmic stratigraphy: a critique of the methodology with particular reference to the North Sea Jurassic record: American Association of Petroleum Geologists Bulletin, v. 70, p. 131–137.

MOORE, W. S., 1982, Late Pleistocene sea level history, *in* Ivanovich, M., and Harmon, R., eds., Uranium-Series Disequilibrium Applications to Environmental Problems in the Earth Sciences: Oxford University Press, New York, p. 481–496.

MORRISON, L. 1985, The day time stands still: New Scientist, no. 1462, p. 20–21.

NUNN, J. A., SLEEP, N. H., AND MOORE, W. E., 1984, Thermal subsidence and generation of hydrocarbons in Michigan basin: American Association of Petroleum Geologists Bulletin, v. 68, p. 296–315.

PARKINSON, N., AND SUMMERHAYES, C., 1985, Synchronous global sequence boundaries: American Association of Petroleum Geologists Bulletin, v. 69, p. 658–687.

ROYDEN, L., SCLATER, J. G., AND VON HERZEN, R. P., 1980, Continental margin subsidence and heat flow; important parameters in formation of petroleum hydrocarbons: American Association of Petroleum Geologists Bulletin, v. 64, p. 173–187.

SEIGLIE, G. A., AND BAKER, M. B., 1984, Relative sea-level changes during the Middle and Late Cretaceous from Zaire to Cameroon (central West Africa), *in* Schlee, J. S., ed., Interregional Unconformities and Hydrocarbon Accumulation: American Association of Petroleum Geologists Memoir 36, p. 81–88.

———, AND MOUSSA, M. T., 1984, Late Oligocene-Pliocene transgression-regression cycles of sedimentation in northwestern Puerto Rico, *in* Schlee, J. S., ed., Interregional Unconformities and Hydrocarbon Accumulation: American Association of Petroleum Geologists Memoir 36, p. 89–96.

SHACKLETON, N. J., AND OPDYKE, N. D., 1973, Oxygen isotope and paleomagnetic stratigraphy of Equatorial Pacific core V28-238: oxygen isotope temperature and ice volumes on a 10^5 year and 10^6 year scale: Quaternary Research, v. 3, p. 39–55.

SLOSS, L. L., 1963, Sequences in the cratonic interior of North America: Geological Society of America Bulletin, v. 74, p. 93–113.

———, 1972, Synchrony of Phanerozoic sedimenty-tectonic events of North American craton and Russian platform: 24th International Geological Congress, Section 6, p. 24–32.

———, AND SPEED, R. C., 1974, Relationships of cratonic and continental margin tectonics episodes, *in* Dickinson, W. R., ed., Tectonics and Sedimentation: Society of Economic Paleontologists and Mineralogists Special Publication 22, p. 98–119.

———, 1984, Comparative anatomy of cratonic unconformities, *in* Schlee, J. S., ed., Interregional Unconformities and Hydrocarbon Accumulation: American Association of Petroleum Geologists Memoir 36, p. 7–36.

SOUTHAM, J., AND WHITMAN, J. M., 1981, Sea level variations, global sedimentation rates and the hypsographic curve: Earth and Planetary Science Letters, v. 54, p. 1–16.

STEINEN, R. P., HARRIS, R. S., AND MATTHEWS, R. H., 1973, Eustatic low stand of sea level between 125,000 and 105,000 B.P. Evidence from the subsurface of Barbados, West Indies: Geological Society of America Bulletin, v. 84, p. 63–70.

THORNE, J., AND WATTS, A. B., 1984, Seismic reflectors and unconformities at passive continental margins: Nature, v. 311, p. 365–367.

VAIL, P. R., AND HARDENBOL, J., 1979, Sea level changes during Tertiary: Oceans, v. 22, no. 3, p. 71–79.

———, ———, AND TODD, R. G., 1984, Jurassic unconformities, chronostratigraphy and sea level changes from seismic and biostratigraphy, *in* Schlee, J. S., ed., Interregional Unconformities and Hydrocarbon Accumulation: American Association of Petroleum Geologists Memoir 36, p. 347–363.

———, MITCHUM, R. M., JR., TODD, R. G., WIDMIER, J. M., THOMPSON, S., III., SANGREE, J. B., BUBB, J. N., AND HATLEILID, W. G., 1977, Seismic stratigraphy and global changes of sea level, *in* Payton, C. E., ed., Seismic Stratigraphy—Applications to Hydrocarbon Exploration: American Association of Petroleum Geologists Memoir 26, p. 49–212.

———, AND TODD, R. G., 1981, Northern North Sea Jurassic unconformities, chronostratigraphy and sea level changes from seismic stratigraphy, *in* Illing, L. V., and Hobson, G. D., eds., Proceedings Petroleum Geology of the Continental Shelf of North West Europe Conference, London, England: Heydon and Son, London, p. 216–235.

VAN HINTE, J. E., 1978, Geohistory analysis—application of micropaleontology in exploration geology: American Association of Petroleum Geologists Bulletin, v. 62, p. 201–222.

VELLA, P., 1961, Terms for real and apparent height changes of sea level and parts of the lithosphere: Transactions of the Royal Society of New Zealand, v. 1, p. 101–109.

WALCOTT, R. I., 1970, Flexural rigidity, thickness, and viscosity of the lithosphere: Journal of Geophysical Research, v. 75, p. 3941–3954.

WARD, W. T., AND CHAPPEL, J., 1975, Geology of coral terraces, Huon Peninsula, New Guinea: a study of Quaternary tectonic movements and sea level changes: discussion and reply: Geological Society of America Bulletin, v. 86, p. 1482–1486.

WATTS, A. B., 1982, Tectonic subsidence, flexure and global changes of sea level: Nature, v. 297, p. 469–474.

———, AND STECKLER, M. S., 1979, Subsidence and eustasy at the continental margin of eastern North America, *in* Talwani, M., Hay, W., and Ryan, W. B. F., eds., Deep Drilling Results in the Atlantic Ocean: Continental Margins and Paleoenvironment: American Geophysical Union Ewing Series, v. 3, p. 218–234.

WEIMER, R. J., 1984, Relation of unconformities, tectonics and sea level changes, *in* Schlee, J. S., ed., Interregional unconformities and hydrocarbon accumulation: American Association of Petroleum Geologists Memoir 36, p. 7–36.

WHEELER, H. G., 1958, Time stratigraphy: American Association of Petroleum Geologists Bulletin, v. 42, p. 1047–1063.

WILLIAMS, D. F., MOORE, W. S., AND FILLON, R. H., 1981, Role of glacial Arctic Ocean ice sheets in Pleistocene oxygen isotope and sea level records: Earth and Planetary Science Letters, v. 56, p. 157–166.

WYATT, A. R., 1984, Relationship between continental area and elevation: Nature, v. 311, p. 370–372.

INTRAPLATE STRESSES: A TECTONIC CAUSE FOR THIRD-ORDER CYCLES IN APPARENT SEA LEVEL?

SIERD CLOETINGH*

Vening Meinesz Laboratory, Institute of Earth Sciences, University of Utrecht, Budapestlaan 4, 3584 CD Utrecht, The Netherlands

ABSTRACT: Thermo-mechanical modeling demonstrates that tectonically induced vertical motions of the lithosphere may provide an explanation for third-order cycles in apparent sea level deduced from the seismic stratigraphic record of passive margins. The interaction of fluctuations in intraplate stresses and the deflection of the lithosphere caused by sedimentary loading can produce apparent sea-level changes of as much as 100m at the flanks of passive margins.

In general, stress variations of a few hundred bars associated with local adjustment of stresses at passive margins suffice to explain a significant part of the stratigraphic record associated with short-term variations in sea level on the order of a few tens of meters. To induce short-term apparent sea-level fluctuations with magnitudes on the order of 50m or more, which occur less frequently in the record, changes in stress level in excess of one kbar are required. These larger fluctuations in apparent sea level could be related to major reorganizations of lithospheric stress fields due to rifting and fragmentation of plates, dynamic changes at convergent plate boundaries, or collision processes. A fluctuating horizontal stress field in the lithosphere can explain contemporaneous changes in apparent sea level in neighboring depositional environments. In principle, it implies the possibility of regional correlations in different basin settings. Specific short-term fluctuations in the curves of Vail and others, (1977; 1984) can be associated with particular plate tectonic reorganizations of lithospheric stress fields. The seismic stratigraphic record may provide a new source of information on paleo-stress fields which can be correlated with results of independent numerical modeling of intraplate stresses.

INTRODUCTION

During the last decade, major advances in quantitative analysis of the sea-level record in sedimentary basins have resulted from studies by Vail and his coworkers at Exxon (Vail and others, 1977; Vail and Todd, 1981; Vail and others, 1984). At the same time, there has been a continuous debate over the mechanisms that cause third-order cycles in sea level (e.g., Bally, 1980, 1982). Recently, Cloetingh and others (1985a,b) have proposed a new tectonic mechanism for apparent sea-level variations of about 1-10 cm/1,000 yr, with a maximum magnitude of a few hundred meters. Their model can explain these third-order cycles in sea level, provided that horizontal stresses exist in the lithosphere and that changes in these stress fields occur on time scales of a few million years and longer. In this paper, we will concentrate on the consequences of the action of intraplate stresses for the stratigraphic record. We shall then investigate the evidence for changes in the intraplate stress field. This will be followed by a discussion of some new elements in passive margin and sedimentary basin research that follow from this work. Mechanisms for long-term changes in sea level (e.g., Kominz, 1984; Heller and Angevine, 1985) fall beyond the scope of the present paper.

A TECTONIC MECHANISM FOR SHORT-TERM SEA-LEVEL FLUCTUATIONS

Pitman and Golovchenko (1983) pointed out that glacially induced fluctuation in sea level is the only known mechanism that can cause changes in sea level at rates in excess of 1 cm/1,000 yr and with magnitudes in excess of 100 m. Glacial-eustatic changes may, therefore, explain the Oligocene unconformities, but cannot explain those third-order cycles of sea level where glacial events are thought to have been insignificant (Thorne and Watts, 1984). For example, with the exception of the Oligocene event, there is no evidence in the geological and geochemical records for significant Mesozoic and Cenozoic glacial events prior to middle Miocene (Frakes, 1979). Alternatively, the sea-level record has been attributed to tectonic processes (e.g., Bally, 1980, 1982; Watts, 1982; Hallam, 1984; Veizer, 1985; Hallam, this volume), although the nature of these mechanisms for explaining rapid short-term fluctuations remained unclear (Watts, 1982; Pitman and Golovchenko, 1983; Hallam, 1984; Miall, 1986).

We interpret apparent fluctuations in sea level, or onlap/offlap sequences, as expressions of regional, and possibly global, tectonic processes rather than in terms of eustatic changes. It is recognized, however, that the latter are always potentially present and that a separation of the two processes may not be possible without independent supporting evidence concerning, for example, climatic change.

The tectonic model proposed by Cloetingh and others (1985a,b) represents the interaction between intraplate stresses and the deflections of the lithosphere caused by sedimentary loading (Fig. 1). This interaction can produce apparent sea-level changes of more than 100 m within a few million years along the flanks of sedimentary basins. Therefore, Cloetingh and others (1985a,b) argued that glacial fluctuations (Pitman and Golovchenko, 1983) are not the only mechanisms capable of producing third-order cycles in apparent sea level with a magnitude and rate comparable to those inferred from the stratigraphic record. The model of Cloetingh and others can explain contemporaneous fluctuations in apparent sea level in neighboring basin settings. The action of changing horizontal stresses is not restricted to passive margins but also modifies the vertical movements within intracratonic basins and foreland basins. This tectonic mechanism may provide an explanation for some of the observed correlations between the timing of the sea-level changes in oceanic and intracontinental regions noted by Sloss (1979) and Bally (1980).

The basin stratigraphy and apparent sea-level record can exhibit considerable variations associated with changes in the stress field. The model of Cloetingh and others (1985a,b)

*Present address: Department of Sedimentary Geology, Institute of Earth Sciences, Free University, P.O. Box 7161, 1007 Mc Amsterdam, The Netherlands

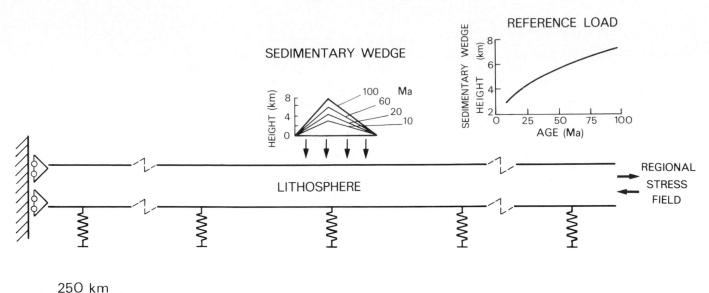

Fig. 1.—Tectonic model for fluctuations in apparent sea level. Variations in intraplate stress field affect the vertical displacements at a passive margin evolving through time, due to the thermal evolution of the lithosphere and the loading of this lithosphere by a wedge of sediments (see inset for reference model of sediment loading). A model is adopted with uniform elastic properties; rheological differences between continental and oceanic lithosphere are neglected for simplicity.

requires that intraplate stress fields be of sufficient magnitude to modulate the deflections of the lithosphere caused by other subsidence processes and that they have variations on time scales of a few million years and longer. In what follows, we review evidence for the existence of such intraplate stress fields.

Calculated apparent sea-level changes.—

The total deflection of the lithosphere at rifted margins is dominated by sediment loading and thermal contraction (Sleep, 1971). Subsidence at a rifted margin is controlled by the stretching of crust or lithosphere and by subsequent cooling and sediment loading. The subsidence curves will therefore be characterized by a decreasing subsidence rate with time, the magnitude of subsidence being a function of lithospheric properties, driving force parameters, the rate of cooling, and the rate of sedimentation. To model passive-margin processes, we consider, for convenience, an elastic oceanic lithosphere that cools with time according to the boundary layer model of thermal conduction. In our modeling, the sediment load is represented by two adjacent triangular wedges, one on the continental shelf, the other on the continental rise. Sedimentation is assumed to be sufficiently rapid to keep up approximately with subsidence (Turcotte and Ahern, 1977). Because thermal subsidence is proportional to the square root of age, the maximum thickness of the sedimentary wedges grows with the same time dependence. This leads to a maximum sedimentary thickness of 7.3 km at 100 Ma (Fig. 1). This reference model gives a fair representation of the sediment-loading histories and thicknesses at rifted margins and agrees with the observation that sediment accumulation rates tend to decrease with time after the initial rifting phase (Southam and Hay, 1981). Although certainly an oversimplification (see Pit-

man, 1978, and Pitman and Golovchenko, 1983, for a discussion), the assumption that sedimentation keeps up with subsidence allows us to equate onlap and offlap with rises and falls in sea level, respectively.

Analyses of the flexural response of oceanic lithosphere to tectonic processes (Bodine and others, 1981; McAdoo and others, 1985) as well as seismo-tectonic studies (Wiens and Stein, 1983), show an increase in the effective flexural rigidity, or equivalent elastic thickness, of the oceanic lithosphere with age of the crust. Thus, the response of the oceanic lithosphere to the sediment load (Watts and others, 1982), and the interaction between this response and the inplane stress, is time dependent not only because the sediment load builds up with time, but also because of the changing mechanical properties of the lithosphere.

The interactions between sediment loading, lithospheric thermal evolution, and intraplate stresses are calculated here using finite element techniques. We investigate the response of the lithosphere, adopting an equivalent elastic layer whose thickness increases with age according to a square root of age function (Bodine and others, 1981). Of particular interest is the modification of the basin shape by variations in intraplate stress fields. Figure 2 (see inset) shows the effect for a sedimentary basin underlain by 30-Ma oceanic lithosphere with a corresponding elastic plate thickness. A transition from tension of 1 kbar (1 kbar = 100 MPa) to compression of the same magnitude produces a net uplift (or apparent sea-level fall) of as much as about 40 m at the edge of the basin, immediately landward of the principal sediment load. The deflections at other stress levels are proportionally scaled to the magnitude of the applied horizontal stresses, with a stress field of a few hundred bars corresponding to a deflection of about 10 m. Evidence for the existence of intraplate stress fields with magnitudes as much as a few kilobars is presented later.

The dependence of deflection at the edge of the basins on the age of the underlying lithosphere is demonstrated in Figure 2. The differential uplift $\triangle w$, defined as the difference in deflection for a change in stress or in-plane force from tension to compression of equal magnitude, is computed for the edge of the basin as a function of variations in the in-plane force and stress. Curves I−V illustrate the deflection for changes in stress from 0.25 to −0.25 kbar, from 0.5 to −0.5 kbar, from 1 to −1 kbar, from 2 to −2 kbar, and from 3 to −3 kbar, respectively, in all cases with the same time-dependent reference sediment load (negative stress denotes compression). In these models, the in-plane force increases with increasing lithospheric age because of the associated thickening. Curve VI illlustrates the deflection, under the same reference load, for a change in the in-plane force from $5 \times 10^{12} \ Nm^{-1}$ to $-5 \times 10^{12} \ Nm^{-1}$, irrespective of lithospheric thickness.

The deflection of the lithosphere at the edge of the basin is of particular stratigraphic interest. External compression (or, equivalently, relaxation of tension) narrows the basin, causes relative uplift at the edge, and manifests itself as stratigraphic offlap. An apparent fall in sea level results, possibly exposing the sediments to produce an erosional or weathering horizon. External tension (or equivalently, relaxation of compression) widens the basin, causes relative subsidence at the edge, and manifests itself as stratigraphic onlap. This situation is characterized by an apparent rise in sea level, and renewed deposition, with a corresponding facies change, is possible. Figure 3 displays in an idealized fashion the stratigraphy for the edge of a sedimentary basin underlain by 40-Ma lithosphere, as predicted on the basis of the calculations shown in Figure 2. The synthetic stratigraphy at the basin edge is schematically shown for the following three situations: (a) widening of the basin with cooling in the absence of an intraplate-stress field (Fig. 3a), (b) the same case with a superimposed transition to 500-bar compression at 30 Ma (Fig. 3b), and (c) the case of a stress change to 500-bar tension at 30 Ma (Fig. 3c). Watts and others (1982) have pointed out the differences in stratigraphy predicted at the basin edge for elastic and viscoelastic models of the lithosphere in the absence of intraplate stresses. These authors showed that for elastic models, the width of the basin increases, inducing progressive overstepping of sediments caused by the increase in the thickness of the elastic layer since basin formation. They demonstrated that viscoelastic models for the lithosphere predict narrowing of the infill of the basin, with younger sediments restricted to the basin center. Care should be taken, however, in rigidly classifying basins as elastic or viscoelastic strictly on the basis of this criterion. The stratigraphy we predict for a basin located on an elastic lithosphere and under the influence of a compressional-stress regime during deposition of younger strata will be very similar to the stratigraphy characteristicly attributed (Watts and others, 1982) to the response of viscoelastic lithosphere to surface loading alone.

Our modeling demonstrates that the incorporation of intraplate stresses in elastic models of basin evolution can, in principle, predict a succession of onlap and offlap patterns such as observed along the flanks of passive margin

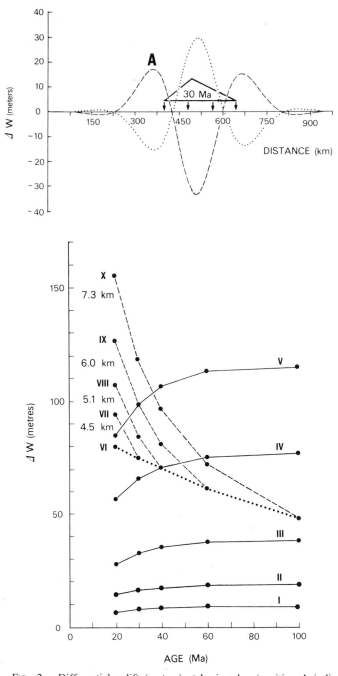

FIG. 2.—Differential uplift (meters) at basin edge (position A indicated in inset) due to superposition of variations in regional stress field on flexure caused by sediment loading, plotted as a function of the age of the underlying lithosphere. Solid lines (I−V): results for a fixed stress transition from 0.25, 0.5, 1, 2, and 3 kbar tension to 0.25, 0.5, 1,2, and 3 kbar compression, respectively, superimposed on sediment loading according to the reference model. Dotted curve (VI): results for a transition from $5 \times 10^{12} \ Nm^{-1}$ tension to $5 \times 10^{12} \ Nm^{-1}$ compression, corresponding to stresses varying between ±1.25 kbar at 100 Ma and ±2.8 kbar at 20 Ma, superimposed on sediment loading according to the reference model. Dashed lines (VII−X): results from sediment-loading models that are different from the reference model. Heights of the sedimentary wedges are specified in kilometers. Inset: differential subsidence or uplift (meters) from the deflection due to sediment loading and thermal contraction of 30-Ma oceanic lithosphere caused by an intraplate stress field of 1 kbar compression (dashed curve) and tension (dotted curve). Sign convention for deflections: uplift is positive, subsidence is negative.

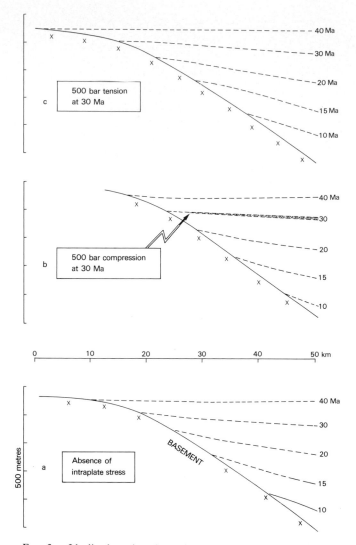

FIG. 3.—Idealized stratigraphy at the edge of a basin underlain by 40-Ma lithosphere predicted on the basis of the calculations shown in Figure 2. (a) Onlap associated with cooling of the lithosphere in the absence of an intraplate stress field. (b) A transition to 500-bar compression at 30 Ma induces a short-term phase of offlap at that time. (c) A transition to 500-bar tension causes an additional short-term phase of onlap at 30 Ma.

of the United States (Fig. 4). Such stratigraphy can be interpreted as the natural consequence of the mechanical widening and narrowing of basins by fluctuations in intraplate stresses, superimposed on the longer term broadening of the basin with cooling since its formation.

The calculations show that the effectiveness of external forces in producing deflections increases with increasing thickness of deposited sediments but decreases with the increasing age of the lithosphere. Hence, external forces are most effective at those young passive margins that are subject to rapid sediment loading. External forces are least effective at old margins subjected to low rates of sediment loading. This is most clearly seen when a constant sediment load is placed on an oceanic lithosphere of increasing effective elastic thickness with age and when sediment loads greater than the reference model are placed on the lithosphere (curves VII–X, Fig. 2).

Similar changes in subsidence occur within the basins, although the relative subsidence there, on the order of a few hundred meters, is small in comparison with the total subsidence, which is on the order of several kilometers. The sign and magnitude of the apparent sea-level change will be a function of the sampling point, which may provide a means of testing the model or distinguishing this mechanism from eustatic contributions. The estimates of the magnitude of the vertical displacement from our modeling of the deflection of a uniform elastic lithosphere are conservative. Introduction of a depth-dependent rheology with a brittle-ductile transition (Goetze and Evans, 1979) in the modeling would enhance the effectiveness of the action of variations in intraplate stress (Cloetingh and others, 1982). The same is true for rheological weakening of the lithosphere due to flexural stresses (Cloetingh and others, 1982, 1983) induced by sediment loading. In areas that combine the presence of thick sedimentary loads and a high level of intraplate stress, such as the Bay of Bengal (Cloetingh and Wortel, 1985), this can even cause folding of the lithosphere (McAdoo and Sandwell, 1985). Using Seasat altimeter data and a depth-dependent rheology of the lithosphere, McAdoo and Sandwell (1985) showed that present compressional stress levels in the northeastern Indian Ocean (Fig. 5) can cause the observed basement undulations with wavelengths of roughly 200 km and amplitudes as much as 3,000 m. In general, however, for deformation not reaching the buckling limit, the effect of depth-dependent lithospheric rheology will be less dramatic (Cloetingh, in prep.).

The vertical movements within the basins become more complex when the effective flexural thickness is varied across the continental margin (Cloetingh and others, 1985a). In this case, an additional tilting of the crust is induced at the transition from oceanic to continental lithosphere. This tilting is a consequence of the change in the thickness of the layer that carries the intraplate stress, possibly due to changes in lithospheric thinning for rifted basins, and it occurs even in the absence of sediment loading. These tilts amplify the vertical displacement at the basin edge (shown in Fig. 2) when the effective elastic thickness of the continental lithosphere is less than that of the oceanic lithosphere. Studies of rifting processes in oceanic and continental lithosphere (Vink and others, 1984) indicate that continental lithosphere at passive margins might be mechanically weaker than the adjacent oceanic lithosphere. The effect is, however, highly dependent on the assumed rheological contrast between oceanic and continental lithosphere, and the degree of amplification or reduction will therefore vary for different basins.

Figure 2 demonstrates that variations in relative sea level of at least 10 m can be caused by regional changes in the in-plane stresses on the order of a few hundred bars. If these stress changes occur on a time scale of 10^6 years, then the associated sea-level change rates are on the order of at least 1 cm/1,000 yr. The actual sea-level variation for a given change in stress is controlled by the magnitude of the perturbation or deflection of the lithosphere at the time that the in-plane stress is applied or, in the context of basin evolution, by the rate of sedimentation and the response of the lithosphere to this sediment load.

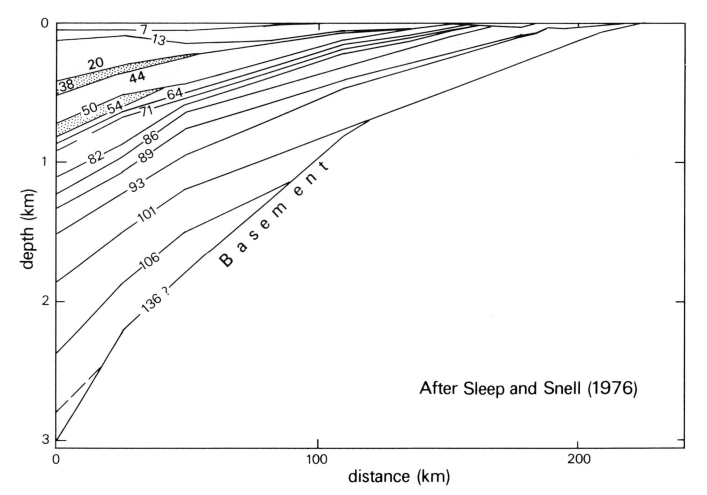

FIG. 4.—Stratigraphic cross section of the shelf of the United States Atlantic margin at Cape Hatteras. The shelf break is about 40 km from the left of the figure. Ages of formation boundaries are given in million years (after Sleep and Snell, 1976).

Independent estimates for the magnitude of apparent sea-level changes. —

Different estimates exist for the magnitude of relative sea-level changes (e.g., Bond, 1978; Pitman, 1978; Watts and Steckler, 1979). Pitman (1978) has scaled the curves of Vail and others (1977) from calculations based on changes in the volume of mid-ocean ridges caused by changes in spreading rates and ridge lengths since Cretaceous times. He estimates a maximum change of 350 m in about 70 million years. Watts and Steckler (1979) examined borehole records for the eastern margin of North America and proposed an average long-term fall in sea level of about 100–150 m since Cretaceous time, a value largely consistent (see also Kominz, 1984) with estimates derived from studies of continental flooding (Bond, 1978). Hardenbol and others (1981) used a method similar to the one employed by Watts and Steckler (1979) on well data from offshore northwest Africa. They estimated mid-Cretaceous sea level to be about 300 m above present sea level. As noted by Falvey and Deighton (1982), it is unreasonable to expect that the well sequence of Hardenbol and others (1981) with 95 million years of missing section, should provide quantitatively better estimates than the well sequences analyzed

by Watts and Steckler (1979), which had only a few tens of millions of years of missing sections. Thorne and Bell (1983) derived a eustatic sea-level curve from histograms of North Sea subsidence that is consistent with lower estimates of the amplitude of the sea-level changes.

New evidence from studies of Oligocene-Miocene carbon isotope cycles and abyssal circulation changes (Miller and Fairbanks, 1985) and from modeling subsidence at the passive margin of the United States (Watts and Thorne, 1984) has provided revised quantitative estimates for the magnitude of the mid-Oligocene fall in sea level. The magnitude of this fall in sea level, by far the largest shown in the curves of Vail and others (1977), is estimated by Miller and Fairbanks (1985) and Watts and Thorne (1984) to be at most 50–60 m. The modified curve of Vail and others (1984; see also Vail and Todd, 1981) for Jurassic sea levels retains the same overall form as the original coastal onlap and off-lap curve, but exhibits a general reduction in magnitude, with some of the corresponding sea-level changes more nearly symmetrical. These findings lead us to explore here the consequences of the possibility that a significant number of the short-term sea-level fluctuations inferred from seismic stratigraphy have a characteristic magnitude of only a few

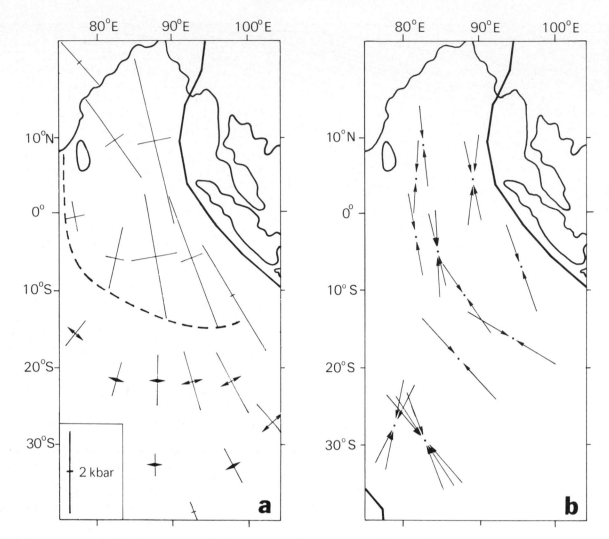

Fig. 5.—Intraplate stress field in the northeastern Indian Ocean. (a) Calculated stress field (after Cloetingh and Wortel, 1985). Principal horizontal non-lithostatic stresses averaged over a uniform elastic plate with a reference thickness of 100 km are plotted. Symbols (⟵⟶) and (⟍⟋) denote tension and compression, respectively. The length of the arrows is a measure of the magnitude of the stresses. The dashed line is the southern limit of the observed deformation in the northeastern Indian Ocean (Geller and others, 1983). (b) The orientation of maximum horizontal compressive stress inferred from a focal mechanism study by Bergman and Solomon (1985).

tens of meters within a time interval of a few million years (Aubry, 1985; S.O. Schlanger, pers. commun., 1986). We propose that such fluctuations may reflect the magnitude and rate of the underlying variation in intraplate stresses.

EVIDENCE FOR FLUCTUATIONS IN INTRAPLATE STRESS FIELDS

Independent studies of lithospheric deformation in active continental margin and intraplate tectonic settings have provided strong evidence for the existence of horizontal stresses in the lithosphere and for stress levels that may reach magnitudes of as much as a few kilobars. Folding of the oceanic lithosphere in the southern part of the Bay of Bengal has been interpreted in terms of deformation caused by compressive forces on the order of several kilobars (Geller and others, 1983; McAdoo and Sandwell, 1985). Departures from isostatic equilibrium in several tectonic provinces within Australia have been attributed to in-plane stresses at the

level of a few kilobars (Lambeck and others, 1984). The formation of sedimentary basins by lithospheric stretching also requires horizontal (tensional) stresses on the order of a few kilobars (Houseman and England, 1986).

Recent numerical modeling of stresses induced by plate-tectonic forces in the lithosphere (Wortel and Cloetingh, 1981, 1983; Cloetingh and Wortel, 1985) has yielded greater insight into the underlying causes of the variations in stress level observed in different plates. The driving plate-tectonic forces are the ridge push, which results from the elevation of the spreading ridge above the adjacent ocean floor and the thickening of the lithosphere with cooling, and the pull acting on the downgoing slab in a subduction zone. The incorporation into the modeling of the dependence of slab pull and ridge push on the age of the oceanic lithosphere provided a quantitative basis for analyzing, explaining, and predicting various deformational processes in lithospheric plates from the resulting stress field.

Of particular interest is the comparison of stress fields in various plates. Modeling of the stress field in the Indo-Australian plate (Cloetingh and Wortel, 1985) has shown that the joint occurrence in this single plate of an exceptionally high level of compressive deformation in the plate's interior (Bergman and Solomon, 1985; McAdoo and Sandwell, 1985) and normal faulting in the near-ridge areas (Wiens and Stein, 1984) is a transient feature unique to the present dynamic situation of the Indo-Australian plate. The high level of intraplate seismicity in the Indo-Australian plate enables the intraplate stress field to be estimated through the determination of earthquake focal mechanisms (Bergman and Solomon, 1985). Stress orientation data from Bergman and Solomon (1985) given in Figure 5b show a rotation of the observed stress field in the northern Indian Ocean from North-South-oriented compression in the north to a more Northwest-Southeast-directed compression in the southeastern part of the Bay of Bengal region. A similar pattern is found in the calculated stress model (Fig. 5a). Furthermore, the stress field as calculated gives a consistent explanation for the observed concentration of seismic activity and significant deformation in the oceanic crust (Geller and others, 1983; McAdoo and Sandwell, 1985) in the northeastern Indian Ocean segment of the Indo-Australian plate. The stresses in the Indo-Australian plate are an order of magnitude greater than those we have calculated using the same technique for the Nazca plate, which is characterized by a low level of intraplate seismicity. There, stresses are on the order of 500 bars (Wortel and Cloetingh, 1985), a stress level more characteristic for plates not involved in continental collision or fragmentation processes.

An important feature of the stress field in the Indo-Australian plate is the occurrence of strong lateral *spatial variations* in the stress field along its trench systems (Fig. 6). The lateral stress variations along the Java-Sumatra trench (Cloetingh and Wortel, 1985), and similar variations along the Peru-Chile trench (Wortel and Cloetingh, 1985) are in excellent quantitative agreement with marine geophysical data on the structure of the trench regions. There is a close

correspondence between the extent to which grabens are developed on the seaward trench slope, and hence the style of accretionary tectonics in the trench region (Hilde, 1983), and the calculated regional stress field. The transition from a compressive-stress field off and parallel to Sumatra to a tensional-stress field normal to Java is the result of the contrast in age of the subducted lithosphere under Sumatra (40–70 Ma) and Java (140 Ma). Following the Sunda arc in an easterly direction, under-thrusting of continental shelf occurs from just west of Flores onward, which results in a compressive-stress field off this trench segment. Rapid *temporal variations in stress* have occurred here, at the onset of the Banda arc collision, where a period during which the stress field was controlled by the slab pull associated with subduction of old oceanic lithosphere has been followed by a phase of net compressive resistance due to the arrival of buoyant continental lithosphere at the subduction zone. A stress field dominated by tension changed to a stress field of predominantly compressional character 5 Ma. As demonstrated by observed paleo-stress orientations in western Europe (Letouzey and Trémolières, 1980), such changes induced at the convergent boundaries propagate into the interiors of the plates, where they affect passive margins and intracratonic basins.

Rapid temporal changes in stress field are not limited to collision processes. Geologic evidence also points to episodic tectonic events on time scales of a few million years (Megard and others, 1984), possibly caused by the response of individual plates to longer term global readjustments of the plate motion patterns. Rifting or fragmentation of plates is also associated with drastic changes in stress state. An example of such an event is the breakup of the Farallon plate into the Cocos and Nazca plates. We have demonstrated (Wortel and Cloetingh, 1983) that fragmentation of the Farallon plate took place under the influence of tensional stresses induced by age-dependent slab pull forces. Strong support for the stress modeling procedure and the fragmentation hypothesis formulated for the Farallon plate is provided by Warsi and others (1983). In their marine geophysical studies of the Nazca plate off Peru, these authors discovered that the Mendana Fracture zone, oriented perpendicular to the Peruvian Trench, is in an incipient stage of spreading, very much along the lines suggested for the origin of the Cocos-Nazca spreading center (Wortel and Cloetingh, 1981). Concentration of slab pull forces dominates the plate-tectonic stress field (Wortel and Cloetingh, 1981; Patriat and Achache, 1984; Cloetingh and Wortel, 1985). Passive margins in plates not involved in collision or subduction processes are not subject to the influence of these slab pull forces. In such circumstances, other sources of stress (e.g., membrane stresses; Turcotte and Oxburgh, 1976) can be much more important than the regional stress field induced by the remaining plate-tectonic forces. Thus, for passive margins located in the interiors of the American plates, stresses induced by sediment loading are an order of magnitude greater than the regional stress field associated with ridge push forces (Cloetingh and others, 1982). The latter is typically on the order of a few hundred bars (Richardson and others, 1979). Under such circumstances, local adjustment of stresses at passive mar-

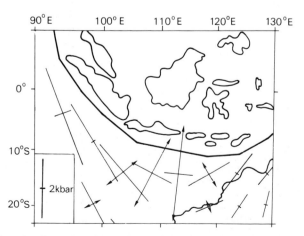

FIG. 6.—Lateral variations in the intraplate stress field in the Indo-Australian plate segment close to the Sunda arc. Figure conventions as in Figure 5a.

gins (e.g., by initiation of spreading in adjacent oceans) rarely involves changes of more than a few hundred bars.

DISCUSSION

In the foregoing, we have argued for a tectonic cause for short-term apparent changes in sea level. Others (e.g., Bally, 1980; 1982; Watts, 1982) have proposed a tectonic explanation but were unable to identify a specific mechanism for short-term lowering in sea level. In particular, Bally (1982) has pointed out the strong correlation in timing of plate-tectonic reorganizations and some lowerings in sea level shown in the curves of Vail and others (1977).

The tectonic mechanism discussed here can explain contemporaneous fluctuations in apparent sea level in neighboring basinal settings. It is important to realize that the degree of correlation between the timing of sea level changes induced by fluctuations in intraplate stress fields will depend primarily ont he dimensions of the stress province. Numerical modeling (Cloetingh and Wortel, 1985; Wortel and Cloetingh, 1985) and observation of lithospheric deformation (Zoback and Zoback, 1980; Wiens and Stein, 1984; Bergman and Solomon, 1985; McAdoo and Sandwell, 1985) show that the stress provinces can vary in size from that of an entire lithospheric plate to that of a small part of a plate. Boundaries of plate-tectonic stress provinces in plates do not necessarily coincide with the transition of oceanic to continental lithosphere at passive margins. Examples are found in the North American and Indo-Australian plates. The eastern and central parts of North America form a compressional-stress province together with the northern Atlantic, while the western part of the continent is subject to a tensional regime (Zoback and Zoback, 1980). Western and central Australia together with the adjoining Indian Ocean share a regional compressional-stress field, while eastern Australia and the northern part of the Tasman Sea are under the influence of a tensional regional stress field (Cloetingh and Wortel, 1985).

Ziegler (1982) and Sclater and Christie (1980) have pointed out the existence of a correlation between tectonic phases in the North Sea basin and changes in the history of opening of the Atlantic and Tethyan oceans (see also Schwan, 1985, for a more general discussion). Changes in spreading rates in the Atlantic and Tethyan regions are probably caused by or associated with changes in plate-tectonic forces. Both modeling of lithospheric stress fields and measurements of paleo-stress directions (Letouzey and Trémolières, 1980; J. Letouzey, pers. commun. 1986) strongly suggest that the induced stress changes are propagated over great distances from the plate boundaries into the northwestern European basins, which provides a dynamic explanation for the observed correlations. Furthermore, a causal relation might exist between these stress changes and the timing of salt diapirism in the North Sea basin (Thorne, 1986).

Small-scale variations in stress can occur superimposed on long-wavelength stress patterns. Examples of these are found in oblique-slip zones (Ballance and Reading, 1980), where alternations of compressional and tensional segments occur. The effects of such alternation in stress fields are reflected in the seismic stratigraphic onlap and offlap patterns along the eastern margin of Canada (A.J. Tankard, pers. commun., 1985) and are consistent with the tectonic mechanism discussed here.

The regional character of the tectonic mechanism sheds new light on some observed deviations from the global sea-level curve of Vail and others (1977). These include examples from well-studied areas in both the northern (Hallam, 1984; Harris and others, 1984) and southern (Chaproniere, 1984; Carter, 1985) hemispheres. While such deviations from a global pattern are a natural consequence of the character of the tectonic mechanism, they do not preclude the presence of global events in the stratigraphic record. These are to be expected when major reorganizations in stress fields occur simultaneously in more than one plate, as conjectured for the early Cenozoic global plate reorganization (Rona and Richardson, 1978), or when glacio-eustatic changes dominate.

An interesting outcome of our analysis is the dependence of the magnitude of the changes in sea level on the age of the margin. Thorne and Watts (1984) showed that relatively slow changes in sea level can cause unconformities on old passive margins, while relatively rapid changes are required to form unconformities in a subsiding young margin. Note that the tectonic mechanism discussed herein is most effective for young passive margins that are subject to rapid sediment loading.

Relative changes in sea level documented for the eastern North American margin concommitant with the breakup of the eastern American continental margin (Sheridan, 1983) and early Cretaceous volcanism on the northeastern American margin (Jansa and Pe-Piper, 1985) may reflect the adjustment of stress associated with these tectonic events. Similarly, high-frequency oscillations in Cretaceous sea levels having characteristic periods of a few million years, which cannot be correlated with fluctuations in spreading rates and ridge lengths (S.O. Schlanger, pers. commun., 1986), could be the result of adjustment of stresses in the process of the opening of the South Atlantic. Other areas where the timing of tectonic events and associated stress changes correspond to short-term changes in apparent sea level include the south Pyrenean basin (Atkinson and Elliott, pers. commun., 1985), the Illinois basin (De Vries Klein, 1986), the late Cenozoic basins of the Mediterranean region (Meulenkamp, 1982; Meulenkamp and Hilgen, 1986), and the Australian passive margins (Cloetingh and others, 1985a). Here, changes from transgression to regression (Steele, 1976) at the onset of the Banda-arc and Himalayan collison events, associated with stress changes from tension to compression (Cloetingh and Wortel, 1985), have been noted.

In general, stress variations of a few hundred bars associated with local adjustment of stress at passive margins and intracratonic basins would suffice to explain a significant number of the short-term variations in sea level inferred from the seismic stratigraphic record. Stress changes of more than 1 kbar are required in order to induce sea-level fluctuations with magnitudes on the order of 50 m such as inferred for the mid-Oligocene event (Miller and Fairbanks, 1985; Watts and Thorne, 1984). These must be related to major reorganizations in lithospheric-stress fields. As pointed out by Engebretson and others, (1985), major plate reorganizations generally occur with characteristic intervals of a few tens of millions of years. The duration of the individual events, and hence the associated stress changes, is, however, much shorter and on the order of at most a few million years (Engebretson and others, 1985). In this context, it is interesting to note that the mid-Oligocene fall in sea level is coincident with a global-plate reorganization, presumably with a concomitant change in the paleo-stress state, in which the breakup of the Farallon plate into the Cocos and Nazca plates (Wortel and Cloetingh, 1981) played a major part. The superposition of the effect of the tectonically induced fall in sea level and an important glacio-eustatic event might explain the exceptional magnitude of the mid-Oligocene fall in apparent sea level (see also Schlanger and Premoli-Silva, 1986).

Although we have concentrated on the relation between tectonics and stratigraphy at passive margins, the tectonic mechanism discussed is applicable in a wider range of sedimentary environments. Other settings where lithosphere is flexed under the influence of sediment loading occur in foreland basins (Quinlan and Beaumont, 1984) and in the vicinity of volcanic islands (Ten Brink and Watts, 1985). Despite its height of only a few hundred meters, the peripheral bulge flanking foreland basins is of particular stratigraphic interest (Quinlan and Beaumont, 1984). The action of intraplate stresses of tensional or compressional character, of which the latter is more natural in this tectonic setting, can reduce or amplify the height of the peripheral bulge and, consequently, greatly influence the stratigraphic record of foreland basins. Intraplate stresses may also affect the postulated interaction (Quinlan and Beaumont, 1984) of adjacent foreland arches.

The detailed study by Ten Brink and Watts (1985) of the seismic stratigraphy of the flexural moat flanking the Hawaiian Islands provides an example of flexural deformation and associated onlap and offlap patterns in a distinctly different intraplate tectonic setting. The observed pattern of initial onlap followed by offlap and the occurrence later of rapid variations in onlap and offlap in the upper part of the stratigraphic record may be an expression of relaxation of tensional stresses during the formation of this part of the Hawaiian Island chain, followed by an ensuing period of fluctuations in intraplate stress fields. The seismic stratigraphic record at Hawaii and other volcanic islands provides a potential source of information on fluctuations in the paleo-stress field in the Pacific plate, which lacks the stratigraphic record found at passive margins and foreland basins of other plates.

From the above, we anticipate that the seismic stratigraphic record may provide a useful new source of information for paleo-stress fields. As has been shown recently (Cloetingh, 1986; Lambeck and others, 1987), a synthetic paleo-stress curve derived from the seismic stratigraphic record mirrors the tectonic evolution of northwestern Europe and the North Sea basin: rifting episodes correspond to relaxation of tensional paleo-stresses. Examination of the stratigraphic record of individual basins in a wide range of

tectonic settings in connection with independent numerical modeling of paleo-stresses, such as that carried out by Wortel and Cloetingh (1981), is required to exploit fully the potential of this new approach.

CONCLUSIONS

The interaction of intraplate stress with flexural loading forms a new element in basin stratigraphy. Stress variations in the lithosphere of a few hundred bars can explain a significant part of the seismic stratigraphic record at passive margins. The tectonic mechanism for regional sea-level fluctuations proposed by Cloetingh and others (1985a,b) explains observed correlations (Bally, 1980, 1982) between the timing of lowerings in sea level inferred from the seismo-stratigraphic record and tectonic events. The model also explains contemporaneous fluctuations in apparent sea level in neighboring basin settings. The action of changing intraplate stresses is not restricted to passive margins but also modifies the veritcal movements within intracratonic basins, foreland basins, and flexural moats flanking intraoceanic volcanic complexes. For this reason, the model may provide a tectonic explanation for some of the observed correlations between the timing of apparent sea-level changes in oceanic and intracontinental regions.

ACKNOWLEDGMENTS

K. Lambeck and H. McQueen, and M.J.R. Wortel made significant contributions to the work on tectonic causes of changes in sea level and to the modeling of intraplate stress fields, respectively. M.J.R. Wortel, J. Thorne, and J.E. Meulenkamp offered valuable suggestions. J. Veizer, S. Schlanger, and G. de Vries Klein furnished preprints of their papers.

REFERENCES

AUBRY, M.-P., 1985, Northwestern European Paleogene magnetostratigraphy, biostratigraphy and paleogeography: calcareous nannofossil evidence: Geology, v. 13, p. 198–202.

BALLANCE, P. F. AND READING, H. G., eds., 1980, Sedimentation in Oblique Slip Mobile Zones: International Association of Sedimentologists Special Publication, v. 4, 236 p.

BALLY, A. W., 1980 Basins and subsidence—a summary: American Geophysical Union Geodynamics Series, v. 1, p. 5–20.

———, 1982, Musings over sedimentary basin evolution: Philosophical Transactions of the Royal Society of London, v. A 305, p. 325–338.

BERGMAN, E. A., AND SOLOMON, S. C. 1985, Earthquake source mechanisms from body-waveform inversion and intraplate tectonics in the northern Indian Ocean: Physics of Earth and Planetary Interiors, v. 40, p. 1–23.

BODINE, J. H., STECKLER, M. S., AND WATTS, A. B., 1981, Observations of flexure and the rheology of the oceanic lithosphere: Journal of Geophysical Research, v. 86, p. 3695–3707.

BOND, G., 1978, Speculations on real sea-level changes and vertical motions of continents at selected times in the Cretaceous and Tertiary periods: Geology, v. 6, p. 247–250.

CARTER, R. M., 1985, The mid-Oligocene Marshall paraconformity, New Zealand: coincidence with global eustatic sea-level fall or rise: Journal of Geology, v. 93, p. 359–371.

CHARPRONIERE, G. C. H., 1984, Oligocene and Miocene larger foraminiferida from Australia and New Zealand: Bureau of Mineral Resources, Geology and Geophysics Bulletin, v. 188, p. 1–98.

CLOETINGH, S., 1986, Intraplate stresses: a new tectonic mechanism for relative fluctuations of sea level: Geology, v. 14, p. 617–620.

————, McQueen, H., and Lambeck, K., 1985a, On a tectonic mechanism for regional sea level variations: Earth and Planetary Science Letters, v. 75, p. 157–166.

————, ————, and ————, 1985b, Intraplate stresses as a mechanism for relative sea level fluctuations: Symposium on Cycles and Periodicity in Geologic Events, Evolution and Stratigraphy, Program and Abstracts, Princeton University, Princeton, New Jersey, p. 6.

————, and Wortel, R. 1985, Regional stress field of the Indian plate: Geophysical Research Letters, v. 12, p. 77–80.

————, ————, and Vlaar, N. J., 1982, Evolution of passive continental margins and initiation of subduction zones: Nature, v. 297, p. 139–142.

————, ————, and ————, 1983, State of stress at passive margins and initiation of subduction zones: American Association of Petroleum Geologists Memoir 34, p. 717–723.

De Vries Klein, G., 1986, Mechanisms of tectonic subsidence and sedimentary response in the Illinois basin and other cratonic basins: Symposium on New Perspectives in Basin Analysis, Program and Abstracts, Minneapolis, Minnesota, p. 21.

Engebretson, D. C., Cox, A., and Gordon, R. G., 1985, Relative motions between oceanic and continental plates in the Pacific Basin: Geological Society of America Special Paper, v. 206, p. 1–56.

Falvey, D. A., and Deighton, I. 1982, Recent advances in burial and thermal geohistory analysis: Australian Petroleum Exploration Association Journal, v. 22, p. 65–81.

Frakes, L. A., 1979, Climates Through Geologic Time: Elsevier, Amsterdam, 310 p.

Geller, C. A., Weissel, J. K. and Anderson, R. N., 1983, Heat transfer and intraplate deformation in the central Indian Ocean: Journal of Geophysical Research, v. 88, p. 1018–1032.

Goetze, C., and Evans, B., 1979, Stress and temperature in the bending lithosphere as constrained by experimental rock mechanics: Geophysical Journal of the Royal Astronomical Society, v. 59, p. 463–478.

Hallam, A., 1984, Pre-Quarternary sea-level changes: Annual Reviews, Earth and Planetary Sciences, v. 12, p. 205–243.

Hardenbol, J., Vail, P. R., and Ferrer, J., 1981, Interpreting palaeoenvironments, subsidence history and sea-level changes of passive margins from seismic and biostratigraphy: Oceanologica Acta, v. 4, p. 33–44.

Harris, P. M., Frost, S. H., Seiglie, G. A., and Schneiderman, N., 1984, Regional unconformities and depositional cycles, Cretaceous of the Arabian peninsula: American Assoication of Petroleum Geologists Memoir 36, p. 67–80.

Heller, P. L. and Angevine, C. L., 1985, Sea level cycles during the growth of Atlantic type oceans: Earth and Planetary Science Letters, v. 75, p. 417–426.

Hilde, T. W. C., 1983, Sediment subduction versus accretion around the Pacific: Tectonophysics, v. 99, p. 381–397.

Houseman, G. A., and England, P. C., 1986, A dynamical model of lithosphere extension and sedimentary basin formation: Journal of Geophysical Research, v. 91, p. 719–729.

Jansa, L. F., and Pe-Piper, G., 1985, Early Cretaceous volcanism on the northeastern American margin and implications for plate tectonics: Geological Society of America Bulletin, v. 96, p. 83–91.

Kominz, M. A., 1984, Oceanic ridge volumes and sea-level change—an error analysis: American Assoication of Petroleum Geologists Memoir 36, p. 109–127.

Lambeck, K., Cloetingh, S., and McQueen, H., 1987, Intraplate stresses and apparent changes in sea level: The barriers of northwestern Europe: Canadian Society of Petroleum Geologists, Memoir 12, p. 259–268.

Lambeck, K., McQueen, H. W. S., Stephenson, R. A., and Denham, D., 1984, The state of stress within the Australian continent: Annale Geophysica, v. 2, p. 723–741.

Letouzey, J., and Trémolières, P., 1980, Paleo-stress fields around the Mediterranean since the Mesozoic from microtectonic, comparison with plate tectonic data: Rock Mechanics Supplement, v. 9, p. 173–192.

McAdoo, D. C., Martin C. F., Poulouse, S., 1985, Seasat observation of flexure: evidence for a strong lithosphere; Tectonophysics, v. 116, p. 209–222.

————, and Sandwell, D. T., 1985, Folding of oceanic lithosphere: Journal of Geophysical Research, v. 90, p. 8563–8568.

Megard, F., Noble, D. C., McKee E. H., and Bellon, H., 1984, Multiple pulses of Neogene compressive deformation in the Ayacucho intermontane basin, Andes of Central Peru: Geological Society of America Bulletin, v. 95, p. 1108–1117.

Meulenkamp, J. E., 1982, On the pulsating evolution of the Mediterranean: Episodes, v. 1, p. 13–16.

————, and Hilgen, F. J., 1986, Event stratigraphy, basin evolution and tectonics of the Hellenic and Calabro-Sicilian arcs: Developments in Geotectonics, v. 21, p. 327–350.

Miall, A. D., 1986, Eustatic sea level changes interpreted from seismic stratigraphy: a critique of the methodology with particular reference to the North Sea Jurassic record: American Association of Petroleum Geologists Bulletin, v. 70, p. 131–137.

Miller, K. G., and Fairbanks, R. G., 1985, Oligocene—Miocene global carbon and abyssal circulation changes: American Geophysical Union Geophysical Monograph, v. 32, p. 469–486.

Patriat, P., and Achache, J. 1984, India-Eurasia collision chronology has implications for crustal shortening and driving mechanisms of plates: Nature, v. 311, p. 615–621.

Pitman, III, W. C., 1978, Relationship between eustacy and stratigraphic sequences at passive margins: Geological Society of American Bulletin, v. 89, p. 1389–1403.

————, and Golovchenko, X., 1983, The effect of sea level change on the shelfedge and slope of passive margins: Society of Economic Paleontologists and Mineralogists Special Publication 33, p. 41–58.

Quinlan, G. M., and Beaumont, C., 1984, Appalachian thrusting, lithospheric flexure and the Paleozoic stratigraphy of the eastern Interior of North America: Canadian Journal of Earth Sciences, v. 21, p. 973–996.

Richardson, R. M., Solomon, S. C., and Sleep, N. H., 1979, Tectonic stress in the plates: Reviews of Geophysics and Space Physics, v. 17, p. 981–1019.

Rona, P. A., and Richardson, E. S., 1978, Early Cenozoic global plate reorganization: Earth and Planetary Science Letters, v. 40, p. 1–11.

Schlanger, S. O., and Premoli-Silva, I., 1986, Oligocene sea level falls recorded in mid-Pacific atoll and archipelagic apron settings: Geology, v. 14, p. 392–395.

Schwan, W., 1985, The worldwide active middle/late Eocene geodynamic episode with peaks at ±45 and ±37 m.y.b.p., and implications and problems of orogeny and sea-floor spreading: Tectonophysics, v. 115, p. 197–234.

Sclater, J. G., and Christie, P. A. F., 1980, Continental stretching: an explanation of the post-mid-Cretaceous subsidence of the central North Sea basin: Journal of Geophysical Research, v. 85, p. 3711–3739.

Sheridan, R. E., 1983, Phenomena of pulsation tectonics related to break-up of the eastern North American continental margin: Tectonophysics, v. 94, p. 169–185.

Sleep, N. H., 1971, Thermal effects of the formation of Atlantic continental margins by continental break-up: Geophysical Journal of the Royal Astronomical Society, v. 24, p. 325–350.

————, and Snell, N. S., 1976, Thermal contraction and flexure of mid-continent and Atlantic marginal basins: Geophysical Journal of the Royal Astronomical Society, v. 45, p. 125–154.

Sloss, L. L., 1979, Global sea level change: a view from the craton: American Assoication of Petroleum Geologists Memoir 29, p. 461–467.

Southam, J. R., and Hay, W. W., 1981, Global sedimentary mass balance and sea level changes, in Emiliani, C., ed., The Oceanic Lithosphere, The Sea, v. 7: Wiley, New York, p. 1617–1684.

Steele, R. J., 1976, Some concepts of seismic stratigraphy with application to the Gippsland basin: Australian Petroleum Exploration Association Journal, v. 16, p. 67–71.

Ten Brink, U. S., and Watts, A. B., 1985, Seismic stratigraphy of the flexural moat flanking the Hawaiian Islands: Nature, v. 317, p. 421–424.

Thorne, J. R., 1986, A quantitative analysis of North Sea subsidence: Symposium on New Perspectives in Basin Analysis, Program and Abstracts, Minneapolis, Minnesota, p. 31.

————, and Bell, R., 1983, A eustatic sea level curve from histograms of North Sea subsidence: EOS Transactions, American Geophysical Union, v. 64, p. 858.

————, and Watts, A. B., 1984, Seismic reflectors and unconformities at passive continental margins: Nature, v. 311, p. 365–368.

TURCOTTE, D. L., AND AHERN, J. L., 1977, On the thermal and subsidence history of sedimentary basins: Journal of Geophysical Research, v. 82, p. 3762–3766.

————, AND OXBURGH, E. R., 1976, Stress accumulation in the lithosphere: Tectonophysics, v. 35, p. 183–199.

VAIL, P. R., HARDENBOL J., AND TODD, R. G., 1984. Jurassic unconformities, chronostratigraphy, and sea level changes from seismic stratigraphy and biostratigraphy: American Association of Petroleum Geologists Memoir 36, p. 129–144.

————, MITCHUM R. M., Jr., AND THOMPSON, S., III, 1977, Global cycles of relative changes of sea level: American Association of Petroleum Geologists Memoir 26, p. 83–97.

————, AND TODD, R. G., 1981, Northern North Sea Jurassic unconformities, chronostratigraphy and sea level changes from seismic stratigraphy, *in* Illing, L. V., and Hobson, G. D., eds., Petroleum Geology of the Continental Shelf of North-West Europe: Heyden, London, p. 216–235.

VEIZER, J., 1985, Carbonates and ancient oceans: isotopic and chemical record on time scales of $10^7 - 10^9$ years: American Geophysical Union Geophysical Monograph, v. 32, p. 595–601.

VINK, G. E., MORGAN, W. J., AND ZHAO, W.-L., 1984, Preferential rifting of continents: a source of displaced terranes: Journal of Geophysical Research, v. 89, p. 10072–10076.

WARSI, W. E. K., HILDE, T. W. C. AND SEARLE, R. C., 1983, Convergence structures of the Peru-Chile Trench between 10°S and 14°S: Tectonophysics, v. 32, p. 331–351.

WATTS, A. B., 1982, Tectonic subsidence, flexure and global changes of sea level: Nature, v. 297, p. 469–474.

————, KARNER, G. D., AND STECKLER, M. S., 1982, Lithospheric flexure and the evolution of sedimentary basins: Philosophical Transactions of the Royal Society of London, v. A 305, p. 249–281.

————, AND STECKLER, M. S., 1979, Subsidence and eustasy at the continental margin of eastern North America: American Geophysical Union, Ewing Series, v. 3, p. 218–234.

————, AND THORNE, J., 1984, Tectonics, global changes in sea level and their relationship to stratigraphical sequences at the US Atlantic continental margin: Marine Petroleum Geology, v. 1, p. 319–339.

WIENS, D. A., AND STEIN, S., 1983, Age dependence of oceanic intraplate seismicity and implications for lithosphere evolution: Journal of Geophysical Research, v. 88, p. 6455–6468.

————, AND ————, 1984, Intraplate seismicity and stresses in young oceanic lithosphere: Journal of Geophysical Research, v. 89, p. 11442–11464.

WORTEL, R., AND CLOETINGH, S., 1981, On the origin of the Cocos-Nazca spreading center: Geology, v. 9, p. 425–430.

————, AND ————, 1983, A mechanism for fragmentation of oceanic plates, American Association of Petroleum Geologists Memoir 34, p. 793–801.

————, AND ————, 1985, Accretion and lateral variations in tectonic structure along the Peru-Chile trench: Tectonophysics, v. 112, p. 443–462.

ZIEGLER, P. A., 1982, Geological Atlas of Western and Central Europe: Shell Internationale Petroleum Maatschappij, The Hague, 130 p.

ZOBACK, M. L., AND ZOBACK, M. 1980, State of stress in the conterminous United States: Journal of Geophysical Research, v. 85, p. 6113–6156.

EVIDENCE FOR AND AGAINST SEA-LEVEL CHANGES FROM THE STABLE ISOTOPIC RECORD OF THE CENOZOIC

DOUGLAS F. WILLIAMS

Isotope Stratigraphy Group, Department of Geological Sciences, University of South Carolina, Columbia, South Carolina 29208

ABSTRACT: The stable oxygen isotope record for the Cenozoic is characterized by a series of large third-order steps of +1 per mil superimposed on a long-term second-order trend. This second-order trend accounts for a $\delta^{18}O$ change of nearly +4 per mil from the early Eocene into the Neogene. The second- and third-order changes in the $\delta^{18}O$ signal are driven primarily by a combination of glacio-eustatic sea-level and ocean paleotemperature changes. These changes are global responses to evolving circulation and climate patterns. Timing of the $\delta^{18}O$ events is in good agreement with the seismically defined changes in the coastal-onlap curve (Vail and others, 1977). Agreement in the timing of events supports a common mechanism, perhaps that glaciation is apparent throughout much of the record and certainly intensified beginning in the Neogene. Agreement is not good between the magnitudes of apparent changes in sea level using the EXXON onlap record and oceanic $\delta^{18}O$ events. Consideration of the $\delta^{18}O$, ice volume, and sea-level relationships during the Pleistocene suggests that sinusoidal eustatics, i.e., the rise and fall of sea level being equal, is not a good assumption at fourth- and fifth-order sea-level events. Although interpretation of the $\delta^{18}O$ record is not without its assumptions and limitations, it offers an independent geochemical check on seismically defined changes in stratal patterns.

INTRODUCTION

The integration of seismic stratigraphy and biostratigraphy (Fig. 1) has significantly advanced our knowledge of regional and perhaps global-unconformity surfaces (Vail and others, 1977; Pitman, 1978; Watts, 1982; Thorne and Watts, 1984; see also this volume). Translation of these seismically defined stratal patterns into regional-unconformity surfaces, and then into a global record of eustatic sea level, is not without controversy. The primary reasons for some of the controversy involve complex geoidal considerations (Morner, this volume), uncertainties in the timing of unconformity surfaces in a precise manner, and difficulties in quantifying the magnitude and rates of sea-level changes using the seismic approach.

I suggest that the stable oxygen isotope stratigraphies of exploration wells in passive-margin settings can add an important new dimension concerning the timing, magnitude, and rates of sea-level change. The chemical record of marine carbonates contains important information about Cenozoic, Mesozoic, and perhaps Paleozoic sea levels, particularly when integrated with seismic and biostratigraphic interpretations. Stable carbon isotopic records also provide important new information regarding the burial of organic carbon as a function of sea-level-influenced sedimentation (Scholle and Arthur, 1980; Arthur, 1983; Shackleton, 1985). In addition, the combined $\delta^{18}O$ and $\delta^{13}C$ records provide a global stratigraphic framework which is independent of local or regional biostratigraphic zonations. This framework can be used to refine the chronostratigraphy of exploration wells and to enhance the detection of unconformity surfaces that are transparent to biostratigraphic resolution. This enhanced precision will improve the correlation of unconformity surfaces.

The purpose of this paper is to compare the information about Cenozoic sea levels contained in the oxygen isotope record with that in the relative sea-level record determined from seismic stratal patterns (Vail and others, 1977; Vail and Hardenbol, 1979). As shown in Figure 2, the Cenozoic oxygen isotope record should provide an independent evaluation of the seismically defined sea-level record, because the major variables controlling the sequence-boundary patterns and the $\delta^{18}O$ signals of foraminifera are considerably different. The types of stratal patterns found on continental margins are determined by the volume and supply rates of sediments, subsidence rates, tectonism, and eustatic sea-level changes.

The primary variables which determine the $\delta^{18}O$ signal in marine carbonates are water temperature, postdepositional diagenesis, and the $^{18}O/^{16}O$ ratio of sea water. The effects of temperature on the global $\delta^{18}O$ record of the Cenozoic can be reduced, but not eliminated, by utilizing benthic foraminifera from deep-sea cores where water temperatures are less variable than in surface waters. Matthews and Poore (1980) and Matthews (1984) have attempted to model the sea-level versus temperature components of the $\delta^{18}O$ record by using the signal in tropical planktonic foraminifera. Their conclusions opened for discussion the possibility of significant volumes of ice on the earth prior to the middle Miocene (Shackleton and Kennett, 1975; Woodruff and others, 1981; Savin and others, 1981; among others). The effects of diagenesis on the $\delta^{18}O$ record will vary as a function of time, temperature, and depth of burial. For the most part, these effects can be checked by using specimens that are well preserved and whose isotopic compositions are well understood. Certainly, the potential of diagenetic effects becomes more likely in the Mesozoic and Paleozoic and may vary significantly in carbonate versus siliciclastic facies.

Lastly, by establishing global patterns for the $\delta^{18}O$ and $\delta^{13}C$ records for various time periods and ocean basins, it should be possible to determine the eustatic component controlling the character and amplitude of the isotopic signal. To a first approximation, the amount of sea water removed from the oceanic reservoir and deposited in ice sheets during a drop in sea level is translated into an increase in the $^{18}O/^{16}O$ ratio of the remaining sea water. The best estimate for this effect in the Pleistocene is a 0.11 per mil change in $\delta^{18}O$ for every 10 m of sea water removed (Fairbanks and Matthews, 1978; see Matthews, 1984, for a review). While it is not possible to separate unequivocally the effects of temperature and seawater compositional changes in the past, the $\delta^{18}O$ signal is currently the best

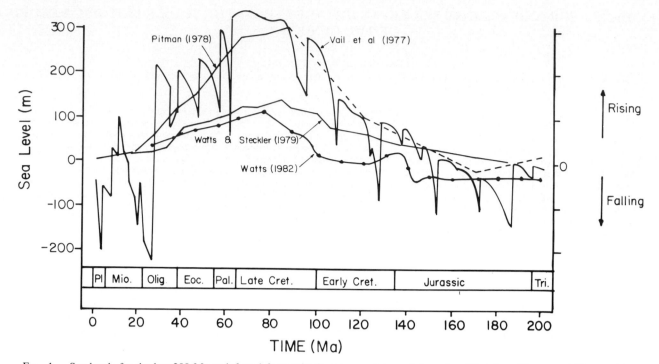

FIG. 1.—Sea levels for the last 200 Ma as inferred from seismic-sequence interpretations (modified from Thorne and Watts, 1984).

parameter available for estimating the magnitude and rate of change in sea level independently of the seismic-onlap records.

This paper examines the $\delta^{18}O$ evidence for the timing, magnitude, and rate change involved in second-, third-, and fifth-order sea-level events in the Cenozoic. The definition of the second-, third-, and fifth-order events is the same as that used by the EXXON group, i.e., events that occur on the order of 36 to 9 Ma, 5 to 1 Ma, and 100 ka (Vail and others, 1977).

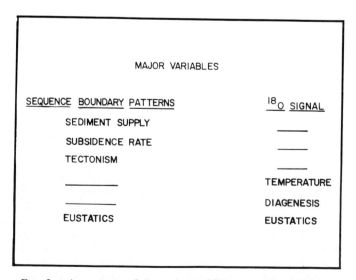

FIG. 2.—A summary of the major variables controlling seismic-sequence patterns and the $\delta^{18}O$ signal in Cenozoic marine carbonates.

OXYGEN ISOTOPIC MODEL FOR CENOZOIC SEA LEVELS

Figure 3 presents a composite $\delta^{18}O$ record for deep-sea benthic foraminifera from Atlantic and Pacific Deep Sea Drilling Project sites. The data have been time-averaged at 1-million-year increments to filter out high-frequency events of the fourth-order and higher (with periods <400 ka, Miller and Fairbanks, 1985). No attempts have been made in compiling Figure 3 to account for the effects of data from different species and biostratigraphic uncertainties of different sites. This record serves as the best approximation of the global $\delta^{18}O$ signal for the Cenozoic and provides a basis for the ensuing comparisons with the seismic-sequence record.

Timing of global $\delta^{18}O$ events.—

The overall change in $\delta^{18}O$ through the Cenozoic is nearly 4 per mil from the lightest $\delta^{18}O$ value in the early Eocene to the very positive values which characterize the Pleistocene (Fig. 3). Positive 1-per-mil $\delta^{18}O$ events are centered at approximately 3, 15, 38, 41–46, and 48–53 Ma. Other significant events, defined as being >0.2 per mil, are present at 5.5, 7–9, 10, 12, 18, 24–26, and 29.5–30.5 Ma and possibly at 41, 44, 58 and 61 Ma, although the paucity of data prior to 40 Ma makes these excursions less evident (Fig. 3). Other significant changes are present in individual records for individual sites (Williams and others, in prep.), but these events are unresolvable in Figure 3 because the data are averaged to filter out events of less than 400 ka duration.

Also shown on the left-hand margin of Figure 3 are the positions of type 1 and 2 unconformities defined by seismi-

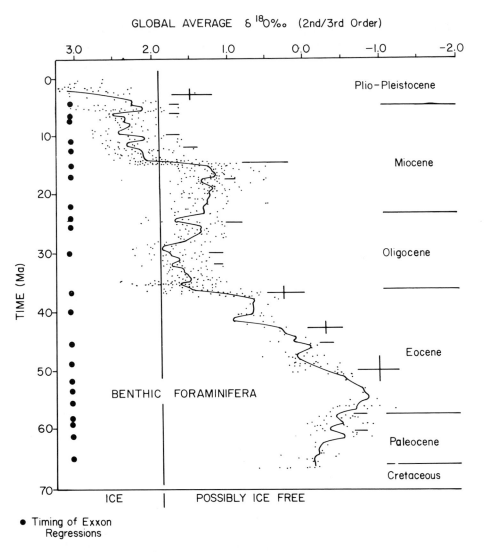

GLOBAL AVERAGE δ^{18}O‰ (2nd/3rd Order)

FIG. 3.—A composite benthic δ^{18}O record from Miller and Fairbanks (1985) compared to the timing of major regressions inferred from the offlap/onlap record of Vail and others (1977). The positions of major δ^{18}O events are indicated by the horizontal lines or crosses to the right of the δ^{18}O curve.

cally defined stratal patterns (van Wagoner and others, this volume). Fairly good agreement exists between the timing of some of the unconformities defined by the EXXON group and most of the significant δ^{18}O events that are resolvable within the bias introduced by averaging the δ^{18}O data at 1-million-year increments (Fig. 3). Despite the statistical bias of the composite δ^{18}O record, the agreement between these two independent methods suggests that eustatic sea level is the primary control of the isotope signal and seismic patterns throughout much of the Cenozoic. While some temperature effect is present in the δ^{18}O signal, the agreement in timing also supports the basic arguments of Matthews and Poore (1980) for the existence of polar ice in the early Cenozoic prior to the major advance of ice in west Antarctica (Shackleton and Kennett, 1975; Savin, 1977; Woodruff and others, 1981).

Magnitude of δ^{18}O-inferred sea-level events.—

To estimate the magnitude of sea-level events of the Cenozoic from the δ^{18}O record, I have chosen to use the Pleistocene calibration of 0.11 per mil δ^{18}O per 10 m of sealevel change (Fairbanks and Matthews, 1978). This cali-

bration is the only one currently available for a time period that has radiometrically dated isotope and sea-level records. This approximation predicts an overall drop in sea level through the Cenozoic of nearly 364 m. Clearly, a drop of this magnitude is not reasonable, and a significant component of the δ^{18}O record must be due to temperature and climatic deterioration through the Cenozoic (Savin, 1977). The magnitude of the drops in sea level at the 1-per-mil events would be equal to over 90 m. These estimates compare with estimates of lowered sea levels during the late Wisconsin glacial maximum 18 ka of 80–163 m (Shackleton and Opdyke, 1973; Duplessy, 1978; Hecht, 1976; Matthews, 1984; Williams, 1984).

Figure 4 and Table 1 compare the magnitude of changes in sea level for selected third-order events. Agreement is good for the sea-level falls at event 2 (24 Ma) and event 6 (40 Ma). The estimated magnitudes are not in agreement for the other six third-order events, particularly for the mid-Oligocene event 4 at 30 Ma. In almost all cases, the inferred change in sea level using the sequence-boundary patterns is larger than that predicted by the δ^{18}O signal. An important exception to this relationship exists for the long-

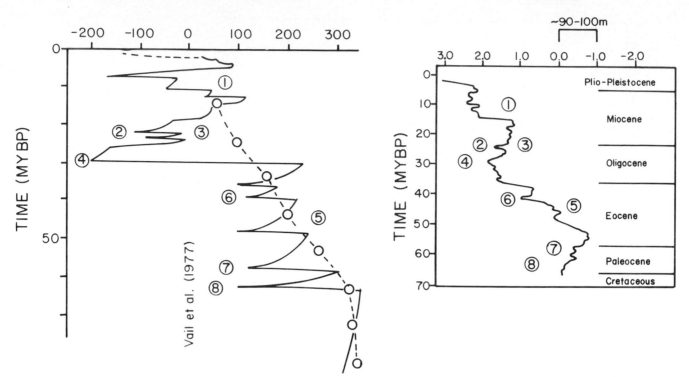

Fig. 4.—Comparison of the magnitude of particular sea-level events of the Cenozoic, as inferred from the seismic stratigraphy (Vail and others, 1977) and composite benthic $\delta^{18}O$ record (Miller and Fairbanks, 1985). The circled numbers refer to particular rises and falls examined in Table 1.

term fall spanning 52.5 to 37 Ma (example 5, Fig. 3). At event 5, the $\delta^{18}O$ signal would predict a sea-level fall of over 250 m. A fall of this magnitude is unreasonable and a temperature change is most likely a significant part of the $\delta^{18}O$ signal in this part of the early Eocene.

Rates of $\delta^{18}O$ and sea-level change.—

Applying the above interpretation to the rapid 1-per-mil $\delta^{18}O$ events of Figure 3 implies that typical long-term rates of sea-level change (of the second and third orders) occurred at an average rate of 10 m and 90–100 m per million years throughout the Cenozoic. These rate estimates represent gross long-term averages due to the broad sample intervals in the $\delta^{18}O$ data sets, and the subsequent averaging of the data to construct Figure 3 (Miller and Fairbanks, 1985).

TABLE 1.—COMPARISON OF THE MAGNITUDE OF SEA-LEVEL EVENTS BASED ON THE $\delta^{18}O$ AND ONLAP RECORDS

Event*	Type	Timing ma	Agreement	Seismic (m)	$\delta^{18}O$ (m)
1	fall	15.5–6.6	−	~300	<50
2	fall	24	+	<50	<50
3	rise	30–15.5	−	>300	<100
4	fall	30	−	>400	<50
5	fall	52–37	−	<100	~250**
6	fall	40	+	~100	~100
7	fall	59	−	<150	<50
8	fall	62.5	−	~200	<50

*As shown in Figure 4.
**A strong temperature component likely.

This approach places constraints on the estimated magnitude of the changes in sea level from seismic stratigraphy (Vail and others, 1977).

We know from detailed $\delta^{18}O$ studies of high-accumulation-rate Deep Sea Drilling Project (DSDP) sequences that 50-to-100-m sea-level changes of the fifth order can occur with frequencies of less than 100,000 years throughout the Neogene (Woodruff and others, 1981; Shackleton, 1982; Williams and others, in prep.) $\delta^{18}O$ records for the last 1.8 Ma (Fig. 5) exhibit significant changes in the frequency and amplitude of the signal (Shackleton and Opdyke, 1976; Pisias and Moore, 1981; Prell, 1982; Thunell and Williams, 1983). Changes over the last 0.8–0.9 ma have a periodicity of 0.1 million years and an average amplitude of over 1.5 per mil (a sea-level equivalent of >130 m). The character of the $\delta^{18}O$ signal in many parts of this time period suggests that the rates of sea-level rise and fall are not necessarily equal (Fig. 6). Using late Pleistocene $\delta^{18}O$ records, Broecker and van Donk (1970) defined the rapid transition from glacial lowstand to interglacial highstands as *terminations*. The rapidity of these changes approaches 10 to 15 m/1,000 yr. The terminations are periods of rapid climatic, oceanographic, and biotic events (Berger and Labeyrie, 1986). In sharp contrast, the subsequent return to full glacially induced lowered sea levels may take an order of magnitude longer (1 to 1.5 m/1,000 yr; Fig. 6). This asymmetry during the late Pleistocene suggests that ice sheets decay more rapidly than they accrete on 100,000 years or less time scales.

This pattern of unequal rates for changes in sea level (rise

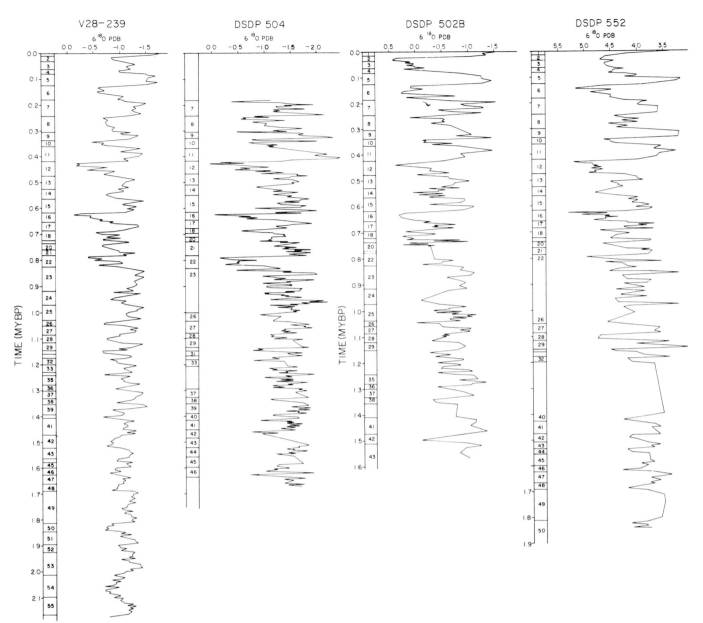

FIG. 5.—Empirical correlation and isotope stage zonation of four detailed Pleistocene oxygen isotope records from the Pacific (V28-239, DSDP Site 504), Caribbean (DSDP Site 502B), and the North Atlantic (DSDP Site 552) (Williams and others, 1984; Williams and Trainor, 1986).

and fall) of the fifth order suggests the following: the assumption that eustatic changes are curvilinear, approaching a sinusoidal function, may not be justified in all cases. The potential effects of this assumption on the stratal geometries of continental margins with differing subsidence rates and tectonic histories should be evaluated in models using sequence-boundary patterns to estimate changes in accommodation of continental margins and eustatic sea levels. The $\delta^{18}O$ signal, even with its limitations regarding the relative temperature-ice-volume effects, provides the only presently known, independent check on this important parameter of eustatic sea-level change.

CONCLUSIONS

Oxygen isotope stratigraphy offers an independent test of some of the inferences of sea level made from onlap records and sequence-boundary interpretations. The future integration of geochemical and geophysical tools, with improved biostratigraphic zonations and spectral processing of the $\delta^{18}O$ (and $\delta^{13}C$) signals, offers possibilities of improved chronostratigraphy in exploration areas of passive continental margins. These results will lead to a more accurate picture of global eustatic sea-level records of the Cenozoic and possibly Mesozoic and older sequences. Such work is in progress at several academic and industrial laboratories.

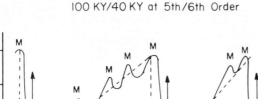

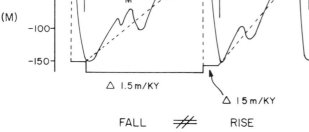

FIG. 6.—A schematic representation of the asymmetry in the rate of sea-level fall or rise during fifth- and sixth-order sea-level/$\delta^{18}O$ changes of the late Pleistocene. The 'M' events represent the possible positions of meltwater events in a basin like the Gulf of Mexico. These 'M' events are schematically superimposed on the global $\delta^{18}O$ change.

ACKNOWLEDGMENTS

The ideas expressed in this paper benefited from the participants at the SEPM Research Workshop, "Sea Level: an Integrated Approach," organized by Cheryl Wilgus, Christopher Kendall and Walter Pitman. I thank the organizers for including me in the workshop. I also thank T. C. Moore and R. B. Dunbar for their thoughtful reviews and the members of the Isotope Stratigraphy Group at the University of South Carolina for their assistance. This paper was written in 1985, revised in 1986, and thus does not include much recent literature of relevance.

REFERENCES

ARTHUR, M. A., 1983, The carbon cycle—controls on atmospheric CO_2 and climate in the geologic past, in, Climate in Earth History: National Academy Press, Washington, D.C., p. 55–67.

BERGER, W. H., AND LABEYRIE, L. D., eds., 1986, The Book of Abstracts and Reports from the Conference on Abrupt Climatic Change: Scripps Institute of Oceanography, Reference 85–8, La Jolla, California, 274 pp.

BROECKER, W. S., AND VAN DONK, J., 1970, Insolation changes, ice volumes, and the ^{18}O record in deep sea cores: Review of Geophysics and Space Physics, v. 8, p. 169–198.

DUPLESSY, J. C., 1978, Isotope studies, in Griffin, J., ed., Climatic Change: Cambridge University Press, New York, p. 46–67.

FAIRBANKS, R. G., AND MATTHEWS, R. K., 1978, The marine oxygen isotopic record in Pleistocene coral, Barbados, West Indies: Quaternary Research, v. 10, p. 181–196.

HECHT, A., 1976, The oxygen isotope record of foraminifera in deep-sea sediment, in Hedley, R. H., and Adams, C. G., eds., Foraminifera: Academic Press, London, v. 2, p. 1–43.

MATTHEWS, R. K., 1984. Oxygen isotope record of ice-volume history: 100 million years of glacio-eustatic sea-level fluctuation, in Schlee, J. S., ed., Interregional Unconformities and Hydrocarbon Accumulation: American Association of Petroleum Geologists Memoir 36, p. 97–107.

———, AND POORE, R. A., 1980. Tertiary ^{18}O record and glacio-eustatic sea-level fluctuations: Geology, v. 8, p. 501–504.

MILLER, K. G., AND FAIRBANKS, R. G., 1985, Cainozoic $\delta^{18}O$ record of climate and sea level: South African Journal of Science, v. 81, p. 248–249.

PISIAS, N. G., AND MOORE, J. C., Jr., 1981, The evolution of Pleistocene climate: a time series approach: Earth and Planetary Science Letters, v. 52, p. 450–458.

PITMAN, W. C., III, 1978, Relationships between eustasy and stratigraphy sequences of passive margins: Geological Society of America Bulletin, v. 89, p. 1389–1403.

PRELL, W. L., 1982, Oxygen and carbon isotope stratigraphy for the Quaternary of Hole 502B: evidence for two modes of isotopic variability: Initial Reports of the Deep Sea Drilling Project, v. 68, p. 455–464.

SAVIN, S. M. 1977, The history of the Earth's surface temperature during the past 100 million years: Annual Review of Earth and Planetary Sciences, v. 5, p. 319–355.

———, DOUGLAS, R. G., KELLER, G., KILLINGLEY, J. S., SHAUGHNESSY, L., SOMMER, M. A., VINCENT E., AND WOODRUFF, F., 1981, Miocene benthic foraminifera isotope records: a synthesis: Marine Micropaleontology, v. 6, p. 423–250.

SCHOLLE, P. A., AND ARTHUR, M. A., 1980, Carbon isotope fluctuations in Cretaceous pelagic limestones: potential stratigraphic and petroleum exploration tool: American Association of Petroleum Geologists Bulletin, v. 64, p. 67–87.

SHACKLETON, N. J., 1982, The deep-sea record of climate variability: Progress in Oceanography, v. 11, p. 199–218.

———, 1985, Oceanic carbon isotope constraints on oxygen and carbon dioxide in the Cenozoic atmosphere, in Sundquist, E., and Broecker, W. S., eds., Natural Variations in CO_2: Past and Present Time: American Geophysical Union Monograph Series, No. 32, p. 412–417.

———, AND KENNETT, J. P., 1975, Paleotemperature history of the Cenozoic and the initiation of Antarctic glaciation: oxygen and carbon isotope analyses in DSDP Sites 277, 279 and 281, in Kennett, J. P., Houtz, R. E., and others, eds., Initial Reports of the Deep Sea Drilling Project: U.S. Government Printing Office, Washington, D.C., v. 29, p. 743–755.

———, AND OPDYKE, N. D., 1973, Oxygen isotope and paleomagnetic stratigraphy of equatorial Pacific core V28-238: oxygen isotope temperatures and ice volumes on a 10^T year and 10^Y year scale: Quaternary Research, v. 3, p. 39–55.

———, AND OPDYKE, N. D., 1976, Oxygen isotope and paleomagnetic stratigraphy of Pacific core V28-239: Late Pliocene to latest Pliocene, in Cline, R. M., and Hays, J. D., eds., Investigations of Late Quaternary Paleoceanography and Paleoclimatology: Geological Society of America Memoir 145, p. 449–464.

THORNE, J., AND WATTS, A. B., 1984, Seismic reflectors and conformities at passive continental margins: Nature, v. 311, p. 365–367.

THUNELL, R. C., AND WILLIAMS, D. F., 1983, The stepwise development of Pliocene—Pleistocene paleoclimatic and paleoceanographic conditions in the Mediterranean: Utrecht Micropaleontological Bulletin, v. 30, p. 111–127.

VAIL, P. R., AND HARDENBOL, J., 1979, Sea level changes during the Tertiary: Oceanus, v. 22, p. 71–79.

———, MITCHUM, R. M., Jr., AND THOMPSON, S., III, 1977, Seismic stratigraphy and global changes of sea level, Part 4: Global cycles of relative changes of sea level, in Payton, C. E., ed., Seismic Stratigraphy—Applications to Hydrocarbon Exploration: American Association of Petroleum Geologists Memoir 26, p. 83–97.

WATTS, A. B., 1982, Tectonic subsidence, flexure and global changes of sea level: Nature, v. 297, p. 469–474.

———, AND STECKLER, M. S., 1979, Subsidence and eustacy at the continental margin of eastern North America, in Talwani, M., Hay, W. F., and Ryan, W. B. F., eds., Deep Drilling Results in the Atlantic Ocean: Continental Margins and Paleoenvironment: American Geophysical Union Ewing Series No. 3, p. 218–234.

WILLIAMS, D. F., 1984, Correlation of Pleistocene marine sediments of the Gulf of Mexico and other basins using oxygen isotope stratigraphy, in Healy-Williams, N., ed., Principles of Pleistocene Stratigraphy Applied to the Gulf of Mexico; International Human Resources Development Corporation Press, p. 67–118.

———, THUNELL, R. C., AND MUCCIARONE, D., 1984, Toward a new oxygen isotope chronostratigraphy of early to middle Pleistocene deep-sea sediments: Abstracts with Programs, Geological Society of America, v. 16, p. 694.

———, AND TRAINOR, D., 1986, Application of isotope chronostratigraphy in the northern Gulf of Mexico: Gulf Coast Association of Geological Societies, Transactions, v. XXXVI, p. 589–600.

WOODRUFF, F., SAVIN, S. M., AND DOUGLAS, R. G., 1981, Miocene stable isotopic record: a detailed deep Pacific Ocean study and its paleoclimatic implications: Science, v. 212, p. 665–668.

PART II
SEA LEVEL CHANGES AND SEQUENCE STRATIGRAPHY

AN OVERVIEW OF THE FUNDAMENTALS OF SEQUENCE STRATIGRAPHY AND KEY DEFINITIONS

J. C. VAN WAGONER, H. W. POSAMENTIER,[1] R. M. MITCHUM,
P. R. VAIL,[2] J. F. SARG, T. S. LOUTIT, AND J. HARDENBOL
Exxon Production Research Company, P.O. Box 2189, Houston, Texas 77252-2189

The objectives of this overview are to establish fundamental concepts of sequence stratigraphy and to define terminology critical for the communication of these concepts. Many of these concepts have already been presented in earlier articles on seismic stratigraphy (Vail and others, 1977). In the years following, driven by additional documentation and interaction with co-workers, our ideas have evolved beyond those presented earlier, making another presentation desirable. The following nine papers reflect current thinking about the concepts of sequence stratigraphy and their applications to outcrops, well logs, and seismic sections. Three papers (Jervey, Posamentier and Vail, and Posamentier and others) present conceptual models describing the relationships between stratal patterns and rates of eustatic change and subsidence. A fourth paper (Sarg) describes the application of sequence stratigraphy to the interpretation of carbonate rocks, documenting with outcrop, well-log, and seismic examples most aspects of the conceptual models. Greenlee and Moore relate regional sequence distribution, derived from seismic data, to a coastal-onlap curve. The last four papers (Haq and others; Loutit and others; Baum and Vail; and Donovan and others) describe application of sequence-stratigraphic concepts to chronostratigraphy and biostratigraphy.

Sequence stratigraphy is the study of rock relationships within a chronostratigraphic framework of repetitive, genetically related strata bounded by surfaces of erosion or nondeposition, or their correlative conformities. The fundamental unit of sequence stratigraphy is the **sequence,** which is bounded by unconformities and their correlative conformities. A sequence can be subdivided into **systems tracts,** which are defined by their position within the sequence and by the stacking patterns of **parasequence sets** and **parasequences** bounded by marine-flooding surfaces. Boundaries of sequences, parasequence sets, and parasequences provide a chronostratigraphic framework for correlating and mapping sedimentary rocks. Sequences, parasequence sets, and parasequences are defined and identified by the physical relationships of strata, including the lateral continuity and geometry of the surfaces bounding the units, vertical and lateral stacking patterns, and the lateral geometry of the strata within these units. Absolute thickness, the amount of time during which they form, and interpretation of regional or global origin are not used to define sequence-stratigraphic units.

Sequences and their stratal components are interpreted to form in response to the interaction between the rates of eustasy, subsidence, and sediment supply. These interactions can be modeled and the models verified by observations to predict stratal relationships and to infer ages in areas where geological data are limited.

The following paragraphs define and briefly explain the terms important for the communication of sequence stratigraphy concepts. Each term will be discussed more fully in the nine papers previously mentioned.

Parasequences and parasequence sets are the fundamental building blocks of sequences. A **parasequence** is a relatively conformable succession of genetically related beds or bedsets bounded by **marine-flooding surfaces** and their correlative surfaces (Van Wagoner, 1985). Siliciclastic parasequences are progradational and therefore shoal upward. Carbonate parasequences are commonly aggradational and also shoal upward. A **marine-flooding surface** is a surface that separates younger from older strata, across which there is evidence of an abrupt increase in water depth. This deepening is commonly accompanied by minor submarine erosion (but no subaerial erosion or basinward shift in facies) and nondeposition, and a minor hiatus may be indicated. Onlap of overlying strata onto a marine-flooding surface does not occur unless this surface is coincident with a sequence boundary. Marine-flooding surfaces are planar and commonly exhibit only very minor topographic relief ranging from several inches to tens of feet, with several feet being most common. The marine-flooding surface commonly has a correlative surface in the coastal plain and a correlative surface on the shelf. The correlative surface in the coastal plain is not marked by significant subaerial erosion due to stream rejuvenation, a downward shift in coastal onlap, a basinward shift in facies, nor onlap of overlying strata. The correlative surface in the coastal plain may be marked by local erosion due to fluvial processes and minor subaerial exposure. Facies analysis of the strata across the correlative surfaces usually does not indicate a significant change in water depth; often, the correlative surfaces in the coastal plain or on shelf can be identified only by correlating updip or downdip from a marine-flooding surface.

A **parasequence set** is a succession of genetically related parasequences which form a distinctive stacking pattern that is bounded, in many cases, by major **marine-flooding surfaces** and their correlative surfaces (Van Wagoner, 1985). Parasequence set boundaries (1) separate distinctive parasequence stacking patterns; (2) may be coincident with sequence boundaries; and (3) may be downlap surfaces and boundaries of systems tracts. Stacking patterns of parasequences in parasequence sets (Fig. 1) are progradational, retrogradational, or aggradational, depending upon the ratio of depositional rates to accommodation rates. These stacking patterns are predictable within a sequence.

A **sequence** is a relatively conformable succession of genetically related strata bounded by unconformities and their correlative conformities (Mitchum, 1977). An **unconformity** is a surface separating younger from older strata, along

[1]Present addresses: Esso Resources Canada Ltd., 237 4th Avenue SW, Calgary, Alberta T2P 0H6; [2]Department of Geology, Rice University, Houston, Texas 77251.

Sea-Level Changes—An Integrated Approach, SEPM Special Publication No. 42

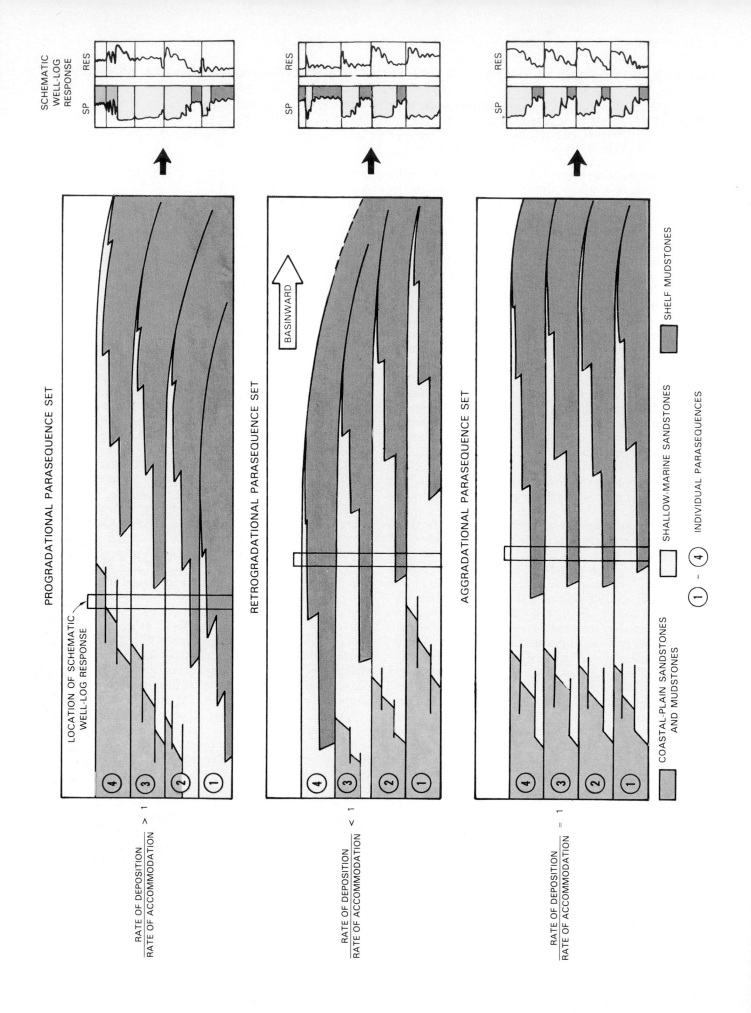

SCHEMATIC WELL-LOG RESPONSE

PROGRADATIONAL PARASEQUENCE SET

RETROGRADATIONAL PARASEQUENCE SET

AGGRADATIONAL PARASEQUENCE SET

BASINWARD

LOCATION OF SCHEMATIC WELL-LOG RESPONSE

$$\frac{\text{RATE OF DEPOSITION}}{\text{RATE OF ACCOMMODATION}} > 1$$

$$\frac{\text{RATE OF DEPOSITION}}{\text{RATE OF ACCOMMODATION}} < 1$$

$$\frac{\text{RATE OF DEPOSITION}}{\text{RATE OF ACCOMMODATION}} = 1$$

COASTAL-PLAIN SANDSTONES AND MUDSTONES

SHALLOW-MARINE SANDSTONES

SHELF MUDSTONES

1 – 4   INDIVIDUAL PARASEQUENCES

which there is evidence of subaerial erosional truncation (and, in some areas, correlative submarine erosion) or subaerial exposure, with a significant hiatus indicated. This definition restricts the usage of the term unconformity to significant subaerial surfaces and modifies the definition of unconformity used by Mitchum (1977). He defined an unconformity as "a surface of erosion or **nondeposition** that separates younger strata from older rocks and represents a significant hiatus" (p. 211). This earlier, broader definition encompasses both subaerial and submarine surfaces and does not sufficiently differentiate between sequence and parasequence boundaries. Local, contemporaneous erosion and deposition associated with geological processes, such as point-bar development, distributary-channel erosion, or dune migration, are excluded from the definition of unconformity used in this paper.

A **conformity** is a bedding surface separating younger from older strata, along which there is no evidence of erosion (either subaerial or submarine) or nondeposition, and along which no significant hiatus is indicated. It includes surfaces onto which there is very slow deposition, with long periods of geologic time represented by very thin deposits.

Type 1 and type 2 sequences are recognized in the rock record. A type 1 sequence (Figs. 2, 3) is bounded below by a type 1 sequence boundary and above by a type 1 or a type 2 sequence boundary. A type 2 sequence (Fig. 4) is bounded below by a type 2 sequence boundary and above by a type 1 or a type 2 sequence boundary. A **type 1 sequence boundary** (Figs. 2, 3) is characterized by subaerial exposure and concurrent subaerial erosion associated with stream rejuvenation, a basinward shift of facies, a downward shift in coastal onlap, and onlap of overlying strata. As a result of the basinward shift in facies, nonmarine or very shallow-marine rocks, such as braided-stream or estuarine sandstones above a sequence boundary, may directly overlie deeper water marine rocks, such as lower shoreface sandstones or shelf mudstones below a boundary, with no intervening rocks deposited in intermediate depositional environments. A typical well-log response produced by a basinward shift in facies marking a sequence boundary is illustrated in Figure 2. A type 1 sequence boundary is interpreted to form when the rate of eustatic fall exceeds the rate of basin subsidence at the **depositional-shoreline break,** producing a relative fall in sea level at that position. The **depositional-shoreline break** is a position on the shelf, landward of which the depositional surface is at or near base level, usually sea level, and seaward of which the depositional surface is below base level (Po-

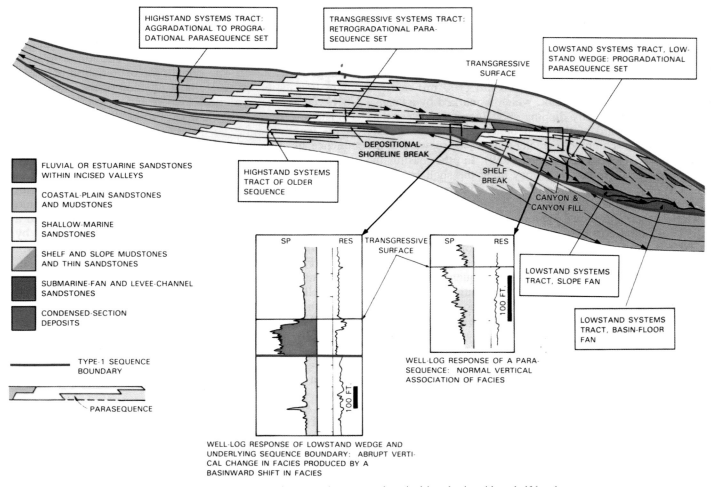

FIG. 2.—Stratal patterns in a type 1 sequence deposited in a basin with a shelf break.

cised-valley fill (Figs. 2, 3), which commonly onlaps onto the sequence boundary, and on the slope by progradational fill with wedge geometry overlying and commonly downlapping onto the basin-floor fan or the slope fan. Lowstand wedge deposition is not coeval with basin-floor deposition. Lowstand wedges are composed of progradational to aggradational parasequence sets. The top of the lowstand wedge, coincident with the top of the lowstand systems tract, is a marine-flooding surface called the **transgressive surface** (Figs. 2–4). The **transgressive surface** is the first significant marine-flooding surface across the shelf within the sequence. Lowstand wedge deposition is interpreted to occur during a slow relative rise in sea level.

The **lowstand systems tract,** if deposited in a basin with a ramp margin (Fig. 3), consists of a relatively thin **lowstand wedge** that may contain two parts. The first part is characterized by stream incision and sediment bypass of the coastal plain interpreted to occur during a relative fall in sea level during which the shoreline steps rapidly basinward until the relative fall stabilizes. The second part of the wedge is characterized by a slow relative rise in sea level, the infilling of incised valleys, and continued shoreline progradation, resulting in a lowstand wedge composed of incised-valley-fill deposits updip and one or more progradational parasequence sets downdip. The top of the lowstand wedge is the transgressive surface; the base of the lowstand wedge is the lower sequence boundary.

The **shelf-margin systems tract** (Fig. 4) is the lowermost systems tract associated with a type 2 sequence boundary. This systems tract is characterized by one or more weakly progradational to aggradational parasequence sets; the sets onlap onto the sequence boundary in a landward direction and downlap onto the sequence boundary in a basinward direction. The top of the shelf-margin systems tract is the transgressive surface, which also forms the base of the transgressive-systems tract. The base of the shelf-margin systems tract is a type 2 sequence boundary.

The **transgressive-systems tract** (Figs. 2–4) is the middle systems tract of both type 1 and type 2 sequences. It is characterized by one or more retrogradational parasequence sets. The base of the transgressive-systems tract is the transgressive surface at the top of the lowstand or shelf-margin systems tracts. Parasequences within the transgressive-systems tract onlap onto the sequence boundary in a landward direction and downlap onto the transgressive surface in a basinward direction. The top of the transgressive-systems tract is the **downlap surface.** The **downlap surface** is a marine-flooding surface onto which the toes of prograding clinoforms in the overlying highstand systems tract downlap. This surface marks the change from a retrogradational to an aggradational parasequence set and is the surface of maximum flooding. The **condensed section** (Figs. 2–4) occurs largely within the transgressive and distal highstand systems tracts. The **condensed section** is a facies consisting of thin marine beds of hemipelagic or pelagic sediments deposited at very slow rates (Loutit and others, this volume). Condensed sections are most extensive during the time of regional transgression of the shoreline.

The **highstand systems tract** (Figs. 2–4) is the upper systems tract in either a type 1 or a type 2 sequence. This systems tract is commonly widespread on the shelf and may be characterized by one or more aggradational parasequence sets that are succeeded by one or more progradational parasequence sets with prograding clinoform geometries. Parasequences within the highstand systems tract onlap onto the sequence boundary in a landward direction and downlap onto the top of the transgressive or lowstand systems tracts in a basinward direction. The highstand systems tract is bounded at the top by a type 1 or type 2 sequence boundary and at the bottom by the downlap surface.

Systems tracts are interpreted to be deposited during specific increments of the eustatic curve (Jervey and Posamentier and others, this volume).

- lowstand fan of lowstand systems tract—during a time of rapid eustatic fall;
- slope fan of lowstand systems tract—during the late eustatic fall or early eustatic rise;
- lowstand wedge of lowstand systems tract—during the late eustatic fall or early rise;
- transgressive-systems tract—during a rapid eustatic rise;
- highstand systems tract—during the late part of a eustatic rise, a eustatic stillstand, and the early part of a eustatic fall.

The subdivision of sedimentary strata into sequences, parasequences, and systems tracts provides a powerful methodology for the analysis of time and rock relationships in sedimentary strata. Sequences and sequence boundaries subdivide sedimentary rocks into genetically related units bounded by surfaces with chronostratigraphic significance. These surfaces provide a framework for correlating and mapping. Interpretation of systems tracts provides a framework to predict facies relationships within the sequence. Parasequence sets, parasequences, and their bounding surfaces further subdivide the sequence and component systems tracts into smaller genetic units for detailed mapping, correlating, and interpreting depositional environments.

REFERENCES

BROWN, L. F., AND FISHER, W. L., 1977, Seismic-stratigraphic interpretation of depositional systems: examples from Brazil rift and pull-apart basins, *in* Payton, C. E., ed., Seismic Stratigraphy—Applications to Hydrocarbon Exploration: American Association of Petroleum Geologists Memoir 26, p. 213–248.

FISHER, W. L., AND McGOWAN, J. H., 1967, Depositional systems in the Wilcox Group of Texas and their relationship to occurrence of oil and gas: Gulf Coast Association of Geological Societies, Transactions, v. 17, p. 213–248.

HEEZEN, B. C., THARP, M., AND EWING, M., 1959, The floors of the ocean, I. The North Atlantic: Geological Society of America Special Paper 65, 122 p.

MITCHUM, R. M., 1977, Seismic stratigraphy and global changes of sea level, Part 1: Glossary of terms used in seismic stratigraphy, *in* Payton, C. E., ed., Seismic Stratigraphy—Applications to Hydrocarbon Exploration: Association of Petroleum Geologists Memoir 26, p. 205–212.

VAIL, P. R., MITCHUM, R. M., AND THOMPSON, S., III, 1977, Seismic stratigraphy and global changes of sea level, Part 3: Relative changes of sea level from coastal onlap, *in* Payton, C. W., ed., Seismic Stratigraphy—Applications to Hydrocarbon Exploration: American Association of Petroleum Geologists Memoir 26, p. 83–97.

———, AND TODD, G. R., 1981, North Sea Jurassic unconformities, chronostratigraphy and sea-level changes from seismic stratigraphy: Petroleum Geology of the Continental Shelf, Northwest Europe, Proceedings, p. 216–235.

———, HARDENBOL, J., AND TODD, R. G., 1984, Jurassic unconformities, chronostratigraphy and sea-level changes from seismic stratigraphy and biostratigraphy, *in* Schlee, J. S., ed., Interregional Uncon-

formities and Hydrocarbon Accumulation: American Association of Petroleum Geologists Memoir 36, p. 129–144.

VAN WAGONER, J. C., 1985, Reservoir facies distribution as controlled by sea-level change: Abstract and Poster Session, Society of Economic Paleontologists and Mineralologists Mid-Year Meeting, Golden, Colorado, p. 91–92.

QUANTITATIVE GEOLOGICAL MODELING OF SILICICLASTIC ROCK SEQUENCES AND THEIR SEISMIC EXPRESSION

M. T. JERVEY

Canadian Hunter Exploration Ltd., 700, 435 4th Avenue SW, Calgary, Alberta T2P 3A8

ABSTRACT: In order to clarify the principles that govern the development of siliciclastic sequences and their bounding surfaces, a mathematical model of progradational basin filling was created for Atlantic-type continental margins. This paper discusses the model and its implications with respect to depositional facies, sandstone geometry, and seismic stratigraphic interpretation.

Basin filling is modeled as the interaction of subsidence, change in sea level, and sediment influx. The simulations show that seismic-sequence boundaries are located, in time, near inflection points of eustatic sea-level fluctuation, where rates of fall or rise are maximized. Changes in the rate of accommodation development, both in time and space, are believed to play a dominant role in shaping the internal facies distribution, the geometry, and the nature of the bounding surfaces of depositional sequences. The pattern of coastal onlap and offshore condensed sections displayed by global-cycle charts are shown to develop in the context of smoothly fluctuating eustatic and relative sea level.

INTRODUCTION

An assessment of probable reservoir facies and sand continuity patterns is one of the important objectives of predevelopment basin studies. These depositional reservoir properties determine, to a large degree, the production potential of prospective basins. I hypothesize that these properties are, in considerable measure, a function of the interaction of three major geologic variables: (1) eustasy; (2) subsidence; and (3) sediment influx, within a time/space framework. If this hypothesis is true, then an understanding of these variables can aid predrill prediction of reservoir quality.

Reservoir sands are contained within depositional sequences. A rather substantial literature has developed (Payton, 1977) which incorporates the notion that changes in relative sea level control the position of sequence boundaries, the geometry of the sequence, and its relationship to other sequences. The term relative sea level refers to the elevation of sea level relative to some previous depositional surface. Changes in relative sea level reflect the combined effect of eustatic sea-level fluctuation and basin subsidence. The precise relationship between changes in relative sea level, sediment influx, and sequence development in a time/space framework are difficult to conceptualize. Consequently, a mathematical analysis was developed to investigate the interaction of these major sequence-controlling variables and their effect on sequence geometry, depositional facies, and seismic stratigraphy.

The following discussion focuses on the geologic principles that govern the development of siliciclastic sequences and their bounding surfaces. The concept of sediment accommodation, in conjunction with assumed (or defined) rates of sediment influx, forms the basis for mathematical simulations of progradational basin filling on an Atlantic-type continental margin. These simulations, and their implications with respect to reservoir facies, continuity, and seismic-sequence development are the subject of this report. The principles developed for siliciclastic sequences are also relevant for carbonate sequences, although some of the fundamental assumptions of the simulations must be modified to reflect the specifics of carbonate sedimentation.

ACCOMMODATION

In order for sediments to accumulate, there must be space available below base level (the level above which erosion will occur). On the continental margin, base level is controlled by sea level and, at first approximation, is equivalent to sea level. For the purposes of this discussion, I shall also consider sea level to limit the accumulation of marine sediments, although, in fact, a secondary marine profile of equilibrium is attained that reflects the marine-energy flux in any region. This space made available for potential sediment accumulation is referred to as accommodation.

Accommodation is a function of both sea-level fluctuation and subsidence. In this paper, sea-level changes are considered to be eustatic and independent of subsidence at the continental margin. This assumption may not be completely valid for certain divergent margins near ridge axes, where the rate of lithospheric cooling and subsidence can govern long-term sea-level variations. Subsidence refers to the sinking of the basin floor by any process, the most important of which (on divergent margins) are lithospheric cooling and sediment loading. Compaction of the sediment column, while creating additional space for sediment to accumulate below base level, is regarded as a secondary effect and is not included in the concept of accommodation used there.

Figure 1 illustrates accommodation developed at three points in a basin under differing conditions of subsidence. In this example, subsidence is represented as a straight line, the slope of which is equal to the rate of subsidence at each point. Sea-level fluctuation is represented as the same smooth curve in all three cases. The curve describing the development of accommodation with time in each of these three cases is found simply by addition of the subsidence and sea-level curves. Changes in accommodation are equivalent to changes in relative sea level.

At locations at which slow subsidence occurs, maximum accommodation is achieved near the sea-level maximum. When sea level falls to its original position, accommodation declines to a value that reflects only subsidence. With increased rates of subsidence, the time of maximum accommodation is progressively later, as measured from time zero on the diagrams in Figure 1. At points in the basin where

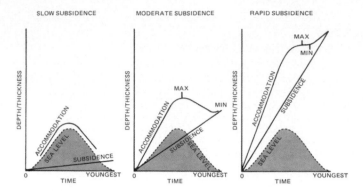

FIG. 1.—Relationship of sediment accommodation to sea-level fluctuation and subsidence.

extremely rapid subsidence prevails, no decrease in accommodation is experienced, even though sea level is falling. These three cases may be regarded as representing three locations near, intermediate, and far from the basin margin, if subsidence rate increases toward the basin center. Thus, the variation in accommodation development shown in Figure 1 is geographically controlled and predictable. Such variations have a profound influence on accumulating sedimentary sequences.

SEDIMENT INFLUX

The amount of sediment supplied to basin locations is a function of the general rate of sediment influx into the basin and proximity to zones of active transport. Let us consider the relationship of facies, accommodation, and rates of sediment accumulation at three specific points in the basin (Fig. 2) with identical accommodation curves. These might represent three positions along strike at varying distances from a point source of sediment. At the location with low rates of sediment influx, we find that accommodation always ex-

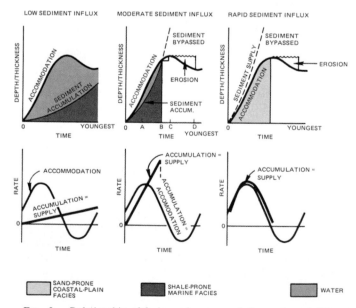

FIG. 2.—Relationship of facies and accommodation under conditions of varying sediment of influx rates.

ceeds sediment accumulation, so that a considerable depth of water is developed. Shale-prone marine facies are deposited at some distance from a coastline located marginward of the figured depositional site. Since these marine sediments accumulate below base level (sea level), the rate of accumulation is controlled by rate of supply and does not reflect fluctuations in the rate of accommodation development.

With a moderate rate of sediment influx, the sea floor can aggrade to sea level, and the sequence of facies development is more complex. At time zero, the rate of increase of accommodation exceeds the ability of sedimentation to maintain the sediment surface at sea level, and a transgression begins while the coastline migrates marginward of the figured location. During this transgressive period, water depths increase, as represented by the height of the accommodation curve above the sediment accumulation curve, and marine shales are deposited on the continental shelf.

As the rate of sea-level rise diminishes, with accompanying decrease in rate of accommodation development, regression of the coastline commences. The first evidence of this regression at the figured basin location is the deposition of shale-prone prodelta marine facies. Regression continues with rapid deposition of marine facies until marine sediments have aggraded to sea level and the coastline is again found at the location shown in Figure 2. Thereafter, sediment supply exceeds accommodation, the sediment surface is maintained at sea level, and sandy coastal- or deltaic-plain sediments accumulate. Excess sediment, which cannot be accommodated, is bypassed basinward. The rate of accommodation development declines to zero, and falling sea level causes a decrease in accommodation with a potential for erosion of previously deposited sediments at the figured location.

During deposition of shale-prone marine facies, the rate of accumulation is a function of supply. When base level is reached, however, the rate of accommodation development controls the rate of accumulation. Erosion is possible when the rate of accommodation is negative and accumulation ceases.

At the location with a rapid rate of sediment influx, sediment supply always exceeds accommodation. Sand-prone coastal- or deltaic-plain facies accumulate at or near sea level during times of accommodation development. Here, the rate of accumulation is a function of accommodation. Excess sediment is bypassed basinward. Erosion of previously deposited coastal-plain facies is favored during accommodation decrease, with the development of a subaerial-erosion surface.

The previous discussion suggests a relationship between accommodation development and the timing of subaerial erosion. The erosion surfaces so formed are important stratigraphic- and seismic-sequence boundaries. Marine nondepositional or hiatal surfaces are also important unconformities in the geologic record, and their development can be related to rapid increases in accommodation. Figure 3 illustrates the timing of a period of nondeposition with respect to accommodation under conditions of moderate-sediment influx at the figured basin location.

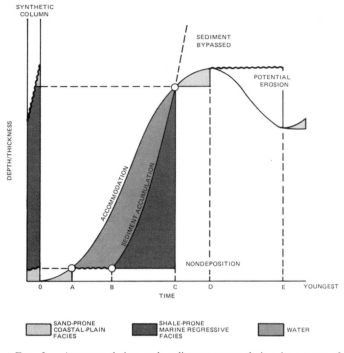

FIG. 3.—Accommodation and sediment accumulation in a case of moderate rates of sediment influx.

At time zero, a beach of the low sea level occurs at the site of Figure 3, and coastline facies are deposited. Subsequently, sea level begins to rise, and coastal-plain sand-prone sediments accumulate, as accommodation increases, until time A. At time A, the rate of increase of accommodation exceeds the ability of sedimentation to maintain the sediment surface at sea level, and a transgression begins while the coastline migrates marginward of the figured location. During this transgression, water depths increase and nondeposition prevails at this location. Sediments such as glauconitic relict sands, phosphorites, black organic shales, or carbonates may accumulate at this basin position during the time of greatly reduced clastic influx (A to B). Clastic deposition is restricted to nearshore and coastal-plain sites marginward of the figured location.

Nondeposition ceases at time B as decreasing rates of accommodation development in coastal areas permit progradation at the coastline, and marine regressive facies begin to accumulate. The sequence of regressive sedimentation and potential subaerial erosion is thereafter similar to that described for the case of moderate-sediment influx in Figure 2. If the rate of sediment influx is low, the period of nondeposition will be extended. On the other hand, if it is high, no period of nondeposition may occur under the same conditions of accommodation development.

Rapid increases in accommodation favor the generation of marine hiatal surfaces of regional significance. Local hiatal surfaces are more likely to reflect changes in sediment influx caused by shifts in drainage patterns or marine-current systems. The regional and global controls of subsidence and eustasy on hiatal surface development are the subject of further discussion in this paper.

These sediment data, discussed in a time framework, may be projected to the depth/thickness axis to yield a synthetic stratigraphic column. The regressive sequence deposited between time B and D is a depositional unit bounded below by a hiatal marine unconformity and above by a subaerial unconformity of variable truncation. Figure 3 thus contains a progradational depositional sequence formed during a period of diminishing accommodation increase.

MULTIPLE CYCLES

Sea-level fluctuations and subsidence combine to produce fluctuations in accommodation through time. This was illustrated for a single eustatic cycle in previous examples. Figures 4 and 5 illustrate the development of several sedimentary units during six cycles of accommodation fluctuation related to six cycles of eustatic sea-level rise and fall in a steadily subsiding basin. Figure 4 represents a case of low-sediment influx at some specific basin location. Following a transgression near time zero, which is related to the rapid rate of accommodation development, three cycles of marine-sediment deposition occur, separated by nondepositional periods. These depositional events reflect progradation of the coastline marginward of the figured location and represent bottomset units of the distal clinoform slope or prodelta apron. The hiatal surfaces develop during nondepositional transgressive phases when the coastline is migrating marginward. In this model, the development of hiatal surfaces occurs under conditions of rapid accommodation increase, whereas progradation is related to declining rates of accommodation development.

At the end of cycle three, a profound decrease in accommodation exposes the marine sediments of the cycle to subaerial erosion (Fig. 4). The rate of erosion is indicated by the slope of the dashed line. Increasing accommodation in cycle four allows sediments to accumulate again on the erosion surface. Coastal-plain sediments are deposited during this cycle until the rate of accommodation development exceeds the ability of sedimentation to maintain the sediment surface at sea level at the site of Figure 4. Transgression and nondeposition then commence, followed by progradation as the rate of accommodation increase declines. The deposition of marine sediments continues until declining accommodation and falling sea level expose the sediments to subaerial erosion. This cyclic pattern of nondeposition, deposition, and erosion is repeated in cycles five and six.

The situation in a case of higher rate of sediment supply is illustrated in Figure 5. Here, the initial rapid rate of accommodation development is accompanied by deposition of a thick sequence of coastal-plain sediments. A brief transgression during the time of most rapid rise in sea level is followed by regression and deposition of prodelta or slope sediments at the figured location. Sediment accumulation is sufficiently rapid that the marine sequence is exposed to subaerial erosion during declining accommodation. Sediment accumulation in cycle two is similar to that described for cycle one. In cycle three, however, rapid accumulation of prodelta sediments aggrades the sediment surface to sea level, and a thick sequence of coastal- or deltaic-plain sand-prone sediments is deposited as accommodation continues to increase.

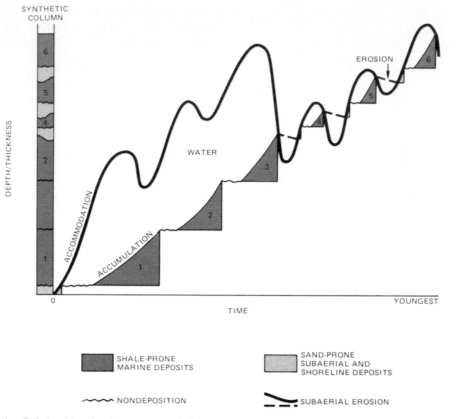

FIG. 4.—Relationship of sediment accumulation to accommodation with a low rate of sediment supply.

With the subsequent decline in accommodation in cycles three and four, a prolonged period of erosion commences, during which the deposits of cycle three are deeply truncated. Base level, represented by the accommodation curve, remains below the actively eroding land surface until cycle five, at which time rising sea level raises base level to the erosion surface and coastal-plain sediments can accumulate. This episode of deposition is followed by truncation and deposition of additional sand-prone sediments during cycle six.

The synthetic columns of Figures 4 and 5 represent vertical sequences produced during model time. It is apparent that continuous sections tend to be preserved at sites of low-sediment influx in contrast to areas of high rates of influx where profound erosional truncation may accompany declines in accommodation.

The cycles of marine deposition are coarsening-upward sequences of the prodelta clinoform slope. The textural gradient reflects the approach of the deltaic coastline during progradation. Where these prodelta sequences aggrade to sea level during periods of increasing accommodation (as in Fig. 5, cycle three), they are overlain by a complex of stream-mouth bar and/or beach facies and subsequently by deltaic- or coastal-plain facies in the classic pattern of deltaic/coastal progradation. These are the deltas of high sea-level stands. We may speculate that because accommodation is increasing, the individual deltaic-lobe sequences within the deltaic complex will be thick and well defined by intervening prodelta shales.

Where the prodelta sequences aggrade to sea level during periods of decreasing accommodation, as in many of the illustrated cycles in Figures 4 and 5, the situation of low stand deltaic sedimentation occurs. Sediment influx from the basin margin and eroded material are channeled to the coast through incised-valley systems. Little or no deltaic-plain sediments can accumulate, and sands are funneled down the subaqueous depositional slope to accumulate as mounded turbidite facies or as prodelta sand wedges.

Following truncation and incision of previous highstand and lowstand deposits, increased accommodation due to relative sea-level rise results in the accumulation of a variety of fluvial, estuarine, and shoreline deposits that may be very thin, although shown to be of substantial thickness in figured cycles. These deposits may fill incised topography or may simply be a veneer of transgressive facies. Hiatal surfaces that develop during transgressions, indicated as periods of nondeposition in Figures 4 and 5, are sites of reworking and accumulations of various nonclastic sediment types mentioned previously.

These remarks are introductory to a more thorough discussion of the influence of accommodation fluctuation on large-scale patterns of environmental-facies development. They serve to indicate the scope of investigation and some of the speculations that may be documented by further study.

THE MODEL

The preceding discussion has presented qualitative relationships between accommodation development, sediment

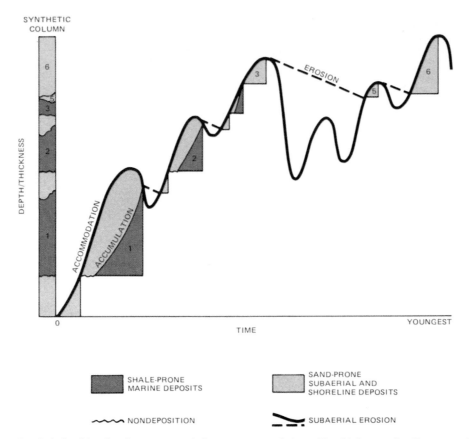

FIG. 5.—Relationship of sediment accumulation to accommodation with a high rate of sediment supply.

accumulation, and environmental facies at specific basin locations. To clarify these interactions, a mathematical simulation of basin filling was developed for a progradational continental margin. The program models the development of sedimentary sequences under given conditions of sea-level fluctuation, subsidence, and sediment influx in a time/space framework. Simulated seismic profiles are generated that illustrate certain fundamental relationships between accommodation, sedimentation, and seismic sequences.

Like all mathematical models, the simulations are approximations of reality involving a number of simplifying assumptions. Their purpose is to clarify principles of sedimentation and seismic stratigraphic interpretation relative to major geologic processes. The mathematics of the model encompass progradation and transgression within the basic geometry of a wedge-shaped cross section. Other forms of deposition, such as marine-drape or onlapping deep-water sequences, are not included in the model *per se* but can be accommodated conceptually within the model framework where appropriate.

Depositional geometry.—

Most siliciclastic depositional sequences that contain prospective reservoir sands have a fundamental wedge-shaped geometry in dip-oriented cross sections (Fig. 6). The sequence is bounded below by an erosional or nondepositional unconformity or correlative conformity that usually reflects a hiatus in sediment influx or change in base level.

During its accumulation, the sequence is bounded above by a depositional surface, although erosion, nondeposition, and sediment bypass may accompany sequence development. The depositional surface of the sequence can be divided into two segments: (1) the aggradational portion, which is maintained at or near sea level (base level); and (2) the progradational portion, which slopes in a seaward direction and approaches the lower bounding surface. Former depositional surfaces within the sequence are chronostratigraphic surfaces or time lines whose configuration reflects depositional geometry and subsequent compaction and subsidence.

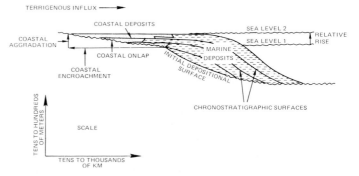

FIG. 6.—Wedge-shaped depositional sequence developed in a subsiding basin (from Vail and others, 1977).

Aggradation refers to the vertical building up of the sedimentary sequence. Aggradation occurs under conditions of relative rise in sea level, related to subsidence and/or eustatic sea-level rise, where the rate of sediment influx is sufficient to maintain the depositional surface at or near sea level. The rate of aggradation therefore depends on the rate of accommodation development. Coastal onlap and encroachment of the aggradational deposits upon the underlying unconformity accompany aggradation. The zone of active sediment accumulation thereby expands toward the basin margin.

Progradation refers to the seaward or lateral building of the subaqueous depositional slope by the addition of marine sediments. The formation of the depositional slope is a result of declining rates of sedimentation seaward of the coastline or shelf edge. At first approximation, the coastline on an actively prograding continental margin forms the break in slope between the aggradational and progradational segments of the depositional surface, although a narrow marine-shelf and sediment-bypass zone may be developed across which sediment is transported to the depositional slope. The inclination of the progradational depositional slope is maintained in the range of 1° to 5° by a variety of submarine processes, including sediment redistribution by currents and mass transport mechanisms. Progradation results in the basinward expansion of the zone of rapid coastal sedimentation.

The rate of progradation varies inversely with the rate of accommodation development and the initial basin bathymetry. This is true because, with fixed-sediment influx, more rapid accommodation development requires that more sediment be stored in aggradational deposits. Also, deep water requires a long depositional slope, with the result that sediment is distributed over a broad area with low overall rates of sedimentation.

Aggradational deposits consist of coastal and deltaic-plain sand-prone facies, such as fluvial and crevasse deposits and associated shaley facies, including flood-plain, marsh, and interdistributary-bay/lagoonal sediments. Shoreline and nearshore marine sandy facies can often be regarded as aggradational at the scale of this investigation, although they are transitional to progradational facies and, under conditions of active regression (see the following discussions), would be considered a part of that latter group.

Progradational facies are generally shale-prone, with the bulk of sedimentation occurring via transport of suspended slit and clay away from the coast. Most sand brought to the coast is trapped there by wave residual currents. Tidal and other wind-generated currents may transport sand to the shelf edge, where density currents can redistribute these sediments to the slope. Certain circumstances (to be discussed further) favor the formation of density currents, and sand-prone facies may dominate on the slope. The formation of turbidite fans, resulting from long-term density-flow transport, is here considered part of the progradational process, although the fans themselves can be recognized as depositional sequences.

Because sands are concentrated along the coast, the position of the shoreline within the sequence is of importance. Figure 7 illustrates the role of the rate of sediment influx

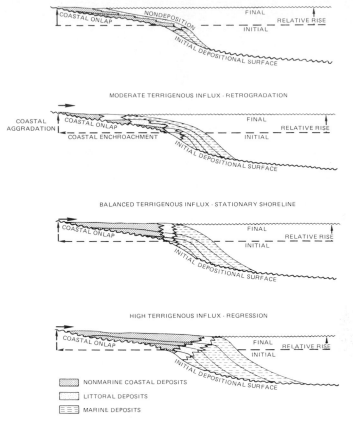

FIG. 7.—Transgression, retrogradation, regression, and coastal onlap during relative rise of sea level. Rate of terrigenous influx determines which situation is produced during relative rise of sea level (modified from Vail and others, 1977).

in determining the geometry of coastal deposits under given conditions of relative rise in sea level. The term transgression, as used in this paper, refers to the landward or marginward movement of the coastline accompanied by nondeposition and the development of a broad, starved shelf, such as is characteristic of many modern shelf areas. This situation is induced by low-sediment influx concurrent with a relative rise in sea level. A transgression is distinguished from a retrogradation of the coastline, in which the landward or marginward movement of the coast is accompanied by progradation of the depositional slope and significant aggradation on the continental shelf. Transgression is associated with the generation of coarse relict-shelf, reworked sands or nonclastic sediments, whereas retrogradation is characterized by normally graded clastic-shelf deposits (i.e., coarsest at the coast and becoming finer and more calcareous offshore).

Sediment influx may also be balanced against relative sea-level rise, such that the shoreline remains stationary through time while progradation continues; or, with high-sediment influx rates, regression or seaward movement of the coast may occur so that more landward facies overlie more seaward facies. With changes in the rate of sediment influx or relative sea-level rise (accommodation increase), any one of these forms of sequence development may pass into any of the others during the deposition of the sequence.

Model geometry and parameters.—

The mathematical model simulates the development of depositional sequences under defined conditions of sediment influx and accommodation development. The assumption is made that the modeled continental margin consists of a sedimentary prism deposited on a basement of premodel rocks. If the length of the prism parallel to the basin margin can be considered fixed by basin size, then the cross-sectional area of the prism taken perpendicular to the basin margin (dip-oriented) is a linear function of prism volume. Therefore, sediment influx into the basin from the surrounding uplands can be modeled by defining a rate of increase of prism cross-sectional area, and the model cross section is representative of any prism cross section. Chronostratigraphic lines in model prism cross sections represent surfaces and are referred to as such in the following discussion.

In reality, a sedimentary prism consists of laterally coalesced and stacked lobate bodies that are deposited by stream systems as deltas composed of aggradational and progradational facies. At the scale of simulation (see next section), the complexity of these small-scale features can be generalized to yield a simplified presentation of sequence geometry. In its simplest form, this cross-sectional geometry is one of a sedimentary wedge (Fig. 8). The base of the wedge is defined by the underlying premodel basement rocks. The depositional surface consists of: (1) a coastal or deltaic plain under which sand-prone facies aggrade to a base level defined as sea level; and (2) a seaward-sloping depositional clinoform slope.

The model coastline is defined as the top of the clinoform slope at sea level when progradation is occurring, as shown in Figure 8. This assumption seems reasonable considering the lack of significant sedimentation on modern slopes except where major rivers have developed deltas close to the shelf edge. It should be emphasized, however, that the sand-prone aggradational facies may include intercalated neritic and coastal deposits at a level of detail finer than the simulation described here. During model transgression, the coastline moves in a landward direction and a broad shelf is formed. The shelf-slope break becomes progressively deeper as subsidence continues.

Modeled deposition of the sequence begins at a time of eustatic sea-level lowstand based on a sinusoidal sea-level fluctuation curve. A smooth, symmetrical function was used in simulations reported here because part of the impetus for

the work was to determine whether symmetrical rises and falls of sea level could account for observed sequence geometry.

The period and amplitude of eustatic sea-level fluctuation were selected to be consistent with estimates of ranges of these variables based on seismic stratigraphic observation (Payton, 1977). For models presented in this paper, sea level was assumed to fluctuate by about 61 m (200 ft) over a period of 4 million years. The development of the sequence(s) occurs in a three-dimensional framework of time and space. Model space is defined by a coordinate system of distance and depth thickness, with an origin at a structural hinge at the basin margin. The hinge is a position of no subsidence of the basement surface when movement is measured against some fixed datum. Subsidence occurs basinward of the hinge with rate of subsidence increasing toward the basin center. The rate of subsidence is constant at any given distance from the hinge. The mathematical function describing subsidence allows the basement surface to collapse from an initial configuration to achieve a convex-down profile of the modeled basement surface.

The modeling procedure is initiated by specifying appropriate values for key parameters that control the mathematical operation. These include the amplitude and period of eustatic changes in sea level, the initial shoreline position, the rate of subsidence at the initial shoreline, the initial elevation of the basement at the shoreline, and the rate of sediment influx as specified by a rate of increase in model cross-sectional area. For models presented in the following sections, the same profile of basin subsidence was used to facilitate comparisons of the effects of other variables. The modeled subsidence of 3 to 5 cm/1000 yr conforms to the range of subsidence rates reported by Pitman and Golovchenko (1983) for the Atlantic margin of the east coast of the United States in the earlier stages of rifting. The slope of the clinoforms is also specified, but variations in clinoform slope have little influence on model results reported in this paper.

These parameters may be varied in any combination in the modeling procedure, with the exception that eustatic sea level may not rise above the elevation of the hinge. Modeling results suggest that the mathematics are robust: that is, small changes in model assumptions do not result in large changes in outcomes. The models shown here were selected to illustrate major conclusions.

Figure 8 illustrates a dip-oriented cross section through a modeled progradational clastic sequence on a subsiding continental margin where sea level remains constant. In this case, accommodation is achieved by subsidence alone. An initial coastline position was defined at distance $X = 20$ for time $t = 0$, where initial sea level intersected the basement surface. The sequence was allowed to develop through four time units with a constant rate of sediment influx (constant rate of cross-sectional area increase). The size of the clastic wedge formed at any time, and therefore coastline position, is determined by the requirement that the areas under the depositional chronostratigraphic surfaces must equal the integral of the rate of areal increase as defined for the model for that time period. Since, in this case, the rate was constant, the areas under surfaces 1, 2, 3, and 4 are simple

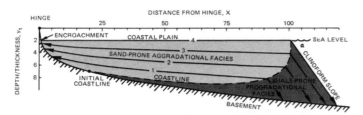

FIG. 8.—Modeled sedimentary wedge simulating a prograding continental margin. Units are arbitrary.

multiples of the given rate. For instance, the area under surface 2 is exactly twice the area under 1 and 2/3 of that under 3.

Figure 8 shows that, during the development of the sequence, sand-prone aggradational facies encroach hingeward upon the basement surface and progradational facies expand in a basinward direction. The coastline is traced as the juncture of these two major types of facies. Although coastal plain depositional surfaces 1, 2, 3, and 4 are horizontal at sea level when deposited, they are deformed to curved surfaces during subsidence because of the differential rate of subsidence along the profile. Depositional surfaces of the clinoform slope are also deformed, but to a lesser degree, because they occupy a narrow zone of slowly changing rates.

Figure 8 also illustrates the influence of accommodation development and bathymetry on the rate of progradation. Progradation is initially quite rapid, because sediment influx is distributed across a shallow basin. With continued subsidence, however, the basin deepens. In addition, the prograding wedge encounters ever-increasing rates of subsidence as the coastline shifts toward the basin center. As the coastal plain widens, more and more of the available sediment must be stored in aggradational facies, and the amount that can be contributed to the progradation of the clinoform slope diminishes. The rate of progradation de-

clines, and the coastline reaches a stable position dominated by aggradation.

When fluctuation of eustatic sea level accompanies subsidence, the geometry of the depositional sequence can become considerably more complex. Figure 9 traces the development of a depositional sequence during one cycle of sea-level rise and fall on a subsiding margin. A constant rate of sediment influx begins at time $t = 0$. An initial, rapid progradation slows, and retrogradation begins near $t = 2.5$. This transition from progradation to retrogradation reflects an increase in rate of accommodation development due to increase in rate of eustatic sea-level rise (see sea-level curve, Fig. 9). Progradation ceases at $t = 3$ because the rate of accommodation development is sufficiently rapid to require that all sediment be stored in aggradational facies. This consequence of a rapid rise in sea level can be appreciated by referring to Figure 9 at $t = 3$. A calculation of coastline position based on a wedge-shaped profile at time $t = 4$ would place the clinoform slope marginward of the slope at $t = 3$, as shown by the dashed line. This is geologically unreasonable since, from the point of view of volume, it would require the erosional redistribution of sediment deposited prior to $t = 3$ and its incorporation in the wedge at $t = 4$. Submarine erosion of the clinoform slope may be important locally, but it is thought to be of minor importance in determining large-scale continental margin

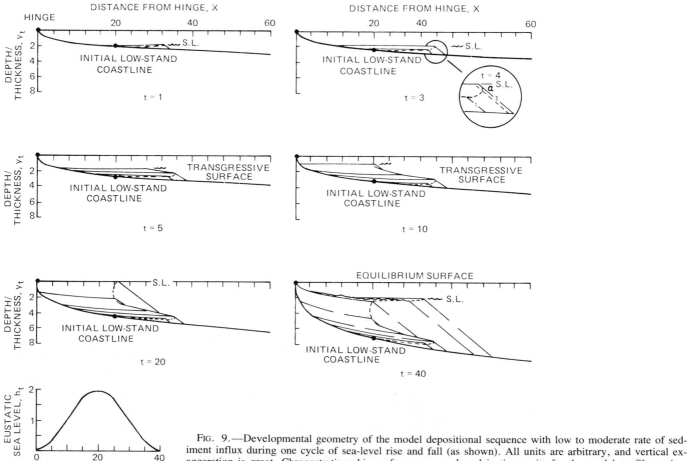

Fig. 9.—Developmental geometry of the model depositional sequence with low to moderate rate of sediment influx during one cycle of sea-level rise and fall (as shown). All units are arbitrary, and vertical exaggeration is great. Chronostratigraphic surfaces are numbered in time units for the model profile at time $t = 40$.

geometry. Rather, the modeling assumption is that transgression, as defined previously, begins when progradation ceases under conditions of increasing rate of accommodation development.

The geologic situation during initial progradation over the basement is shown in Fig. 10. Progradation is accompanied by development of a broad coastal plain of laterally coalesced meandering-stream deposits. The coast is a complex of deltaic deposits near the mouths of major rivers, and interdeltaic-beach, bar, shallow-marine, and bay sediments. Shale-prone sediments of the clinoform slope are silts and clays transported in suspension away from river mouths of shelf-edge deltas or bypassed across the narrow shelf by wind-driven or tidal currents.

It seems reasonable that increased accommodation development due to rise in sea level would result in an embayed coastline as a prelude to transgression. These embayments would be the accumulation sites of estuarine facies and marsh peats lateral to belts of active fluvial deposition.

The transgression beginning at time $t = 3$ (Fig. 9) leads to the development of a transgressive hiatal surface that, in the assumption of the simulation, is nondepositional with respect to clastics. The rapid rate of increase in accommodation requires that all sediment be stored in aggradational facies. As transgression proceeds, coastal-plain facies are deposited hingeward of the coast, while no sediment is supplied to the transgressive surface. The transgressive surface is, in effect, a broad continental shelf. In Figure 9, the growth of this shelf area continues until the rate of relative sea-level rise begins to decline at time $t = 10$.

Figure 11 illustrates model geography and geometry during transgression. Estuarine sediments accumulate in large bays and drowned areas of the coastal plain behind the transgressive coastline. Coastal-plain facies continue to aggrade under conditions of increasing accommodation. On the continental shelf, coastal deposits are reworked by waves and currents to form thin, relict-shelf, sand bars and shoals. Glauconitic sands, phosphorites, and organic-rich shales may accumulate on the shelf, as well as thick carbonates that may form a significant nonclastic aggradational/progradational component not included in the model.

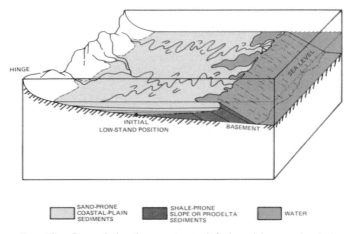

FIG. 10.—Progradational geometry and facies; rising sea level, increasing accommodation.

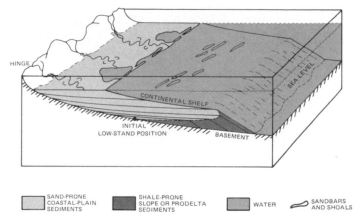

FIG. 11.—Geometry and facies during transgression and development of a broad continental shelf. Rising sea level, increasing accommodation.

As the rate of accommodation increase begins to decline, the inclination of the transgressive surface increases in response to the expansion of the coastal plain. As some point in time ($t = 10$, Fig. 9), the inclination of this surface reaches the defined slope of the model clinoforms, and regression and progradation across the hiatal transgressive surface begin. The slowing of accommodation development means that less and less sediment can be stored in aggradational facies, while the volume contributed to the clinoform slope increases. The rate of progradation across the hiatal surface therefore increases after time $t = 10$.

At time $t = 20$ (Fig. 9), the eustatic sea-level maximum is achieved; as sea level begins to fall, accommodation declines in hingeward areas, where the rate of sea-level fall is greater than the rate of subsidence. At the same time, accommodation continues to develop basinward where the rate of subsidence is greater than the rate of fall in sea level. Since the coastal plain is, by definition, at sea level, it can be shown mathematically that as the rate of sea-level fall increases, the hingeward margin of the coastal plain must migrate basinward. This margin is located where the rate of fall in sea level equals the rate of subsidence. The model assumes that an equilibrium depositional surface will form hingeward of the coastal plain; and that the elevation of this surface will be an exponential function of distance from the hinge, equalling sea level at the inner coastal-plain margin. Modeling has shown that the equilibrium surface is depositional as long as any coastal plain is developed.

Figure 12 shows depositional geometry during sea-level fall. Decreasing accommodation in hingeward areas results in the basinward migration of the coastal-plain margin as the coastal plain becomes narrower. Little aggradation is possible under these conditions, and most of the sediment is stored in the clinoforms. This situation has important consequences for development of facies (to be discussed in a following section). Alluvial sediments accumulate under the equilibrium surface, and broad alluvial valleys, braided-channel systems, and flood plains are created. These aggradational facies are very thin, if shown at true scale, and cannot be depicted in Figure 9, even though vertical exaggeration is considerable there.

In Figure 9, the rate of sea-level fall increases from time

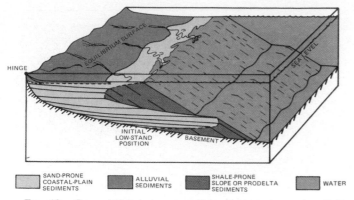

FIG. 12.—Progradational geometry subsequent to transgression. Falling sea level, decreasing accommodation in hingeward areas where alluvial sediments are deposited.

$t = 20$ to $t = 30$. During such conditions of falling sea level, with moderate rates of sediment influx, the coastal plain ceases to exist, and accommodation decreases at the coastline. This situation results in an erosional equilibrium surface as previously deposited coastal-plain and alluvial sediments are stripped away. Since no coastal plain is developed, no aggradational facies accumulate, and all sediment is stored on the clinoform slope. Figure 13 illustrates this circumstance. Stream systems draining the alluvial plain become incised as base level is lowered at the coast. Small-fan deltas may form very near the shelf edge as the rapid lowering in base level reduces the shelf area. Alluvial and stranded-shoreline terraces are probably common geomorphic features that reflect the relative uplift of coastal zones. Because of the incision of the topography, point sources are favored along the shoreline, and sands may be funneled down the clinoform slope by density-flow mechanisms to accumulate as onlapping deep-water sequences. Sand may also accumulate on the clinoform slope as prodelta ramp facies. Sand-prone clinoform and turbidite fan

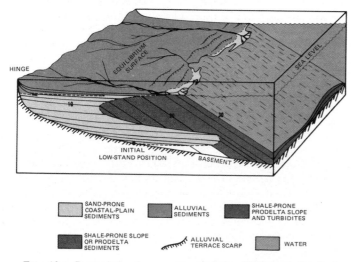

FIG. 13.—Progradational geometry and facies. Falling sea level, decreasing accommodation at the coastline with erosion of the land surface and development of turbidite facies.

deposits can therefore be related to storage capacity reduction on the coastal plain during a fall in sea level.

As the rate of eustatic sea-level fall declines from time $t = 30$ to $t = 40$ (Fig. 9), accommodation begins to increase at the coastline, because the rate of subsidence becomes greater than the rate of sea-level fall. Coastal-plain facies encroach hingeward upon the erosion surface, and the area of coastal-plain development is accompanied by progradation of the clinoform slope. The tendency for density-flow transport of sand to the basin floor is reduced as sand is stored in the coastal plain and along the coastline. Sand is also trapped in estuaries of drowned valleys. Where onlapping deep-marine turbidite facies occur, these are buried beneath the progradational clinoforms. Onlapping marine sequences are not included in the developmental geometry of the models. Sand-prone turbidite facies of Figure 13, however, may be conceptually regarded as possessing an onlapping relationship with the previous shale-prone clinoform slope.

Model scale.—

The models of basin filling simulate the large-scale geometry of progradational continental margins at the supercycle level of global change in sea level (Vail and others, 1977). Although the units of distance and depth/thickness are arbitrary in the following models (Figs. 15–22), they can be approximately related to tens of kilometers and hundreds of meters, respectively. Time units approximate millions of years. The models therefore represent sedimentary wedges on the order of 160 to 960 km (100 to 600 mi) across with single-cycle thickness of 150 to 240 m (500 to 800 ft). Figure 14 illustrates the relationship of model geometry and facies to depositional-sequence development. The simplified progradational geometry of the models (Fig. 14A) encompasses a great deal of complexity within the corresponding seismic sequence (Fig. 14B). Internal onlap and offlap patterns reflect major shifts in the focus of active-sediment influx through time and define the boundaries of subsequences. Subsequences of this scale are visible on seismic records and are caused by growth and abandonment of major portions of the deltaic/coastal plain. Changes in inclination of clinoforms and onlap of prodelta facies at subsequence boundaries are related to the orientation of the profile with respect to the oblique progradational axes of the overlapping subsequences.

The model coastline, separating sand-prone coastal-plain facies from shale-prone prodelta/clinoform slope facies, is the average position of the intertonguing of aggradational and progradational deposits. Shallow-marine shelf sediments, deposited between major subsequences, may be included in coastal-plain aggradational facies in the models. The nature of intertonguing is determined by small-scale events along the coastline. Figure 14C shows that, at the finest level, individual shoreline progradational units comprise the larger scale intertonguing pattern. Stream-mouth bar and beach sands interfinger with marine shales, which thicken toward the nearby clinoform slope. The deposition of these fundamental coastal/deltaic units is governed by local geomorphology and changes in the pattern of distributary channels. Compaction, an effect not incorporated in

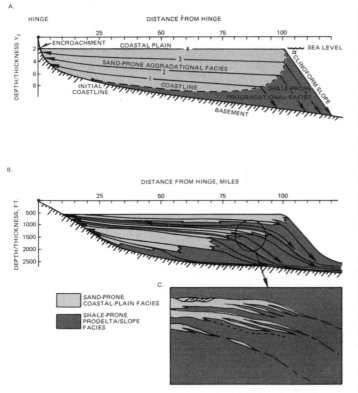

FIG. 14.—Comparison of model geometry and facies to a hypothetical depositional sequence.

the models, becomes at least as important as regional subsidence or eustatic sea-level change in the control of base level at this scale of investigation. We hypothesize, however, that regional controls will significantly influence sand continuity patterns in spite of and in conjunction with local compaction effects.

Selected models.—

Selected simulations of continental-margin progradation are presented to illustrate fundamental interrelationships between the geometry and facies of depositional sequences and rates of sediment influx, subsidence, and sea-level change. These models represent a preliminary effort; other combinations of parameters may yield further interesting results.

We may begin our discussion by considering simulations of three depositional sequences representing low-, moderate-, and rapid-sediment influx, where parameters that control subsidence and sea-level fluctuation remain constant among the three models (Fig. 15). Note that an initial lowstand coastline position is found at a distance of 20 units from the hinge, where the coast was located at time $t = 0$. Four time units were involved in the formation of the simulated profiles, during which time one cycle of sea-level rise and fall occurred, and subsidence of the basin floor depressed the initial coastline elevation from -2 to about -7 units below the elevation of the hinge. By time $t = 4$, sea level has fallen to its initial elevation.

The model profiles of Figure 15 all show rapid initial

progradation, reflecting the initial shallow nature of the basin and the slow rate of sea-level rise. Encroachment and onlap of the basement surface by sand-prone coastal-plain facies are indicated during this initial phase, whereas progradation results in downlapping-cycle terminations. As accommodation development increases, the rate of sediment influx controls the time and position at which initial progradation ceases and transgression begins. In the case of rapid sediment influx (Figure 15A), the coastline regresses to a distance of about 48 units from the hinge by time $t = 0.4$. At that time, retrogradation of the coast occurs as the increasing rate of accommodation development causes more and more sediment to accumulate in aggregational facies. Slower rates of sediment influx (Figure 15B, C) cause earlier termination of progradation, with regressive coastlines located nearer the hinge.

More profound transgressions also characterize models B and C as compared to model A. These transgressions occur during the time of most rapid accommodation development prior to time $t = 1$, when sea-level rise is most rapid. If the rate of sediment influx is extreme, no transgression will take place, but the rate of progradation will decline during this period in response to development of thicker aggradational facies.

In the figured simulations, transgression ends at about time $t = 1$ or at the inflection point during sea-level rise, when the rate of sea-level rise begins to decline. The time when transgression ceases is not very sensitive to sediment influx rate. In cases of low-sediment influx, the transgressive coastline is pushed near the hinge, where rates of subsidence are low. In cases of high-sediment influx, the transgressive coast is far from the hinge, where subsidence is high. Thus, a balance is achieved between sediment influx and accommodation development, such that any decrease in rate of accommodation increase results in progradation regardless of sediment influx rate. Extremely, low-sediment influx rates do result, however, in a longer period of transgression.

During rapid accommodation development, much or all of the sediment is trapped on the coastal plain and along the coast. We may speculate that this situation leads to a decrease in overall sand:shale ratio as silts and clays accumulate on deltaic and coastal plains. Cycles of sand deposition, whether they result from the lateral migration of fluvial channel systems or from small-scale transgressions and progradations, may exhibit a low degree of vertical and possibly lateral continuity as thick shales become interbedded. On the delta plain, the formation of large interdistributary bays may be favored during such conditions. These bays are the sites of deposition of shaley sediments. Rapid accommodation development in the three simulations (Fig. 15) occurs until time $t = 1$. The decline in the rate of sea-level rise following time $t = 1$ causes a decline in the rate of accommodation increase, and progradation across the transgressive surface begins. The basement surface continues to be encroached upon by coastal-plain facies. The rate of progradation increases in the three cases as the rate of sea-level rise declines to zero at time $t = 2$, when sea-level maximum is achieved. We may again speculate that as silt and shale are preferentially stored in clinoform units, the

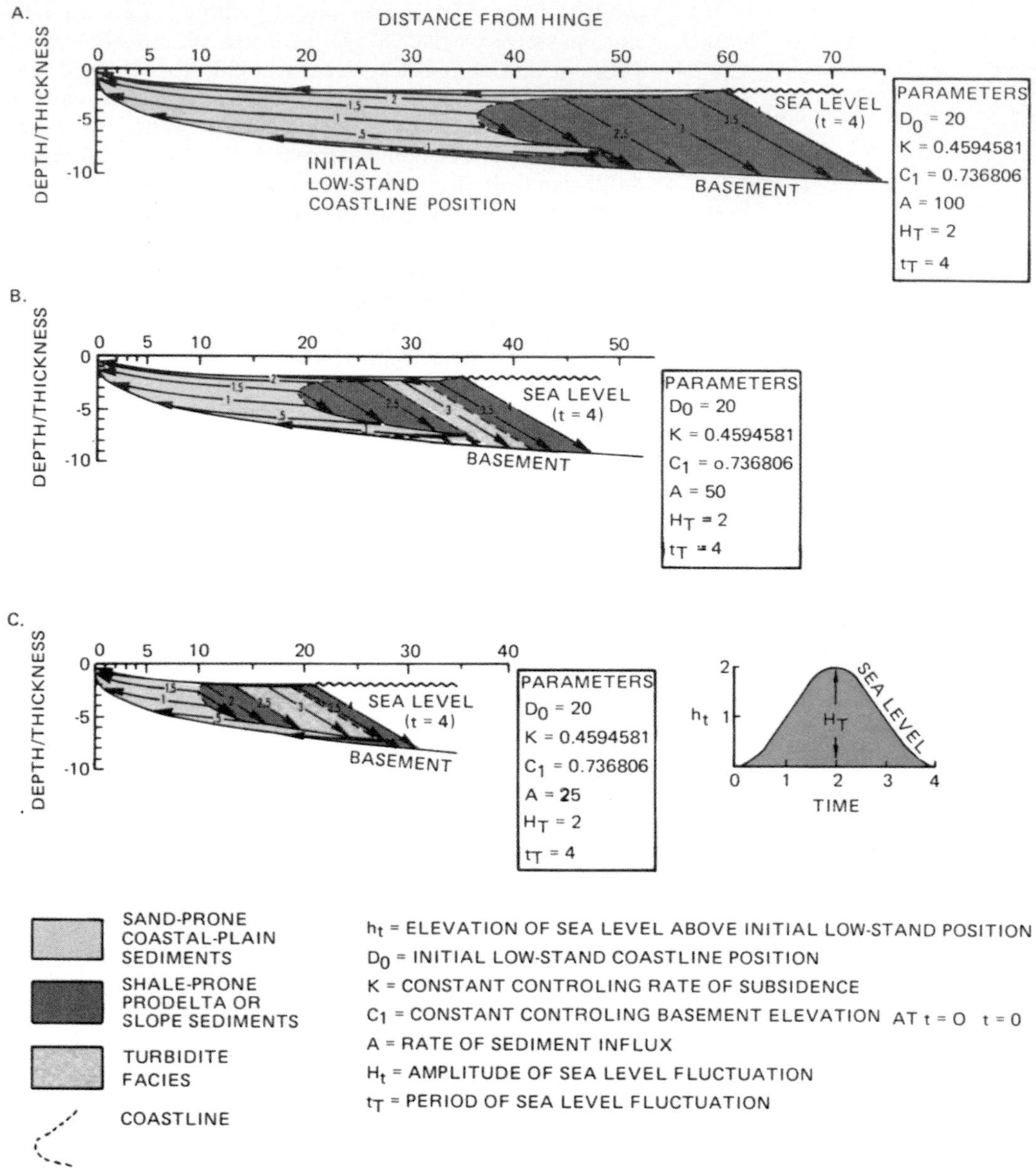

FIG. 15.—Simulations of depositional sequences developed during one cycle of sea-level rise and fall. Sediment influx varies as shown from rapid (A), to moderate (B), to slow (C). All units are arbitrary and vertical exaggeration is great. Chronostratigraphic surfaces, numbered in time units, simulate seismic reflections, and cycle terminations are indicated by arrows.

sand:shale ratio may improve, and continuity may be enhanced on the coastal plain.

Progradation initiated during the period of declining rate of sea-level rise continues during sea-level fall. As sea level begins to fall, accommodation declines in hingeward areas, and the lower margin of the coastal plain, where the rate of sea-level fall equals the rate of subsidence, moves seaward. An equilibrium surface is established hingeward of the coastal plain, as described previously. When sediment influx is rapid relative to rates of sea-level fall, as shown in Figure 15A, the rate of coastal regression (or progradation) is high relative to the rate at which the inner-coastal-plain margin moves basinward; and a coastal plain is always developed during sea-level fall. In this case, deposition of alluvial sediments occurs under the equilibrium surface, and deltaic/coastal plain facies continue to aggrade. As the coastal plain becomes narrower and as ag-

gradation is reduced by the increased rate of sea-level fall, it is probable that sand-prone prodelta or slope deposits accumulate. It seems that since a large proportion of the incoming sediment is stored in the clinoforms at this time, an increase in the sand:shale ratio on the slope must result. The alluvial equilibrium surface and the coastal plain are sites of highly continuous, although possibly thin, accumulations of fluvial sands, as the slow rate of aggradation permits continued reworking and removal of fines to the coast.

Figure 15 shows that the rate of sea-level fall reaches a maximum at time $t = 3$ and thereafter declines to zero at time $t = 4$. The inflection point at $t = 3$ marks the beginning of encroachment and onlap of the equilibrium surface by the expanding coastal-plain facies (model A) while progradation continues. Encroachment is brought about by increasing accommodation due to the decline in the rate of sea-level fall subsequent to time $t = 3$. With the expansion of the coastal plain, more sediment is stored in aggradational facies, and the rate of progradation declines. Presumably, the sand:shale ratio declines on the slope as the rate of accommodation increases.

The sequence of events during sea-level fall is somewhat different in models B and C, where sediment influx is lower relative to rates of sea-level fall. In those cases, the equilibrium surface becomes erosional. The rate of regression of the coastline is slow, so that the rate of sea-level fall exceeds the rate of subsidence at the coast, and no coastal plain is developed. Sediments are brought to the coast through incised valleys, and the formation of turbidite facies is enhanced as sediment is dumped at the top of the clinoform slope. These turbidites may be incorporated into the clinoforms as sand-prone lobes, as shown, or they may form a mounded sequence at the base of the slope. The decline in the rate of sea-level fall subsequent to time $t = 3$ results in encroachment and onlap of the erosion surface. The time of encroachment is later, however, than in model A (Fig. 15), because the rate of subsidence is slow at the coastline (near the hinge), so that the rate of sea-level fall must decline considerably before accommodation will increase there.

The timing of events during the development of the simulated depositional sequences and the possible relationship of reservoir facies to those events can better be investigated with time/space diagrams (Fig. 16). These diagrams contain the same data as Figure 15 but use axes of distance (from the hinge) and time (increasing toward present). Cumulative sediment thickness at any time t is represented by contour units. With reference to Figure 16, an initial coastline position at a distance of 20 units from the hinge is indicated at time $t = 0$. The initial rapid progradation is more pronounced and encompasses more time as rate of sediment influx increases. Progradation is accompanied by encroachment and onlap of the basement surface, which continues until time $t = 2$ of the sea-level maximum. Thereafter, the hingeward margin of the coastal plain moves in a seaward (basinward) direction in response to decreasing accommodation in hingeward areas (sea-level fall). In the case of rapid sediment influx (Fig. 16A), the coastline is located sufficiently basinward during sea-level fall that a coastal plain is always developed. This happens because the rate of subsidence at the coast exceeds the rate of sea-level fall, and accommodation for aggradational facies is available. The inner coastal-plain margin approaches but does not intersect the coastline as the coastal plain narrows. The possible consequence of the narrowing of the coastal plain and the development of alluvial facies with regard to reservoir quality was discussed previously. At time $t = 3$, encroachment and onlap of the depositional equilibrium surface begin as the rate of sea-level fall declines, and accommodation increases along the inner coastal-plain margin.

In the cases of lower sediment influx (Fig. 16B,C), the coastline is located nearer the hinge during negative encroachment subsequent to time $t = 2$. Figure 16 shows that in these cases, the coastal plain narrows and disappears as the rate of subsidence at the coast is exceeded by the rate of sea-level fall. Narrowing of the coastal plain is accompanied by deposition of alluvial sediments under the equilibrium surface. When the equilibrium surface intersects the coastline, however, erosion ensues, because the rate of sea-level fall exceeds the rate of subsidence at the coast and base level is lowered across the land surface. Alluvial sediments and coastal-plain facies are removed; Figure 16 shows that the degree of erosion is related to the rate of sediment influx and coastline position, or, more fundamentally, to the degree to which the rate of sea-level fall exceeds the rate of subsidence at the coast.

A probable consequence of the development of an erosion surface is the deposition of turbidite facies, as mentioned previously. Figure 16 illustrates the relationship between rate of sediment influx and timing of turbidite facies accumulation. Slow rates of sediment influx favor extended periods of turbidite development as the period of erosion is prolonged. Turbidites are most likely to occur when erosion is most rapid, that is, when rate of sea-level fall is greatest. The turbidite-prone time/space window is reduced as the rate of sediment influx increases and the coast regresses farther basinward, where rates of subsidence are higher relative to sea-level fall. Conversely, if rate of influx were held constant, high rates of sea-level fall relative to subsidence rates would enhance the possibility of turbidite development, since a coastal plain would be narrow or lacking entirely.

In the simulations, the assumption is made that erosion ceases when base level is raised and encroachment on the erosion surface begins. Encroachment and onlap of the erosion surface begins when the rate of sea-level fall declines. Figure 16 shows that the time of encroachment is later when the coast is located near the hinge, where rates of subsidence are low. The time of encroachment is at the inflection point on the falling sea-level curve in cases of high-sediment influx where a coastal plain is always present, because in that situation (Fig. 16A) the margin of the coastal plain is found where the rate of subsidence is equal to the rate of sea-level fall. Any decline in the rate of sea-level fall, therefore, results in hingeward encroachment.

Initial progradation, in the three cases of Figure 16, is followed by transgression and nondeposition. Relict-shelf sands and nonclastic facies (mentioned previously) may then accumulate. The time/space window for deposition of these facies is shown to increase as the rate of clastic-sediment influx decreases. Reduction of the rate of sea-level rise at time $t = 1$ results in initiation of progradation across the hiatal shelf surface. Turbidite facies may be included within the progradational sequence or may form distinct depositional sequences during falling sea level. Encroachment of the equilibrium surface is accompanied by clinoform progradation. Sands are stored in the coastal-plain and regressive-shoreline facies, and the likelihood of significant sand accumulation on the clinoform slope is reduced.

Let us now consider simulations involving two cycles of sea-level rise and fall. Figure 17 illustrates model profiles of siliciclastic sequences with different rates of sediment influx. Eight time units were required for their formation. The parameters controlling sediment influx, subsidence, and sea-level fluctuation in Figure 17A and B are identical to those of Figure 15A and B. The patterns of first-cycle pro-

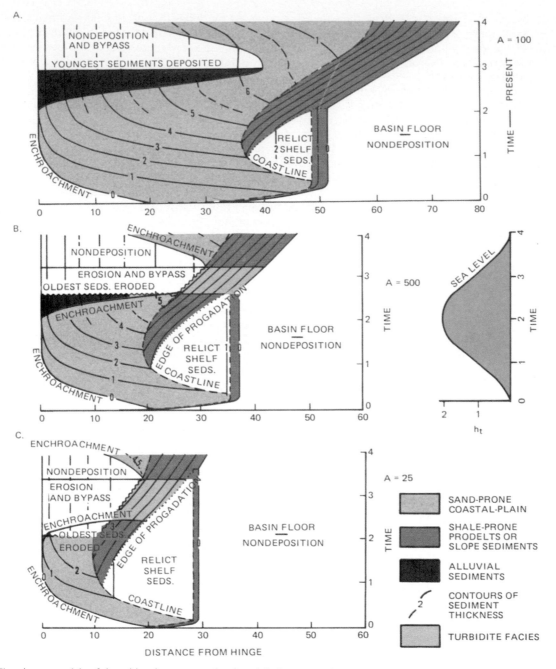

FIG. 16.—Time/space models of depositional sequences developed during one cycle of sea-level rise and fall. Sediment influx varys from rapid (A), to moderate (B), to slow (C).

gradation, transgression, and encroachment are therefore identical in the two-cycle models to those described for one-cycle models. The second cycle of sea-level fluctuation results in a cycle of transgression and subsequent progradation. Encroachment and progradation during first-cycle sea-level fall ($t = 3$ to $t = 4$) continue during the initial slow second-cycle rise. Progradation is terminated as the rate of rise increases, and transgression begins. The coastline retreats until time $t = 5$, the inflection point of rising sea-level, when the rate of rise begins to decline. Progradation then ensues and, in the case of rapid-sediment influx (Fig. 17A), continues throughout subsequent sea-level fall. In the

case of moderate-sediment influx rate (case B), progradation is accompanied by erosion and turbidite facies deposition. Encroachment and onlap of the second-cycle equilibrium surface begins at or near the inflection point on the falling sea-level curve ($t = 7$). Previous generalizations regarding the possible relationship between facies, continuity, and accommodation development apply during second-cycle sedimentation and need not be repeated. It should be pointed out, however, that in multicyclic deposition, the clinoform slope becomes extended as the sediment column grows, and progradation of the continental margin is thereby limited. The margin, as a whole, will therefore aggrade at

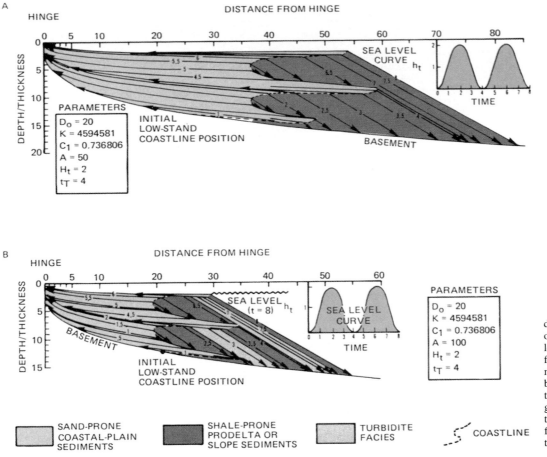

FIG. 17.—Simulations of depositional sequences developed during two cycles of sea-level rise and fall. Sediment influx varys from rapid (A) to moderate (B). All units are arbitrary, and vertical exaggeration is great. Chronostratigraphic surfaces, numbered in time units, simulate seismic reflections, and cycle terminations are indicated by arrows.

a basin position in equilibrium with long-term sediment supply rates. Extremely high rates of sediment supply, probably tectonically controlled, are required to prograde the continental margin once significant shelf-to-basin topography is established.

Model profiles can be regarded as simulated seismic profiles in which chronostratigraphic surfaces represent seismic-reflection horizons. Developmental geometry results in discontinuities in the pattern of time lines that define subunits within the model. These subunits simulate seismic sequences bounded by unconformities or correlative conformities. Figure 18 illustrates the two-cycle models discussed previously with seismic sequences lettered and colored. Sequence boundaries are identified by cycle terminations at onlap and downlap (offlap). A study of Figure 18A will show that five sequences can be defined by onlapping relationships in hingeward areas and offlap down to the basin. In the case of moderate rate of sediment influx (Figure 18B), additional sequences of mounded turbidites or low-stand deltaic deposits may also be included (sequences B and D).

The time/space framework of seismic-sequence development and its relationship to facies is better displayed in Figures 19 and 20. In the case of both moderate and high rates of sediment influx, the time envelope that includes the seismic sequence reflects the processes of encroachment and progradation. Sequence boundaries defined by encroach-

ment and onlap become younger toward the basin margin, while those reflecting progradation and offlap become younger toward the basin center. Sequence boundaries defined by transgressive hiatal surfaces become younger toward the hinge as a result of the diachronous nature of transgression. Where sequence boundaries are concordant, that is, conformable with continuous deposition, they are correlative with discordant boundaries in other areas either hingeward or basinward. Concordant boundaries are time-synchronous and parallel to reflections. The equilibrium surface defines a boundary for the underlying sequence. Where the surface is depositional, as in Figure 19B, it is time-synchronous. Where the surface is erosional, however, as in Figure 20B, it is diachronous. In the latter case, flat-lying coastal-plain facies may be deeply truncated basinward, whereas dipping strata of the clinoforms are only superficially modified by truncation at the coast. Relict-shelf sediments in the models are thin veneers of reworked clastics without seismic expression and are included in sequences A and C of Figures 19 and 20. If significant nonclastic deposition occurs in this time/space window, additional sequences may be generated. Shelf bypass during transgression may also result in marine onlap of the slope, a process not simulated in the models.

The timing of sequence development with respect to sea-level fluctuation is important as we attempt to understand the global distribution of seismic stratigraphic sequences in re-

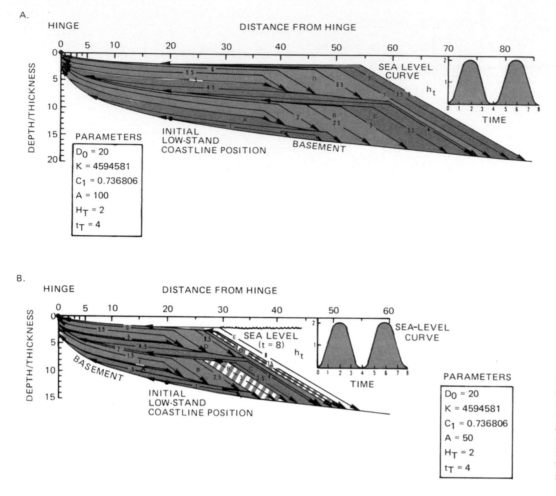

FIG. 18.—Simulations of seismic sequences developed during two cycles of sea-level rise and fall. Sediment influx varying from rapid (A) to moderate (B). All units are arbitrary, and vertical exaggeration is great. Seismic sequences lettered and colored; cycle terminations indicated by arrows.

lation to eustatic cycles of sea-level change. Figures 19 and 20 illustrate the fundamental relationship between sequence-boundary age and eustatic sea level. If seismic sequences are regarded as time-stratigraphic units, then the maximum age of the lower boundary and the minimum age of the upper boundary define the time represented by the sequence as a whole. An examination of Figures 19 and 20 will show that the ages of sequence-bounding surfaces correspond to times and maximum rate change in rising or falling phases of sea-level fluctuation. The development of these time-stratigraphic units is therefore controlled by rate relationships and not by sea-level maximum and minimum. The magnitude of sea-level rise and fall may quantitatively control sediment distribution, but the fundamental qualitative control in the timing of sequence-boundary development lies in the interaction of *rates* of sediment influx, sea-level fluctuation, and subsidence.

Let us examine these statements in detail. Referring to Figures 19 and 20, the age of the upper boundary of sequence A is approximately the time of inflection on the sea-level rise curve. Offlap across the hiatal transgressive surface, by which this unconformable boundary is recognized, occurs in response to declining accommodation as the rate of sea-level rise diminishes at time $t = 1$. Progradation and offlap in sequence B continues through the time of sea-level maximum. In the case of rapid rate of sediment influx (Fig.

19B), the upper boundary of sequence B is defined by onlap of the depositional equilibrium surface. The time at which this onlap begins is the age of the upper sequence boundary and is located at the inflection point on the falling sea-level curve. Onlap begins at this time because the rate of sea-level fall begins to decline, and accommodation, produced by subsidence, increases at the inner coastal-plain margin. In the case of moderate or slow rate of sediment influx (Fig. 20B), the upper boundary of sequence B is defined by onlap of an erosional equilibrium surface. Highstand progradation continues until the rate of sea-level fall exceeds the rate of subsidence at the coast. This situation is reached as the rate of sea-level fall increases. "Lowstand" deltaic deposition or turbidite fan development (sequence B) occurs concurrently with the most rapid rate of sea-level fall. "Lowstand" deposition terminates with a sequence boundary defined by onlap of the erosion surface. This occurs somewhat younger than the inflection point on the sea-level fall curve because of relatively low rates of subsidence at the coast. Onlap, as in the case of rapid-sediment influx, results from declining rates of sea-level fall and increase in accommodation.

Sequence C of Figure 19B and Figure 20B spans the time of sea-level minimum as progradation and offlap continue. The age of the upper bounding surface of C is defined, as for A, by the time of initiation of offlap in sequence D at

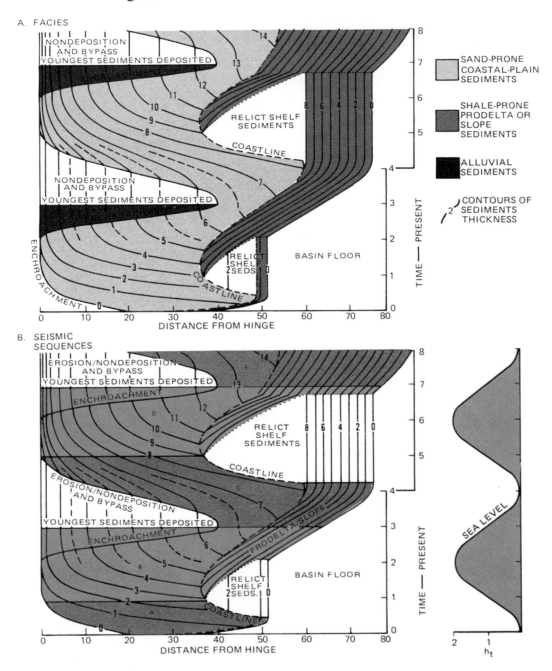

FIG. 19.—Space/time diagrams of depositional sequences with rapid rate of sediment influx, showing facies types (A) and seismic sequences (B).

the inflection point of rising sea level, where rate of rise begins to decline. A sequence of events identical to those discussed for first-cycle seismic sequences is repeated during second-cycle rise and fall. Clearly, then rate relationships control the timing of seismic-sequence development.

Models with stillstands.—

Models discussed previously have involved sinusoidal fluctuation in sea level. We will now examine a model in which sea-level rises and falls are interrupted by stillstands. Figure 21 is a model profile of sequences developed during eight time units, with sea level fluctuating as shown. Initial rapid progradation and regression during the slow initial sea-level rise is followed by retrogradation or near-stability as

the rate of sea-level rise increases. No transgression occurs during the initial rise, because subsidence rate is relatively low at the regressive coastline. Stillstand during sea-level rise results in a decline in accommodation for aggradational facies, leading to an increase in the rate of progradation and regression. Regression continues until shortly after time $t = 2$, when the rate of sea-level rise increases, and the concurrent increase in accommodation development causes a transgression to commence. Transgression occurs following stillstand because, in contrast to the situation during initial rise, prior regression has moved the coastline basinward where rates of subsidence are high. The combined effect of high-subsidence rate and rapid sea-level rise are responsible for the transgression.

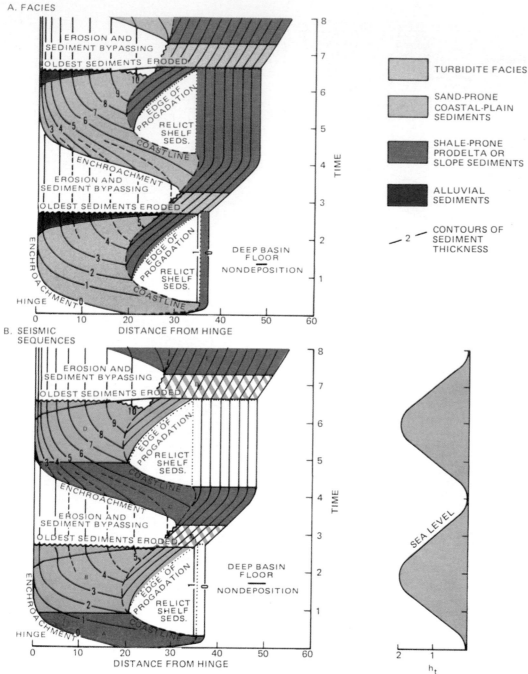

A. FACIES

ENCHROACHMENT

EROSION AND
SEDIMENT BYPASSING

OLDEST SEDIMENTS ERODED

EDGE OF
PROGADATION

RELICT
SHELF
SEDS.

COASTLINE

ENCHROACHMENT

EROSION AND
SEDIMENT BYPASSING

OLDEST SEDIMENTS ERODED

ENCHROACHMENT

EDGE OF
PROGADATION

RELICT
SHELF
SEDS.

COASTLINE

HINGE

ENCHROACHMENT

DEEP BASIN
FLOOR
NONDEPOSITION

TIME

DISTANCE FROM HINGE

TURBIDITE FACIES

SAND-PRONE
COASTAL-PLAIN
SEDIMENTS

SHALE-PRONE
PRODELTA OR
SLOPE SEDIMENTS

ALLUVIAL
SEDIMENTS

─ 2 ─   CONTOURS OF
SEDIMENT
THICKNESS

B. SEISMIC
SEQUENCES

EROSION AND
SEDIMENT BYPASSING

OLDEST SEDIMENTS ERODED

EDGE OF
PROGADATION

RELICT
SHELF
SEDS.

COASTLINE

ENCHROACHMENT

EROSION AND
SEDIMENT BYPASSING

OLDEST SEDIMENTS ERODED

ENCHROACHMENT

EDGE OF
PROGADATION

RELICT
SHELF
SEDS.

COASTLINE

HINGE

ENCHROACHMENT

DEEP BASIN
FLOOR

NONDEPOSITION

TIME

DISTANCE FROM HINGE

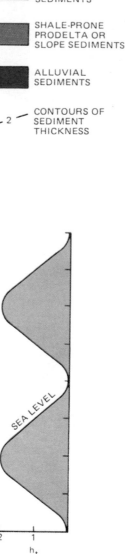

SEA LEVEL

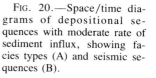

h_t

FIG. 20.—Space/time diagrams of depositional sequences with moderate rate of sediment influx, showing facies types (A) and seismic sequences (B).

Transgression continues until time $t = 2.5$, the inflection point on the sea-level rise curve, following stillstand, when the rate of rise begins to decline. Regression and progradation occur as the rate of accommodation development decreases and more sediment is stored in clinoforms. As sea level begins to fall, accommodation is reduced in hingeward areas, and the coastal plain narrows. As the rate of sea-level fall declines at the inflection point (time $t = 4.5$) prior to stillstand, accommodation begins to increase in hingeward areas, and encroachment and onlap of the depositional equilibrium surface begins. Encroachment is accompanied by regression, but near-stability in coastal position is achieved during stillstand as accommodation increases on the coastal plain and the rate of progradation declines. The rate of regression increases following stillstand as the rate of sea-level fall increases. Another cycle of encroachment and onlap begins when the inflection point, following stillstand, is reached at time $t = 6.5$ and the rate of sea-level fall begins to decline.

Four seismic sequences are simulated by the model shown in Figure 21. As in previous models, they are defined by unconformities of onlap or offlap and their correlative con-

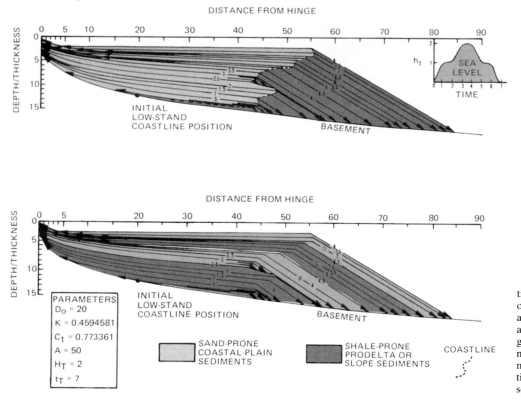

FIG. 21.—Simulations of depositional and seismic sequences developed during one cycle of sea-level rise and fall with stillstands. All units are arbitrary, and vertical exaggeration is great. Chronostratigraphic surfaces, numbered in time units, simulate seismic reflections, and cycle terminations are indicated by arrows. Seismic sequences are lettered and colored.

formities. The sequences are recognized from downlapping relationships on transgressive surfaces and basement and onlapping cycle terminations on basement and equilibrium depositional surfaces. The space/time framework of the simulated sequences and their relationship to facies can be better seen in Figure 22.

Fluctuations in rate of sea-level change, here incorporated as stillstands, have a profound effect in the generation of sequence boundaries. Increases in rate of sea-level rise may cause transgression, whereas decreases lead to progradation and regression. Sequence boundaries are produced at inflection points, where the rate of rise declines, as at the upper boundary of sequence A (Fig. 22). These are recognized by the unconformity of the transgressive surface with the offlapping progradational strata above. Changes in the rate of encroachment accompany fluctuations in rate of sea-level rise, but onlapping boundaries between depositional sequences do not occur during sea-level rise, because accommodation is increasing everywhere. Apparent onlap due to oblique progradational axes, however, is a commonly observed feature on seismic profiles under conditions of increasing accommodation.

Fluctuations in the rate of sea-level fall produce sequence boundaries of onlap and encroachment on depositional surfaces. Boundaries are located at inflection points, as at the upper boundaries of sequences B and C (Fig. 22), where the rate of sea-level fall begins to decline and accommodation increases. These surfaces of onlap may be mistaken as indications of sea-level fluctuation rather than simply as the result of rate changes during a general sea-level fall.

Accommodation/accumulation diagrams.—

Let us return briefly to Figure 19 and examine the modeled relationship between accommodation, sediment accumulation, and facies at selected basin locations, namely $X = 20$ and $X = 46$. Figure 23 presents the data at these locations; we may regard these as well sites.

Close to the basin margin at $X = 20$, sand-prone coastal-plain facies accumulate in response to increased accommodation, and the rate of accumulation is controlled by rate of accommodation development. The sequence rests on basement, which it onlaps. The boundary between sequences A and B is concordant, and therefore no evidence for it is seen at this location. Following the sea-level maximum at time $t = 2$, accommodation begins to decrease at $X = 20$ and, under the assumption of the model, equilibrium surface is established. Alluvial sediments are deposited on this surface until time $t = 3$, when the rate of sea level declines and a period of nondeposition and bypass ensues at $X = 20$, while encroachment begins basinward. At time $t = 3.8$, accommodation begins to increase at $X = 20$, and deposition of coastal-plain facies of sequence C is initiated with onlap onto the depositional surface of sequence B. The second cycle of sea-level rise begins at time $t = 4$, and the sequence of events is repeated for seismic sequences C, D, and E. We may speculate that rate changes on the accommodation curve will be reflected in facies type and continuity. Increased reworking of near-surface and surface deposits may occur during periods of slow accommodation increase, resulting in a high sand:shale ratio as

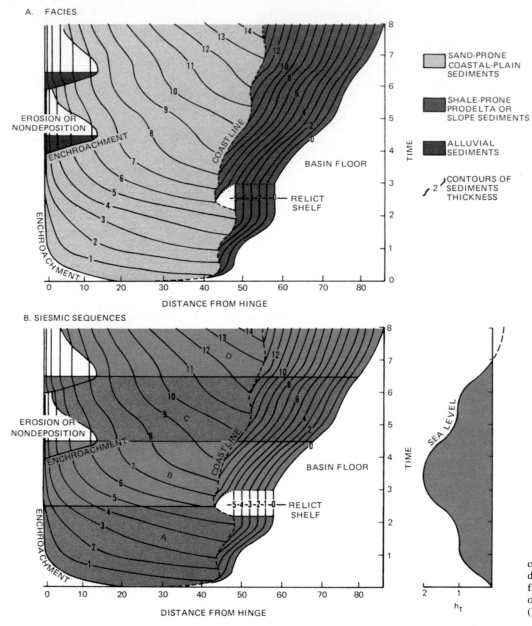

A. FACIES

SAND-PRONE COASTAL-PLAIN SEDIMENTS

SHALE-PRONE PRODELTA OR SLOPE SEDIMENTS

ALLUVIAL SEDIMENTS

CONTOURS OF SEDIMENTS THICKNESS

EROSION OR NONDEPOSITION

ENCHROACHMENT

COASTLINE

BASIN FLOOR

RELICT SHELF

ENCHROACHMENT

TIME

DISTANCE FROM HINGE

B. SIESMIC SEQUENCES

EROSION OR NONDEPOSITION

ENCHROACHMENT

COASTLINE

BASIN FLOOR

RELICT SHELF

ENCHROACHMENT

TIME

SEA LEVEL

DISTANCE FROM HINGE

h_t

FIG. 22.—Space/time diagrams of depositional sequences developed during one cycle of sea-level rise and fall with stillstands, showing facies of types (A) and seismic sequences (B).

fines are bypassed seaward. Thus, lenticular point-bar sands of low-reservoir continuity may give way upward, within the major cycles of accommodation development, to more continuous, sheetlike, meander-belt and braided-stream sands.

Farther from the hinge at $X = 46$, the history of deposition is quite different. Here, accommodation is always increasing because of high-subsidence rate, although the rate of accommodation increase varies in response to sea-level fluctuation. The base of the sequence is defined by offlap of shale-prone marine facies. Sedimentation rates are sufficient to aggrade the sea floor to sea level during the deposition of sequence A. Sand-prone coastal-plain facies are then deposited until time $t = 0.5$, at which time the rate of accommodation development exceeds the rate at which sediments may accumulate under model constraints. A trans-

gressive period of nondeposition ensues and continues until time $t = 2$ at $X = 46$, when bottomset marine facies are deposited at the toes of prograding clinoforms. Note that the hiatal period lasts almost until time $t = 2$, even though progradation and regression began hingeward of $X = 46$ at time $t = 1$. Progradation sequence B is deposited with offlap on the hiatal surface and aggrades to sea level. At location $X = 46$, coastal-plain facies deposition continues uninterrupted across the concordant B/C boundary to time $t = 4.5$. At this time, the high rate of sea-level rise during the second eustatic cycle causes another transgression, and the sequence of depositional events is repeated. Coarsening- or thickening-upward cycles of prodelta/slope and coastal deposition characterize the model synthetic column. Hiatal surfaces are overlain by progradational shales and silts which grade upward into shoreline sands. Relict, re-

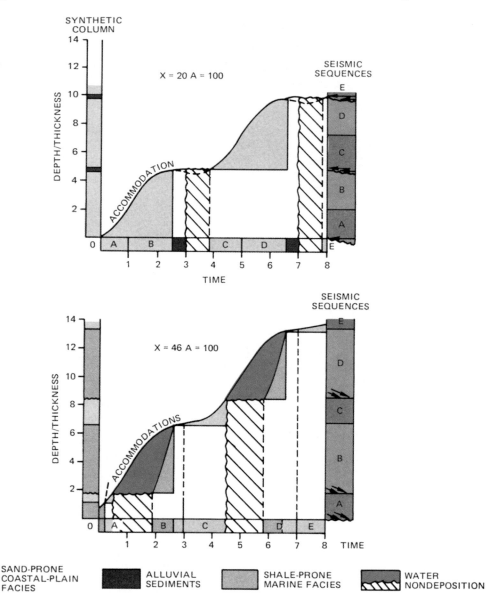

FIG. 23.—Accommodation, sediment accumulation, and facies at two basin locations for a simulation with rapid sediment influx (see Fig. 19). Distance from the hinge (X) as shown.

worked sand, glauconites or other nonclastic rocks mark the hiatal period and are underlain by coastal and deltaic facies.

Figure 23 emphasizes the obvious fact that depositional events within sequences are greatly dependent on basin position. Relative changes in sea level are also a function of basin location. A relative fall in sea level, for instance at $X = 20$, $t = 3.5$, is concurrent with a relative rise in sea level at basinward locations, $X = 46$, $t = 3.5$. Nondeposition and transgression in rapidly subsiding areas are synchronous with continuous continental or deltaic development in other more stable regions.

Implications for seismic stratigraphy.—

Since the completion of the research presented in this paper, a number of concepts derived from the basin-filling

simulations have been incorporated into the seismic stratigraphic literature (e.g., Vail and others, 1984). Figure 24 illustrates relationships currently used in seismic stratigraphic analysis. These concepts can be related directly to modeling results and predictions.

Three sequences are shown in Figure 24A, each of which is bounded by unconformities. The simulations suggest that these unconformities form during sea-level fall when accommodation is declining toward the basin margin. The type 1 unconformity (bounding-sequence 1) of Vail and others (1984) arises from processes described in this paper in which the rate of sea-level fall exceeds the rate of subsidence at the coast (shelf edge). Erosion then prevails, and sediment is bypassed to deep water through incised valleys. Note the mounded turbidite fan sequence at the base of sequence 1 clinoforms during this period of erosion. In contrast, se-

A) STRATIGRAPHIC CROSS SECTION

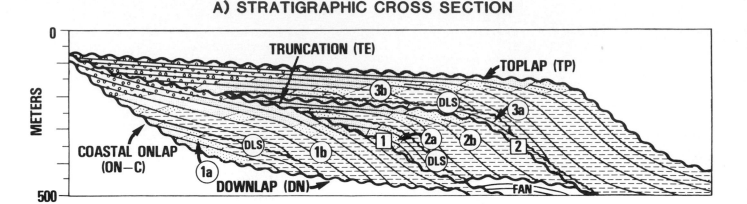

B) PARAMETER CLASSIFICATION CHARTS

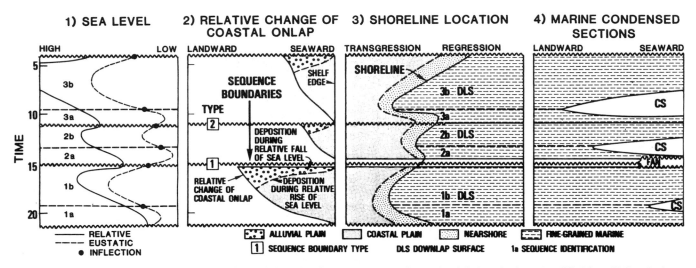

FIG. 24.—Diagrammatic stratigraphic cross section and chart showing parameters used to make global-cycle charts (modified from Vail and others, 1984; sea-level curves added). The stratigraphic cross section (A) shows the distribution of boundary types, downlap surfaces (condensed sections), and facies of three idealized sequences. The parameter classification chart (B) shows the relation between the parameters of the three sequences and relative and eustatic sea level, in time. CS means condensed section.

quence 2 is bounded above by a type 2 unconformity. Modeling shows that this type of unconformity is likely when the rate of subsidence exceeds the rate of sea-level fall at the coast (shelf edge). In this situation, accommodation for sediment continues to develop under the coastal plain, and the equilibrium surface is dominantly depositional. The tendency for sediment bypass to deep water is therefore reduced, and major fan development is less likely.

The cyclic model pattern of progradation/transgression/progradation, which arises in response to changes in relative sea level, is quite similar to observed geometries in seismic sequences, diagramatically represented in Figure 24A. Figure 24A shows that each major sequence (1, 2, 3) includes subsequences (e.g., 1A, 2A and so on) that correspond in genesis and timing to model sequences shown in Figures 18, 19, and 20. For instance, the pretransgressive progradational units of the simulations (sequences A, C, E; Figs. 18, 19, 20) are equivalent to subsequences 1A, 2A, and 3A of Figure 24A. These subsequences are bounded

above by downlap surfaces (DLS) that are the transgressive hiatal surfaces of this paper. The downlap pattern is a result of progradation across these surfaces as the rate of relative sea-level rise declines. Subsequences 1A, 2A, and 3A are examples of lowstand deposits as they span the time from the inflection point on the falling sea-level curve, when the rate of fall begins to decline, to the inflection point during rising sea level.

Subsequences 1B, 2B, and 3B of Figure 24A are progradational units deposited during accommodation increase and decline as modeled in Figures 18, 19, and 20 (sequences B and D). These sequences are "highstand" deposits, as they represent the time from the rising sea-level inflection point, through eustatic highstand, to the falling sea-level inflection point.

Figure 24B illustrates the relationship, suggested by the simulations, between eustatic and relative sea-level fluctuation and patterns of coastal onlap, facies development, and condensed interval occurrence. Coastal onlap and encroachment of coastal-plain facies upon preexisting surfaces are shown to occur during periods of relative sea-level rise (Figure 24B, 1 and 2). Alluvial facies are most prevalent near the hingeline during falling relative sea level, when base level is lowered and fluvial gradient and energy are increased. Encroachment and onlap of alluvial facies continues during relative sea-level fall until either (1) relative sea level begins to rise at the inflection point on the eustatic curve, in the case of a type 2 unconformity; or (2) erosion begins to truncate the previously deposited sediments, in the case of type 1 unconformity. Eustatic sea level continues to fall as encroachment and onlap begin across sequence-bounding unconformities, as for example at the base of sequence 2 illustrated in Figure 24B.

Changes in shoreline position, shown in Figure 24B, 3, simply reflect the transgressive or regressive nature of the coast during sequence development. Modeling predicts that the maximum incursion during transgression will occur near the inflection point on rising eustatic and relative sea-level curves when the rate of rise (accommodation development) is maximized. Modeling also suggests that condensed sections (Fig. 24B, 4), or hiatal, nondepositional periods in the terminology of this paper, occur during this transgressive time frame. Condensed sections represent periods of greatly reduced clastic influx, when most sediment is stored in aggradational facies of the coastal plain. The time included in the condensed section expands in a basinward direction because of continued low-sedimentation rates far from the coast, even during periods of regression. These relationships are illustrated in simulations of Figures 19 and 20 in which the condensed or hiatal intervals are identified as times of relict shelf deposition.

Global-cycle Charts (see for example, Vail and others, 1984) display world-side onlap and encroachment patterns through geologic time. Each major cycle or supercycle is shown to be characterized by slow encroachment during the time period of the cycle (varying in rate, as in Fig. 24B) followed by abrupt basinward movement of the encroachment point and termination of the cycle. This encroachment geometry was thought to be caused by and directly to reflect strong asymmetry in relative sea-level fluctuation. Such an abrupt fall in relative sea level, as suggested by the shift in encroachment location, can only occur, in the face of continued subsidence, in conjunction with an equally abrupt eustatic sea-level fall. Such abrupt sea-level falls have been difficult to explain. The simulations presented in this paper have shown that the observed encroachment geometry can result from the interaction of subsidence and sinusoidal sea-level fluctuations. The apparent patterns of onlap and relative sea-level change displayed by global cycle charts reflect the geometry of the preserved sediment packages. These patterns can be explained as the natural consequence of the interaction of a smoothly fluctuating eustatic sea level and basin subsidence.

SUMMARY AND CONCLUSIONS

Several generalizations can be drawn from basin-filling simulations. These are offered here as speculations to be documented by further research.

(1) The pattern of encroachment and onlap displayed by global-cycle charts can develop in the context of a smoothly fluctuating or sinusoidal eustatic sea level.

(2) Seismic-sequence boundaries, where the development of the sequence is controlled by factors of accommodation, are located at or near inflection points on the sea-level fluctuation curve, where rates of fall or rise are maximized.

(3) Sand-prone "lowstand" deltaic and turbidite sequences are most likely to develop when rate of sea-level fall exceeds the rate of subsidence at the coast. This situation is most probable when rate of fall is highest near the inflection point on the eustatic sea-level curve.

(4) Rates of aggradation of coastal plain sediments are controlled by accommodation development. Periods of rapid accommodation increase may result in storage of clay in low-energy coastal-plain environments and an overall reduction in reservoir continuity. Periods of slow accommodation increase may allow continual reworking of surface sediments by the fluvial systems and may result in an increase in sand continuity as fines are removed to the coast.

(5) Under conditions of constant sediment influx, decrease in accommodation development is reflected in increased rates of progradation. Sand content may increase in shelf and upper clinoform slope units at times of reduced accommodation, leading to the deposition of potential reservoir units.

(6) Glauconitic sands, phosphorites, and organic-rich shales in the geologic record may indicate periods of rapid accommodation increase due to rapid eustatic sea-level rise, when most clastics were stored in coastal-plain and shoreline facies.

ACKNOWLEDGMENTS

The author thanks Exxon Production Research Company for permission to publish this work. Also, the support and encouragement offered by many associates is gratefully acknowledged. In particular, Peter Vail provided much of the impetus for the further development of ideas presented in this paper and for their publication in the public domain.

REFERENCES

PAYTON, C. E., ed., 1977, Seismic Stratigraphy—Applications to Hydrocarbon Exploration: American Association of Petroleum Geologists Memoir 26, 516 p.

PITMAN, W. C., III, AND GOLOVCHENKO, X., 1983, The Effect of sea level change on the shelfedge and slope of passive margins: Society of Economic Paleontologists and Mineralogists Special Publication 33, p. 41–58.

VAIL, P. R., HARDENBOL, J., AND TODD, R. G., 1984, Jurassic unconformities, chronostratigraphy, and sea-level changes from seismic stratigraphy and biostratigraphy, in Schlee, J. S., ed., Interregional Unconformities and Hydrocarbon Accumulation: American Association of Petroleum Geologists Memoir 36, p. 129–144.

———, MITCHUM, R. M., AND THOMPSON, S., III, 1977, Seismic stratigraphy and global changes of sea level, Part 3: Relative changes of sea level from coastal onlap, in Payton, C. E., ed., Seismic Stratigraphy—Applications to Hydrocarbon Exploration: American Association of Petroleum Geologists Memoir 26, p. 83–97.

MESOZOIC AND CENOZOIC CHRONOSTRATIGRAPHY AND CYCLES OF SEA-LEVEL CHANGE

BILAL U. HAQ,[1] JAN HARDENBOL, AND PETER R. VAIL[2]

Exxon Production Research Company, P.O. Box 2189, Houston, Texas 77252-2189

In collaboration with

L. E. Stover, J. P. Colin, N. S. Ioannides, R. C. Wright, G. R. Baum, A. M. Gombos, Jr., C. E. Pflum, T. S. Loutit, R. Jan du Chêne, K. K. Romine, J. F. Sarg, H. W. Posamentier, and B. E. Morgan

ABSTRACT: Sequence-stratigraphic concepts are used to identify genetically related strata and their bounding regional unconformities, or their correlative conformities, in seismic, well-log, and outcrop data. Documentation and age dating of these features in marine outcrops in different parts of the world have led to a new generation of Mesozoic and Cenozoic sea-level cycle charts with greater event resolution than that obtainable from seismic data alone. The cycles of sea-level change, interpreted from the rock record, are tied to an integrated chronostratigraphy that combines state-of-the-art geochronologic, magnetostratigraphic and biostratigraphic data. In this article we discuss the reasoning behind integrated chronostratigraphy and list the sources of data used to establish this framework. Once this framework has been constructed, the depositional sequences from sections around the world, interpreted as having been formed in response to sea-level fluctuations, can be tied into the chronostratigraphy.

Four cycle charts summarizing the chronostratigraphy, coastal-onlap patterns, and sea-level curves for the Cenozoic, Cretaceous, Jurassic, and Triassic are presented. A large-scale composite-cycle chart for the Mesozoic and Cenozoic is also included (in pocket). The relative magnitudes of sea-level falls, interpreted from sequence boundaries, are classified as major, medium, and minor, as are the condensed sections associated with the intervals of sediment starvation on the shelf and slope during the phase of maximum shelf flooding during each cycle. Generally, only the sequence boundaries produced by major and some medium-scale sea-level falls can be recognized at the level of seismic stratigraphic resolution; detailed well-log and/or outcrop studies are usually necessary to resolve the minor sequences.

INTRODUCTION

The usefulness of a chronostratigraphic eustatic framework in exploration geology lies in the fact that it provides predrill estimates of geologic parameters from seismic data. This enhances interregional correlations, particularly in frontier areas, leading to more accurate stratigraphic, structural, and facies interpretations. The first such eustatic stratigraphic framework (Vail and others, 1977; Vail and Hardenbol, 1979) was an outcome of developments in seismic stratigraphy that were based on the realization that primary seismic reflections had chronostratigraphic significance. The cycles of sea-level change on the cycle chart (Vail and others, 1977) were derived from seismic data, aided by paleontologic age control from wells. Now, a new generation of cycle charts has been constructed, representing a considerable improvement on the earlier cycle charts constructed from seismic data alone. The new cycle charts were made possible by the recognition of depositional patterns interpreted as having formed during various phases of the sea-level cycle, coupled with an enhanced ability to recognize accurately genetically related sediment packages in outcrops and well logs.

Recent advances in magnetobiostratigraphy, especially for the Late Cretaceous and Cenozoic, have been important for the refinement of the global cycle charts. In numerous sections of marine strata in different parts of the world, magnetic-polarity reversals have been correlated directly to biohorizons (fossil first and last occurrences) that are the most readily available source of age dating. Over the past several years, researchers at Exxon Production Research Company have attempted to develop a sequence-keyed global

stratigraphic framework that is based on up-to-date magneto-, bio-, and sequence chronostratigraphies in sedimentary basins in different parts of the world. Updated and integrated chronostratigraphies were seen as a prerequisite to the construction of cycle charts. Existing time scales for the Mesozoic and Cenozoic were disparate and were based on differing criteria and approaches. We felt that an internally consistent approach was needed and that existing time scales could be improved. This led to an integrated chronostratigraphic framework to which the sea-level changes of the Mesozoic and Cenozoic have been tied. An early summary of this work (Haq and others, 1987) disseminated the results expediently so that the testing of the model of sea-level change and constructive feedback could ensue. In the present paper we include details of the approach used to construct the chronostratigraphic framework that could not be included in the summary, and we discuss the results of this effort. The cycle charts presented here include some modifications and corrections to our earlier versions presented with the summary. One important modification is the inclusion of the prominent 2.4-Ma event in the latest Pliocene, which was inadvertently left off the Cenozoic cycle chart (Haq and others, 1987, fig. 2).

The ultimate objective of stratigraphy is to identify and arrange successions of events that aid in increasingly precise correlations. From a practical standpoint, the best stratigraphic correlations are those that have a high degree of reproducibility within increasingly narrower temporal limits. Toward this end, it is important to identify and employ multiple stratigraphic criteria concurrently to maximize precision. As additional, independent, correlative systems are found and used parallel to the existing systems, a greater degree of consistency and confidence in the results is to be expected. An excellent example of a new system that has only recently become available is Sr-isotopic stratigraphy, based on the relatively slow but significant vari-

---

[1] Present addresses: Marine Geology/Geophysics Program, National Science Foundation, 1800 G Street, NW, Washington D.C., 20550.

[2] Department of Geology, Rice University, Houston, Texas 77251.

Sea-Level Changes—An Integrated Approach, SEPM Special Publication No. 42

ations in the ratios of ^{87}Sr to ^{86}Sr of sea water with time, which holds much promise for the future as an age-dating tool (Burke and others, 1982, DePaolo and Ingram, 1985).

Our chronostratigraphic framework has been aided significantly by the recently developed sequence-stratigraphic models that assist in the recognition of depositional sequences in outcrops and well logs. This not only helps in the precise dating of the sea-level events, but also in the more accurate integration of standard chronostratigraphic units (stages).

THE CALIBRATION OF GEOLOGIC TIME

Because units of time are not unique in themselves, we perceive the passing of time by association with unique events. The transition of geologic time is similarly conceptualized by either the sequence of geologic events that are distinctive in themselves, or by cyclic events that are distinguished from each other by association with other unique events.

The only strictly direct method of measuring geologic time is through the built-in clock of isotopic decay, the proportion of radiogenic parent-to-daughter elements determining the elapsed time since the genesis (or closure) of the measured system, according to a known rate of decay; however, radiochronology is the most time-consuming and thus the least practical tool for routine age determination. Radiometric dates, nevertheless, can provide the basis to elevate other systems of gauging relative temporal order to practical but indirect time-measuring tools.

The most readily available method of detecting the relative antiquity of sediments is through biostratigraphy, i.e., the unique biologic events (the first and last occurrence of distinctive fossil taxa) that occur during the course of biologic evolution. Although biostratigraphy discriminates older from younger rocks, it cannot measure absolute time. To accomplish this, biostratigraphic events have to be age dated, either through correlation with radiometrically dated horizons, or indirectly through association with magnetic-polarity or climatic events that have been previously dated.

Magnetic-polarity reversal events have the distinctive attribute of being relatively numerous and globally synchronous. They are, therefore, most appropriate as precise global correlation criteria. Because polarity reversals are not unique events, however, they must be used in association with other unique events, such as biostratigraphic datums and/or radiometric dates.

Climatic events, as depicted by oxygen isotope variations, and other environmental events, such as productivity fluctuations as reflected in carbon isotope variations, are increasingly used for regional and global correlations. Like magnetic-polarity reversals, however, the isotopic variations can only be used in association with other unique events and, so far, detailed stable isotope "stage" stratigraphy has been developed only for the past 2 million years or so of the stratigraphic record.

Here, we must add a note about the potential of Sr-isotopic stratigraphy. Although Sr-isotopic stratigraphy holds much promise as a semi-independent system for age dating of both fossiliferous and nonfossiliferous marine sediments,

much work needs to be done to calibrate accurately Sr-isotopic variations to the chronostratigraphic schemes. The present state of the art (DePaolo and Ingram, 1985, Elderfield, 1986) does not permit the routine use of Sr-isotopic stratigraphy.

These limitations to various direct and indirect age-dating methods suggest that, within the framework of the tools presently available, a widely applicable chronologic framework has to be based on the most practical and readily available criteria of correlation (biostratigraphy), constrained as broadly as possible by absolute dates (radiochronology), and the precision of interregional correlations ensured by a system of globally synchronous events (magnetostratigraphy).

INTEGRATION PROCEDURE FOR THE CONSTRUCTION OF CYCLE CHARTS

The construction of the sea-level cycle charts is an integration process that involves many iterative steps (Fig. 1). Recent improvements in direct correlations between microplanktonic biohorizons and magnetic reversals (magnetic-polarity chrons) are of prime importance for the construction of accurate chronostratigraphy. Irrespective of the numerical calibration of magnetic-polarity reversals, such direct correlations between fossil datums and polarity chrons provide a reliable indication of the true ranges of fossil taxa, especially after the relationships have been ascertained in a series of sections from several different areas. Range stacking from different latitudinal locations, when possible, serves to document the maximum (chronostratigraphic) extent of the taxa and to elevate *biozones* into *biochronozones*. This magnetobiostratigraphic reconciliation is the first essential step in the construction of the global-cycle charts and the basis of all subsequent integration.

The magnetobiostratigraphic framework in turn allows the assignment of stratigraphically constrained radiometric dates to appropriate polarity chrons. The criteria for the selection of radiometric dates are that they are analytically sound, geochemically reliable, and biostratigraphically constrained. Through paleontologic correlation, they can then be placed within the polarity reversal framework, along with their limits of uncertainties (stratigraphic and analytical errors).

Following the assignment of radiometric dates to polarity chrons, these dates can be plotted against seafloor magnetic-anomaly profiles from different ocean basins (i.e., the distance of anomalies from the spreading center) to arrive at best-fit age models for the anomalies, constrained by inflection points preferably where seafloor spreading rate changes are predicted by a comparison of the seafloor magnetic-anomaly profile from different basins (Fig. 2). The resulting anomaly ages from various basins are then stacked to enhance the signal-to-noise ratio. In this way, the stacked mean ages were used to calibrate the late Cretaceous through Cenozoic magnetic-polarity scale on the cycle charts presented in this paper. This procedure reduces the noise from radiometric analytical techniques, as well as minor irregularities in seafloor spreading.

The next step in the development of the cycle charts in-

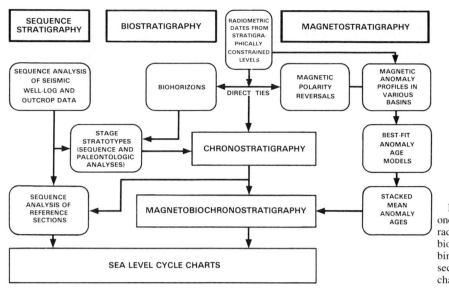

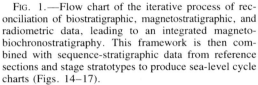

FIG. 1.—Flow chart of the iterative process of reconciliation of biostratigraphic, magnetostratigraphic, and radiometric data, leading to an integrated magnetobiochronostratigraphy. This framework is then combined with sequence-stratigraphic data from reference sections and stage stratotypes to produce sea-level cycle charts (Figs. 14–17).

volves the incorporation of the standard chronostratigraphic units (stages) in the magnetobiochronologic scheme. This integration involves two separate but related activities. First, the detailed biostratigraphic analyses of the stage stratotypes (and magnetostratigraphy or radiometric dates, when available) help determine the biochronostratigraphic extent of the stratotype. Second, the depositional-sequence analysis of the same sections assists in a more accurate positioning of the stages within a sequence-stratigraphic framework. The significance of the latter in the refinement of chronostratigraphy will become clear from the examples of field applications discussed in the following sections, especially the example from the neostratotype of the Lutetian Stage.

The final step in the construction of the cycle charts involves the integration of sea-level events with the magneto- and biochronostratigraphies. The sea-level changes interpreted through stratigraphic analyses of sequences in both subsurface and outcrop data from sections in various parts of the world, dated through their paleontologic and physical/stratigraphic relationships, are incorporated into the chronostratigraphic scheme to produce the sea-level cycle charts that are the ultimate purpose of this exercise.

In the ensuing sections, various aspects of this integration will be discussed in more detail. To communicate these concepts clearly, it is first necessary to review briefly the stratigraphic terminology utilized on the cycle charts.

Stratigraphic terminology.—

Table 1 summarizes the stratigraphic terminology employed in the text and on the cycle charts. As explained in the International Stratigraphic Guide (Hedberg, 1976), the distinction between rock (lithostratigraphic) and time (geochronologic) units is quite obvious. Time-rock (chronostratigraphic) units are often confused in the literature, however, and their nature needs to be clarified, especially when applied to depositional sequences. The cycle charts represent both time and rock data tied together by standard

chronostratigraphic units (stages). Depositional sequences are first identified in lithostratigraphic sections, and later, indirectly, tied to a linear time scale through their biostratigraphic and/or magnetostratigraphic relationships to establish their chronostratigraphic identity.

This underscores the need for a consistent and distinctive chronostratigraphic terminology for biostratigraphic, magnetostratigraphic, and sequence-stratigraphic units. For example, the biochronostratigraphic equivalent of a locally or regionally defined biostratigraphic zone (biozone) is a biochronozone. It represents the maximum temporal extent of the biozone world-wide, based on the total temporal ranges (or maximum first- or last-occurrence datums in the case of zones based on concurrent or partial ranges of taxa), as opposed to local ranges of the taxa on which the zone is defined. Because most taxa are environmentally, and thus areally, restricted, their geographic distribution at different times reflects the changes in their preferred environmental factors. A locally defined biostratigraphic zone (biozone) may therefore only span a part of the true temporal extent (biochronozone) of the taxa world-wide (see Table 1). Generally, most biostratigraphically useful taxa of the low and middle latitudes are environmentally excluded from the higher latitudes. Even in the tropical and temperate latitudes, however, considerable differences between biozonal and biochronozonal limits are possible as a result of circulation patterns, or other environmentally induced fluctuations (see also Loutit and others, this volume, for further discussion of biozones vs. biochronozones).

The distinction between magnetic-rock, time-rock, and time units has been made by Harland and others (1982), and we have adopted their terminology. We have also suggested a distinction between sequence-stratigraphic, sequence-chronostratigraphic and sequence-chronologic units (see Table 1). This terminology subdivides sequence-stratigraphic units into first-order (megasequence), second-order (supersequence), and third- or fourth-order (sequence) units, as well as into shorter term flooding events within se-

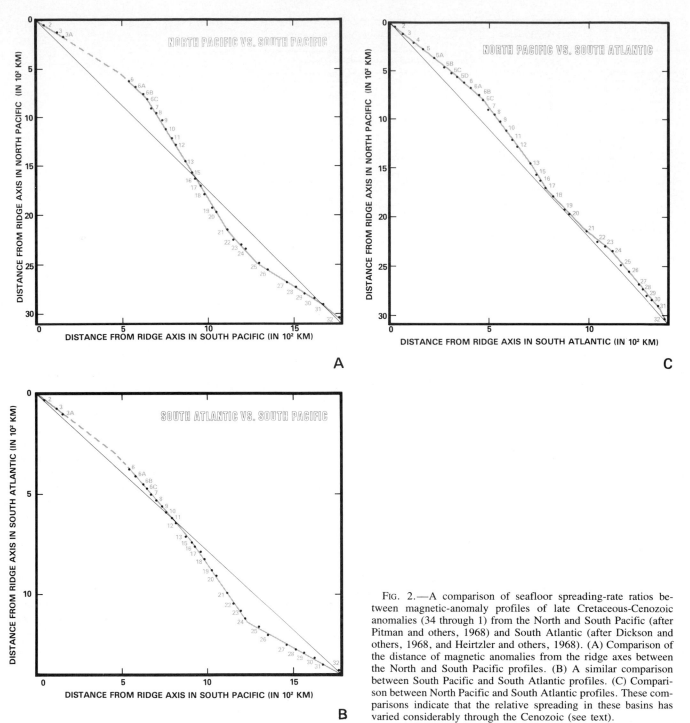

FIG. 2.—A comparison of seafloor spreading-rate ratios between magnetic-anomaly profiles of late Cretaceous-Cenozoic anomalies (34 through 1) from the North and South Pacific (after Pitman and others, 1968) and South Atlantic (after Dickson and others, 1968, and Heirtzler and others, 1968). (A) Comparison of the distance of magnetic anomalies from the ridge axes between the North and South Pacific profiles. (B) A similar comparison between South Pacific and South Atlantic profiles. (C) Comparison between North Pacific and South Atlantic profiles. These comparisons indicate that the relative spreading in these basins has varied considerably through the Cenozoic (see text).

quences that produce parasequences (Van Wagoner and others, this volume) and their chronostratigraphic and geochronologic equivalents.

BIOCHRONOSTRATIGRAPHIC AND MAGNETOCHRONOSTRATIGRAPHIC TIES

The progress in the calibration of biostratigraphic events to magnetostratigraphy in the last decade was first due to the availability of undisturbed piston cores of Neogene sediments, and more recently through the use of the Hydraulic Piston Corer (HPC) by the Deep Sea Drilling Project (DSDP) and its successor, the Ocean Drilling Program (ODP). Prior to the deployment of the HPC, the upper 200 m of softer sediments was usually too severely disturbed, as a result of the rotary action of the drill bit, to be suitable for magnetic or refined biostratigraphic analyses. HPC coring has provided the basis for direct ties between fossil datums and magnetic-polarity reversal events of pre-Neogene sedi-

TABLE 1.—STRATIGRAPHIC TERMINOLOGY SHOWING ROCK, TIME-ROCK, AND TIME UNITS.

Lithostratigraphy (Rock)	Chronostratigraphy (Time-Rock)	Geochronology (Time)
Group	Erathem	Era
Formation	System	Period
Member	Series	Epoch
Bed	Stage	Age
Biostratigraphy	Biochronostratigraphy	Biochronology
Superzone (Superbiozone)	Superbiochronozone	
Zone (Biozone)	Biochronozone	Biochron
Biohorizon	Biochronohorizon	
Magnetostratigraphy	Magnetochronostratigraphy	Magnetochronology
Polarity Superzone	Polarity Superchronozone	Polarity Superchron
Polarity Zone	Polarity Chronozone	Polarity Chron
Sequence Stratigraphy	Sequence Chronostratigraphy	Sequence Chronology
Megasequence (First Order)	Megasequence Chronozone	Megasechron-Megacycle
Supersequence (Second Order)	Supersequence Chronozone	Supersechron-Supercycle
Sequence (Third or Fourth Order)	Sequence Chronozone	Sechron-Cycle
Parasequence (Fourth and Higher Order)	Parasequence Chronozone	

The time-rock terminology (Chronostratigraphy, biochronostratigraphy, magnetochronostratigraphy, and sequence chronostratigraphy) adopted in the cycle charts (Figs. 14–17) is shown in the center column.

ments. In addition, magnetic/biostratigraphic studies in carbonate outcrops have also been critical in establishing a magnetobiostratigraphic framework for late Cretaceous and Paleogene intervals.

The desirability of assembling a magnetobiochronologic scheme of wide geographic applicability is quite obvious. Consequently, it is important to distinguish between the need to adopt *biochronozones,* wherever possible, as opposed to locally defined *biozones,* as previously mentioned.

To utilize the biozones in the amplified sense of biochronozones, it is essential that the biostratigraphic events on which the zones are based be compared to magnetic-polarity reversal stratigraphy in sections from different latitudes and hemispheres. So far, this is only possible for a very few biostratigraphic events. For instance, many of the late Cretaceous and early Cenozoic planktonic-datum events have been tied directly to polarity reversals in one or two sections and, rarely, in more than two sections. The optimization of the total-range information of the less well-constrained planktonic-datum events can sometimes be accomplished by cross-correlations of datums between taxa of the same group and between different groups. An illustrative example of such cross-correlations is provided by two early Oligocene datum events.

The last occurrence of the planktonic foraminifers *Pseudohastigerina* spp. and the first occurrence of nannofossil *Sphenolithus distentus* have been tied to the reversed-polarity interval of magnetic chron C12. Both of these are important datum events for the recognition of the Early Oligocene. Authors working in different areas, however, have recorded both a distinct overlap in the ranges of these taxa and a lack of such overlap. The cross-comparison of these taxa in all tropical and subtropical DSDP sites, where they have been jointly recorded, showed that a definite overlap existed in only 25 percent of the sites, a slight overlap was recorded in 56 percent of the sites, and a definite lack of overlap was recorded in 19 percent of the sites compared.

Whereas subtropical sites consistently show just a slight overlap, tropical sites show both a distinct overlap and a lack of overlap. Such biogeographic comparisons help ascertain the true maximum or biochronozonal extent of the taxa.

It is apparent that the temporal extent of many of the fossil zones as yet may not approximate their chronozonal limits, and ranges of marker taxa may have to be refined as more magneto- and biostratigraphic and paleobiogeographic data are compiled from different latitudes. This aspect of biostratigraphy, i.e., documentation of the true biochronozonal extent of marker taxa, needs a strong emphasis in the future, and major advances are to be expected in this area.

Recent paleomagnetic-biostratigraphic studies furnishing direct correlations between late Cretaceous and younger calcareous plankton datum events and polarity reversal events that form the basis of our magneto- and biochronostratigraphic framework include the following: Gartner (1973), Ryan and others (1974), Alvarez and others (1977), Haq and others (1977), Premoli-Silva (1977), Thierstein and others (1977), Haq and others (1980), Channell and Medizza (1981), Lowrie and others (1982), Poore and others (1982), Stradner and Allram (1982), Backman and others (1983), Napoleone and others (1983), Poore and others (1983), Berggren and others (1984), Hsü and others (1984), Shackleton and others (1984), Miller and others (1985), and Monechi and Thierstein (1985). In addition, recent publications by Berggren and others (1985 a,b) are a compilation of a wealth of data and reference sources on the state of the art of Cenozoic biostratigraphic magnetic polarity-reversal correlations.

For siliceous plankton, direct magnetobiostratigraphic correlations are limited to the Neogene. These studies include Burckle (1978), Theyer and others (1978), Burckle and Trainer (1979), and Barron (1985).

For the Late Jurassic and Early Cretaceous, direct ties

between fossil occurrences and magnetic reversals are provided by a few isolated studies. These include Alvarez and others (1977), Channell and others (1979), Lowrie and others (1980), Marton (1982), Marton and others (1980), Hörner and Heller (1983), Ogg (1983), Galbrun and Rasplus (1984), Ogg and others (1984), Steiner and others (1985), and Bralower (1987). These studies tie magnetic reversals to calcareous plankton datums, and occasionally ammonite biozones. It is obvious that for much of the Jurassic and Cretaceous considerable work needs to be done on magnetic-biostratigraphic relationships before the results can be considered adequate. Until then, such correlations should be considered more as provisional working schemes rather than as refined magneto-biostratigraphies.

TUNING OF THE NUMERICAL TIME SCALE

Ideally, a chronostratigraphic system should be based on the aggregate of reliable empirical data and must be consistent with known earth historical facts. It must also seek to minimize the elements of uncertainty in the data base and avoid unnecessary assumptions. In the past it has been customary to construct geochronologic schemes on the basis of a few selected radiometric dates. These dates are used as isolated constraints on the stratigraphic column, and the intervening segments are then subjected to interpolation, either according to sediment thickness or the distance of magnetic anomalies from spreading ridges. Inherent in such interpolation is the assumption of constancy of rates, whether sediment accumulation or seafloor spreading. These assumptions may not be always warranted, as will become clear in later discussion.

POTENTIAL SOURCES OF ERROR

Since radiometric, biostratigraphic, and magnetostratigraphic data supply the building blocks on which the global chronostratigraphic framework must be based, it may be instructive to review the scope of errors and the ultimate limits of refinement that each type of data offers.

Sources of error in radiometric data.—

In spite of the obvious advantages of basing a time scale on well-constrained radiometric dates, in practice radiometric data are severely limited by numerous sources of error inherent in the samples, as well as in the various radiometric techniques (see Odin, 1982b, for a summary of the sources of error in various techniques).

For example, unaltered plutonic rocks, which probably yield the most reliable radiometric dates because of their greater chemical stability, are often the most difficult to place stratigraphically. Moreover, the numbers obtained from plutonics will depend a great deal on the temperatures at which the various isotopes used to date the rock become 'fixed' in the crystal lattice (its closure temperature), as well as rate of cooling of the rock (Jäger and Hunziker, 1979). When this closure temperature is relatively low, as may often be the case, the subsequent reheating above this tem-

perature will reset the isotopic clock, and any results obtained from such samples only indicate the time of resetting (Gale, 1982). Thus, a careful petrological scrutiny to detect alteration and to ensure freshness of samples becomes an important prerequisite. Several analyses of the minerals in question may also be necessary for accurate age estimates.

Volcanic extrusives, such as lava flows, may suffer from similar sources of error as intrusives, but the presence of excessive radiogenic ^{40}Ar may be especially problematic in these rocks, which would lead to older apparent K/Ar age estimates. For example, in deep-sea lava flows, even when the rock is relatively fresh, excess radiogenic ^{40}Ar trapped during their formation often yields older apparent ages (Seidemann, 1977). Conversely, the presence of a recrystallized glassy component, common in whole-rock volcanic samples, would give younger age estimates (Odin 1982b). There is also the problem of inhomogeneity in samples of volcanic rocks, which could be a serious limitation. In general, however, fresh-acid volcanics that lack post-extrusive alteration can yield reliable Rb/Sr ages.

Bentonites, which are clay deposits formed by devitrification and chemical alteration of tuffs and volcanic ash (see Person, 1982), are generally better constrained stratigraphically than plutonics, but they have their own set of inherent sources of inaccuracies. For example, contaminations resulting from inherited crystal nuclei, and material from the volcanic vent and detrital material, as well as contamination introduced during hydrous alteration, combined with the effects of alteration from glass to clay (Baadsgaard and Lerbekmo, 1982), can all introduce significant uncertainties in the dates obtained from bentonites. For bentonites as well, pooled dating of many samples from a single horizon may be needed to obtain better age estimates (Baadsgaard and Lerbekmo, 1982).

Glauconites (or glaucony), marine authigenic minerals related to micas, are generally more abundant than plutonic rocks or bentonites and are best constrained stratigraphically; however, they require great caution in the choice of samples and in analysis. Glauconite grains are fraught with syngenetic and diagenetic sources of error that could alter the apparent ages obtained from them. The behavior of the relevant nuclides during formation and stabilization of the system, as well as the alteration of nuclide ratios since the time of the closure, will affect the ultimate age values (Keppens and Pasteels, 1982). In addition to such post-genetic factors as burial diagenesis, tectonism and reheating, and alteration due to weathering, which affect all types of minerals, the glauconites are also more susceptible to early- and late-diagenetic processes operating at low temperatures. Extra caution is therefore needed to ensure the reliability of dates obtained from glauconites. Glauconitic samples need to be chosen carefully to select grains in which the influence of alteration processes can be mineralogically discounted (Keppens and Pasteels, 1982).

In addition, each radiometric technique, depending on the isotopes being measured, has to contend with a string of special sources of uncertainty. These may be due to the presence of relatively high proportions of the original daughter isotope present in the rock sample at the time of the genesis and the addition or removal of parent or daughter isotopes since that time. In plutonic rock samples, the Rb/Sr system is generally considered more stable than in other rocks where radiogenic Sr may be removed more easily because of temperature rise. The presence of detrital contaminants, on the other hand, may mean the presence of excessive radiogenic Sr. There may also be a considerable inhomogeneity with respect to Sr in the rock itself (Clauer, 1982). Post-genetic alterations can be caused by leaching by ground water or by weathering by fresh and/or sea water. Exchange reactions resulting from groundwater circulation, however, seem to alter Rb/Sr ages more than K/Ar ages (Keppens and Pasteels, 1982). In general, K/Ar systems are less stable than Rb/Sr systems because Ar atoms carry no charge, are not well-bound chemically, and are subject to easier removal by alteration processes (Keppens and Pasteels, 1982).

Such common laboratory treatment of samples as ultrasonic cleaning or excessive heating prior to analysis may also significantly alter the apparent age of the rock by removal of Ar or Sr. Interlaboratory variations and instrument errors are additional sources of error that make the assessment of isotopic ages from literature an extremely difficult exercise (Webb, 1981).

Comparison of Rb/Sr and K/Ar analyses on the same plutonic rock show that K/Ar may yield ages that are some 4 to 9 million years younger than Rb/Sr (Shibata and Ishihara, 1979). This is attributed to the Rb/Sr system's representing the age of emplacement system's and K/Ar representing the age of cooling (Webb, 1981). An allowance for the age discrepancies may therefore have to be made when dealing with dates from the two systems in a data set. In glauconites, the Rb/Sr and K/Ar age discrepancies are attributed to the open-lattice mineral grains, which may represent alteration or incomplete glauconitization. Thus, there is a need to ascertain the closed-lattice nature of glauconitic minerals before analysis (Odin, 1982a). Good comparative results are obtained by the two methods on glauconites when it is ensured through careful petrographic/sedimentary examination that the samples have not been affected by temperature or pressure changes or by reworking (Keppens and Pasteels, 1982). When K/Ar analyses are performed on carefully chosen samples, a precision of as much as 2 to 3 percent at a 95 percent confidence level is possible (Webb, 1981).

These characteristics and analytical requirements of radiometric dating severely limit the usefulness of both high- and low-temperature radiometric dates, especially in obtaining reliable dates from stratigraphically constrained samples. Significant uncertainties may be involved with *all* types of dates, which underscores the need to use radiometric data with qualification. In our view, given the existing set of radiometric data, one cannot, as yet, achieve a consistent and widely acceptable resolution of the Paleogene and older linear time scales. A *qualified* use of a large set of analytically acceptable dates with known stratigraphic limits, however, can provide important constraints for chronostratigraphy that will lead to a time scale that, hopefully, better approximates reality than does the use of singular dates, used in isolation to nail down relatively long segments of the stratigraphic column.

Sources of error in biostratigraphic data.—

In biostratigraphic data, the most serious source of error that may go undetected is introduced by variations in the biogeographic distribution of marker taxa. Because distribution of taxa is environmentally controlled, their appearance or disappearance in an area may result from changes in environmental parameters rather than from evolutionary first appearance or extinction (see previous discussion on biozones vs. biochronozones). These biogeographic wrinkles can sometimes be ironed out through the aid of magnetostratigraphy or isotopic stratigraphy, but such documentation over large geographic areas exists for only a relatively small number of biohorizons.

The degree of biostratigraphic resolution and precision depends a great deal on the degree to which biostratigraphers can effectively communicate to other workers the morphologic changes they observe in fossils (Haq and Worsley, 1982). When biostratigraphic data are synthesized from widely different sources, a considerable uncertainty can be introduced as a result of differing taxonomic concepts among workers, differing levels of rigor in biostratigraphic work, and in recognizing and recording absolute first- and last-appearance levels in assemblages. Uncertainties can also be introduced by differing or inconsistent sampling intervals and because of undetected vertical discontinuities in sections, or because of reworking and dissolution (see Haq and Worsley, 1982, for further discussion).

Sources of error in magnetic data.—

Because the signal of the earth's individual magnetic-polarity reversal events is not unique, the potential of interpretive error is significant if the magnetostratigraphic data are not stratigraphically well constrained. The interpretation of polarity chrons is commonly dependent on radiometric or biostratigraphic control. Thus, all sources of error inherent in radiometric and biostratigraphic methodologies may ultimately bias the magnetostratigraphic data as well.

Like biostratigraphic data, the magnetostratigraphic data are also subject to sampling limitations and stratigraphic inconsistencies, which determine the ultimate limits of the temporal precision of polarity-reversal boundaries within any given section.

Magnetic-anomaly recognition on the sea floor may be hampered by a number of factors. Seafloor magnetic profiles are complicated by fracture zones, ridge jumps, changes in spreading centers, asymmetric spreading along opposite flanks of the spreading centers, and major plate reorganizations that punctuate the seafloor record. The low amplitude of some anomalies and anomaly skewness can also lead to misidentification of magnetic anomalies (Klitgord and Schouten, 1986). This heavy dependence on interpretation and synthesis may introduce a considerable element of uncertainty in the measurement of distances between anomalies.

In view of this array of sources of error and uncertainties in various types of techniques used in calibrating geologic time, the debates concerning the merits of using one technique over the other, and the criteria for selection of chronologic data, seem spurious. One reaches the inescapable conclusion that a chronologic framework based on any singular technique or unduly restrictive-selection criteria could inherit considerable bias. Instead, an approach that is based on multiple correlative systems, and attempts to reconcile earth historic facts, without ignoring a large body of empirical and analytical data, has the potential of reducing inconsistencies introduced by any singular technique and of producing a more realistic and robust chronostratigraphy.

MAGNETO-BIOCHRONOSTRATIGRAPHY

As mentioned earlier, in the construction of our cycle charts, we have integrated radiometric, biostratigraphic and magnetostratigraphic data in an iterative manner, which produces a reconciliatory magneto-biochronostratigraphic model whose object is that it does not contradict known earth historic facts. For example, one assumption that is often made in the construction of time scales is the constancy of seafloor spreading rates for fairly long intervals of time. Does this assumption agree with the known spreading histories of the major ocean basins? This premise can be evaluated with a relatively simple test.

In Figures 2A–C we have compared the seafloor spreading profiles (i.e., the distances of magnetic anomalies 32 through 2 from the spreading centers) of three different ocean basins. Figure 2A shows a comparison between two North and South Pacific profiles (both after Pitman and others, 1968). Figure 2B shows a similar comparison between North Pacific and South Atlantic profiles (the latter after Dickson and others, 1968, and Heirtzler and others, 1968), and Figure 2C compares the South Atlantic profile with those of the South Pacific. If the spreading rates in these basins had been consistently uniform, these comparisons would have produced linear relationships (i.e., straight lines). Instead, the comparisons reveal that the relationships have varied considerably over time; therefore, spreading rates cannot be assumed to be uniform over the long term in at least two, and perhaps all three, of these basins.

Curry (1985) has also discussed the perils of the assumption of constant spreading rates over long periods of time in constructing time scales. He cites the widely used South Atlantic profile as an example and demonstrates the variability in the Cenozoic spreading rates between and within regions. Pitman and others (1968, fig. 8) made a similar comparison between the various profiles in the South Pacific and the composite North Pacific magnetic-anomaly profile and came to the conclusion that the ratio of spreading in the two basins varied during the Cenozoic; however, the similarities in the shapes of the spreading curves representing three different regions in the South Pacific suggested to them that spreading rates over these regions varied in a similar fashion, even though each region is bounded by different sets of fracture zones. Similarly, Klitgord and Schouten (1986, fig. 8) compared relative spreading between the magnetic-anomaly profile of the Cenozoic mid-Atlantic Ridge and a profile from the western flank of the Pacific-Antarctic Ridge. This comparison revealed distinct changes in the relative spreading of these two basins around anomalies 24/25, 21, 13, 5C, and 2.

Spreading rate comparisons in Figure 2 also indicate where the rate changes are most likely. For example, a consistent change in spreading rates is centered between anomalies 24 and 25, and additional changes may have occurred between anomalies 20 and 21, and 6C and 5C, which is consistent with the findings of Klitgord and Schouten (1986). Changes may also have occurred around anomaly 17 and between 13 and 15 in at least one of the basins. If a linear time scale is to reconcile spreading-rate data, it should be able to reproduce at least the most consistent of the spreading-rate changes (K. Klitgord, pers. commun., 1986). This provides us with important constraints in the integration of magnetostratigraphic data into a chronostratigraphic scheme. The comparison of seafloor spreading rates (Fig. 2) also demonstrates that, over the long term, constant spreading rates cannot be assumed for any one basin, and time scales based on such assumptions could inherit considerable error.

Because seafloor magnetic anomalies form the basis of magnetostratigraphy, the ultimate temporal fidelity of a chronostratic scheme that includes magnetostratigraphy depends on a meaningful reconciliation of the magnetic-anomaly data from various basins. (See Ness and others, 1980, for a discussion and critique of various late Cretaceous-Cenozoic marine magnetic-anomaly scales.)

As mentioned earlier, our approach has been to assign analytically acceptable and biostratigraphically constrained radiometric dates to magnetic anomalies, through the known relationships of biostratigraphic events to magnetic-polarity reversals. The radiometric dates for the Cenozoic are listed in Appendix A. These dates have been plotted against the North Pacific marine magnetic-anomaly profile of Pitman and others (1968) in Figure 3. The same set of dates are plotted against the South Pacific profile (also after Pitman and others, 1968) in Figure 4, and against the South Atlantic profile of Dickson and others (1968) and Heirtzler and others (1968) in Figure 5. In each case the ages of the younger ends (tops) of the magnetic anomalies are calculated from the "best-fit" solution of the high- and low-temperature dates, which is weighted in favor of the older ranges of low-temperature dates when no high-temperature dates are available. Our rationale for using this weighted approach is that, when reliable high- and low-temperature dates are available for the same stratigraphic interval, the older range ends of low-temperature dates overlap with high-temperature dates (Figs. 3–6; see also the discussion on sources of error in radiometric data). As discussed previously, the anomalies where consistent changes in spreading rates are predicted (Fig. 2) can provide potential inflection points on the best-fit curves. Ages (Table 2) calculated from these solutions are then stacked to obtain a mean from the three profiles, which provides us with an integrated chronology of the magnetic anomalies. By stacking the curves, we have, in effect, attempted to reduce the noise due both to uncertainties in the radiometric analytical techniquecs and to minor irregularities in seafloor spreading patterns in any single basin. It also means that any individual spurious datum will have little effect on the overall validity of the stacked best-fit solution. In view of the inherent analytical uncertainties in radiometric techniques, however, we estimate that an average limit of uncertainty of about ±1.4 million years can

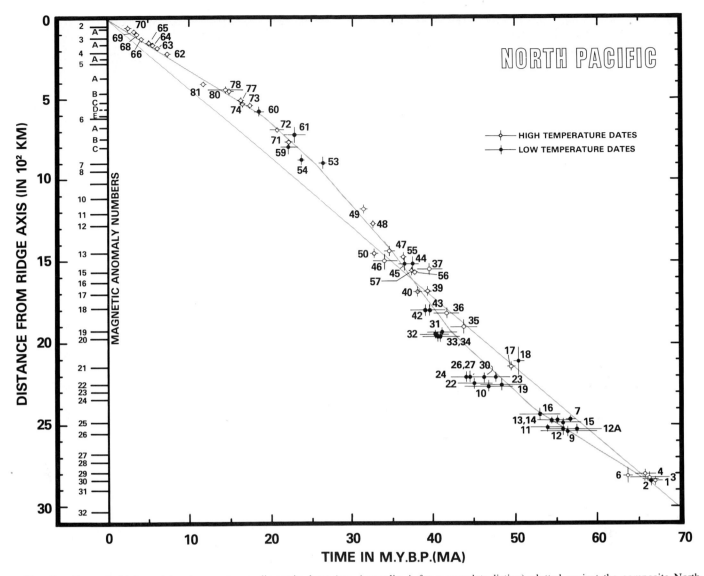

FIG. 3.—Cenozoic high- and low-temperature radiometric dates (see Appendix A for a complete listing) plotted against the composite North Pacific seafloor magnetic-anomaly profile of Pitman and others (1968). The dates are plotted both with their analytical limits of uncertainty (horizontal error bars) and their stratigraphic limits of uncertainty (vertical error bars). (Radiometric dates would align along the straight lines if the spreading rate in this basin were uniform over the Cenozoic.) The ages of the anomalies according to the segmented best-fit solution are given in Table 2 (see text for further explanation).

be ascribed to the Early Tertiary stage boundaries. The limits of uncertainty are considerably narrower for the Miocene stage boundaries (±0.6 million years) and even more so for the Pliocene (in the range of ±1.5 thousand years).

Because of the general lack of magneto-biostratigraphic constraints for the Mesozoic, a direct approach, similar to that used for the Cenozoic, cannot be employed for the Mesozoic. Existing chronologic schemes can be tested and improved, however, by using available stratigraphically constrained radiometric dates.

For example, the set of radiometric dates from the Jurassic and Cretaceous (Appendix B) can be compared to the time scale of Harland and others (1982) or to that of Kent and Gradstein (1985). Since the latter is a modified version of the former, and both are essentially similar (with the

exception of the post-Bathonian Jurassic), we have plotted the Jurassic-Cretaceous radiometric dates against the stage boundaries suggested by Harland and others (Fig. 6). Both high- and low-temperature dates have been plotted with their stratigraphic and analytical limits of uncertainty. It is obvious that the majority of the radiometric data older than Cenomanian implies consistently younger ages for the stage boundaries than that suggested by the time scale of Harland and others (1982) for this interval, which was based on the interpolation between two dates at the Aptian/Albian and Anisian/Ladinian boundaries. We propose a chronology of stage boundaries based on the best-fit solution (Fig. 6) through the entire set of data, again weighted in favor of older ends of the low-temperature dates when no high-temperature dates are available. Within each stage, where di-

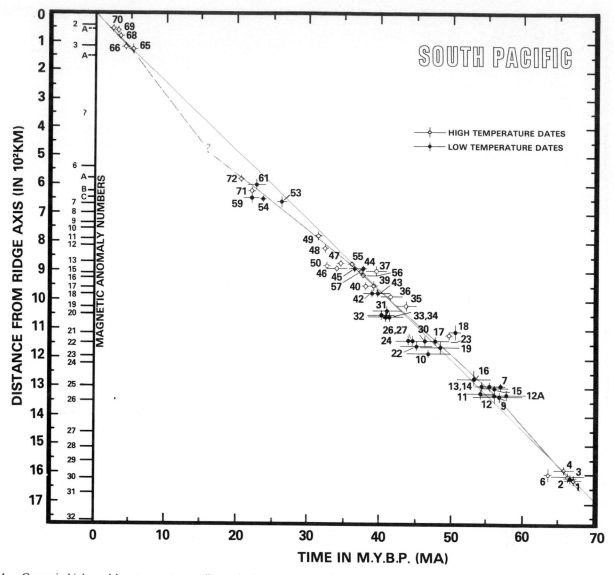

FIG. 4.—Cenozoic high- and low-temperature radiometric dates (same set of dates as in Fig. 3 and Appendix A) plotted against a South Pacific (EL-19S) seafloor magnetic-anomaly profile (after Pitman and others, 1968). Position of anomalies 4 through 5D is uncertain in this profile (see caption of Fig. 3 for explanation).

rect ties between biostratigraphic zonal boundaries and magnetic-polarity reversal events have not been established, we are forced to assign equal duration to biozones of such commonly used fossil groups as ammonites. Hallam and others (1985) have discussed the reason for assigning equal duration to ammonite zones (subzones) at some length. Our approach differs in that we have assigned equal duration to zones within each stage after assigning ages to stage boundaries from the radiometric best-fit solution, and only where no first- or second-order ties with magnetic reversals existed. This avoids the assumption of uniform evolutionary rates for ammonites over a very long period of time and restricts this element of uncertainty to shorter durations within the stage boundaries. The estimated average limits of uncertainty ascribed are ±3.5 million years to the Jurassic stage boundaries, ±3.0 million years to the Early

Cretaceous, and ±1.75 million years to the Late Cretaceous.

The magnetic-polarity scale for the Callovian through Aptian interval is based on the marine magnetic-anomaly sequence that has been developed since the early 1970s. Larson and Chase (1972) and Larson and Pitman (1972) first published the late Jurassic-early Cretaceous M anomaly data based on the Hawaiian lineations in the Pacific. This sequence has since been added to and modified by Larson and Hilde (1975) and Cande and others (1978). The North Atlantic Mesozoic magnetic anomalies have been identified and were subsequently reinterpreted by Schouten and Klitgord (1977), Vogt and Einwich (1979), and Sundvik and others (1984), among others (see also Klitgord and Schouten, 1986, for a discussion of North and Central Atlantic magnetic-anomaly sequences). The correlations be-

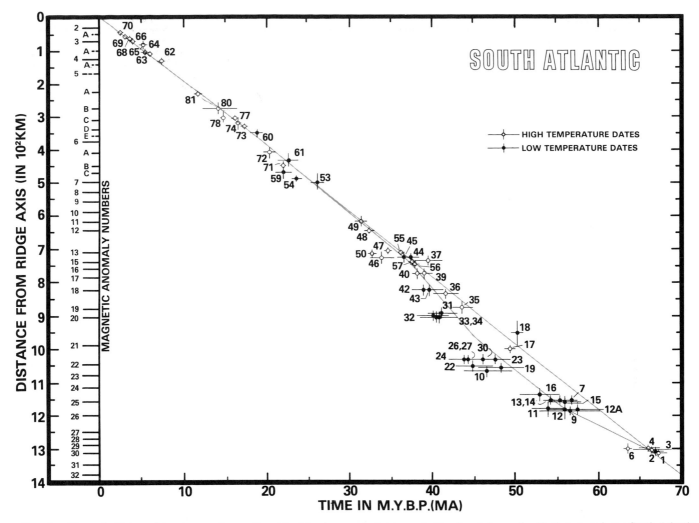

FIG. 5.—Cenozoic high- and low-temperature radiometric dates (same set of dates as in Fig. 3 and Appendix A) plotted against a South Atlantic (V20 S.A.) seafloor magnetic-anomaly profile (after Dickson and others, 1968, and Heirtzler and others, 1968). (See caption of Fig. 3 for explanation.)

tween stage boundaries and magnetic anomalies for the Callovian through Aptian interval have been discussed by Kent and Gradstein (1985).

How does the best-fit Jurassic-Cretaceous time scale of Figure 6 compare with the known magnetic-anomaly profiles in the Pacific and the North Atlantic? For such a comparison (Fig. 7), the ages of M anomalies as predicted by our best-fit solution of Figure 6 have been plotted against the profiles (distance of anomalies from anomaly M) in the North Pacific (after Hilde and others, 1975) and the western North Atlantic (after Klitgord and Schouten, 1986). The North Pacific basin, which has been spreading since Late Triassic-Early Jurassic time (Fig. 7A) shows a relatively constant spreading rate over this interval, with a slight change around anomaly M10N. The western North Atlantic, on the other hand, which began its consistent opening around Blake Spur Magnetic Anomaly (BSMA) time, sometime in the Bathonian-early Callovian, shows considerable variation in spreading rates. The early opening of the basin was relatively fast (Fig. 7B), then slowed down somewhat around

anomalies M25, M21, and M15 times, and accelerated slightly between anomalies M10N and M4. The changes at M21, M15, and around M10N are quite consistent with the changes in the plate kinematics of the North Atlantic as outlined by Klitgord and Schouten (1986; and M. Sundvik, pers. commun. 1984). The change predicted by our time scale at M25 may be entirely due to the relatively young age usually given to BSMA. Extrapolation of the trend between anomalies M22 and M25 to the position of the BSMA (dashed line), predicts a middle Bathonian age for this event and for the initiation of the seafloor spreading in this basin. This older age is more consistent with the middle Bathonian age for the opening of the Central and the North Atlantic as indicated by faunal and paleobiogeographic data from Europe, North America, and the southern Andes (Hallam, 1983). The slight acceleration in spreading around anomaly M10N in both the North Pacific and North Atlantic basins (Fig. 7) suggests that a possible plate reorganization event may have influenced both these basins at the same time.

A similar comparison of the late Cretaceous magnetic

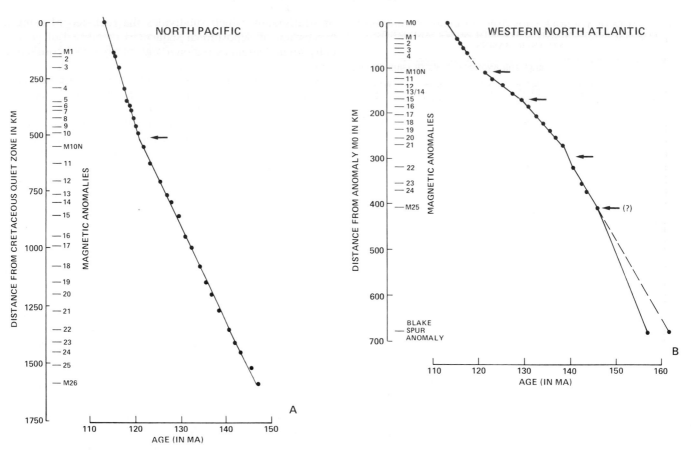

FIG. 7.—The ages of M-series magnetic anomalies, as predicted by the best-fit solution time scale plotted against the seafloor magnetic-anomaly profiles in the North Pacific (after Hilde and others, 1975) and western North Atlantic (after Schouten and Klitgord, 1982, and Klitgord and Schouten, 1986). (A) Anomalies M26 through M1 are shown with their distance from the Cretaceous Magnetic Quiet Zone. (B) The distance of Blake Spur and M25 through M1 anomalies is shown from anomaly M0. Points represent ages on the anomalies as estimated from the best-fit time scale of Figure 6. In the North Pacific, this time scale predicts an essentially uniform seafloor spreading rate, with a slight acceleration between anomalies M10 and M10N. In the western North Atlantic, the same time scale suggests changes at anomalies marked by arrows. The dashed line between M25 and Blake Spur anomalies represents the continuation of the spreading-rate trend of the interval between M22 and M25 to Blake Spur. This would date the Blake Spur anomaly and the beginning of the opening of the North Atlantic at about 162 Ma (mid-Bathonian). (See text for discussion.)

During the course of a complete cycle of sea-level change, lowstand, transgressive, and highstand facies are all well defined, although lowstand and highstand facies are generally represented by relatively thicker sections, and transgressive facies by thinner sections.

In outcrop sections, the lowstand, shelf margin, transgressive-, and highstand systems tracts can be identified between three prominent depositional surfaces. The first is the "transgressive surface" (Fig. 10), which occurs at the top of the lowstand systems tract. The transgressive surface is the first flooding surface marking the beginning of the more rapid sea-level advance over the shelf. This surface is usually marked by conspicuous lithologic changes, for example, from marginal marine below to nearshore marine above, or from more terrigenous to less terrigenous sediments on the outer shelf and slope. The lowstand deposits below the transgressive surface are characterized by the sediments of the most regressive phase of the sequence. When lowstand deposits are lacking, the transgressive surface may coincide with the underlying unconformable portion of the sequence boundary.

The second readily recognizable surface in outcrops of shelf strata is the "surface of maxium flooding" of the shelf. This surface manifests itself as a "downlap surface" on seismic profiles. It is associated with the period of depositional starvation on the outer shelf and slope, when depocenters have moved landward in response to rapidly rising sea level. The sediment starvation produces a physically condensed section (see Loutit and others, this volume, for a detailed discussion of condensed sections and their stratigraphic importance). The condensed section forms partly within the transgressive- and partly within the highstand systems tracts of the sequence. Because of the lack of terrigenous input, the condensed section may compromise a zone of high pelagic-fossil concentration, rich glauconitic or phosphatic layers, or a hardground caused by lithification. The relative duration of sediment starvation within a condensed section increases basinward, until beyond the area of terrigenous influence, the biogenic deposition in the deeper basins occurs as a series of stacked condensed sections. The downlap surface separates the transgressive- from the highstand systems tracts, and the accompanying prominent faunal

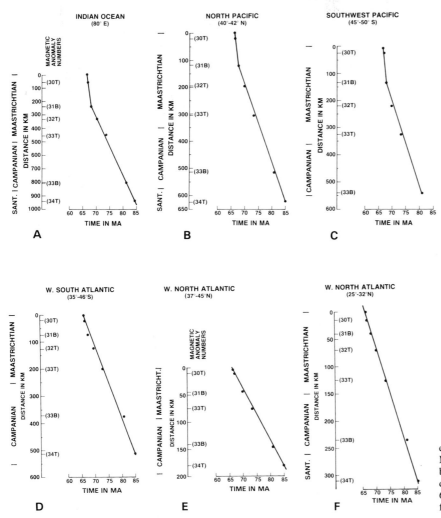

Fig. 8.—Tops (younger ends) and bottoms (older ends) of the latest Cretaceous (younger than Cretaceous Magnetic Quiet Zone) magnetic anomalies in different basins plotted against the ages of the anomalies predicted by the Cretaceous time scale adopted here (Fig. 6). All profiles are after Cande and Kristoffersen (1977, fig. 5).

and lithologic changes at this surface also often lead to confusion between the downlap surface and the sequence boundary.

The third surface is the sequence boundary, expressed as the downward (basinward) shift of the coastal-onlap pattern, or by truncation on seismic profiles. In outcrops the sequence boundary may be an obvious unconformity or its correlative conformity, depending on the position of the section along the shelf-to-basin profile and the rate of sea-level fall. For example, if the position of a section is proximal (landward) on the shelf, the probability of lowstand deposition in the area is reduced or precluded. In such a case, the sequence boundary may be an unconformity that coincides with the transgressive surface (Fig. 10). Basinward, the sequence boundary becomes conformable and may be typified by a change from interbedded progradational deposits to more massive, aggradational, deposits.

The outcrop interpretation of the Cenozoic and Mesozoic sea-level cycles, using the sequence-stratigraphic principles outlined previously (and detailed in Jervey, and Posamentier and others, this volume), has been made on continental margin and interior sections in various parts of the world. The bulk of the documentation so far, however, comes from sections in western Europe, the United States Gulf and Atlantic coasts, and the United States Western Interior. Sections in New Zealand, Australia, and Pakistan, and in the Arctic islands of Bjørnøya and Svalbard have also contributed to the documentation. Lists of sections that have contributed to the documentation of the sequences on the cycle charts can be found in Appendix C for Cenozoic strata, Appendix D for Cretaceous strata, Appendix E for Jurassic strata, and Appendix F for Triassic strata.

Some examples of outcrop studies.—

To illustrate the sequence-stratigraphic analysis of outcrop sections in different settings, we include three examples of outcrop studies. The first example is from the neostratotype of the Lutetian stage in the Paris Basin, the second from the type area of the Cenomanian, and the third from a reference section of Scythian age from the Salt Range in Pakistan. Other examples of identification of systems tract boundaries in outcrop sections are enumerated in this volume by Sarg; Donovan and others; Greenlee and others (and in prep.); Loutit and others; and Baum and others. Examples; discussed in the above studies and those given below provide a broad overview of the application of sequence-

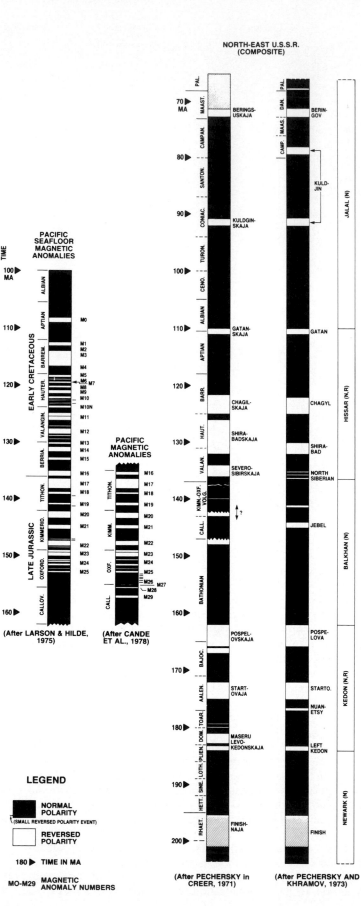

FIG. 9.—A paleomagnetic composition of known magnetic-polarity reversal data from sections in the U.S.S.R. and central Europe (France, Italy, Switzerland, and Hungary). The Pacific seafloor magnetic-anomaly data (Larson and Hilde, 1975; Cande and others, 1978) and a global composite (of McElhinny and Burek, 1971) are also plotted. The provisional pre-Callovian polarity reversal model (Fig. 16) is based on a synthesis of the paleomagnetic data of the same interval shown here.

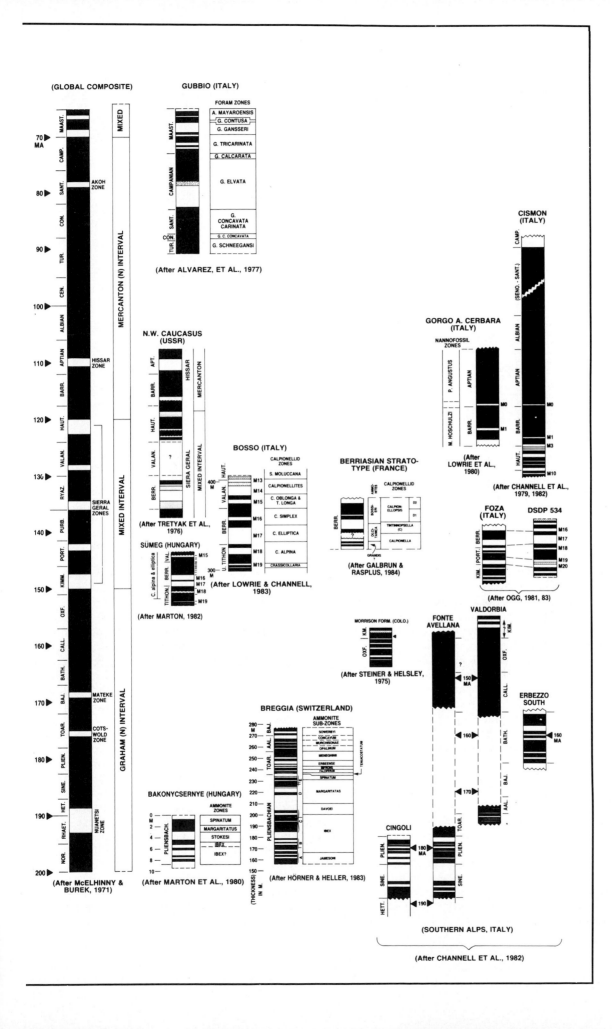

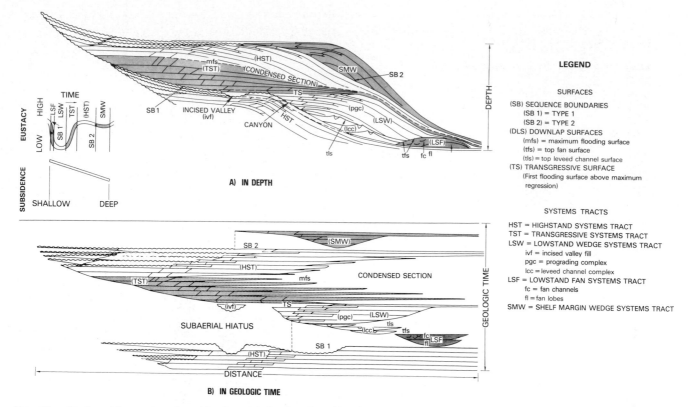

FIG. 10.—Stratigraphic sequence depositional model showing depositional-systems tracts and their bounding surfaces. The model shows systems tracts associated with type 1 sequence boundary (i.e., lowstand fan, lowstand wedge, transgressive-, and highstand systems tracts). Shelf margin wedge systems tract develops following a type 2 sequence boundary (see text). (A) The geometry of the systems tracts with depth. (B) The same features as in A in relationship to geologic time.

stratigraphic concepts in identifying cycles of sea-level change.

The identification of depositional surfaces and sequence boundaries in the outcrop is based on the premise that changes in relative sea level affect facies relationships in the sediments. A correct analysis of the depositional facies is easiest in areas with pronounced changes in lithofacies. Facies changes in shallow-water clastic settings are often more obvious than those in carbonate settings. The preliminary sequence interpretation in the three examples given here is based on available facies analyses from published information and subsequent visits to the outcrops. It must be emphasized, however, that a more precise positioning of the depositional surfaces needs to be confirmed by more detailed litho- and biofacies analyses of such sections on a broader, regional basis.

The first example, the neostratotype of the Lutetian, the "standard" stage of the lower Middle Eocene, is based on a composite section from the lower Lutetian at St. Leu d'Esserent and the middle and upper Lutetian at St. Vaast-les-Mello in the Paris Basin (Blondeau, 1980; Blondeau and Renard, 1980). The lower Lutetian is composed of approximately 12 m of coarse sands, dolomitic and glauconitic limestone, and the middle and upper Lutetian of 33 m of limestones and alternating limestones, marls, and clays, overlain by Quaternary sands (Fig. 11). Four sequence boundaries can be identified.

The lowest sequence boundary (49.5 Ma) is placed between micaceous sands attributed to the Cuisian Stage and coarse glauconitic sands with bryozoans and shark teeth which form the transgressive base of onlapping lower Lutetian deposits. The sequence boundary and the transgressive surface are interpreted to coincide at the St. Leu d'Esserent locality, as indicated by the presence of an unconformity between the Cuisian and Lutetian in this area. The sandy transgressive deposits of the lower Lutetian become more calcareous upward. The downlap surface is not well exposed at this locality. Thus, the base of the Lutetian in the type section falls between the sequence boundary at 49.5 Ma and the downlap surface at 49.0 Ma.

The next sequence boundary (48.5 Ma) can be identified in the upper part of the lower Lutetian, where sandy, calcareous deposits are overlain by glauconitic sands containing *Nummulites laevigatus*. The transgressive surface is marked by a highly glauconitic sandy limestone with very abundant *N. laevigatus*. This transgressive surface with its abundant fossil content has traditionally been considered the boundary between the lower and middle Lutetian (Blondeau, 1980). The sandy glauconitic deposits between the sequence boundary and the transgressive surface are interpreted here as a thin shelf margin wedge. The transgressive deposits overlying the transgressive surface consist of coarse calcarenites with abundant molluscan molds. The downlap surface can best be seen at St. Vaast-les-Mello, at the base

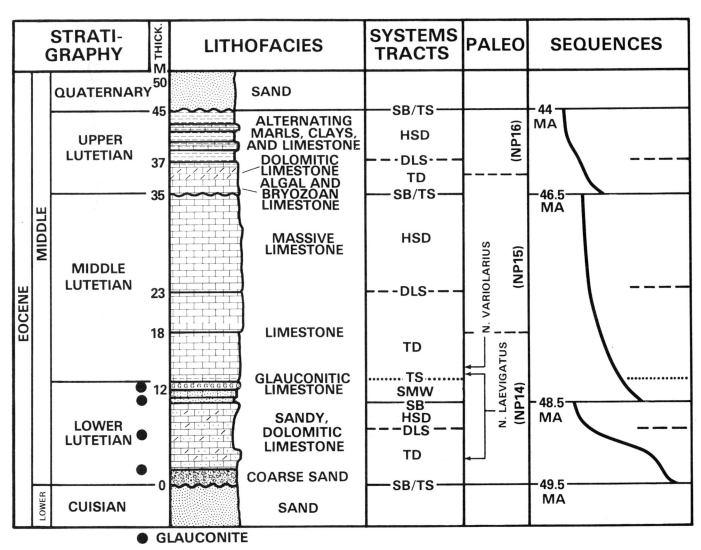

FIG. 11.—Lithofacies, biostratigraphy, and sequence stratigraphy of the neostratotype sections of the Lutetian (middle Eocene) Stage in the Paris Basin. The section is a composite of the lower Lutetian from St. Leu d'Esserent and middle and upper Lutetian from St. Vaast-les-Mello (stratigraphic subdivisions and lithologic sections after Blondeau, 1980, and Blondeau and Renard, 1980, and indirect nannofossil zonal assignments in the paleo column after Bouche, 1962, and Aubry, 1983). Symbols in the systems tracts column are as follows: SB: Sequence boundary; TS: Transgressive surface; TD: Transgressive deposits; HSD: Highstand deposits; SMW: Shelf-margin wedge. (See text for further explanation.)

of a massive 12-m-thick, fine calcarenite bed with abundant miliolids ("Banc Royal"), which represents the highstand deposits of this sequence.

The 46.5-Ma sequence boundary is placed above the "Banc Royal" at the base of a coarser grainstone high in the middle Lutetian. The downlap surface in this sequence is marked by the transition from higher energy grainstones in the transgressive-systems tract to interbedded marls and fine limestones in the highstand systems tract. The fourth sequence boundary, at 44 Ma, lies between the alternating marls, clays, and limestones of the upper Lutetian and the transgressive sands of the Auversian. The later sequence boundary is not exposed at St. Vaast because Quaternary sediments unconformably overlie the upper Lutetian.

Indirect calcareous nannofossil information from the Lutetian elsewhere in the Paris Basin (Bouché, 1962; Au-

bry, 1983) permits the correlation and positioning of these four sequences relative to the cycle chart.

The second outcrop study example comes from the region of the type Cenomanian in the southeastern Paris Basin. In sections in the Le Mans-Ballon and the Theligny-St. Calais areas, the uppermost Albian rests unconformably on Oxfordian limestone, which marks the lowest sequence boundary in the area (98 Ma). In Figure 12, composite sections (compiled by P. Juignet) from the two areas are shown along with lithostratigraphy, biostratigraphy (ammonite zones after Amédro, 1980, 1986; Wright and Kennedy, 1984) and sequence-stratigraphic interpretations based on broader, regional considerations of the depositional facies relationships.

Both the Le Mans-Ballon and Theligny-St.Calais sections are composed of from 115 to 120 m of glauconitic

The excellent fossil record of the Salt Range sections includes conodonts, ammonites, ostracodes, and dinoflagellates. These groups have been extensively studied and tied to global stratigraphy (e.g., Matsuda, 1985; Nakazawa and others, 1985) and permit the accurate dating of the transgressive and downlap surfaces and sequence boundaries.

DESCRIPTION OF THE CYCLE CHARTS

Figures 14 and 17 present the chronology of sea-level fluctuations in the Cenozoic, Cretaceous, Jurassic and, Triassic. A large-scale composite chart outlining the chronostratigraphy and eustatic cycles of the Mesozoic and Cenozoic is also included (in pocket). The cycle charts combine the linear time scale (in Ma, or millions of years before present, repeated on the left, center, and right of the charts for convenience) with magnetochronostratigraphy, standard chronostratigraphy, biochronostratigraphy, and sequence chronostratigraphy.

Magnetochronostratigraphy includes the polarity reversal sequence, seafloor magnetic-anomaly numbers (from Oxfordian to Recent), and the numeric terminology of polarity chronozones (*sensu* Harland and others, 1982). The late Cretaceous and Cenozoic chron numerology is adopted from Tauxe and others (1983) and the Callovian through Aptian from LaBrecque and others (1983). The old system of Neogene polarity "epochs" (LaBrecque and others, 1977, 1983) is also included, for correlation purposes with the older literature. In the present version, however, the correlation of epochs 7 through 14 with magnetic anomalies has been updated following Barron and others (1985).

The magnetostratigraphy adopted for the cycle charts is a combination of four types of paleomagnetic data of varying quality. For the past 6.5 Ma, an accurate geomagnetic-polarity scale tied to reliable radiometric dates of lavas is available (McDougall and others, 1976, 1977; Mankinen and Dalrymple, 1979). The Cenozoic polarity reversal scale (older than 6.5 Ma and up to magnetic anomaly 32) is based on the stacked mean ages of marine magnetic anomalies from three ocean basins, as discussed earlier. The composite-polarity scale for this interval has developed from the marine magnetic-anomaly scale of Heirtzler and others (1968) with later refinements suggested by LaBrecque and others (1977) and Lowrie and Alvarez (1981). The Oxfordian through Barremian polarity scale has been adopted by calibrating the M-series magnetic anomalies and standard stages against a best-fit numerical scale based on available Jurassic-Cretaceous radiometric dates (see Fig. 6). The M-series magnetic-anomaly scale has developed through the studies of Larson and Pitman (1972), Larson and Hilde (1975), and Cande and others (1978). An update of the relationship of these anomalies to stage boundaries has been recently discussed by Kent and Gradstein (1985). Here, however, we have included a minor reversed-polarity event near the top of the Aptian (above MO), following the finding of Lowrie and others (1980). For the pre-Callovian Jurassic and the Triassic, for which no seafloor magnetic anomaly data are available, the polarity reversal sequence has been synthesized from the paleomagnetic studies of Helsley (1969), Burek (1970), McElhinny and Burek (1971),

Perchesky (in Creer, 1971), Perchesky and Khramov (1973), Helsley and Steiner (1974), Marton and others (1980), Channell and others (1982), and Hörner and Heller (1983). The magnetic zonal terminology for this interval is largely after Burek (1970), McElhinny and Burek (1971), and Perchesky and Khramov (1973).

The chronostratigraphic section includes hierarchical subdivisions of system, series, and stage. We have adopted the commonly used European stage nomenclature because it has become the "standard" for world-wide correlations. The suprastage designations, such as Buntsandstein, Muschelkalk, Keuper, Lias, Dogger, Malm, Neocomian, and Senonian, that are still frequently used in regional stratigraphic literature, are also included in the stage columns, but no formal status is implied. The Jurassic/Cretaceous boundary has been placed between the Portlandian and Ryazanian stages (at 131 Ma), following the common North Sea usage. Alternative boundary placement is between the Tithonian and Berriasian stages (at 134 Ma), which is also common in the continental European literature.

The third section of the cycle charts includes biochronostratigraphy. This section incorporates two types of information: zonal schemes, where zones have been formally defined, and first- and last-occurrence events (biohorizons) of some fossil taxa whose formal zonal schemes have not been established. The included zonal schemes vary for each cycle chart, depending on the relative usefulness of the fossil groups for each interval. All four cycle charts include palynomorph biohorizons (mostly dinoflagellates, with the exception of some spore and pollen in the early Triassic) that have been synthesized and compiled by palynologists of Exxon Production Research Company (L. E. Stover, M. Millioud, N. S. Ioannides, R. Jan de Chêne, Y. Y. Chen, J. D. Shane, and B. E. Morgan). These data are based mostly on information from western Europe and the North Sea area.

On the Cenozoic-cycle chart (Fig. 14) the biochronostratigraphic data include planktonic foraminiferal zones after Blow (1969), Berggren (1972), Stainforth and others (1975); calcareous nannofossil zones after Martini (1971), Okada and Bukry (1980) and Bukry (1981); radiolarian zones after Sanfilippo and others (1981, 1985); and diatom zones after Gombos (1982), Gombos and Ciesielski (1983), Fenner (1985), and Barron (1985).

The Cretaceous biochronostratigraphic data (Fig. 15) include zonal schemes of planktonic foraminifera after Premoli-Silva and Bolli (1973), Premoli-Silva and Boersma (1977), van Hinte (1976), Robaszynski and others (1979, 1983), and Caron (1985); calpionellids after Alleman and others (1971) and Remane (1978, 1985); calcareous nannofossils after Thierstein (1976), Sissingh (1977), Manivit and others (1977), and Roth (1978, 1983). Cretaceous boreal macrofossil zones (mostly ammonoids and belemnoids, but also some echinoids in the Late Cretaceous) are after Rawson and others (1978) and Kennedy (1984). Tethyan ammonoid zones are after various authors in Cavalier and Rogers (1980), Busnardo (1984), Amedro (1980, 1981, 1984), Kennedy (1984), Robaszynski and others (1983), and Clavel and others (1986).

The Jurassic biochronostratigraphy (Fig. 16) includes radiolarian events and preliminary zones after Pessagno (1977),

Pessagno and Blome (1980), Baumgartner (1985), and Pessagno and others (1987a,b) and calcareous nannofossil zones after Barnard and Hay (1974), Medd (1982), Hamilton (1982), and Roth and others (1983). Boreal ammonoid zones and subzones are from Cope and others (1980).

The Triassic-cycle chart (Fig. 17) includes radiolarian zones after Blome (1984) and conodont zones after Mosher (1970), Sweet and others (1971), and Matsuda (1985). The Tethyan ammonoid zones are after Tozer (1984) and North American ammonoids after Silbering and Tozer (1968).

The fourth section of the cycle charts consists of sequence chronostratigraphy, which includes sequence chronozones. The first-order sequence chronozones include megasequence and megasequence set chronozones. The Phanerzoic is divided into two megasequence set chronozones (lower and upper), and all of the interval from the Late Permian through the Recent fits within the Upper Phanerozoic megasequence set. The Mesozoic and Cenozoic, however, are composed of three megasequence chronozones (the upper part of the Absaroka, the Zuni, and the Tejas). The sequence chronozone terminology adopted here is based on the sequence terminology of Sloss (1963). The megasequences are subdivided into second-order supersequence and supersequence set chronozones. Seven supersequence sets include 27 super-sequences, all of which terminate with sequence boundaries of major magnitude. The sequence chronozone terminology is followed by scaled relative change in coastal onlap associated with each third-order sequence chronozone. The third-order sequence chronozones include sequences that terminate with major, medium, and minor boundaries, and these boundaries are distinguished as such on the cycle charts. A total of 120 sequence chronozones has been identified from the base of the Triassic to Recent. Of these, 19 begin with major-magnitude falls of sea level, 43 with relatively medium-magnitude falls, and 58 with minor sea-level falls. Generally, only the major- and some medium-magnitude cycles can be identified at seismic stratigraphic resolution level. Detailed well-log and/or outcrop studies are necessary to resolve the minor cycles.

Alongside the coastal-onlap column, the ages of the sequence boundaries and downlap surfaces are indicated in separate columns, as are the depositional systems tracts. Boundaries where lowstand fans have been observed are indicated by "F." The unshaded triangle within each coastal-onlap cycle represents the condensed sections within that sequence (see Loutit and others, this volume). The shape of the triangle depicts the interval of slow deposition on the shelf and slope, following a rapid rise of sea level, the relative duration of which increases basinward. The relative magnitude of the condensed sections (major, medium, or minor) is also identified by the relative thickness of the lines representing the downlap surface associated with each condensed section.

The long- and short-term eustatic curves complete the cycle charts. The scale represents our best present estimate of sea-level rises and falls compared to the modern average mondial sea level. The relative magnitude of long-term sea-level variations is estimated following the methods described in Hardenbol and others (1981), with a high value (in the Turonian) adopted from Harrison (1986). The relative magnitude of the short-term sea-level changes is a best estimate from seismic and sequence-stratigraphic data. Although we believe these estimates to be within realistic ranges, they may have to be modified as better, more rigorous methods of determining the magnitude of sea-level changes become available. Today's long- and short-term sea-levels diverge because the long-term eustatic curve is estimated assuming an ice-free world.

In the long term, the low-sea levels of the Late Paleozoic reached their lowest point in Tatarian (Late Permian). This trend continued into the Triassic and Early Jurassic. The Hettangian saw another marked drop, and the levels remained generally low through most of the Early and Middle Jurassic, with a slight rise in the Bajocian. The trend finally reversed itself in the Callovian, and the sea level continued a long-term rise through the Oxfordian, reaching a Jurassic peak in the Kimmeridgian.

After a marked decline in the Valanginian, the sea level began rising once again, reaching its Mesozoic-Cenozoic peak in the Turonian. A gradual decline began in the latest Cretaceous and continued through the Cenozoic. With the exception of short-lived highs in the Danian, Ypresian, Rupelian, Langhian-early Serravallian, and Zanclean, this trend toward lower sea level has continued through to the present.

In the short term, major sea-level falls occurred in the early Portlandian, early Aptian, mid-Cenomanian, latest Ypresian, and latest Bartonian, near the Rupelian/Chattian boundary, in the Burdigalian-Langhian, in the late Serravallian, and throughout the late Pliocene-Pleistocene. These marked sea-level falls are often associated with world-wide major unconformities and frequently with canyon-cutting events on the shelves. At least since the Oligocene, sea-level falls are in part due to the increasing influence of glaciation. This is displayed by the relatively higher amplitude variations of the short-term sea level since the Oligocene.

ACKNOWLEDGMENTS

We are grateful to many colleagues for stimulating discussions about magnetostratigraphy, biostratigraphy, and radiochronology, which convinced us that a truly robust chronostratigraphy requires a broad integration of all reliable empirical and analytical data. In particular, we acknowledge our discussions at various stages of this work with J. A. Barron, W. A. Berggren, D. V. Kent, K. D. Klitgord, J. L. LaBrecque, K. G. Miller, N.-A. Mörner, G. S. Odin, E. A. Pessagno, Jr., R. Z. Poore, P. H. Roth, A. Salvador, H. Schouten, N. J. Shackleton, and E. L. Winterer. This paper was reviewed by J. A. Barron, C. A. Ross, J. Van Wagoner, R. M. Mitchum, and S. R. May, who improved the quality of the text. The following provided invaluable assistance in the field in different parts of the world with the special knowledge of their areas: F. Amédro, G. Badillet, A. Blondeau, R. Busnardo, B. Clavel, C. Collete, M. Delamette, R. Demyttenaere, A. Donovan, P. J. Felder, J. Ferrer, J. C. Fouchet, S. Gartner, R. C. Glenie, A. Hallam, J. Hooker, P. Juignet, E. Kaufman, P. Laga, E. Mancini, N. Morris, A. Mork, S. Nathan, C. Pomerol, L. Pray, K. Akhtar Qureshi, S. Mahmood Raza,

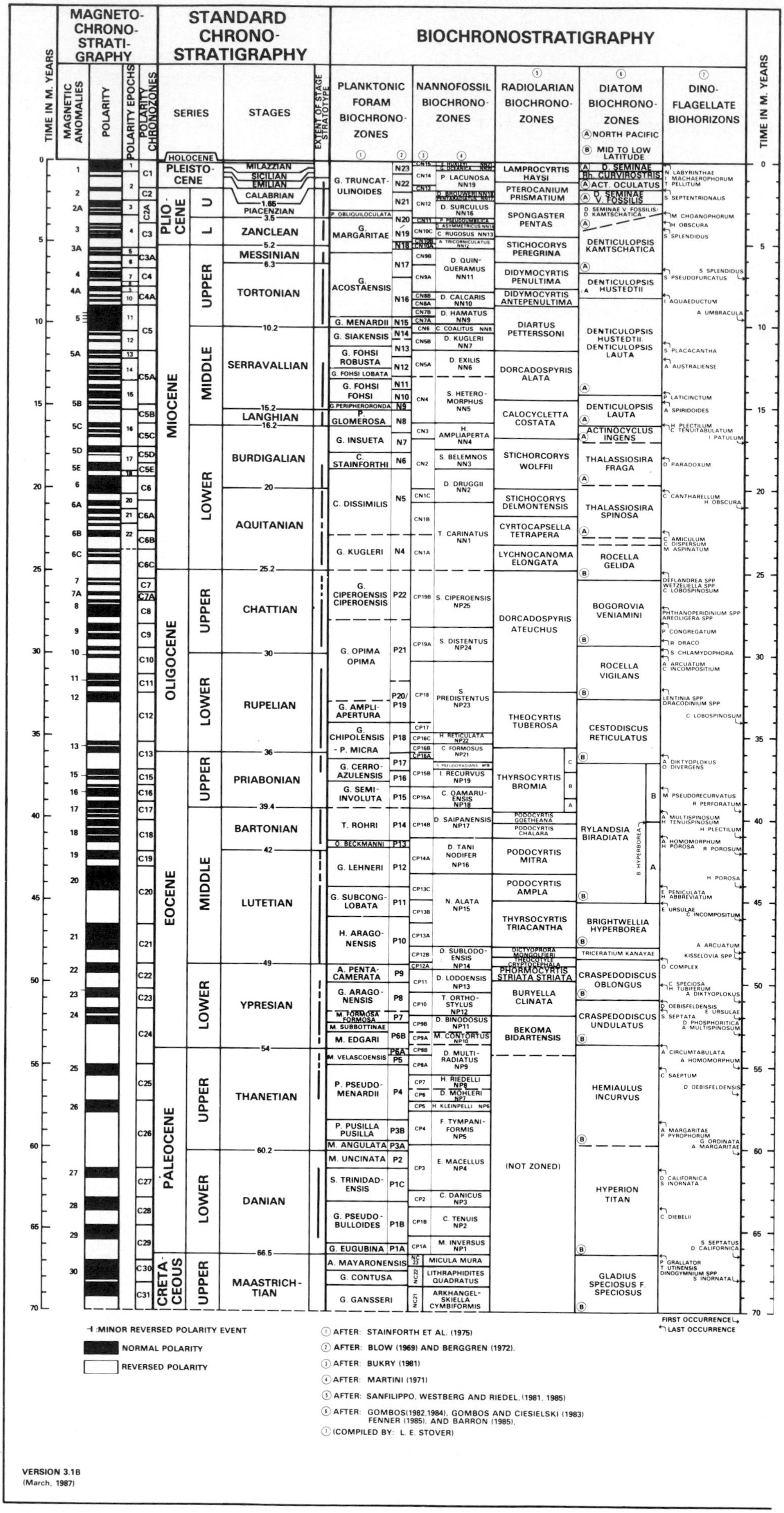

FIG. 14.—Cenozoic chronostratigraphic- and eustatic-cycle chart. Sections of the chart from left to right include magnetochronostratigraphy, standard chronostratigraphic subdivisions, biochronostratigraphy, and sequence chronostratigraphy (see Table 1 for stratigraphic terminology used here). Sequence chronostratigraphy includes cycle terminology (modified after Sloss, 1963), and the observed relative change of coastal onlap associated with various cycles. The unshaded triangles within each coastal-onlap cycle represent the condensed sections, depicting the interval of slow deposition on the outer shelf and slope following rapid sealevel rise. The key at the bottom of the figure explains the sequence boundary types (type 1 and type 2, associated with rapid or slow sea-level falls, respectively), relative magnitude of sequence boundaries and condensed sections (major, medium, and minor) by the relative thickness of the lines, and symbols for the systems tracts. Type 1 boundaries, where lowstand fans have been observed, are indicated by "F" in the systems tracts column. Long- and short-term eustatic curves complete the cycle chart. The 2.4-Ma event that was inadvertently omitted in an earlier version of this chart (Haq and others, 1987) has been added.

Cenozoic chronostratigraphic- and eustatic-cycle chart (Version 3.1B, March 1987).

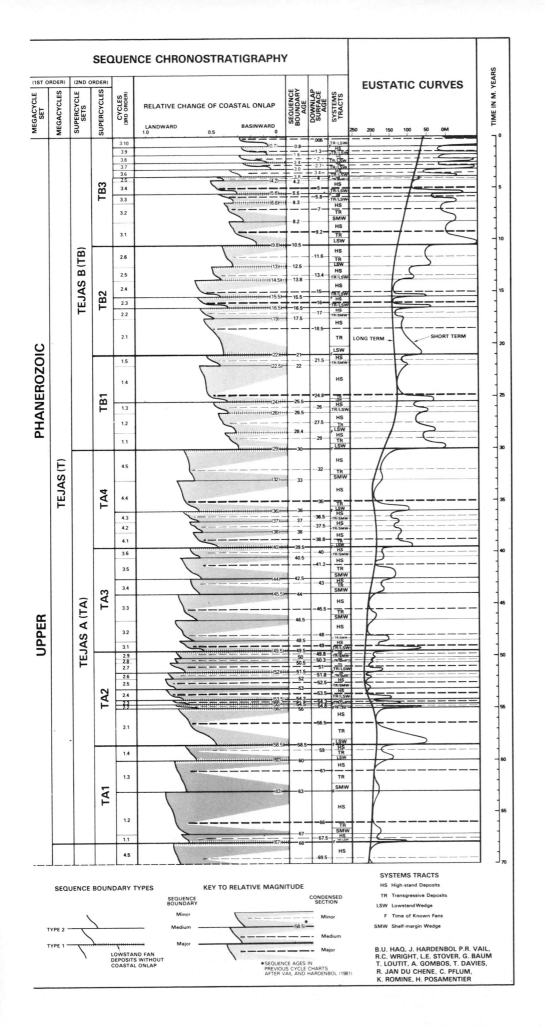

SEQUENCE CHRONOSTRATIGRAPHY

EUSTATIC CURVES

LONG TERM ← → SHORT TERM

SEQUENCE BOUNDARY TYPES

TYPE 2

TYPE 1

LOWSTAND FAN
DEPOSITS WITHOUT
COASTAL ONLAP

KEY TO RELATIVE MAGNITUDE

SEQUENCE BOUNDARY	CONDENSED SECTION
Minor	Minor
Medium	(58.5)
Major	Medium
	Major

● SEQUENCE AGES IN
PREVIOUS CYCLE CHARTS
AFTER VAIL AND HARDENBOL (1981)

SYSTEMS TRACTS

HS   High-stand Deposits

TR   Transgressive Deposits

LSW Lowstand Wedge

F   Time of Known Fans

SMW Shelf-margin Wedge

B.U. HAQ, J. HARDENBOL P.R. VAIL,
R.C. WRIGHT, L.E. STOVER, G. BAUM
T. LOUTIT, A. GOMBOS, T. DAVIES,
R. JAN DU CHENE, C. PFLUM,
K. ROMINE, H. POSAMENTIER

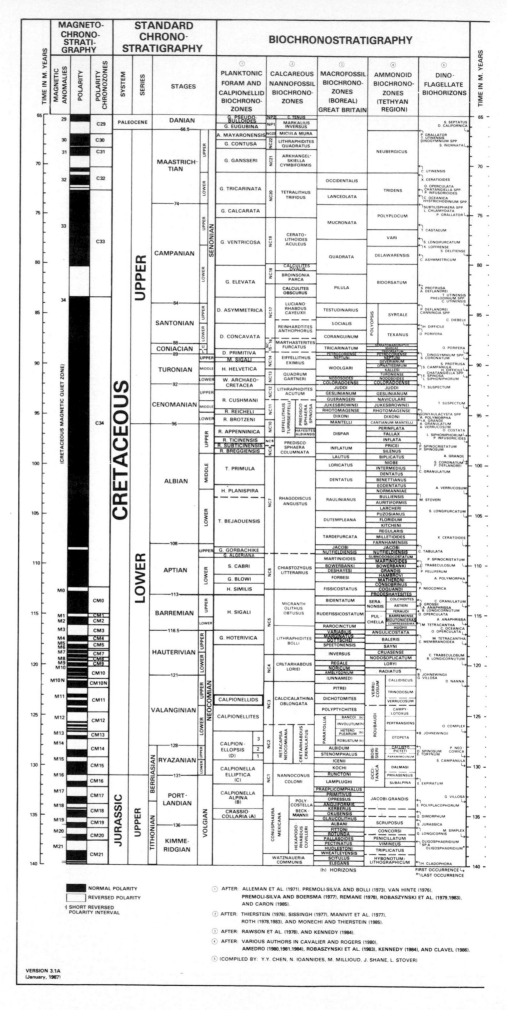

FIG. 15.—Cretaceous chronostrati-graphic- and eustatic-cycle chart. (See Fig. 14 caption and text for explanation.)

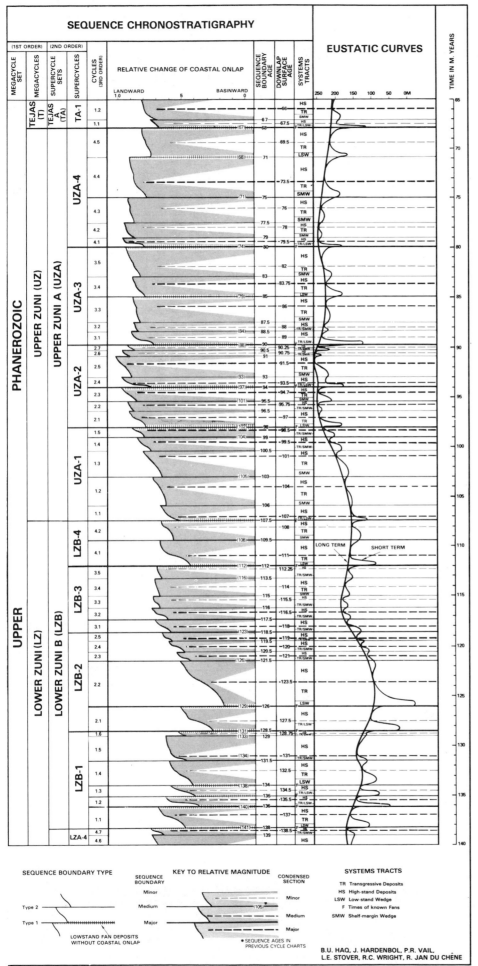

SEQUENCE CHRONOSTRATIGRAPHY

EUSTATIC CURVES

(1ST ORDER)	(2ND ORDER)									TIME IN M. YEARS
MEGACYCLE SET	MEGACYCLES	SUPERCYCLE SETS	SUPERCYCLES	CYCLES (3RD ORDER)	RELATIVE CHANGE OF COASTAL ONLAP	SEQUENCE BOUNDARY AGE	DOWNLAP SURFACE AGE	SYSTEMS TRACTS		

SEQUENCE BOUNDARY TYPE

Type 2

Type 1

LOWSTAND FAN DEPOSITS WITHOUT COASTAL ONLAP

KEY TO RELATIVE MAGNITUDE

SEQUENCE BOUNDARY

Minor

Medium

Major

CONDENSED SECTION

Minor

(105)

Medium

Major

● SEQUENCE AGES IN PREVIOUS CYCLE CHARTS

SYSTEMS TRACTS

TR Transgressive Deposits
HS High-stand Deposits
LSW Low-stand Wedge
F Times of known Fans
SMW Shelf-margin Wedge

B.U. HAQ, J. HARDENBOL, P.R. VAIL,
L.E. STOVER, R.C. WRIGHT, R. JAN DU CHÈNE

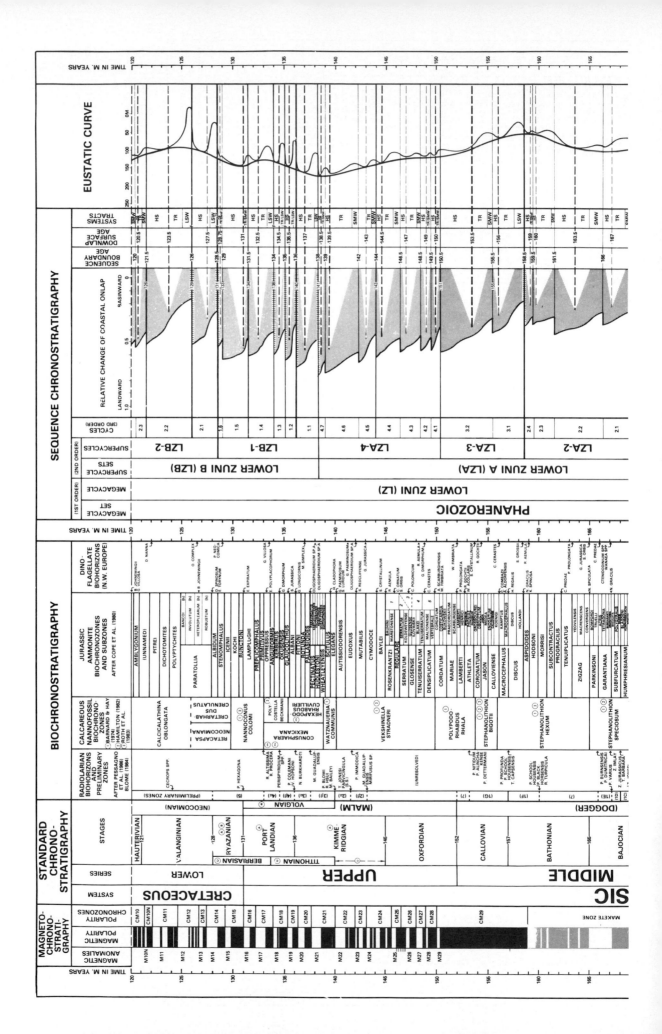

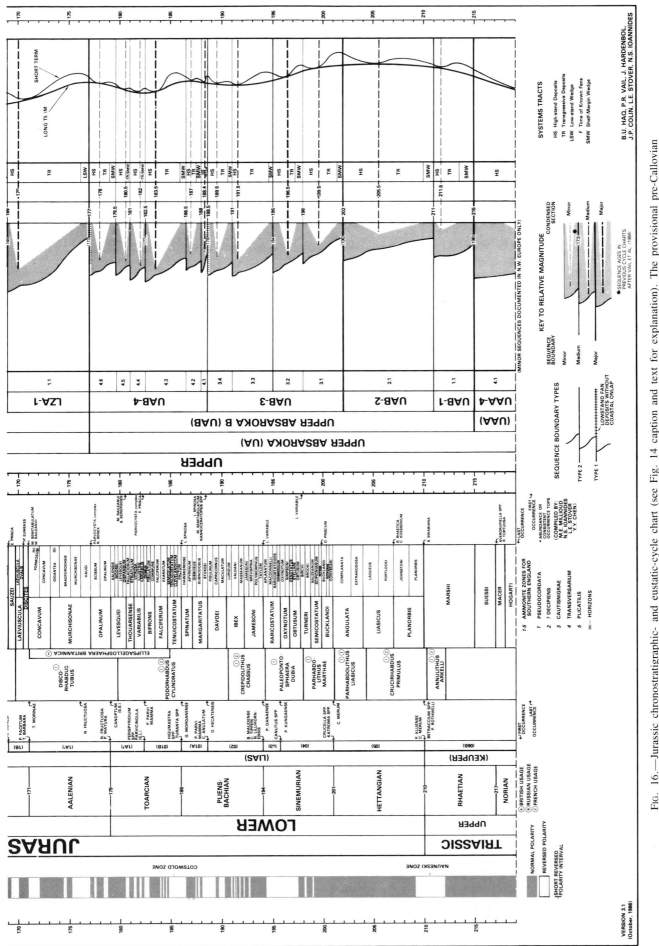

Fig. 16.—Jurassic chronostratigraphic- and eustatic-cycle chart (see Fig. 14 caption and text for explanation). The provisional pre-Callovian magnetic-polarity reversal pattern (shown in gray and white) is synthesized from known paleomagnetic data and may be subject to modification as more data become available from this interval.

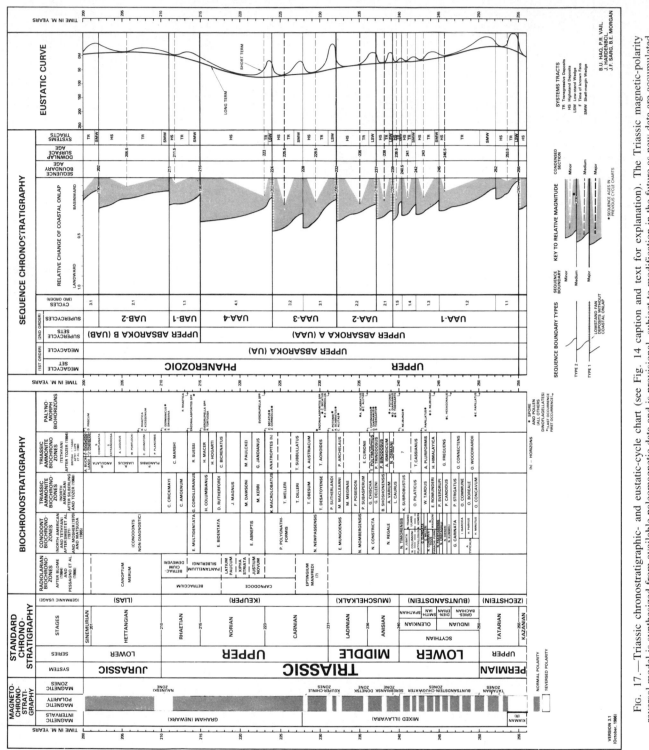

Fig. 17.—Triassic chronostratigraphic- and eustatic-cycle chart (see Fig. 14 caption and text for explanation). The Triassic magnetic-polarity reversal model is synthesized from available paleomagnetic data and is provisional, subject to modification in the future as new data are accumulated from this interval.

F. Robaszynski, M. Seroni-Vivien, A. Strasser, P. Strong, B. Thompson, and N. Vandenberghe. We acknowledge Exxon Production Research Company for permission to publish this article.

REFERENCES

ABELE, C., AND PAGE. R. W., 1974, Stratigraphic and isotopic ages of basalts at Maude and Aureys Inlet, Victoria, Australia: Royal Society of Victoria, Proceedings, v. 86, p. 143–150.

ADAMS, C. J. D., 1975, New Zealand potassium-argon age list-2: New Zealand Journal of Geology and Geophysics, v. 18, p. 443–467.

ALLEMAN, F., CATALANO, R., FARES, F., AND REMANE, J., 1971, Standard calpionellid zonation (Upper Tithonian-Valanginian) of the western Mediterranean Province: 2nd Planktonic Conference, Proceedings, Rome, v. 2, p. 1337–1340.

ALVAREZ, W., ARTHUR, M., FISHER, A., LOWRIE, W., NAPOLEONE, G., PREMOLI-SILVA, I., AND ROGGENTHEN, W., 1977, Upper Cretaceous-Paleocene magnetic stratigraphy at Gubbio, Italy: Geological Society of America Bulletin, v. 88, p. 367–389.

AMEDRO, F., 1980, Synthèse biostratigraphique de l'Aptien et Santonien du Boulonnais a partir de sept groupes paleontologiques: Revue de Micropaleontologie, v. 2, p. 195–321.

———, 1981, Actualisation des zonations d'ammonites dans le Cretace Moyen du Bassin Anglo-Parisien: Cretaceous Research, v. 2, p. 261–269.

———, 1984, Nouvelles donnes paleontologique (ammonites) sur l'Abiende la Bordure Nord-est du Bassin de Paris: Bulletin de la Société de la Geologie Normandie et Amis Museum du Harve, v. 71, p. 17–30.

———, 1986, Biostratigraphie des craies cenomaniennes du Boulonnais par les ammonites: Annales Societé Geologique du Nord, v. CV, p. 159–167.

ARMSTRONG, R. L., 1982, Late Triassic—early Jurassic time scale calibration in British Columbia, Canada, *in* Odin, G. S., ed., Numerical Dating in Stratigraphy: Wiley, New York, p. 509–513.

AUBRY, M.-P., 1983, Biostratigraphie du Paleogene epicontinental de l'Europe du Nord-Quest. Etude fondée sur les nannofossiles calcaires: Documents Laboratoire de Geologie Lyon, 89, 317 p.

BAADSGAARD, H., AND LERBEKMO, J. F., 1982, The dating of bentonite beds, *in* Odin, G. S., ed., Numerical Dating in Stratigraphy: Wiley, New York, p. 423–440.

BACKMAN, J., SHACKLETON, N. J., AND TAUXE, L., 1983, Quantitative nannofossil correlation to open ocean deep-sea sections from Plio-Pleistocene boundary at Vrica, Italy: Nature, v. 304, p. 156–158.

BANDY, O. L., HORNIBROOK, N. de B., AND SCHOFIELD, J. C., 1970, Age relationships of the *Globigerinoides trilobus* zone and the andesite at Muriwai Quarry, New Zealand: New Zealand Journal of Geology and Geophysics, v. 13, p. 980–995.

BARNARD, T., AND HAY, W. W., 1974, On Jurassic coccoliths: a tentative zonation of Jurassic of southern England and North France: Eclogae Geologicae Helvetiae, v. 67, p. 563–585.

BARRON, J. A., 1985, Miocene to Holocene planktic diatoms, *in* Bolli, H. M., Saunders, J. B., and Perch-Nielsen, K., eds., Plankton Stratigraphy: Cambridge University Press, Cambridge, p. 763–809.

———, KELLER, G., AND DUNN, D. A., 1985, A multiple microfossil biochronology for the Miocene: Geological Society of America memoir 163, p. 21–36.

BAUMGARTNER, P. O., 1985, Summary of Middle Jurassic and Early Cretaceous radiolarian biostratigraphy of DSDP Site 534 (Blake-Bahama basin) and correlations to Tethyan sections: Initial Reports of the Deep Sea Drilling Project, v. 76, p. 569–571.

BERGGREN, W. A., 1972, A Cenozoic time-scale—some implications for regional geology and paleobiogeography: Lethaia, v. 5, p. 195–215.

———, HAMILTON, N., JOHNSON, D. A., PUJOL, C., WEISS, W., CEPEK, P., AND GAMBOS, A., 1984, Magnetostratigraphy of DSDP Leg 72 Sites 515–518: Initial Reports of the Deep Sea Drilling Project, v. 72, p. 675–713.

———, KENT, D. V., AND FLYNN, J. J., 1985a, Jurassic to Paleogene: Part 2. Paleogene geochronology and chronostratigraphy, *in* Snelling, N. J., ed., The Chronology of the Geological Record: Blackwell Scientific Publishing, Oxford, and Geological Society of London Memoir 10, p. 141–195.

———, ———, AND VAN COUVERING, J. A., 1985b, The Neogene: Part 2. Neogene geochronology and chronostratigraphy, *in* Snelling, N. J., ed., The Chronology of the Geological Record: Blackwell Scientific Publishing, Oxford and Geological Society of London Memoir 10, p. 211–260.

BLOME, C. D., 1984, Upper Triassic Radiolaria and radiolarian zonation of western North America: Bulletin of American Paleontology, v. 85, p. 5–88.

BLONDEAU, A., 1980, Lutetien, *in* Cavelier, C., and Roger, J., eds., Les étages Français et leurs stratotypes: Bureau de Recherches Géologiques et Minières, Memoire, v. 109, p. 211–223.

———, AND RENARD, M., 1980, Le Lutetien stratotypique de la region de Creil (Oise): Bulletin de Information Géologique du Bassin de Paris, no. h-s, Excursion B-15 du 26eme Congress Géologique Internationale, p. B15-1–B15-11.

BLOW, W. H., 1969, The late Middle Eocene to Recent planktonic foraminiferal biostratigraphy: 1st Planktonic Conference, Proceedings, Geneva, 1967, p. 199–422.

BOUCHÉ, P., 1962, Nannofossiles calcaires du Lutetian du Bassin de Paris: Revue de Micropaleontologie, v. 5, p. 75–103.

BRALOWER, T. J., 1987, Valanginian to Aptian calcareous nannofossil stratigraphy and correlation with the upper M-sequence magnetic anomalies: Marine Micropaleontology, v. 11, p. 293–310.

BUKRY, D., 1981, Cenozoic coccoliths from the Deep Sea Drilling Project: Society of Economic Paleontologists and Mineralogists Special Publication 32, p. 335–353.

BURCKLE, L. H., 1978, Early Miocene to Pliocene diatom datum levels from the equatorial Pacific: 2nd Working Group Meeting, Biostratigraphic datum-planes of the Pacific Neogene, Proceedings: International Geological Correlation Program, Project 114, Bandung, 1977, Special Publication of the Geological Research and Development Center, no. 1, p. 25–44.

———, AND TRAINER, J., 1979, Middle and Late Pliocene diatom datum levels from the central Pacific: Micropaleontology, v. 25, p. 281–293.

BUREK, P. J., 1970, Magnetic reversals: their applications to stratigraphic problems: American Association of Petroleum Geologists Bulletin, v. 54, p. 1120–1139.

BURKE, W. H., DENISON, R. E., HETHERINGTON, E. A., KOEPNICK, R. B., NELSON, H. F., AND OTTO, J. B., 1982, Variation of seawater 87-Sr/86-Sr throughout Phanerozoic time: Geology, v. 10, p. 516–519.

BUSNARDO, R., 1984, Ammonites, *in* Cretace Inferieur: Bureau de Recherches Geologiques et Minières, Memoire, v. 125, p. 292–294.

CANDE, S. C., AND KRISTOFFERSEN, Y., 1977, Late Cretaceous magnetic anomalies in the North Atlantic: Earth and Planetary Science Letters, v. 35, p. 215–224.

———, LARSON, R. L., AND LaBRECQUE, J. L., 1978, Magnetic lineations in the Pacific Jurassic quiet zone: Earth and Planetary Science Letters, v. 41, p. 436–440.

CARON, M., 1985. Cretaceous planktonic foraminifera, *in* Bolli, H. M., Saunders, J. B., and Perch-Nielsen, K., eds., Plankton Stratigraphy: Cambridge University Press, Cambridge, p. 11–86.

CAVALIER, C., AND ROGERS, J., 1980, Les étages Français et leurs stratotypes: Bureau Recherches Geologiques et Minières Memoire, v. 109, 295 p.

CHANNELL, J. E. T., AND MEDIZZA, F., 1981, Upper Cretaceous and Paleogene magnetic stratigraphy and biostratigraphy from Venetian (southern) Alps: Earth and Planetary Science Letters, v. 55, p. 419–432.

———, LOWRIE W., AND MEDIZZA, F., 1979, Middle and early Cretaceous magnetic stratigraphy from the Cismon section, northern Italy: Earth and Planetary Science Letters, v. 42, p. 153–166.

———, OGG, J. G., AND LOWRIE, W., 1982, Geomagnetic polarity in the Early Cretaceous and Jurassic: Philosophical Transactions of the Royal Society of London, Series A, v. 306, p. 137–146.

CLAUER, N., 1982. The rubidium-strontium method applied to sediments: certitudes and uncertainties, *in* Odin, G. S., ed., Numerical Dating in Stratigraphy: Wiley, New York, p. 245–276.

CLAVEL, B., 1986, Précisions stratigraphiques sur le Cretacé inferieur basal du Jura meridional: Eclogae Geologieae Helvetiae, v. 79, p. 319–341.

COPE, J. C. W., GETTY, T. A., HOWARTH, M. K., MORTON, N., TORRENS, H. S., DUFF, K. L., PARSONS, C. F., WIMBLEDON, W. A., AND WRIGHT

J. K., 1980, Jurassic, Parts 1 and 2: Geological Society of London, Special Reports, no. 14, 15, 73 and 109 p.

CREER, K. M., 1971, Mesozoic paleomagnetic reversal column: Nature, v. 233, p. 545–546.

CURRY, D., 1985, Oceanic magnetic lineaments and the calibration of late Mesozoic-Cenozoic time scales, *in* Snelling, N. J., ed., The Chronology of the Geological Record: Blackwell Scientific Publishing, Oxford, and Geological Society of London, Memoir, 10, p. 269–272.

DEPAOLO D. J., AND INGRAM, B. L., 1985, High-resolution stratigraphy with strontium isotopes: Science, v. 227, p. 938–941.

DICKSON, G. O., PITMAN, W. C., AND HEIRTZLER, J. R., 1968, Magnetic anomalies in the South Atlantic and ocean floor spreading: Journal of Geophysical Research, v. 73, p. 2087–2100.

ELDEFIELD, H., 1986, Strontium isotope stratigraphy: Palaeogeography, Palaeoclimatology, Palaeoecology, v. 57, p. 71–90.

EMRY, R. J., 1973, Stratigraphy and preliminary biostratigraphy of the Flagstaff Rim area, Natrona County, Wyoming: Smithsonian Contributions to Paleobiology, v. 18, p. 1–42.

EVERNDEN, J. F., SAVAGE, D. E., CURTIS, G. H., AND JAMES, C. T., 1964, Potassium-argon dates and the Cenozoic mammalian chronology of North America: American Journal of Science, v. 262, p. 145–198.

FENNER, J., 1985, Late Cretaceous to Oligocene planktic diatoms, *in* Bolli, H. M., Saunders, J. B., and Perch-Nielsen, K., eds., Plankton Stratigraphy: Cambridge University Press, Cambridge, p. 713–762.

FITCH, F. J., HOOKER, P. J., MILLER, J. A., AND BRERETON, N. R., 1978, Glauconite dating of Paleocene-Eocene rocks from East Kent and the time-scale of Paleogene volcanism in North Atlantic region: Journal of the Geological Society of London, v. 135, p. 499–512.

FLYNN, J. J., 1981, Magnetic polarity stratigraphy and correlation of Eocene strata from Wyoming and southern California (abs.): EOS, American Geophysical Union Transactions, v. 62, p. 264.

FORSTER S. C., AND WARRINGTON, G., 1985, Geochronology of the Carboniferous, Permian and Triassic, *in* Snelling, N. J., ed., The Chronology of the Geological Record: Blackwell Scientific Publishing, Oxford and Geological Society of London, Memoir 10, p. 99–113.

GALBRUN, B., AND RASPLUS, L., 1984, Magnetostratigraphie du stratotype du Berriasien, premiers resultats: Comptes Rendus, Academie des Sciences, Paris, v. 298, p. 219–222.

GALE, N. H., 1982, The dating of plutonic events, *in* Odin, G. S., ed., Numerical Dating in Stratigraphy: Wiley, New York, p. 441–453.

GARTNER, S., 1973, Absolute chronology of the Late Neogene calcareous nannofossil succession in the equatorial Pacific: Geological Society of America Bulletin, v. 84, p. 2021–2034.

GHOSH, P. K., 1972, Use of bentonites and glauconites in potassium−40/argon−40 dating in Gulf Coast stratigraphy: unpublished Ph.D. Dissertation, Rice University, Houston, Texas, 136 p.

GOMBOS, A. M., Jr., 1982, Early and middle Eocene diatom evolutionary events: Bacillaria, v. 5, p. 225–236.

———, 1984, Late Paleocene diatoms in the Cape Basin: Initial Reports of the Deep Sea Drilling Project, v. 73, p. 495–512.

———, CIESIELSKI, P. F., 1983, Late Eocene to early Miocene diatoms from the southwest Atlantic: Initial Reports of the Deep Sea Drilling Project, v. 71, p. 583–634.

HALLAM, A., 1983, Early and mid-Jurassic molluscan biogeography and the establishment of the central Atlantic seaway: Palaeogeography, Palaeoclimatology, Palaeoecology, v. 43, p. 181–193.

———, HANCOCK, J. M., LaBRECQUE, J. L., LOWRIE, W., AND CHANNELL, J. E. T., 1985, Jurassic to Paleogene: Part 2, Jurassic and Cretaceous geochronology and Jurassic to Paleogene magnetostratigraphy, *in* Snelling, N. J., ed., The Chronology of the Geological Record: Blackwell Scientific Publishing, Oxford, and Geological Society of London, Memoir 10, p. 118–140.

HAMILTON, G. B., 1982, Triassic and Jurassic and Jurassic nannofossils, *in* Lord, A. R., ed., A Stratigraphic Index of Calcareous Nannofossils: Ellis Horwood, Chichester, p. 17–39.

HAQ, B. U., BERGGREN, W. A., AND VAN COUVERING, J. A., 1977, Corrected age of the Pliocene-Pleistocene boundary: Nature, v. 269, p. 483–289.

———, HARDENBOL, J., AND VAIL, P. R., 1987, The chronology of fluctuating sea level since the Triassic: Science, v. 235, p. 1156–1167.

———, AND WORSLEY, T. R., 1982, Biochronology—biological events in time resolution, their potential and limitations, *in* Odin, G. S., ed., Numerical Dating in Stratigraphy: Wiley, New York, p. 19–36.

———, ———, BURCKLE, L. H., DOUGLAS, R. G., KEIGWIN, L. D., Jr., OPDYKE, N. D., SAVIN, S. M., SOMMER, M. A., II, VINCENT, E., AND WOODRUFF F., 1980, The Late Miocene carbon isotopic shift and the synchroneity of some phytoplanktonic biostratigraphic datums: Geology, v. 8, p. 427–431.

HARDENBOL, J., AND BERGGREN, W. A., 1978, A new Paleogene numerical time scale: American Association of Petroleum Geology, Studies in Geology, v. 6, p. 213–234.

———, VAIL, P. R., AND FERRER, J., 1981, Interpreting paleoenvironments, subsidence history, and sea–level changes of passive margins from seismic and biostratigraphy: Oceanologica Acta, Suppl. to v. 3, p. 33–44.

HARLAND, W. B., COX, A. V., LLEWELLYN, P. G., PICKTON, C. A. G., SMITH, D. G., AND WALTERS, R., 1982, A geologic time scale: Cambridge University Press, Cambridge, 131 p.

HARRISON, C. G. A., 1986, Long-term eustasy and epeirogeny in continents: National Academy of Sciences, Studies in Geophysics: p. 111–131.

HEDBERG, H. D., ed., 1976, International Stratigraphic Guide: Wiley, New York, 200 p.

HEIRTZLER J. R., DICKSON, G. O., HERRON, E. M., PITMAN W. C., III, AND LePICHON, X., 1968, Marine magnetic anomalies, geomagnetic field reversals, and motions of ocean floor and continents: Journal of Geophysical Research, v. 73, p. 2119–2136.

HELLSEY, C. E., 1969, Magnetic stratigraphy of the Lower Triassic Moenkopi Formation of western Colorado: Geological Society of America Bulletin, v. 80, p. 2431–2450.

———, AND STEINER, M. B., 1974, Paleomagnetism of the Lower Triassic Moenkopi Formation: Geological Society of America Bulletin, v. 85, p. 457–464.

HILDE, T. W. C., ISEZAKI, N., AND WAGEMAN, J. M., 1975, Mesozoic sea-floor spreading in the North Pacific, *in* Sutton, G. H., Manghnani, M. H., and Moberly, R., The Geophysics of the Pacific Ocean Basin and its Margins: American Geophysical Union Monograph 19, p. 205–226.

HÖRNER, F. AND HELLER, F., 1983, Lower Jurassic magnetostratigraphy at the Breggia Gorge (Tincino, Switzerland) and Alpe Turati (Como, Italy): Journal of the Royal Astronomical Society of London, v. 73, p. 705–718.

HSÜ, K. J., PERCIVAL S. F., Jr., WRIGHT, R. C., AND PETERSEN, N., 1984, Numerical ages of magnetostratigraphically calibrated biostratigraphic zones: Initial Reports of the Deep Sea Drilling Project, v. 73, p. 623–635.

JÄGER, E., AND HUNZIKER, J. C., 1979, Isotope Geology: Springer-Verlag, Heidelberg, 329 p.

KAWAI, N., AND HIROOKA, K., 1966, Some age results on Cenozoic igneous rocks from southwest Japan (abs.): Symposium on Age of Formation for Japanese Acid Rocks by Dating Results: Geological Society of Japan, p. 5.

KENNEDY, W. J., 1984, Ammonite faunas and the 'standard zones' of the Cenomanian to Maastrichtian stages in their type areas, with some proposals for the definition of stage boundaries by ammonites: Bulletin of the Geological Society of Denmark, v. 33, p. 147–161.

KENT, D. V., AND GRADSTEIN, F. M., 1985, A Cretaceous and Jurassic geochronology: Geological Society of American Bulletin, v. 96, p. 1419–1427.

KEPPENS, E., AND PASTEELS, P., 1982, A comparison of rubidium-strontium and potassium-argon apparent ages on glauconies, *in* Odin, G. S., ed., Numerical Dating in Stratigraphy: Wiley, New York, p. 225–239.

KLITGORD, K. D., AND SCHOUTEN, H., 1986, Plate kinematics of the central Atlantic, *in* Vogt, P. R., and Tucholke, B. E., eds., The Geology of North America: Geological Society of America, v. M, p. 351–378.

LaBRECQUE, J. L., HSU, K. J., CARMAN, M., KARPOFF, A.-M., McKENZIE, J., PERCIVAL, S., PETERSEN, N., PISCIOTTO, K., SCHREIBER, E., TAUXE, L., TUCKER, P., WEISSERT, H., AND WRIGHT, R., 1983, DSDP Leg 73: Contributions to Paleogene stratigraphy in nomenclature, chronology and sedimentation rates: Palaeogeography, Paleoclimatology, Paleoecology, v. 42, p. 91–125.

———, KENT, D. V., AND CANDE, S. C., 1977, Revised magnetic polarity time scale for Late Cretaceous and Cenozoic time: Geology, v. 5, p. 330–335.

LARSON, R. L., AND CHASE, C. G., 1972, Late Mesozoic evolution of the western Pacific: Geological Society of America Bulletin, v. 83, p. 3627–3644.

——, AND HILDE, T. W. C., 1975, A revised time scale of magnetic reversals for early Cretaceous, late Jurassic: Journal of Geophysical Research, v. 80, p. 2586–2594.

——, AND PITMAN, III, W. C., 1972, World-wide correlation of Mesozoic magnetic anomalies and its implications: Geological Society of America Bulletin, v. 83, p. 3645–3662.

LOWRIE, W., AND ALVAREZ, W., 1981, 100 million years of geomagnetic polarity history: Geology, v. 9, p. 392–397.

——, ——, NAPOLEONE G., PERCH-NIELSEN, K., PREMOLI-SIVA, I., AND TOUMARKINE, M., 1982, Paleogene magnetic stratigraphy in Umbrian pelagic carbonate rocks: The Contessa sections, Gubbio: Geological Society of America Bulletin, v. 93, p. 414–432.

——, ——, PREMOLI-SILVA, I., AND MONECHI, S., 1980, Lower Cretaceous magnetic stratigraphy in Umbrian pelagic carbonate rocks: Geophysical Journal of the Royal Astronomical Society, v. 60, p. 263–281.

——, AND CHANNELL, J. E. T., 1983, Magnetostratigraphy of the Jurassic-Cretaceous boundary in the Maiolica Limestone (Umbria, Italy): Geology, v. 12, p. 44–47.

MANIVIT, H., PERCH-NIELSEN K., PRINS, B., AND VERBEEK, J. W., 1977, Mid-Cretaceous calcareous nannofossil biostratigraphy: Koninklijke Nederlandse Akademie van Wetenschappen, v. B80, p. 109–181.

MANKINEN, E. A., AND DALRYMPLE, G. B., 1979, Revised late Cenozoic geomagnetic polarity time scale: U.S. Geological Survey Professional Paper 1100, 167 p.

MARTINI, E., 1971, Standard Tertiary and Quaternary calcareous nannoplankton zonation: 2nd Plankton Conference Proceedings, Rome, 1969, v. 2, p. 739–785.

MARTON, E., 1982, Late Jurassic/early Cretaceous magnetic stratigraphy from the Sumeg section, Hungary: Earth and Planetary Science Letters, v. 57, p. 182–190.

——, MARTON, P., AND HELLER, F., 1980, Remnant magnetization of a Pliensbachian limestone sequence at Bakonycsernye (Hungary): Earth and Planetary Science Letters, v. 48, p. 218–226.

MATSUDA, T., 1985, Late Permian to early Triassic conodont paleobiogeography in the 'Tethys Realm', in Nakazawa, K., and Dickins, J.M., eds., The Tethys—Her Paleogeography and Paleobiogeography from the Paleozoic to Mesozoic: Tokai University Press, Tokyo p. 157–170.

McDOUGALL, I., SAEMUNDSSON, K., JOHANNESSON, H., WATKINS, N.D., AND KRISTJANSSON, L., 1977, Extension of the geomagnetic polarity time scale to 6.5 m.y.: K-Ar dating, geological and paleomagnetic study of a 3,500 m lava succession in western Iceland: Geological Society of America Bulletin, v. 88, p. 1–15.

——, WATKINS, N.D., WALKER, G.P.L., AND KRISTJANSSON, L., 1976, Potassium-argon and paleomagnetic analysis of Icelandic lava flows—limits on the age of anomaly 5: Journal of Geophysical Research, v. 81, p. 1505–1512.

McELHINNY, M.W., AND BUREK, P.J., 1971, Mesozoic paleomagnetic stratigraphy: Nature, v. 232, p. 98–102.

McKEE, E.H., 1975, K-Ar ages of deep-sea basalts, Benham Rise, W. Philippine Basin, Leg 31: Initial Reports of the Deep Sea Drilling Project, v. 31, p. 599–611.

MEDD, A.W., 1982, Nannofossil zonation of the English Middle and Upper Jurassic: Marine Micropaleontology, v. 7, p. 73–95.

MILLER, K.G., KHAN, M.J., AUBRY, M.-P., BERGGREN, W.A., KENT, D.V., AND MELILLO, A., 1985, Oligocene-Miocene biostratigraphy, magnetostratigraphy, and isotopic stratigraphy of western North Atlantic: Geology, v. 13, p. 257–261.

MITCHUM, R.M., VAIL, P.R., AND THOMPSON, S., 1977, The depositional sequence as a basic unit for stratigraphic analysis: American Association of Petroleum Geologists Memoir 26, p. 53–62.

MONECHI, S., AND THIERSTEIN, H., 1985, Late Cretaceous—Eocene nannofossil and magnetostratigraphic correlations near Gubbio, Italy: Marine Micropaleontology, v. 9, p. 419–440.

MOSHER, L.C., 1970, New conodont species as Triassic guide fossils: Journal of Paleontology, v. 44, p. 737–742.

NAKAZAWA, K., ALI, S.T., BANDO, Y., ISHII, K., OKIMURA, Y., QURESHI, K.A., AND SHUJA, T.A., 1985, Permian and Triassic Systems in the Salt Range and Surghar Range, Pakistan, in Nakazawa, K., and Dickens, J.M., eds., The Tethys—Her Paleogeography and Paleobiogeog-

raphy from Paleozoic to Mesozoic: Tokai University Press, Tokyo, p. 221–312.

NAPOLEONE, G., PREMOLI-SILVA, I., HELLER, F., CHELI, P., COREZZI, S., AND FISCHER, A. G., 1983, Eocene magnetic stratigraphy at Gubbio, Italy, and its implications for Paleogene geochronology: Geological Society of America Bulletin, v. 94, p. 181–191.

NESS, G., LEVI, S., AND COUCH, R., 1980, Marine magnetic anomaly time scales for the Cenozoic and late Cretaceous: a précis, critique, and synthesis: Reviews of Geophysics and Space Physics, v. 18, p. 753–770.

NEWMAN, K. R., 1979, Cretaceous/Tertiary boundary in Denver Formation at Golden, Colorado, USA, in Christensen, W. K., and Birkelund, T., eds., Cretaceous-Tertiary Boundary Events: University of Copenhagen, Sweden, v. 2, p. 246–248.

OBRADOVICH, J. D., AND COBBAN, W. A., 1975, A time-scale for the Late Cretaceous of the Western Interior of North America, in Caldwell, W. G. E., ed., The Cretaceous System in the Western Interior of North America: Geological Association of Canada, Special Paper 13, p. 31–54.

ODIN, G. S., eed., 1982a, Numerical Dating in Stratigraphy: Wiley, New York, 1040 p.

——, 1982b, The Phanerozoic time scale revisited: Episodes, no. 3, p. 3–9.

OGG, J. G., 1981, Sedimentology and paleomagnetism of Jurassic pelagic limestones ('Ammonitico Rosso' facies): Unpublished Ph.D. Dissertation, University of California, San Diego, California, 203 p.

——, 1983, Magnetostratigraphy of Upper Jurassic and lowest Cretaceous sediments, Deep Sea Drilling Project Site 534, western North Atlantic: Initial Reports of the Deep Sea Drilling Project, v. 76, p. 685–697.

——, STEINER, M. B., OLORIZ, F., AND TAVERA, J. M., 1984, Jurassic magnetostratigraphy, 1. Kimmeridgian-Tithonian of Sierra Gorda and Carcabuey, southern Spain: Earth and Planetary Science Letters, v. 71, p. 147–162.

OKADA, H., AND BUKRY, J., 1980, Supplementary modification and introduction of code numbers to low-latitude coccolith biostratigraphic zonation (Bukry, 1973; 1975): Marine Micropaleontology, v. 5, p. 321–326.

PECHERSKY, D. M., AND KHRAMOV, A. N., 1973, Mesozoic paleomagnetic scale of the U.S.S.R.: Nature, v. 244, p. 499–501.

PERSON, A., 1982, The genesis of bentonites, in Odin, G. S., ed., Numerical Dating in Stratigraphy: Wiley, New York, p. 407–422.

PESSAGNO, E. A., Jr., 1977, Upper Jurassic Radiolaria and radiolarian biostratigraphy of the California Coast Range: Micropaleontology, v. 23, p. 56–133.

——, AND BLOME, C. D., 1980, Upper Triassic and Jurassic Pantanellinae from California, Oregon and British Columbia: Micropaleontology, v. 26, p. 225–273.

——, ——, CARTER, E. S., MacLEOD, N., WHALEN, P. A., AND YEH, K.-Y., 1987b, Part 2. Preliminary radiolarian zonation for the Jurassic of North America: Cushman Foundation for Foraminiferal Research, Special Publication, no. 23, p. 1–18.

——, LONGORIA, J. F., MacLEOD, N., AND SIX, W. M., JR., 1987a, Part 1. Upper Jurassic (Kimmeridgian—Upper Tithonian) Pantanelliidae from the Taman Formation, east-central Mexico: tectonostratigraphic, chronostratigraphic and phylogenetic implications: Cushman Foundation for Foraminiferal Research, Special Publication, no. 23, p. 1–51.

PITMAN, III, W. C., HERRON, E. M., AND HEIRTZLER, J. R., 1968, Magnetic anomalies in the Pacific and sea floor spreading: Journal of Geophysical Research, v. 73, p. 2069–2085.

POORE, R. Z., TAUXE, L., PERCIVAL, S., JR., AND LaBRECQUE, J. L., 1982, Late Eocene—Oligocene magnetostratigraphy and biostratigraphy at South Atlantic DSDP Site 552: Geology, v. 10, p. 508–511.

——, ——, ——, ——, WRIGHT, R., PETERSEN, N., SMITH, C., AND HSU, K., 1983, Late Cretaceous-Cenozoic magnetostratigraphic and biostratigraphic correlations of South Atlantic Ocean (DSDP Leg 73): Palaeogeography, Palaeoclimatology, Palaeoecology, v. 42, p. 127–149.

PREMOLI-SILVA, I., 1977, Upper Cretaceous-Paleocene magnetic stratigraphy at Gubbio, Italy, II. Biostratigraphy: Geological Society of America Bulletin, v. 88, p. 371–374.

——, AND BOERSMA, A., 1977, Cretaceous planktonic foraminifers—

DSDP Leg 39 (South Atlantic): Initial Reports of the Deep Sea Drilling Project, v. 39, p. 615–631.

———, AND BOLLI, H. M., 1973, Late Cetaceous to Eocene planktonic foraminifera and stratigraphy of Leg 15 Sites in the Caribbean Sea: Initial Reports of the Deep Sea Drilling Project, v. 15, p. 499–547.

PROTHERO, D. R., DENHAM, C. R., AND FARMER, H. G., 1982, Oligocene calibration of the magnetic polarity time scale: Geology, v. 10, p. 650–653.

RAWSON, P. F., CURRY, D., DILLEY, F. C., HANCOCK, J. M., KENNEDY, W. J., NEALE, J. W., WOOD, C. J., AND WORSSAM, B. C., 1978, Cretaceous: Geological Society of London Special Report no. 9, 70 p.

REMANE, J., 1978, Calpionellids, in Haq, B. U., and Boersma, A., eds., Introduction to Marine Micropaleontology: Elsevier, New York, p. 161–170.

———, 1985, Calpionellids, in Bolli, H. M., Saunders, J. B., and Perch-Nielsen, K., eds., Plankton Stratigraphy: Cambridge University Press, Cambridge, p. 555–572.

ROBASZYNSKI, F., CARON, M., AND THE EUROPEAN WORKING GROUP ON PLANKTONIC FORAMINIFERA, 1979, Atlas des foraminifères planktoniques du Crétacé Moyen (Mer Boréale et Téthys): Cahiers de Micropaleontologie, no. 1, 2, 185 and 181 p.

———, CARON, M., GONZALES, J. M., AND WONDER, A., 1983, Atlas of Late Cretaceous planktonic foraminifera. Reviews of Micropaleontology, v. 26, p. 145–305.

ROTH, P. H., 1978, Cretaceous nannoplankton biostratigraphy and oceanography of the northwestern Atlantic Ocean: Initial Reports of the Deep Sea Drilling Project, v. 44, p. 731–759.

———, 1983, Jurassic and Lower Cretaceous nannofossils in the western North Atlantic (Site 534): Initial Reports of the Deep Sea Drilling Project, v. 76, p. 587–621.

———, MEDD, A. W., AND WATKINS, D. K., 1983, Jurassic calcareous nannofossil zonations, an overview with new evidence from Deep Sea Drilling Project Site 534: Initial Reports of the Deep Sea Drilling Project, v. 76, p. 573–579.

RYAN, W. B. F., CITA, M. B., DREYFUS RAWSON, M., BURCKLE, L. H., AND SAITO, T., 1974, A paleomagnetic assignment of Neogene stage boundaries and the development of isochronous datum planes between the Mediterranean, the Pacific and the Indian Ocean in order to investigate the response of the world ocean to Messinian Salinity Crisis: Rivista Italiana Paleontologic e Stratigrafia, v. 80, p. 631–688.

SALVADOR, A., 1985, Chronostratigraphic and geochronometric scales in COSUNA stratigraphic correlation charts of the United States: American Association of Petroleum Geologists Bulletin, v. 69, p. 181–189.

SANFILIPPO, A., WESTBERG, M. J., AND RIEDEL, W. R., 1981, Cenozoic radiolarians at Site 462, DSDP leg 61: Initial Reports of the Deep Sea Drilling Project, v. 61, p. 495–505.

———, ———, AND ———, 1985, Cenozoic Radiolaria, in Bolli, H. M., Saunders, J. B., and Perch-Nielsen, K., eds., Plankton Stratigraphy: Cambridge University Press, Cambridge, p. 631–712.

SCHOUTEN, H., AND KLITGORD, K. D., 1977, Map showing Mesozoic magnetic anomalies, western North Atlantic: U.S. Geological Survey Miscellaneous Field Studies, Map MF-915, scale 1:2,000,000.

———, AND ———, 1982, The memory of the accreting plate boundary and the continuity of fracture zones: Earth and Planetary Science Letters, v. 59, p. 255–266.

SEIDEMANN, D. E., 1977, Effects of submarine alteration on K-Ar dating of deep-sea igneous basalts: Geological Society of America Bulletin, v. 88, p. 1660–1666.

SHACKLETON, N. J., AND SHIPBOARD SCIENTIFIC PARTY, 1984, Accumulation rates in Leg 74 sediments: Initial Reports of the Deep Sea Drilling Project, v. 74, p. 621–637.

SHIBATA, K., 1973, K-Ar ages of volcanic rocks from the Hokuriku Group: Geological Society of Japan, Memoir 8, p. 143–149.

———, AND ISHIHARA, S., 1979, Rb-Sr whole rock and K-Ar mineral ages of granitic rocks in Japan: Geochemical Journal, v. 13, p. 113–119.

———, NISHIMURA, S., AND CHINZEI, K., 1984, Radiometric dating related to Pacific Neogene planktonic datum planes, in Ikebe, N., and Tsuchi, R., eds., Pacific Neogene Datum Planes: University of Tokyo Press, Tokyo, p. 85–89.

———, AND NOZAWA, T., 1967, K-Ar ages of granitic rocks from the

Outer Zone of S.W. Japan: Geochemical Journal, v. 1, p. 131–137.

———, SATO, H., AND NAKAGAWA, M., 1981, K-Ar ages of Neogene volcanic rocks from the Noto Peninsula: Japan Association of Petrology, Mineralogy, and Economic Geology, Journal, v. 76, p. 248–252.

———, UCHIUMI, S., AND NAKAGAWA, T., 1979, K-Ar age results-1: Geological Society of Japan Bulletin, v. 30, p. 675–686.

SILBERING, N. J., AND TOZER, E. T., 1968, Biostratigraphic classification of the marine Triassic in North America: Geological Society of America Special Paper, no. 110, 63 p.

SISSINGH, W., 1977, Biostratigraphy of Cretaceous calcareous nannoplankton: Geologie en Mijnbouw, v. 56, p. 335–350.

SLOSS, L. L., 1963, Sequences in the cratonic interior of North America: Geological Society of America Bulletin, v. 64, p. 93–113.

SNELLING, N. J., ed., 1985, The Chronology of the Geological Record: Blackwell Scientific Publishing, Oxford and Geological Society of London, Memoir 10, 343 p.

STAINFORTH, R. M., LAMB, J. L., LUTERBACHER, H., BEARD, J. H., AND JEFFORDS, R. M., 1975, Cenozoic planktonic foraminferal zonation and characteristic index forms: University of Kansas, Paleontological Contributions, v. 62, Lawrence, Kansas, 425 p.

STEINER, M. B., AND HELSLEY, C. E., 1975, Late Jurassic magnetic polarity sequence: Earth and Planetary Science Letters, v. 27, p. 108–112.

———, OGG, J. G., MELENDEZ, G., AND SEQUEIROS, L., 1985, Jurassic magnetostratigraphy, 2. Middle-Late Oxfordian of Aguilon, Iberian Cordillera, northern Spain: Earth and Planetary Science Letters, v. 76, p. 151–166.

STRADNER, H., AND ALLRAM, F., 1982, The nannofossil assemblage of DSDP Leg 66, Middle America Trench: Initial Reports of the Deep Sea Drilling Project, v. 66, p. 589–639.

SUNDVIK, M., LARSON, R. L., AND DETRICK, R. S., 1984, Rough-smooth basement boundary in the western North Atlantic basin: evidence for seafloor-spreading origin: Geology, v. 12, p. 31–43.

SWEET, W. C., MOSHER, L. C., CLARK, D. L., COLLINSON, J. W., AND HASENMUELLER, W. A., 1971, Conodont biostratigraphy of the Triassic: Geological Society of America Memoir, 126, p. 441–465.

TAUXE, L., TUCKER, P., PETERSEN, N. P., AND LABRECQUE, J. L., 1983, The magnetostratigraphy of Leg 73 sediments: Palaeogeography, Palaeoclimatology, Palaeoecology, v. 42, p. 65–90.

THEYER, F., HAMMOND, S. R., AND MATO, Y., 1978, Paleomagnetic and geochronologic calibration of latest Oligocene to Pliocene radiolarian events, Equatorial Pacific: Marine Micropaleontology, v. 3, p. 337–395.

THIERSTEIN, H. R., 1976, Mesozoic calcareous nannoplankton biostratigraphy of marine sediments: Marine Micropaleontology, v. 1, p. 325–362.

———, GEITZENAUER, K. R., MOLFINO, B., AND SHACKLETON, N. J., 1977, Global synchroneity of late Quaternary coccolith datum levels—validation by oxygen isotopes: Geology, v. 5, p. 400–404.

TOZER, E. T., 1984, The Triassic and its ammonoids, evolution of a time scale: Geological Survey of Canada, Miscellaneous Reports, v. 35, 171 p.

TREYAK, A. H., VIGILYANSKAYA, L. I., AND SHEMPELEV, A. G., 1976, Paleomagnitny Razrev Nizhnego mela Severo-Zapadnogo Kavkaza, in Mikhylova, N. P., ed., Paleomagnetizm, Magnetizm, Geomagnitnoye Pole: Izd. Nauk Dumka, Moscow, p. 38–42.

VAIL, P. R., AND HARDENBOL, J., 1979, Sea-level changes during the Tertiary: Oceanus, v. 22, p. 71–79.

———, ———, AND TODD, R. G., 1984, Jurassic unconfomities, chronostratigraphy and sea level changes from seismic stratigraphy and biostratigraphy: Third Annual Conference, Gulf Coast Society and Society of Economic Paleontologists and Mineralogists, Proceedings, p. 347–364.

———, MITCHUM, R. M., JR., TODD, R. G., WIDMIER, J. M., THOMPSON, S., III, SANGREE, J. B., BUBB, J. N., AND HATLELID, W. G., 1977, Seismic stratigraphy and global changes of sea level: American Association of Petroleum Geologists Memoir, 26, p. 49–212.

VAN HINTE, J. E., 1976, A Cretaceous time scale: American Association of Petroleum Geologists Bulletin, v. 60, p. 498–516.

VOGT, P. R., AND EINWICH, A. M., 1979, Magnetic anomalies and seafloor spreading in the western North Atlantic, and a revised calibration of Keathley (M) geomagnetic reversal chronology: Initial Repots of the

Deep Sea Drilling Project, v. 43, p. 857–876.

WEBB, J. A., 1981, A radiometric time scale of the Triassic: Journal of the Geological Society of Australia, v. 28, p. 107–121.

———, 1982, Triassic radiometric dates from eastern Australia, in Odin,

G. S., ed., Numerical Dating in Stratigraphy: Wiley, New York, p. 515–522.

WRIGHT, C. W. AND KENNEDY, W. J., 1983, The Ammonoidea of the Lower Chalk, Part 1: Palaeontgraphical Society, 126 p.

APPENDIX A.—RADIOMETRIC DATES PLOTTED AGAINST SEAFLOOR MAGNETIC-ANOMALY PROFILES (FIGS. 3–5) FOR BEST-FIT SOLUTION OF MAGNETIC-ANOMALY AGES.

Number	Date (in MA)	High-/Low-Temperature Date	Magnetic Anomaly (Assignment Through Paleo)	Reference(s) (*Reference in Odin, 1982a)
1	67.1 ± 1.0	H	30N	Obradovich and Cobban, 1975
2	66.7 ± 1.0	L	Above 30N	*Harris, p. 604
3	66.4 ± 2.5	H	Within 29R	Evernden and others 1964; Obradovich and Cobban, 1975; Newman, 1979
4	65.8 ± 1.4	H	Below 29N	*Odin and Obradovich, p. 768
6	63.5 ± 0.4	H	Between 30N and 29N	*Baadsgaard and Lerbekmo, p. 797
7	56.8 ± 0.8	L	Just above 25N	Fitch and others, 1978
9	56.6 ± 3.4	L	Above 26N	*Harris, p. 604
10	46.7 ± 3.0	L	Above 23N	*Harris, p. 604
11	54.1 ± 2.0	L	Between 26N and 25N	*Curry and Odin, p. 685
12	56.0 ± 2.1	L	Between 26N and 25N	*Curry and Odin, p. 673
12A	57.5 ± 3.0	L	Between 26N and 25N	*Curry and Odin, p. 674
13	54.2 ± 1.9	L	Just above 25N	*Curry and others, p. 694
14	55.2 (?)	L	Just above 25N	*Curry and others, p. 695
15	56.0 ± 1.9	L	Just above 25N	*Curry and Odin, p. 695
16	53.0 ± 2.4	L	Above 25	*Bignot and others, p. 693
17	49.5 ± (?)	H	Top 21N	Flynn, 1981
18	50.3 ± 0.7	L	Between 21N and 20 N	*Kreuzer, p. 932
19	48.3 ± 2.7	L	22N	*Curry and Odin, p. 691
22	45.0 ± 2.2	L	22N	*Curry and Odin, p. 681
23	47.7 ± 1.6	L	Between 22N and 21N	*Curry and Odin, p. 682
24	43.9 ± 2.0	L	Between 22N and 21N	*Curry and Odin, p. 665
26	44.4 ± 2.3	L	Between 22N and 21N	*Curry and Odin, p. 665
27	44.4 ± 2.3	L	Between 22N and 21N	*Curry and Odin, p. 686
30	46.2 ± 1.6	L	Between 22N and 21 N	*Curry and Odin, p.689
31	41.0 ± 1.8	L	20N to 19N	*Curry and Odin, p. 680
32	40.1 ± 1.8	L	20N to 19N	*Curry and Odin, p. 663
33	40.7 ± 1.7	L	20N to 19N	*Curry and Odin, p. 663
34	40.8 ± 2.3	L	20N to 19N	*Curry and Odin, p. 679
35	43.7 ± 1.5	H	19N to 18R	Ghosh, 1972; Hardenbol and Berggren, 1978
36	41.6 ± 1.5	H	18N	Ghosh, 1972; Hardenbol and Berggren, 1978
37	39.5 ± 0.5	H	Between 15N and 13N	Ghosh, 1972; Hardenbol and Berggren, 1978
39	39.1 (?)	H	17N to 16N	McKee, 1975
40	38.0 (?)	H	17N to 16N	McKee, 1975
42	38.9 ± 1.8	L	18N/17R	*Curry and Odin, p. 659
43	39.6 ± 1.8	L	18N/17R	*Curry and Odin, p. 661
44	37.5 ± 0.7	L	13N/R	*Curry and Odin, p. 696
45	36.4 ± 0.7	L	13N/R	*Curry and Odin, p. 696
46	33.9 ± 1.5	H	13N/R	*Baubron and Cavelier, p. 893
47	34.6 (?)	H	Top 13N	Evernden and others, 1964; Prothero and others, 1982
48	32.4 (?)	H	12N	Evernden and others, 1964; Prothero and others, 1982
49	31.3 ± 0.3	H	11N to 10N	*Hartung and others, p. 896
50	32.6 ± 0.3	H	13N	*Hartung and others, p. 896
53	26.2 ± 0.5	L	7N to 6CN	*Kreuzer and Gramann, p. 794
54	23.6 ± 0.2	L	Above 7N	*Kreuzer, p. 793
55	36.1 (?)	H	Mid-13N	Emry, 1973; Prothero and others, 1982
56	37.7 (?)	H	Top 15N	Emry, 1973; Prothero and others, 1982
57	37.4 (?)	H	Above 15N	Emry, 1973; Prothero and others, 1982
59	22.0 ± 1.1	L	6CR to 6BN	*Odin, p. 699
60	18.6 ± 0.7	L	5EN to 5DN	*Odin, p. 702
61	22.6 ± 1.2	L	6BN to 6A	*Odin, p. 703
62	7.24 (?)	H	Top 4N	McDougall and others, 1977
63	5.93 (?)	H	Below 3AN	McDougall and others, 1977
64	5.40 (?)	H	Just below 3AN	McDougall and others, 1977
65	5.07 (?)	H	Lower 3N	McDougall and others, 1977
66	3.99 (?)	H	Upper 3N	McDougall and others, 1977
68	3.55 (?)	H	Just below 2AN	McDougall and others, 1977
69	2.99 (?)	H	Mid-2A	McDougall and others, 1977
70	2.48 (?)	H	Just above 2A	McDougall and others, 1977
71	22.0 ± 0.3	H	6BN	Abele and Page, 1974; Shibata and others, 1984
72	20.4 ± 0.3	H	Mid-6A	Adams, 1975; Shibata and others, 1984
73	17.2 ± 0.4	H	Just below 5C	Bandy and others, 1970; Shibata and others, 1984
74	16.4 ± 0.9	H	Just below 5C	Shibata, 1973
77	16.1 ± 0.4	H	Between 5C and 5B	Shibata and others, 1981
78	14.7 ± 0.4	H	Just above 5B	Kawai and Hirooka, 1966; Shibata and others, 1964
80	14.3 ± 2.1	H	Just above 5B	Shibata and Nozawa, 1967
81	11.6 ± 0.4	H	Below 5AN	Shibata and others, 1979

High-temperature dates are K/Ar or Rb/Sr dates on tuffs, bentonites, lavas, or other high temperature minerals. Low-temperature dates are K/Ar dates on glauconites. See individual references for stratigraphic and analytical details. All dates have been converted to new decay constants (if not already converted).

APPENDIX B.—RADIOMETRIC DATES USED TO OBTAIN A BEST-FIT NUMERICAL TIME SCALE FOR THE JURASSIC-CRETACEOUS (FIG. 6).

Number	Date (in MA)	High-/Low-Temperature Date	Stratigraphic Level	References (In Odin, 1982a)
T13	216.0 ± 2.0	H	Late Norian—early Rhaetian	Priem, Abstract 137, p. 814
J1	206.0 ± 12.0	H	Early Lias	Webb, Abs. 202, p. 876
J2	204.0 ± 5.0	H	Sinemurian—early Pliensb.	Armstrong, Abs. 181, p. 857
J3	202.0 ± 12.0	H	Between early and mid-Lias.	Webb, Abs. 203, p. 877
J4	169.0 ± 10.0	H	Bajocian	Odin and Obradovich, Abs. 102, p. 766
J5	156.0 ± 5.0	H	Bajocian/Bathonian transition	Baubron and Odin, Abs. 214, p. 891
J6	154.0 ± 5.0	H	Bajocian/Bathonian transition	Baubron and Odin, Abs. 214, p. 891
J7	163.0 ± 3.0	H	Bathonian	Kreuzer, Abs. 149, p. 831
J8	155.0 ± 3.0	H	Bathonian	Kreuzer, Abs. 149, p. 831
J9	156.0 ± 3.0	H	Bathonian	Kreuzer, Abs. 149, p. 831
J10	148.0 ± 3.0	L	Oxfordian	Odin and McDowell, Abs. 141, p. 818
J11	140.0 ± 3.0	L	Basal Kimmerid.	Odin and McDowell, Abs. 142, p. 819
J12	136.0 ± 3.0	L	Basal Kimmerid.	Odin and McDowell, Abs. 142, p. 819
J13	134.0 ± 4.0	L	Portlandian	Kennedy and Odin, Abs. 99, p. 764
J14	133.0 ± 4.2	L	Mid-Portland.	Kennedy and Odin, Abs. 76, p. 737
C16	119.0 ± 2.0	L	Latest Valang.	Kreuzer, Abs. 148, p. 830
C17	116.9 ± 2.6	L	Mid-Hauterivian	Odin and Conard, Abs. 74, p. 734
C18	113.7 ± 3.3	L	Mid-Hauterivian	Odin and Conard, Abs. 74, p. 734
C19	110.7 ± 3.6	L	Late Barremian	Conard and Odin, Abs. 73, p. 734
C20	109.3 ± 6.0	H	Late Aptian	Montigny and others, Abs. 188, p. 860
C21	108.0 ± 2.6	L	Late Aptian	Kreuzer and Thiermann, Abs. 146, p. 827
C22	107.3 ± 3.9	L	Late Aptian	Curry and Odin, Abs. 71, p. 732
C26	112.0 ± 3.3	L	Latest Aptian	Kennedy and others, Abs. 98, p. 763
C27	108.6 ± 3.3	L	Early Albian	Kennedy and others, Abs. 70, p. 731
C28	100.5 ± 3.3	L	Early Albian	Elewaut and others, Abs. 78, p. 740
C29	108.3 ± 4.0	L	Early Albian	Elewaut and others, Abs. 78, p. 740
C30	100.5 ± 1.0	L	Early Albian	Kreuzer and others, Abs. 144, p. 821
C31	101.1 ± 3.7	L	Mid-Albian	Elewaut and others, Abs. 79, p. 741
C32	102.4 ± 2.3	L	Mid-Albian	Elewaut and others, Abs. 79, p. 741
C33	96.5 ± 2.7	L	Late Albian	Kreuzer and Thiermann, Abs. 145, p. 826
C34	98.1 ± 3.9	L	Late Albian	Kennedy and Odin, Abs. 65, p. 727
C35	98.6 ± 2.6	L	Late Albian	Kennedy and Odin, Abs. 65, p. 727
C36	97.6 ± 2.0	H	Late Albian	Odin and Obradovich, Abs. 111, p. 774
C37	97.5 ± 2.0	H	Late Albian	Odin and Obradovich, Abs. 111, p. 774
C38	98.4 ± 3.2	L	Latest Albian	Kennedy and Odin, Abs. 63, p. 725
C39	99.5 ± 3.1	L	Latest Albian	Kennedy and Odin, Abs. 63, p. 725
C40	99.7 ± 1.2	H	Late Albian	Obradovich, Abs. 157, p. 838
C41	94.6 ± 0.8	L	Early Cenoman.	Kreuzer and others, Abs. 211, p. 887
C42	94.7 ± 1.1	L	Early Cenoman.	Kennedy and Odin, Abs. 62, p. 722
C43	94.8 ± 3.1	L	Early Cenoman.	Elewaut and others, Abs. 80, p. 742
C44	90.4 ± 2.0	L	Mid-Cenomanian	Elewaut and others, Abs. 81, p. 743
C45	92.9 ± 2.1	L	Mid-Cenomanian	Elewaut and others, Abs. 81, p. 743
C46	91.2 ± 0.5	L	Early Turonian	Kreuzer and others, Abs. 226, p. 903
C47	91.5 ± 1.8	H	Early Turonian	Lanphere, Abs. 118, p. 780
C48	90.5 ± 0.4	L	Early Turonian	Kreuzer and others, Abs. 226, p. 903
C49	88.1 ± 1.5	L	Late Turonian	Kreuzer and Seibertz, Abs. 227, p. 906
C50	85.0 ± 1.0	L	Late Turonian	Kreuzer and Rabitz, Abs. 94, p. 758
C50A	86.0 ± 1.0	L	Late Turonian	Kreuzer and Rabitz, Abs. 94, p. 758
C51	89.0 ± 1.5	H	Turonian/Coniacian boundary	Odin and Obradovich, Abs. 108, p. 772
C52	87.1 ± 2.8	L	Coniacian	Elewaut and others, Abs. 83, p. 745
C53	84.4 ± 1.6	H	Earliest Santonian	Odin and Obradovich, Abs. 107, p. 771
C54	79.3 ± 1.6	H	Earliest Campanian	Odin and Obradovich, Abs. 106, p. 771
C55	77.6 ± 2.5	L	Early Campanian	Kennedy and Odin, Abs. 140, p. 817
C56	74.9 ± 2.3	L	Late Campanian	Kennedy and Odin, Abs. 140, p. 817
C57	74.3 ± 1.4	H	Latest Campanian	Odin and Obradovich, Abs. 105, p. 769
C59	71.0 ± 2.1	L	Late Maastrichtian	Kennedy and Odin, Abs. 139, p. 815
C50	70.1 ± 1.4	H	Late Maastrichtian	Odin and Obradovich, Abs. 104, p. 768
C60	69.0 ± 1.4	H	Late Maastrichtian	Odin and Obradovich, Abs. 104, p. 768

High-temperature dates are Rb/Sr or K/Ar dates on minerals formed at high temperatures, such as those in tuffs, ashes, bentonites, lava flows, biotite, etc. Low-temperature dates are most commonly K/Ar dates on glauconites. See individual references for stratigraphic and analytical details on the dates. All dates were converted to new decay constants in Odin (1982a).

APPENDIX C: CENOZOIC SEQUENCES

Cenozoic sequences were studied in outcrops on the Isle of Wight and in southern England, Belgium, the Paris Basin and southern France, West Germany, Italy, Spain, along the Gulf Coast of the United States, in the Carolinas, New Zealand, and in Australia. Sections studied and the ages of the sequences include the following.

On the Isle of Wight, a relatively complete Upper Paleocene through Lower Oligocene succession of sequences was studied at the White Cliff Bay and Alum Bay sections. In addition, in southern England sections at Herne Bay (type area of the Thanetian Stage), Barton (stratotype of the Bartonian Stage), and Sheppey Island (Ypresian) were studied.

In Belgium, sections in the type area of the Rupelian Stage included those at Sint Niklaas, Kruibeke, and Steendorp. Other outcrops included sections at Pellenberg (Tongrian) and Tongeren (stratotype of the Tongrian Stage); in the type area of the Ypresian Stage, the Kortemark, Egem, and Aalbeke sections; the Balegem section (type Ledian); the

Wemmel section (Wemmelian Stage stratotype); the Ciply section in the Mons Basin (Montian Stage stratotype); in the type area of the Bruxellian Stage, the Zetrud-Lumay and Braine l'Alleud sections; the Gelinden section (Thanetian); and the Vroenhoven section along the Albert Canal (K/T boundary and Danian).

In the Paris Basin, France, outcrops were studied at St. Leu d'Esserent and St. Vaast-les-Mello (Lutetian stratotype). Lutetian sections were also studied at Guitrancourt and Damery. Cuisian sections included those at Cuise-la-Motte (stratotype of the Cuisian) and Gisors. Other outcrops included sections at Auvers-sur-Oise (stratotype of the Auversian), Marine (type Marinesian), Pourcy (Sparnaciana), Cormeilles (Ludian, Sannoisian, and Stampian), and Chalons-sur-Vesle (Thanetian).

In southern France, outcrops in the type area of the Aquitanian Stage included Moulin de Bernachon, Moulin de l'Eglise, L'Ariey (stratotypes of the Aquitanian), and Relevee de Balizac sections. Other outcrops included sections at Grignols (Upper Aquitanian-Lower Burdigalian), Champ du Peloua, Ruisseau de la Coquilleyre, and Pont Pourquey, La Sime (type area of the Burdigalian Stage).

Other European outcrops included the section at Doberg bei Bünde in West Germany (Chattian stratotype); in northern Spain the type Ilerdian section at Tremp, Lerida; in Italy the stratotype of the Priabonian at Priabona. In addition, most literature sources containing measured sections and litho- and chronostratigraphic information on the stratotypes and reference sections of the Late Neogene stages in Italy, i.e., Langhian, Serravallian, Tortonian, Messinian, Zanclean, and Calabrian, were also perused extensively, and sequence-stratigraphic models were proposed on the basis of litho- and biofacies information (field work in these areas is planned for 1987).

Along the United States East Coast, outcrops in North Carolina included the Belgrade quarry near Belgrade (Lower Miocene); Trent River section (Oligocene); New Bern quarry near New Bern (Upper Eocene-Oligocene); and Ideal (Martin Marietta) quarry in New Hannover County (Middle Eocene). Outcrops in South Carolina included quarries in Georgetown and Berkeley counties (Middle Eocene) and the Santee Portland quarry (Upper Eocene).

Along the United States Gulf Coast, outcrops in Alabama included Little Stave Creek (Middle Eocene-Oligocene); St. Stephens quarry (uppermost Eocene through lowest Miocene); sections along the Alabama, Tombigbee, and Chattahoochie rivers (Paleogene); and the Braggs section (K/T boundary).

On South Island, New Zealand, coastal sections were studied from South Canterbury to North Otago (Eocene through Middle Miocene), at North Canterbury (Upper Cretaceous to Oligocene), and from Greymouth to Karamea (Eocene through Middle Miocene) along the west coast.

In Australia, outcrops were studied along the flanks of the Otway Mountains. Along the eastern flanks of the Otways, the Bells Headland, Airey's Inlet, and Soapy Rocks sections (uppermost Eocene through lowest Miocene) were examined. The Browns Creek section (Upper Eocene-lowest Oligocene) at the southwest end of the Otways, and sections along the western flanks of the Otways from Pebble Point (Middle Paleocene through Oligocene) to Port Campbell (Lower Miocene-Pliocene) were also studied.

APPENDIX D: CRETACEOUS SEQUENCES

Cretaceous sequences were studied in the Suisse Romande, Switzerland, in southeastern and northern France, in Belgium and the Netherlands, in the Western Interior of the United States (Colorado, Utah), and in central Texas. Sections studied and their ages include the following.

In the Suisse Romande, Switzerland, the outcrops studied were the section at Châtel-St. Dennis on the right bank of the Veveyse River (Berriasian through Lower Barremian); the section at Montsalvens (Berriasian through Santonian); the section at Valangin (stratotype of the Valanginian Stage); the Cressier and Landeron sections near Hauterive (type area of the Hauterivian Stage); the Borne Valley-Plateau d'Andey sections (Valanginian through Lower Barremian); and the Val de Fier section (Berriasian-Valanginian).

In Ardèche, southern France, the outcrops were the Villeneuve de Berg section (Valanginian-Hauterivian) and the nearby Ibie Valley section (Lower Barremian); the Berrias section (Berriasian Stage stratotype); and the Les Buissieres section near Bessas (Hauterivian). In Vaucluse, southern France, the section at Apt (stratotype of the Aptian Stage) and the Gargas section (Gargasian Stage stratotype) were studied. In Alpes-de-Haute-Provence, southern France, the sections included those near Angles (Valanginian through Barremian, stratotype of Barremian Stage); the La Clue de Vergon section near Castellane (Hauterivian); and the La Rochette section (Turonian). In the Hautes-Alpes, the Bruis section (Upper Aptian-Albian), the Villefranche-le-Château section (Albian) in Drôme, and the La Roudoule section (Albian) in the Alpes Maritime. Along the coast of Bouche-du-Rhone in southern France, the section at Cassis near La Bédoulle (Bedoullian Stage stratotype) was also studied.

In northern France, sequence studies included the type area of the Cenomanian Stage and other Cenomanian-Lower Turonian sections in the Le Mans (Sarthe) area: those at Le Cormier (near Cormes), La Pigalière (St. Ulphace), Les Quatre Chemins (Le Luart), La Gare (Dollon), Les Fosses Blanches (Duneau), Le Moulin Ars (St. Calais), Le Bourgneuf (Mulsanne), Les Acacias (Yvré-l'Evêque), Longueville (Savigné-l'Evêque), Le Chateau (Ballon), Le Sablon and La Goupillerie (near Mézières-sous-Ballon), and at Mercey (Bonnétable).

In the Le Havre area of northern France, sections at Dollemard, Bléville, Octeville, St. Jouin, Le Tilleul, and Fécamp (Upper Albian through Turonian) were studied. Near Touraine, the type area of the Turonian Stage, sections at Amboise (Lower Turonian), Montrichard (mid-Turonian), and Francueil (Upper Turonian and Turonian/Coniacian boundary) were studied along with that at La Bousinière (Upper Cenomanian).

In the Boulonnais area of northern France, outcrops included sections at Wissant (Upper Aptian-Albian) and at Cap Blanc-Nez (Cenomanian-Lower Turonian).

In the Aube, northern France, Albian outcrops were studied at Perchois-Ouest and Perchois-Est in the type area of the Albian Stage, and at Montiéramey, Courcelles, Le Jard, Radonvilliers, and Montreuil-sur-Barse. Other outcrops in northern France included sections at Saintes (stratotype of the Santonian Stage); Cognac (stratotype of Coniacian Stage); and in the type area of the Campanian, sections at Aubeterre (Campanian-Maastrichtian, stratotype of Campanian), Belvès (Santonian-Campanian), and at Bigaroque (Campanian).

In Belgium outcrops included sections at Maisières (Turonian-Coniacian); Obourg (Campanian); Hallembaye (Campanian-Maastrichtian); and Ciply (Maastrichtian). In addition, the stratotype of the Maastrichtian Stage was studied at the Maastricht quarry (ENCI) in the Netherlands.

In the Western Interior of the United States, outcrops in Colorado included the Rock Canyon Anticline near Pueblo (Upper Albian-Santonian); sections near Canon City, Wolcott, Glenwood Springs, New Castle, and Grand Junction (Aptian through Santonian); and sections near Coalville, Utah (Albian through Turonian). In central Texas numerous outcrop sections near the towns of Austin, Temple, Waco, and Waxahachie were studied for mid-Albian through Maastrichtian sequences.

APPENDIX E: JURASSIC SEQUENCES

Jurassic sequences were studied along the Dorset and Somerset coasts of southern England, along the Yorkshire coast of northern England, in west-central France, in southern Germany, and in Switzerland. Sections and the ages of the sequences studied include the following.

In southern England, an almost complete succession of Jurassic sequences can be seen along the Dorset coast. Sections from Pinhay Bay to Eastcliff (Hettangian through Toarcian), at Burton Bradstock and Tidmoor Point (Aalenian through Callovian), and from the Isle of Portland to Swanage (Oxfordian through Purbeckian) were studied. In northern England along the Yorkshire coast, sections were studied between Robin Hood Bay and Staithes (Lower Jurassic), at Blea Wyke Point (Middle Jurassic), and at Cayton Bay (Oxfordian).

In France near the areas of Thouars and St. Jacques de Thouars, the stratotype of the Toarcian Stage was studied at the Rigollier-Vrines sections.

Upper Jurassic (Oxfordian-Tithonian) sequences were studied in the Montsalvens area of Switzerland. Samples were also studied for sequence analysis from the stratotype of the Pliensbachian Stage (from a section near Pliensbach, southern Germany).

APPENDIX F: TRIASSIC SEQUENCES

Triassic sequences were studied in the Dolomites in Italy; in Tyrol, Austria; in the Salt Range, Pakistan; and on the Arctic islands of Svalbard and Bjørnøya. Sections and the ages of the sequences studied include the following.

In the western Dolomites, Italy, the outcrops were the Rosengarten section (Ladinian-Carnian); San Lucano Valley sections (Anisian-Ladinian); Plattaforma del Sella sections (Ladinian-Carnian); and Alpe di Siusi section (Ladinian).

Elsewhere, the Steinplatte sections (Rhaetian) in Tyrol, Austria; the sections in Chiddru Nala, Narmia, and Nammal Gorge (Scythian) in the Salt Range, Pakistan; and the sections in the Festningen area of Isfjorden (Griesbachian through lowest Rhaetian) in Spitsbergen were studied. Triassic sections on Bjørnøya (Bear Island) are exposed on the Miseryfjellet (Sycthian through Carnian).

EUSTATIC CONTROLS ON CLASTIC DEPOSITION I—CONCEPTUAL FRAMEWORK

H. W. POSAMENTIER[1]
Exxon Production Research Company, P.O. Box 2189, Houston, Texas 77252-2189;

M. T. JERVEY,
Canadian Hunter Exploration Ltd., 700, 435 4th Avenue SW, Calgary, Alberta T2P 3A8

AND

P. R. VAIL[2]
Exxon Production Research Company, P.O. Box 2189, Houston, Texas 77252-2189

ABSTRACT: A conceptual framework for understanding the effects of eustatic control on depositional stratal patterns is presented.

Eustatic changes result in a succession of systems tracts that combine to form sequences deposited between eustatic-fall inflection points. Two types of sequences have been recognized: (1) a type 1 sequence, which is bounded at the base by a type 1 unconformity and at the top by either a type 1 or type 2 unconformity and has lowstand deposits at its base, and (2) a type 2 sequence, which is bounded at the base by a type 2 unconformity and at the top by either a type 1 or type 2 unconformity and has no lowstand deposits. Each sequence is composed of three systems tracts; the type 1 sequence is composed of lowstand, transgressive-, and highstand systems tracts, and the type 2 sequence is composed of shelf-margin, transgressive-, and highstand systems tracts. The type 1 sequence is associated with stream rejuvenation and incision at its base, whereas the type 2 sequence is not.

Eustacy and subsidence combine to make the space available for sediment to fill. The results of this changing accommodation are the onlapping and offlapping depositional stratal patterns observed on basin margins. Locally, conditions of subsidence and/or uplift and sediment supply may overprint but usually will not mask the effects of global sea level. Any eustatic variation, however, (e.g., irregular eustatic rise or fall, asymmetric fall, slow or rapid rise or fall, and so on) will be globally effective. The significance of eustatic fall-and-rise inflection points is considered with regard to the occurrence of unconformities and condensed sections, respectively. Type 1 unconformities are related to rapid eustatic falls, and type 2 unconformities are related to slow eustatic falls.

INTRODUCTION

The objective of this paper is to provide a conceptual framework for understanding the relationship between relative sea-level changes and clastic depositional stratal patterns on basin margins. An awareness of this relationship will provide the geologist with a tool to understand better the relationship between depositional sequences and the distribution of lithofacies within these sequences. Key terms used in this paper are defined in Table 1. See Van Wagoner and others (this volume) for additional definitions.

This study builds on previous studies by Vail and others (1977); Vail and Todd (1981); Vail and others (1984); and Jervey (this volume), which established that a relationship exists between relative sea-level change and depositional stratal patterns. This paper analyzes the mechanics of this stratigraphic relationship in the context of sequence, systems tract, and types of unconformity discussed. The paper by Posamentier and Vail (this volume) addresses the sequence and systems-tract models as well as variations on the model.

Conceptual models presented in this report represent an analysis of the effects of accommodation change on clastic sediment deposition and suggest how sedimentary basins will fill. It must be emphasized that the models are *generally* applicable. The effects of local factors such as climate, sediment supply, and tectonics must be incorporated into the models before these models can be applied to a particular basin. Once these considerations have been taken into account, the refined models can then be used predictively to simulate local conditions in order to predict lithologic succession better.

---

[1]Present addresses: Esso Resources Canada Ltd., 237 4th Avenue SW, Calgary, Alberta T2P OH6 (HWP).
[2]Department of Geology, Rice University, Houston, Texas 77251 (PRV).

ASSUMPTIONS

Geologic and seismic observations of accommodation and deposition show that a predictable succession of as many as four systems tracts can be generated by varying global sea level. These are highstand, lowstand (including lowstand fan and lowstand wedge), transgressive-, and shelf-margin systems tracts, and are shown in block diagrams in Figures 1–6. Each systems tract is composed of one or more depositional systems (Brown and Fisher, 1977) and each is characterized by a set of lithofacies. The timing of unconformites or surfaces of nondeposition bounding these systems tracts can also be predicted from the sea-level curve.

Locally, however, the models should first be refined by the incorporation of subsidence and sediment supply information before predictions are made. Each systems tract will be considered in greater detail by Posamentier and Vail (this volume).

In order to develop generally applicable depositional models, it was assumed that the following conditions would be present:

—The rate of seafloor subsidence an any single location on a profile was held constant. Seafloor subsidence is primarily a function of lithospheric cooling and sediment loading (together they compose total subsidence). Geohistory analyses from a variety of sedimentary basins suggest that eustatic variations occur with greater frequency than subsidence variations. Thus, over a limited interval, the assumption of constant subsidence rate seems acceptable. Nonetheless, when the general model is modified to account for local conditions, nonuniform subsidence can be accommodated.
—Total subsidence increases in a basinward direction. This seems to characterize most divergent basin margins.
—Deposition was occurring along a divergent continental margin characterized by a shelf, slope, and basin, where

TABLE 1.—DEFINITION OF KEY TERMS

Sequence Stratigraphy:
The study of rock relationships within a chronostratigraphic framework wherein the succession of rocks is cyclic and is composed of genetically related stratal units (sequences and systems tracts).

Depositional System:
A three-dimensional assemblage of lithofacies, genetically linked by active (modern) or inferred (ancient) processes and environments (delta, river, barrier island, and so on) (Brown and Fisher, 1977).

Systems Tract:
A linkage of contemporaneous depositional systems (Brown and Fisher, 1977). Each is defined objectively by stratal geometries at bounding surfaces, position within the sequence, and internal parasequence stacking patterns. Each is interpreted to be associated with a specific segment of the eustatic curve (i.e., eustatic lowstand—lowstand wedge; eustatic rise—transgressive; rapid eustatic fall—lowstand fan, and so on), although not defined on the basis of this association.

Sequence:
A relatively conformable succession of genetically related strata bounded at its top and base by unconformities and their correlative conformities (Vail and others, 1977). It is composed of a succession of systems tracts and is interpreted to be deposited between eustatic-fall inflection points.

Parasequence:
A relatively comformable succession of genetically related beds or bedsets bounded by marine-flooding surfaces and their correlative surfaces (Van Wagoner, 1985).

Unconformity:
A surface separating younger from older strata, along which there is evidence of subaerial erosional truncation (and, in some areas, correlative submarine erosion) or subaerial exposure, with a significant hiatus indicated.

Condensed Section:
A thin marine stratigraphic interval characterized by very slow depositional rates <1–10 mm/1000 yr (Vail and others, 1984). It consists of hemipelagic and pelagic sediments, starved of terrigenous materals, deposited on the middle to outer shelf, slope, and basin floor during a period of maximum relative sea-level rise and maximum transgression of the shoreline (Loutit, and others, this volume).

Accommodation:
The space made available for potential sediment accumulation (Jervey, this volume).

Equilibrium Point:
The point along a depositional profile where the rate of eustatic change equals the rate of subsidence/uplift. It separates zones of rising and falling relative sea level.

Equilibrium Profile:
The longitudinal profile of a graded stream or of one whose smooth gradient at every point is just sufficient to enable the stream to transport the load of sediment made available to it. It is generally regarded as a smooth, parabolic curve, gently concave to the sky, practically flat at the mouth and steepening toward the source (Gary and others, 1974).

sediment supply remains constant. In the real world, differing rates of sediment supply affect primarily the seaward extent of deposition. In the landward direction, stratal patterns will usually show onlap and aggradation as the space between sea floor and base level is filled, whereas the basinward limit of progradation is a function of sediment supply and basin margin geometry. Thus, in the landward direction, where base level (sea level or graded-stream profile) is the controlling factor, the stratal patterns on the landward side of identical basins will be the same regardless of the sediment supply. When local sediment supply parameters are incorporated into the model, the effect will be observed primarily at the seaward limit of deposition.

—The trend of eustatic change is curvilinear, approaching sinusoidal. Although the actual trend of eustatic change is clearly not sinusoidal, the eustatic curve may nonetheless be resolved by a series of sine curves. It will be shown that depositional stratal patterns are directly related to inflection points on this curve.

Again, it must be emphasized that, although these models are generally applicable, the overprint of local factors must be considered in order to utilize them in a predictive mode for a particular basin. By incorporating local factors, one can readily refine the models to simulate local conditions.

RELATIVE SEA-LEVEL CHANGE AS A FUNCTION OF EUSTASY AND SUBSIDENCE AND ITS EFFECT ON DEPOSITIONAL STRATAL PATTERNS

Stratal patterns and facies distribution depend in part upon (1) the amount of space available for the sediment and (2) the rate of change of new space added. Sediment is deposited in the space between the sea floor and base level (sea level in a marine environment or graded-stream profile in a nonmarine fluvial environment). The space available is referred to as *accommodation,* defined by Jervey (this volume) as "the space made available for potential sediment accumulation, [which] is a function of both sea-level fluctuation and subsidence." In this paper the terms *accommodation* and *new space added* are used. *Accommodation* refers to all the space available for sediment to fill, including old space (leftover space not filled during an earlier time) plus new space added, whereas *new space added* refers only to space contemporaneously being made available. Accommodation may vary as both the upper and lower boundaries of this space move up or down. The result is an *accommodation envelope,* which defines the space available for sediment to fill (Fig. 7).

Relative sea-level change, rather than eustatic change only, controls accommodation change. Eustacy (global sea level) refers only to the position of the sea surface *with reference to a fixed datum,* such as the center of the earth, and is therefore independent of local factors. *Relative sea level* incorporates local subsidence and/or uplift by referring to the position of the sea surface *with respect to the position of a datum (e.g., basement) at or near the sea floor* (Fig. 8). Thus, *relative* sea-level change observed along a profile varies with local subsidence or uplift. A relative rise or fall of sea level determines whether or not new space is being made available for sediment to fill. A relative rise adds space, whereas a relative fall takes space away. Consequently, even during a eustatic stillstand or slow eustatic fall, relative sea level may continue to rise and add new space as a result of local subsidence. This parameter is independent of sediment accumulating above the datum plane and should not be confused with water depth.

Water depth involves the integration of a third parameter—sediment supply—with eustacy and tectonics. Water depth may be described as relative sea level less accumulated sediment, as illustrated in Figure 8. Thus, relative sea level may continue to rise and add new space to accommodate sediment, whereas water depth may nonetheless simultaneously decrease if sediment is accumulating faster than relative sea level is rising.

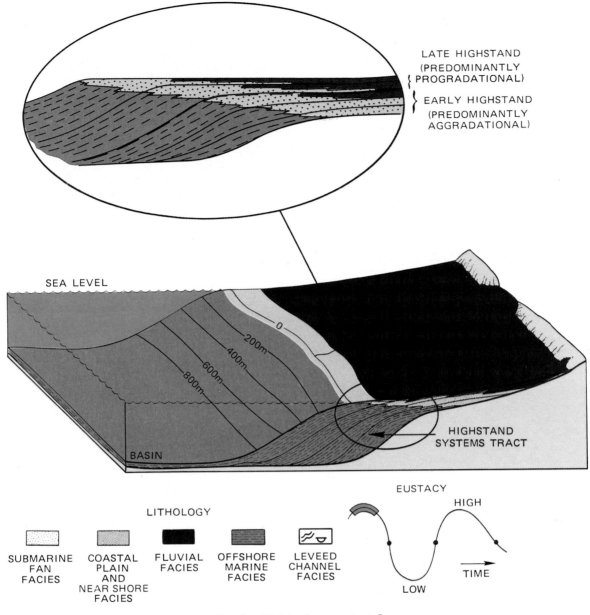

FIG. 1.—Highstand systems tract, I.

Significance of the inflection point.—

Eustatic change is a curvilinear function punctuated by inflection points. These are points on the curve where absolute slope or rate of change is greatest. Figure 9 illustrates a hypothetical sea-level curve with two inflection points. The one on the falling limb will be referred to as the $\underline{F}$ inflection point, and the one on the rising limb will be referred to as the $\underline{R}$ inflection point.

Basin-margin depositional stratal patterns depend in large part on eustacy and seafloor subsidence. Sedimentation on the shelf involves filling the wedge-shaped space between the sea surface and the seaward-dipping sea floor. The stratal pattern which results will depend on the rate at which space has been added and how sediment responds to this

addition of space. If the sediment supply is sufficient to allow continued aggradation to base level, then, as the rate of addition of new shelf space slows, the rate of aggradation will gradually decrease. As a result of the decreased rate of aggradation, progressively less sediment will be required to keep up with slower-rising base level and, consequently, progressively more sediment will be available for progradation.

Figure 10 illustrates how the rate of accommodation change (i.e., dA/dt or rate of new space added) varies with eustacy. At $\underline{F}$ inflection points, the rate at which new shelf space is added is least; at $\underline{R}$ inflection points, the rate is greatest (Fig. 10). Little or no new shelf space is being added at $\underline{F}$ inflection points, so relatively little new sediment can be accommodated there (assuming that sediment

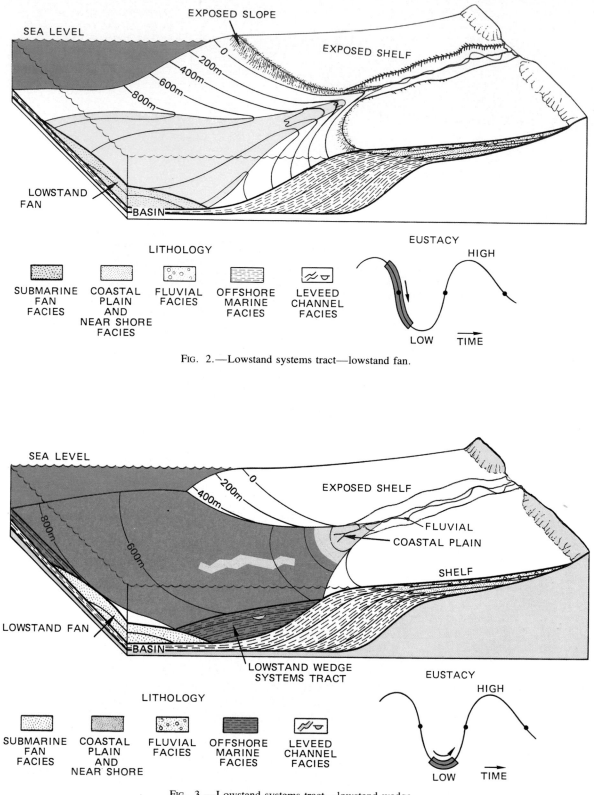

FIG. 2.—Lowstand systems tract—lowstand fan.

FIG. 3.—Lowstand systems tract—lowstand wedge.

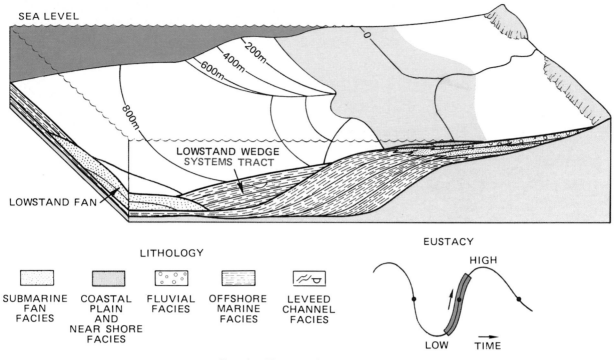

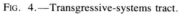

LITHOLOGY

SUBMARINE FAN FACIES COASTAL PLAIN AND NEAR SHORE FACIES FLUVIAL FACIES OFFSHORE MARINE FACIES LEVEED CHANNEL FACIES

EUSTACY

FIG. 4.—Transgressive-systems tract.

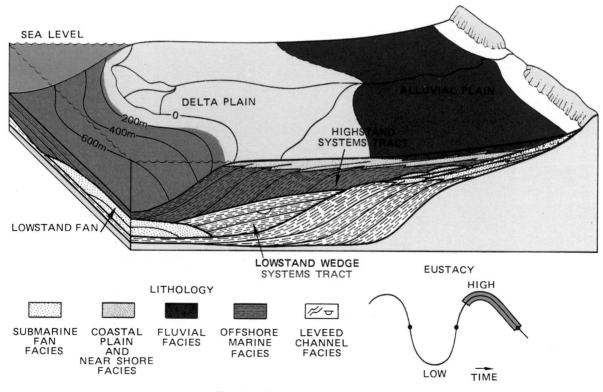

LITHOLOGY

SUBMARINE FAN FACIES COASTAL PLAIN AND NEAR SHORE FACIES FLUVIAL FACIES OFFSHORE MARINE FACIES LEVEED CHANNEL FACIES

EUSTACY

FIG. 5.—Highstand systems tract, II.

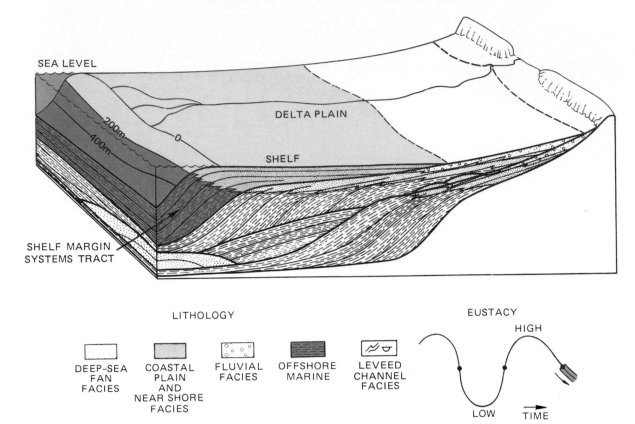

LITHOLOGY   EUSTACY

DEEP-SEA   COASTAL   FLUVIAL   OFFSHORE   LEVEED
FAN   PLAIN   FACIES   MARINE   CHANNEL
FACIES   AND   FACIES
            NEAR SHORE
            FACIES

Fig. 6.—Shelf-margin systems tract.

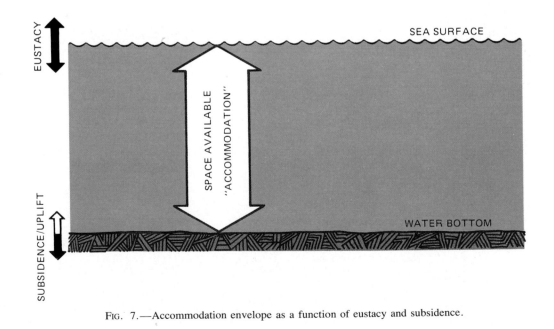

Fig. 7.—Accommodation envelope as a function of eustacy and subsidence.

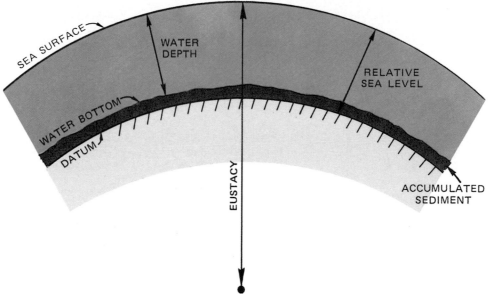

FIG. 8.—Eustacy, relative sea level, and water depth as a function of sea surface, water bottom, and datum position.

builds to base level). Hence, at $\underline{F}$ inflection points, the rate of aggradation will be least and that of progradation will be greatest. At $\underline{R}$ inflection points, the opposite occurs. The thinnest topset beds (per unit time) occur at the $\underline{F}$ inflection point (T6, Fig. 11) and, conversely, the thickest topset beds (per unit time) occur at $\underline{R}$ inflection points. Thus, with constant sediment supply, rates of aggradation and progradation are inversely related. As a result, within successive parasequences, shoreline regression tends to be progressively more rapid approaching the $\underline{F}$ inflection point and gradually less rapid thereafter (Fig. 11). Maximal rate of addition of new space at $\underline{R}$ inflection points commonly re-

sults in transgression and the development of starved or condensed sections.

The maximum landward encroachment of the condensed action or maximum flooding usually occurs sometime after the $\underline{R}$ inflection point during the eustatic rise (Fig. 12). Note that topset (i.e., shelf) beds approach maximum thickness (per unit time) at the eustatic-rise inflection point, and at the same time, approach their minimum areal extent. As these layers onlap progressively farther landward, the position of the basinward pinchout of each layer also migrates landward until T9 is reached. This marks the *time of maximum flooding* (TMF). After time T9, the seaward limit of each time slice migrates progressively basinward as regression resumes.

A basinward shift of coastal onlap characterizes $\underline{F}$ inflection points. *Coastal onlap* may be defined as the landward limit on the shelf or upper slope of sediment distribution—marine or nonmarine. It has been observed that initiation of fluvial erosion resulting in globally synchronous subaerial unconformities is associated with these basinward shifts and apparently is controlled by sea-level change (Vail and others, 1977). This will be considered in Posamentier and Vail (this volume).

One-dimensional model.—

At any point on a continental margin, the rate at which new space is made available for sediment to fill is determined by the rate of relative sea-level change and is equal to the rate of eustatic change minus the rate of subsidence (Fig. 10). For example, if global sea level is falling at a certain rate and the sea floor is subsiding at the same rate, relative sea level remains unchanged and no new space is being made available. If global sea level is falling, but more slowly than the sea floor is subsiding, the net effect will

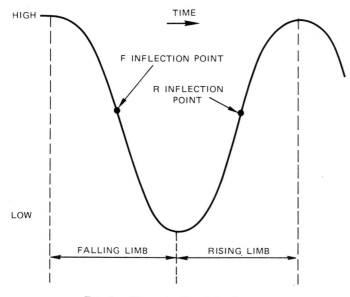

FIG. 9.—Elements of eustatic change.

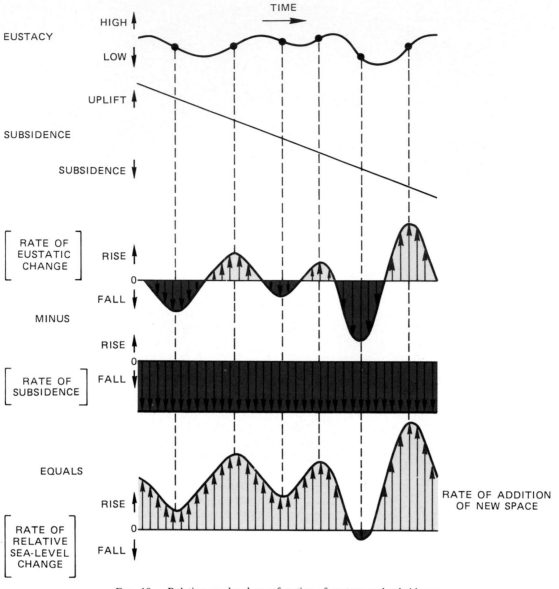

FIG. 10.—Relative sea level as a function of eustacy and subsidence.

be relative sea level rise, and consequently, new space will be added.

In the example shown in Figure 10, most of the interval is characterized by a relative rise of sea level since the rate of subsidence exceeds the rate of eustatic change most of the time. *Thus, new space is being added throughout most of the interval.* The actual accommodation at any given time is the sum of the new space added plus the left-over unfilled space and occurs between base level and the sea floor.

Two-dimensional model.—

On passive continental margins, subsidence gradually increases from shelf to basin, resulting in a basinward increase in the rate of addition of new space. The effect of this differential subsidence is illustrated in Figure 13, where the rate of addition of new space is shown for outer, middle, and inner shelf-edge positions. Note that greatest ac-

commodation occurs on the outer shelf where subsidence is greatest. Conversely, on the inner shelf, where subsidence is least, there are intervals when no new shelf space is being added.

A continental shelf profile may be subdivided into two zones separated by the *equilibrium point.* This point is defined as the point along a profile where the rate of eustatic change equals the rate of subsidence. Seaward of this point, the rate of subsidence is greater than the rate of eustatic fall, resulting in the addition of new space, whereas landward the opposite occurs. Alternatively, the equilibrium point defines two zones: (1) a zone of relative sea-level rise that is seaward of the equilibrium point and, (2) a zone of relative sea-level fall that is landward of the equilibrium point. Figure 14 illustrates the seaward migration of the equilibrium point in response to an increasing rate of eustatic fall. The addition of new space over the whole shelf profile is

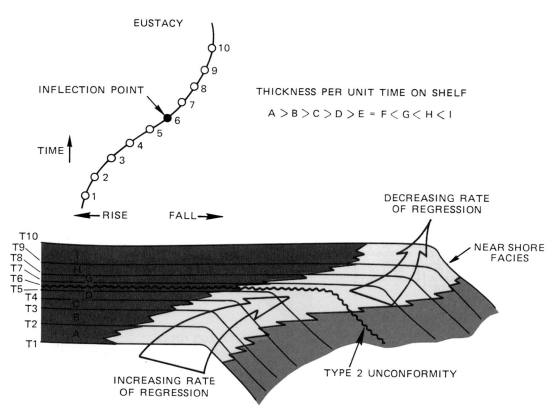

FIG. 11.—Response of topset bed thickness to eustatic fall.

least when the rate of eustatic fall is greatest (at the F inflection point). At this time (T4), the equilibrium point has reached its maximum seaward position. Conversely, the equilibrium point reaches its maximum landward position at the R inflection point.

Figure 15 illustrates the effect of different subsidence rates on the zones where new space is being added. Longitudinal profiles are shown for two basins differing only in their subsidence rates. Given the same eustatic change, basin A,

with a high subsidence rate, is characterized by a greater addition of new shelf space than basin B, where subsidence rate is low. All else being equal, relatively greater aggradation will occur in basin A than in basin B, whereas the latter basin will be characterized by a higher rate of progradation.

The response of sedimentation to an interval of slow eustatic fall is shown in Figure 16. A type 2 unconformity is illustrated in this figure. From T1 to T4, the rate of eustatic

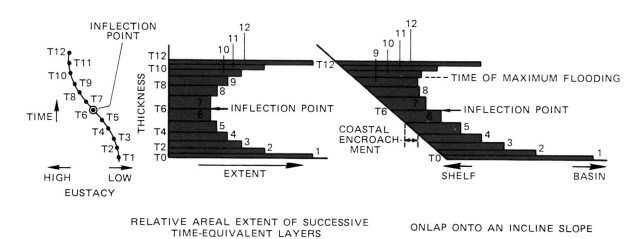

FIG. 12.—Response of topset bed thickness and timing of maximum flooding to eustatic rise.

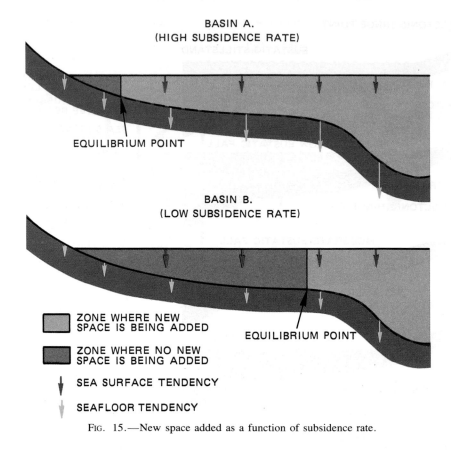

Fig. 15.—New space added as a function of subsidence rate.

ceases, the point of coastal onlap abruptly shifts basinward, occurring again at the bayline (see Posamentier and Vail, this volume). At this time, deposition in the absence of the fluvial component is once more restricted to the wedge-shaped space between the sea floor and the sea surface (T6 to T8, Fig. 16). It should be noted again that a type 2 unconformity is shown here. With a rapid eustatic fall generating a type 1 unconformity, cessation of widespread fluvial deposition is marked by stream rejuvenation and incision. Subsequently, when relative sea-level rise resumes, these incised valleys (which may be quite extensive) fill with fluvial and/or estuarine deposits. Thus, incised-valley fill will overlie the type 1 unconformity in places.

ELEMENTS OF THE COASTAL ONLAP CURVE

The general shape of the coastal onlap curve on the global-cycle chart is based on a set of observations from many sedimentary basins (Vail and others, 1977). It represents the maximum landward limit of terrigenous deposition and comprises either nonmarine or marine sediment. Its specific shape is inferred from models of accommodation and deposition. Figure 18 illustrates a hypothetical coastal onlap curve with two type 2 unconformities. From the $\underline{F}$ inflection point to the $\underline{R}$ inflection point, the rate of addition of new space steadily increases (Fig. 10). As a result, the rate of landward migration of the bayline and thus, coastal onlap, increases. At $\underline{R}$ inflection point time, the rates of relative sea-level rise at the bayline and the rate of landward

bayline migration are greatest. Thereafter, until the equilibrium point reaches the bayline, the rate of relative sea-level rise decreases at the bayline. In addition, the rate of landward migration of the bayline decreases as well. Because the point of coastal onlap is at the bayline during this time, the rate of landward shift of coastal onlap thus also appears to decrease. When the equilibrium point reaches the bayline, the rate of relative sea-level rise there is zero. From this time until $\underline{F}$ inflection point time, the equilibrium point and the bayline move basinward together. The migration rate of the equilibrium point, and hence the bayline, decreases as the equilibrium point gradually reaches its most basinward position.

Concomitant fluvial deposition results in a continued landward shift of coastal onlap until $\underline{F}$ inflection point time. The rate of this landward shift of coastal onlap gradually decreases in response to the decelerating landward shift of stream equilibrium profiles as the $\underline{F}$ inflection point is reached. At this time, the equilibrium point and the bayline change migration direction and once again move landward, resulting in the cessation of widespread fluvial deposition. This, in turn, results in an apparent abrupt basinward shift of coastal onlap back to the bayline. Again, it should be noted that a type 2 unconformity is described here. Type 1 unconformities are often characterized by extensive incised-valley fluvial deposits overlying the unconformity (Posamentier and Vail, this volume).

The chronostratigraphic distribution of the condensed section is also shown in Figure 18. The maximum landward

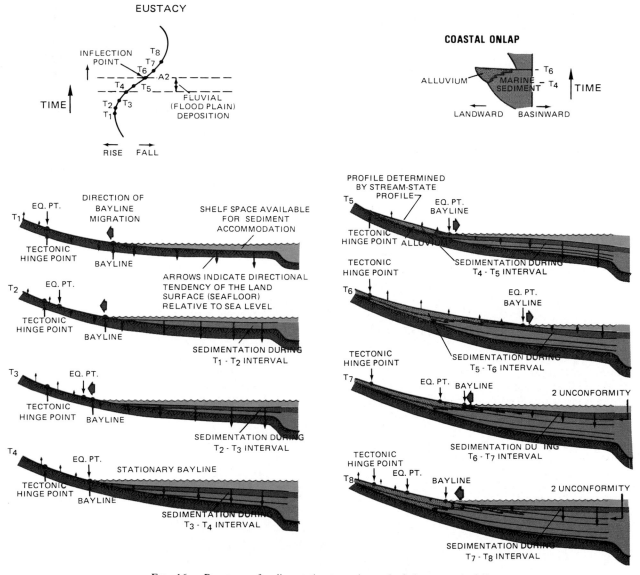

Fig. 16.—Response of sedimentation to an interval of slow eustatic fall.

extent of the condensed section is shown as a dashed line, because the seaward limit of terrigenous deposition may vary significantly as sediment supply varies from basin to basin. The maximum seaward limit of significant terrigenous sediment distribution occurs at the $\underline{F}$ inflection point. At this time, the rate of new shelf space added is at a minimum. The time of maximum regression, however, occurs somewhat later when, with continued increase in rate of new shelf space added, progradation finally gives way to retrogradation. The time represented by the condensed section decreases in a landward direction, with the condensed section reaching its maximum landward position at the time of maximum flooding sometime after the $\underline{R}$ inflection point (see Fig. 12). The surface corresponding to the time of maximum flooding is called the *downlap surface* (DLS) (Vail and others, 1984) or *maximum flooding surface* (MFS). In general, downlap surface is used where seismic data are involved and the downlapping toes of clinoforms can be observed. Maximum flooding surface is more commonly used when only outcrop or well-log data are involved.

STRATAL PATTERNS (PARASEQUENCE SCALE) ASSOCIATED WITH VARYING RATES OF EUSTATIC RISE OR FALL

This discussion addresses perturbations, or "bumps," on the eustatic curve, rather than true higher frequency cycles superimposed upon a lower frequency eustatic curve. The model predicts that all eustatic cycles, regardless of frequency, will result in the deposition of sequences composed of a predictable succession of systems tracts. (Sequence type, i.e., type 1 or 2, will be a function of local subsidence rate.) Unconformities associated with such high-frequency eustatic cycles superimposed on general eustatic rises and falls may correspond to fourth or fifth order (in the sense

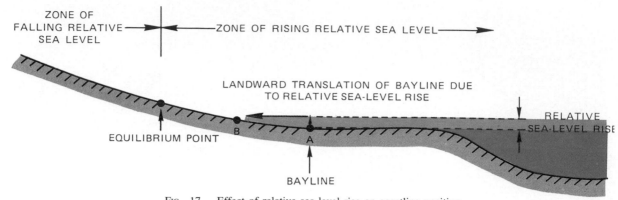

FIG. 17.—Effect of relative sea-level rise on coastline position.

of Vail and others, 1977). Higher frequency eustatic cycles may be associated with parasequence sets or parsequences (Van Wagoner, 1985). Simple perturbations on the eustatic curve, however, where there is no change in eustatic tendency, (i.e., fall to rise or vice versa) generate a different stratal response.

When perturbations in overall eustatic rises or falls occur, additional inflection points are generated. If the rate of eustatic fall changes from decreasing to again increasing following an $\underline{F}$ inflection point on the falling limb of a eustatic curve, another inflection point is generated (Fig. 19). This will be referred to as an F' inflection point. Similarly, on the rising limb of a eustatic curve, another inflection point is generated if the rate of eustatic rise changes from decreasing to again increasing after an $\underline{R}$ inflection point.

This will be referred to as an R' inflection point. R' inflection points are similar to $\underline{F}$ inflection points since both are associated with times of farthest basinward position of the equilibrium point. Similarly, $\underline{F}'$ inflection points are associated with times of farthest landward position of the equilibrium point and are therefore similar to $\underline{R}$ inflection points (Fig. 19).

The effects of perturbations on a eustatic rise may be observed primarily at the seaward limit of terrigenous deposition. Subaerial unconformities, with concomitant basinward shifts of coastal onlap, typical of type 1 or type 2 unconformities, do not occur in association with either $\underline{R}$ or $\underline{R}'$ inflection points. Rather, an uneven eustatic rise is characterized by recurrent condensed sections corresponding to successive intervals of maximum flooding associated

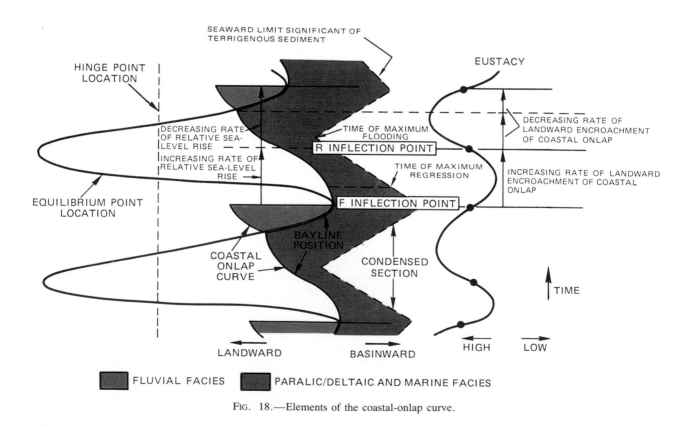

FIG. 18.—Elements of the coastal-onlap curve.

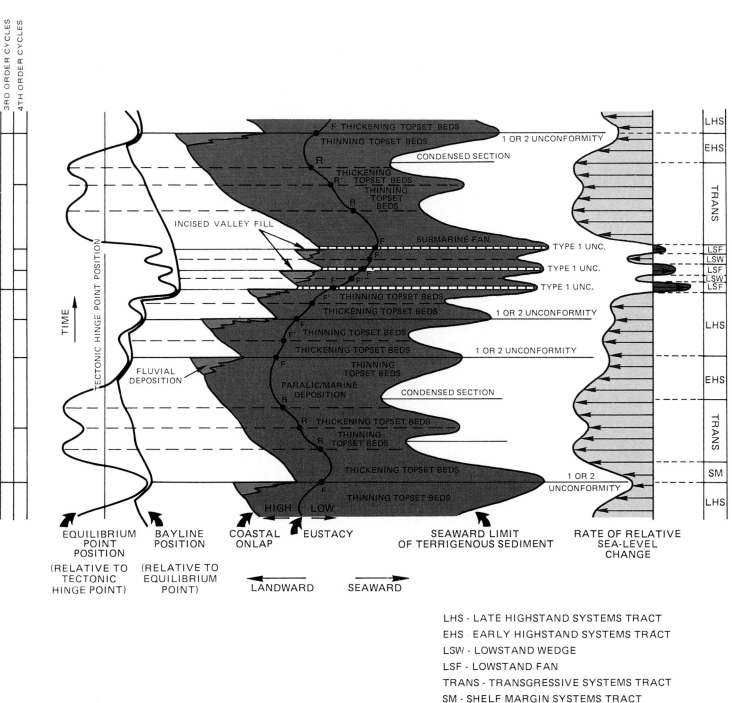

FIG. 19.—Effect of varying rates of eustatic rise and fall.

LHS - LATE HIGHSTAND SYSTEMS TRACT
EHS - EARLY HIGHSTAND SYSTEMS TRACT
LSW - LOWSTAND WEDGE
LSF - LOWSTAND FAN
TRANS - TRANSGRESSIVE SYSTEMS TRACT
SM - SHELF MARGIN SYSTEMS TRACT

with each R inflection point. Figure 19 shows two R inflection points and two condensed sections on each of the general eustatic rises shown. Each condensed section is associated with an R inflection point when the rate of addition of new space is greatest. Note that the second condensed section extends farther landward in each case, even though the rate of relative sea-level rise is the same. (This diagram is drawn so that the slope at each R inflection point is the

same.) During the overall eustatic rise, the equilibrium point remains landward of the tectonic hinge point, never reaching the bayline (Fig. 19). This assumes that the bayline is located seaward of the hinge point. Consequently, the bayline lies in the zone of relative sea-level rise and migrates landward throughout this interval. No fluvial aggradation will occur at this time and, therefore, no subaerial unconformity or basinward shift of coastal onlap characterizes

R′ inflection points. Rather, these points are characterized only by an increased basinward encroachment of terrigenous deposition.

Perturbations on a sea-level fall usually generate the more familiar pattern of a succession of type 1 (or 2) unconformities and condensed sections (Fig. 19). An uneven or stepped sea-level fall is characterized by a succession of F and F′ inflection points. Each F inflection point will generate a type 1 (or 2) unconformity, provided the equilibrium point reaches the bayline prior to that time so as to initiate fluvial deposition. F′ inflection points, similar to R inflection points, are associated with condensed sections. At these inflection points, the equilibrium point is at a maximum landward position, generating conditions analogous to those which generate condensed sections associated with R inflection points. The first basinward shift of coastal onlap occurs at the first F inflection point and separates the highstand systems tract into early and late sections.

The early section of the highstand systems tract is characterized by alternating transgressions and regressions corresponding to R and R′ inflection points, respectively. It is capped by fluvial deposits associated with the first F inflection point. The late section of the highstand systems tract is characterized by recurrent basinward shifts of coastal onlap with associated subaerial exposure surfaces that correspond to successive F inflection points (see Fig. 1). Because the rate of new space added generally decreases throughout the highstand systems tract, the early section of this systems tract is usually characterized more by aggradation than progradation, whereas the opposite usually applies to the late section of the systems tract. The maximum landward position of coastal onlap steps basinward with each successive F inflection point. Each F′ inflection point results in a condensed section. The maximum landward encroachment of the condensed section also steps basinward with each successive F′ inflection point. The dominant F inflection point during an overall eustatic fall is usually characterized by the most pronounced unconformity.

CONCLUSIONS

The observation that similar stratal patterns develop at the same time in widely varied sedimentary basins suggests a globally effective control such as eustatic change. The interaction of eustacy with local tectonics and sediment supply determines local depositional stratal patterns. The concepts discussed here and in Jervey (this volume) serve as the foundation or framework upon which the sequence and systems tract depositional models discussed in the paper by Posamentier and Vail (this volume) are based. Certain simplifying assumptions have been made for the purpose of presenting this model in a straightforward and coherent fashion, but it should be emphasized that these assumptions can and should be modified to conform to the conditions observed in specific basins before the models can be applied.

ACKNOWLEDGMENTS

The authors gratefully acknowledge the support and contributions from their many co-workers at Exxon Production Research Company who contributed to the development of these concepts. We especially thank J. C. Van Wagoner, J. F. Sarg, R. M. Mitchum, and R. A. Hoover for their helpful suggestions and constructive critiques of this manuscript at various stages of its evolution. In addition, we thank G. J. Moir, W. A. Burgis, R. D. Erskine, G. Mirkin, V. Kolla, G. R. Baum, C. G. St. C. Kendall, and T. R. Nardin for their helpful suggestions and comments. Ultimate responsibility for the material presented herein, however, rests with the authors.

REFERENCES

BROWN, L. F., JR., AND FISHER, W. L., 1977. Seismic stratigraphic interpretation of depositional systems: Examples from Brazilian rift and pull apart basins, in Clayton, C. E., ed., Seismic Stratigraphy-Applications to Hydrocarbon Exploration: American Association of Petroleum Geologists Memoir 26, p. 213–248.

GARY, M., McAFEE, R., JR., AND WOLF, C. L., 1974, Glossary of Geology: American Geological Institute, Washington, D.C., 805 p.

VAIL, P. R., HARDENBOL, J., AND TODD, R. G., 1984, Jurassic unconformities, chronostratigraphy and sea-level changes from seismic stratigraphy and biostratigraphy: American Association of Petroleum Geologists Memoir 36, p. 129–144.

VAIL, P. R., MITCHUM, R. M., JR., TODD, R. G., WIDMIER, J. M., THOMPSON, S., III, SANGREE, J. B., BUBB, J. N., AND HATELID, W. G., 1977, Seismic stratigraphy and global changes of sea level, in Clayton, C. E., ed., Seismic Stratigraphy-Applications to Hydrocarbon Exploration: American Association of Petroleum Geologists Memoir 26, p. 49–212.

VAIL, P. R., AND TODD, R. G., 1981, North Sea Jurassic unconformities, chronostratigraphy and sea-level changes from seismic stratigraphy: Petroleum Geology of the Continental Shelf of Northwest Europe, Proceedings, p. 216–235.

VAN WAGONER, J. C., 1985, Reservoir facies distribution as controlled by sea-level change: Abstracts with Programs, Society of Economic Paleontologists and Mineralogists Midyear Meeting, Golden, Colorado, p. 91–92.

EUSTATIC CONTROLS ON CLASTIC DEPOSITION II—SEQUENCE AND SYSTEMS TRACT MODELS

H. W. POSAMENTIER[1] AND P. R. VAIL[2]

Exxon Production Research Company, P.O. Box 2189, Houston, Texas 77252-2189

ABSTRACT: Depositional sequences are composed of genetically related sediments bounded by unconformities or their correlative conformities and are related to cycles of eustatic change. The bounding unconformities are inferred to be related to eustatic-fall inflection points. They are either type 1 or type 2 unconformities, depending on whether sea-level fall was rapid (i.e., rate of eustatic fall exceeded subsidence rate at the depositional shoreline break) or slow (i.e., rate of eustatic fall was less than subsidence rate at the depositional shoreline break). Each sequence is composed of a succession of systems tracts. Each systems tract is composed of a linkage of contemporaneous depositional systems. Four systems tracts are recognized: lowstand, transgressive, highstand, and shelf margin. The lowstand systems tract is divided into two parts: lowstand fan followed by lowstand wedge, where the basin margin is characterized by a discrete physiographic shelf edge, *or* lower followed by upper wedge, where the basin margin is characterized by a ramp physiography. Two sequence types are recognized: a type 1 sequence composed of lowstand, transgressive-, and highstand systems tracts, and a type 2 sequence composed of shelf margin, transgressive-, and highstand systems tracts.

Type 1 and type 2 unconformities are each characterized by a basinward shift of coastal onlap concomitant with a cessation of fluvial deposition. The style of subaerial erosion characterizing each unconformity is different. Type 1 unconformities are characterized by stream rejuvenation and incision, whereas type 2 unconformities typically are characterized by widespread erosion accompanying gradual denudation or degradation of the landscape. Stream rejuvenation and incision are not associated with this type of unconformity. On the slope and in the basin, type 1 unconformities typically are overlain by lowstand fan or lowstand wedge deposits, whereas type 2 unconformities are overlain by shelf margin systems tract deposits. Within incised valleys on the shelf, type 1 unconformities are overlain by either fluvial (lowstand wedge) or estuarine (transgressive) deposits. Type 2 unconformities typically are characterized by a change in parasequence stacking pattern from progradational to aggradational.

Timing of fluvial deposition is also a function of eustatic change insofar as global sea level is the ultimate base level to which streams will adjust. The elevations of stream equilibrium profiles are affected by eustatic change, and, assuming constant sediment supply, streams will aggrade or degrade in response to eustatically induced shifts in these profiles. Fluvial deposition occurs at different times in type 1 and type 2 sequences and is characterized by different geometries within each type of sequence. In type 1 sequences, fluvial deposits occur as linear, incised-valley fill during the time of lowstand wedge and transgressive deposition. Fluvial deposits also may occur during highstand deposition as more widespread floodplain deposits within the late highstand systems tract. Fluvial deposits in type 2 sequences are usually limited to widespread floodplain deposits occurring within the late highstand systems tract.

INTRODUCTION

The following discussion of sequence and systems tract models is based on the conceptual framework presented in Posamentier and others (this volume) and Jervey (this volume). The models presented here are *generally* applicable; before applying the models to specific basins, the models should be modified and adjusted to account for local factors, such as varying sediment supply, tectonics, and climate. Four systems tracts (Figs. 1–6) are identified and described: highstand, lowstand, shelf margin, and transgressive. Each systems tract is associated with a segment of the eustatic curve, although *the exact timing of any given systems tract in any given basin will depend on local subsidence and sediment supply*.

SEQUENCES AND SEQUENCE BOUNDARIES

A *sequence* is defined as a relatively conformable succession of genetically related strata bounded at its top and base by unconformities and their correlative conformities (Vail and others, 1977). It is composed of a succession of systems tracts and is interpreted to be deposited between eustatic-fall inflection points. Two sequence types are recognized: type 1 and type 2. The type 1 sequence is bounded at its base by a type 1 unconformity and at its top by either a type 1 or type 2 unconformity (see discussions of types of unconformities in Vail and Todd, 1981). This sequence is composed of lowstand, transgressive-, and highstand systems tracts. The type 2 sequence is bounded at its base by a type 2 unconformity and at its top by either a type 1 or type 2 unconformity. This sequence is composed of shelf margin, transgressive-, and highstand systems tracts.

The type 1 unconformity is characterized by stream rejuvenation and fluvial incision, sedimentary bypass of the shelf, and abrupt basinward shift of facies and coastal onlap. The type 2 unconformity is more subtle and is not characterized by stream rejuvenation. This unconformity is characterized by a basinward shift of coastal onlap and slow widespread subaerial erosion accompanying gradual denudation or degradation of the landscape. At its correlative conformity, the type 2 unconformity is characterized by a distinct parasequence stacking pattern: increasingly progradational within the underlying highstand systems tract, decreasingly progradational and then aggradational within the overlying shelf margin systems tract, becoming retrogradational within the subsequent transgressive systems (see Fig. 11 in Posamentier and others, this volume).

The occurrence of a type 1 or type 2 unconformity will depend on whether the rate of eustatic fall exceeds or is less than the rate of subsidence at the depositional shoreline break.[1] Thus, a eustatic fall may result in a type 1 unconformity, where low rates of subsidence characterize the ba-

---

[1]Present addresses: Esso Resources Canada Limited 237 4th Avenue SW Calgary, Alberta T2P OH6;

[2]Department of Geology and Geophysics Rice University P.O. Box 1892 Houston, Texas 77251

---

[1]The term *depositional shoreline break* as used here is defined as the concurrent physiographic break, landward of which the sea floor is at or near base level and seaward of which the sea floor is below base level. This break may occur at or seaward of the shoreline. Consequently, it refers to an active depositional physiographic break as opposed to a pre-existing or relict physiographic break at the edge of the continental shelf.

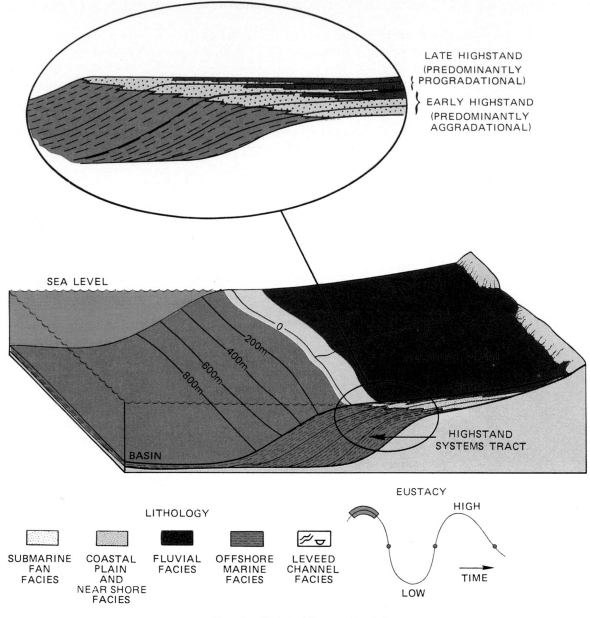

LITHOLOGY

FIG. 1.—Highstand Systems Tract, I.

HIGHSTAND SYSTEMS TRACT

sin margin, and a type 2 unconformity, where high rates of subsidence characterize the basin margin. This underlines the importance of integrating eustacy and local subsidence before applying the sequence-stratigraphic model.

HIGHSTAND SYSTEMS TRACT

The highstand systems tract (Figs. 1, 5) is characterized by an increasingly progradational parasequence stacking pattern and is interpreted to be deposited during the eustatic highstand, generally defined as the interval starting some time after the $\underline{R}$ inflection point and ending before or at the $\underline{F}$ inflection point. In response to gradual slowing of relative sea-level rise after the $\underline{R}$ inflection point (Fig. 7), shoreline transgression eventually gives way to regression,

resulting in the initiation of this systems tract. The actual timing of this event varies with varying sediment supply; however, it typically occurs sometime after the $\underline{R}$ inflection point but before the eustatic peak is reached. The base of this systems tract is associated with a downlap surface or condensed section (Loutit and others, this volume). The top of this systems tract is associated with widespread[2] fluvial

---

[2]In this study, fluvial deposition is described as "widespread" or "incised-valley fill." The distinction used here is that fluvial deposits of the former type are not associated with stream rejuvenation and are not restricted laterally by rejuvenated-valley walls, in contrast with deposits of the latter. Under certain circumstances, however, incised-valley fill deposits can also be quite extensive if stream rejuvenation is accompanied by extensive lateral erosion.

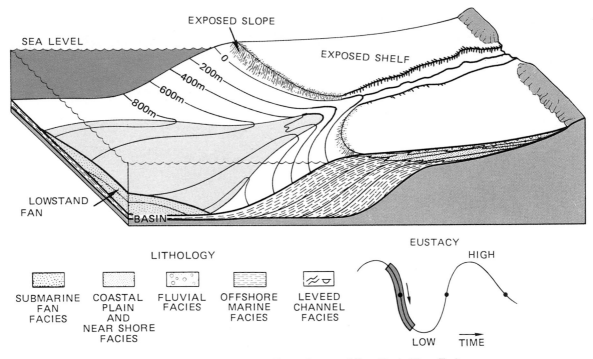

FIG. 2.—Lowstand Systems Tract—Lowstand Fan (Basin Floor Fan).

deposition, which begins some time after the eustatic peak is reached.

The upper boundary of the highstand systems tract is characterized by a type 1 or type 2 unconformity. In the type 1 case, sea level drops below the depositional shoreline break, causing subaerial shelf exposure and stream in-cision. In the type 2 case, sea level never drops below the depositional shoreline break, so that subaerial exposure of the shelf is limited and no stream incision occurs. The principal expression of the type 2 unconformity is an abrupt basinward shift of coastal onlap located landward of the depositional shoreline break because of cessation of fluvial

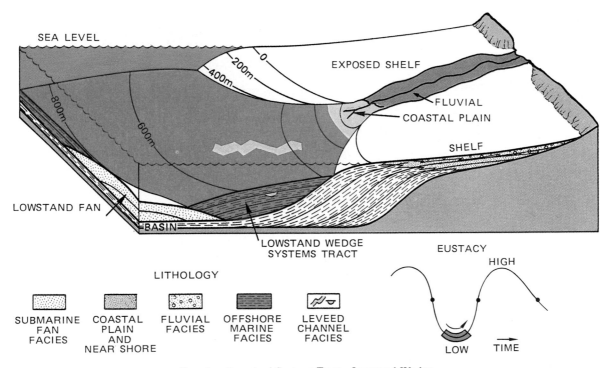

FIG. 3.—Lowstand Systems Tract—Lowstand Wedge.

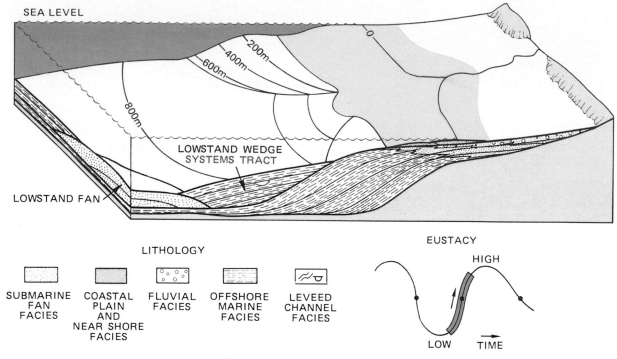

FIG. 4.—Transgressive-Systems Tract.

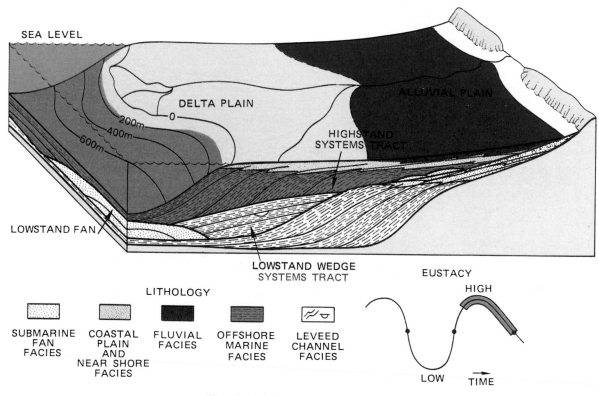

FIG. 5.—Highstand Systems Tract, II.

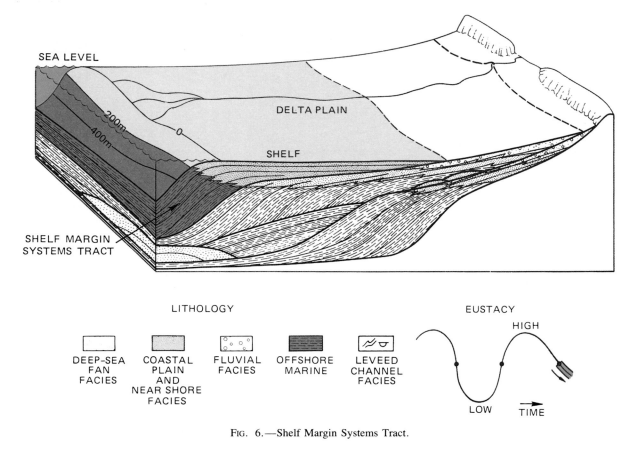

LITHOLOGY

DEEP-SEA FAN FACIES

COASTAL PLAIN AND NEAR SHORE FACIES

FLUVIAL FACIES

OFFSHORE MARINE

LEVEED CHANNEL FACIES

EUSTACY

HIGH

LOW   TIME

Fig. 6.—Shelf Margin Systems Tract.

deposition. Type 2 unconformities are also characterized by a change in parasequence stacking pattern, i.e., from highly progradational to progradational/aggradational.

Different styles of subaerial erosion characterizing type 1 and type 2 unconformities are suggested by the conceptual model. Erosional surfaces associated with type 1 unconformities are characterized by rejuvenated streams and subsequent stream incision as a function of the lowering of stream equilibrium profiles. Given enough time during the ensuing lowstand, the entire landscape could be lowered to this new equilibrium condition, although this is unlikely.

Erosional surfaces associated with type 2 unconformities are characterized by a general degradation of the landscape without significant stream incision. These unconformities are associated with cessation of fluvial deposition without concurrent stream rejuvenation. Rather, stream erosion is more gradual and is associated with a general wearing down of the landscape. This occurs as a result of the gradual reduction of stream equilibrium profile gradients because of the decrease of the grain size of the load as the landscape matures. Thus, slow widespread degradation characterizes type 2 unconformities, in contrast with more restricted stream incision which characterizes type 1 unconformities.

Fluvial depositional system with modern analogues.—

It is suggested that a basinward shift of coastal onlap, with its concomitant subaerial unconformity, occurs at the eustatic-fall inflection point and results from abrupt cessation of fluvial deposition at that time. Figure 8 illustrates this shift that occurs between T6 and T7. The key to explaining the cause of this coastal onlap shift lies in the understanding of the relationship between eustacy and fluvial deposition.

In its natural course of development, a stream will strive to achieve a slope of maximum efficiency, wherein the "slope is delicately adjusted to provide, with available discharge and the prevailing channel characteristics, just the velocity required for transportation of all the load supplied from above" (Mackin, 1948, p. 471). Such a stream is described as graded or in a state of dynamic equilibrium, i.e., there is a constant but balanced flow of material into and out of the system, resulting in no net erosion or deposition.

The actual slope of a graded stream is a function of the discharge or volume of water and the volume and texture of the sediment load (Rubey, 1952). Generally, the slope is inversely proportional to the discharge and directly proportional to the grain size. Thus, slopes gradually decrease in a downstream direction as discharge increases and grain size decreases. This decrease in slope has been described as exponential (Shulits, 1941; Strahler, 1952). The graded river generally has a smooth-curve longitudinal profile (henceforth referred to as the equilibrium profile) that seems to be fairly stable as it ages (Yatsu, 1955).

The elevation at each point on an equilibrium profile is

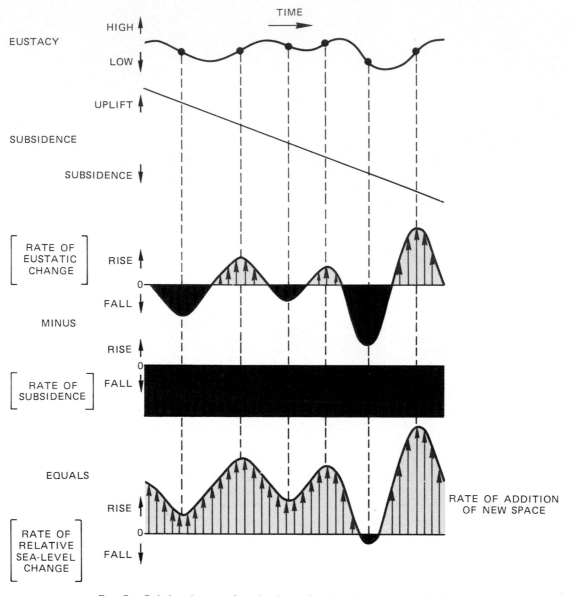

Fig. 7.—Relative changes of sea level as a function of eustacy and subsidence.

determined by base level position at the downstream end of the profile (Mackin, 1948; Rubey, 1952; Figs. 9 and 10), whereas the slope and the actual shape of the equilibrium profile are controlled by factors imposed from upstream. If the point to which the equilibrium profile is graded (i.e., the point where the stream reaches the sea) remains fixed in horizontal and vertical space, then there will be a very slow decrease of slopes along the profile as load-grain size gradually decreases. In the real world, equilibrium profiles are graded to ultimate base level (i.e., sea level). Thus, eustatic fluctuations have a profound effect on the fluvial regime. They cause base-level changes and, consequently, shifts of the points to which equilibrium profiles are graded. If this point is shifted horizontally basinward, the equilibrium profile is also shifted in this direction, resulting in floodplain aggradation (Fig. 11A). If this point is shifted vertically, then the profile is shifted vertically as well (Fig.

11B). An actual example of the case illustrated in Figure 11A is cited by Lane (1955, p.) and is depicted in Figure 12.

A railroad runs up the valley of the Missouri River in the United States and small tributaries of the river pass under the railroad in culverts. When the railroad was built at the point where one of these tributaries entered the river, the river flowed along the side of the valley near the railroad and the culvert was set to conform to the grade established by this situation. The Missouri River shifted to the other side of the valley and the distance from the culvert to base level in the river increased. This caused a raising of the grade of the tributary to such an extent that it was necessary to raise the grade of the railroad where it crossed the tributary.

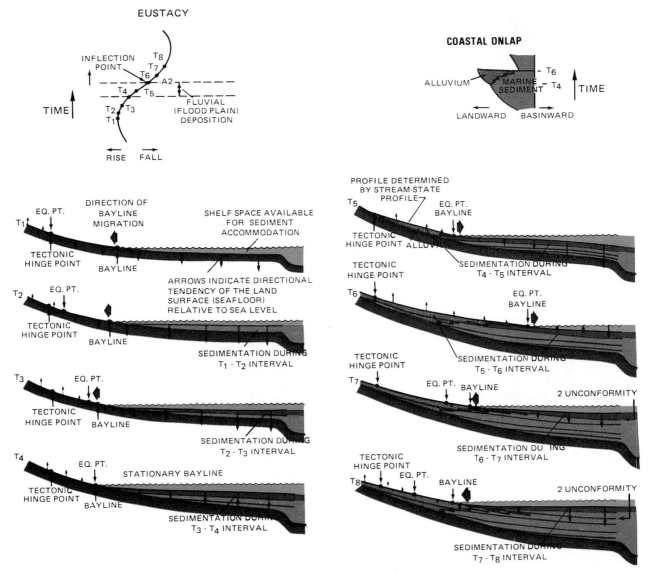

FIG. 8.—Response of sedimentation to an interval of slow eustatic fall.

This example is analogous to the shift of a stream mouth basinward because of a seaward shift of the bayline.[3]

A rapid eustatic rise and then a stillstand of relative sea level at the bayline can be compared to the filling of a man-made reservoir behind a dam. As the reservoir is filled with water, the rate of addition of new space exceeds the rate of sediment supply and "transgression" occurs (Fig. 13A). In effect, the lower reaches of the stream profile are being flooded. Assuming the stream was at grade before construction of the dam, the point to which the profile is graded shifts upstream along the profile from point 1 to point 5 in response to the landward shift of the bayline. During this interval, the equilibrium profile remains stationary and no

widespread fluvial deposition occurs. At T5, the reservoir has been filled with water, and from T5 to Tn the reservoir remains at a constant level. This would be analogous to a basin margin situation, wherein eustatic fall equaled subsidence at the bayline so that relative sea-level rise at the bayline was zero (i.e., when the equilibrium point reached the bayline). During this interval, the point to which the profile is graded shifts from position 5 to 12, resulting in a concomitant basinward shift of the equilibrium profile. This shift produced *subaerial accommodation* for fluvial sediment to fill, at least up to the first knickpoint.

There is substantial evidence that widespread alluviation occurs in response to basinward shifts of equilibrium profiles (Gilbert, 1917; Eakin and Brown, 1939; Lane, 1955; Chitchob and Cowley, 1973; Simons and Senturk, 1977; Leopold and Bull, 1979). The shift of the equilibrium profile is achieved by the upstream propagation of a wave of deposition (Kuiper, 1965; Mackin, 1948). Ultimately, the

---

[3]The *bayline* is defined as the demarcation line between fluvial and paralic/delta plain environments. See discussion in Posamentier and others (this volume).

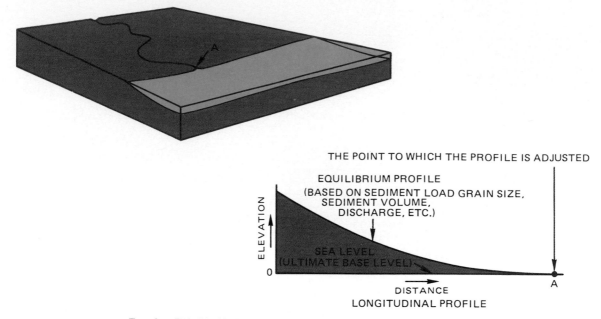

THE POINT TO WHICH THE PROFILE IS ADJUSTED

EQUILIBRIUM PROFILE
(BASED ON SEDIMENT LOAD GRAIN SIZE,
SEDIMENT VOLUME,
DISCHARGE, ETC.)

SEA LEVEL
(ULTIMATE BASE LEVEL)

ELEVATION

0

DISTANCE

A

LONGITUDINAL PROFILE

FIG. 9.—Relationship between base level and stream equilibrium profile.

readjusted profile will be roughly parallel with the original river bed. The portion of the stream affected by these changes is the lower part of the profile, which was originally at grade or equilibrium. As shown in Figure 14, aggradation clearly occurs upstream of Milburn Diversion Dam on the Middle Loup River, Nebraska (from Simons and Senturk, 1977). Some modern studies suggest that the upstream effects of a basinward shift of the point to which stream profiles are adjusted are limited to the upstream end of the backwater curve (Leopold and others, 1964). Nonetheless, Leopold and others (1964) point out that modern streams may have lacked sufficient time to reestablish equilibrium and achieve "ultimate" changes. If, indeed, streams accommodate such base-level changes by straightening and deepening their channels, then coastal plains developing behind prograding shorelines would always be characterized by relatively straight, deep fluvial channels. This, however, generally does not seem to be the case, lending support to the arguments presented here.

Figure 15 illustrates how subaerial accommodation may be filled. Assuming constant sediment supply, the signal that a new, higher equilibrium profile has been established is transmitted upstream from the downstream end of the profile. Thus, a wave of sedimentation gradually propagates in that direction. Initially, sedimentation rates are relatively high because the rate of new space added is high. Renewed sedimentation builds up the profile as the stream strives to reestablish equilibrium. In response to this rapid accommodation increase, rapid aggradation should occur and should be characterized by a vertical stacking pattern of fluvial deposits with relatively low lateral continuity. As the final equilibrium profile position is neared, the rate of new space added decreases and approaches zero. Vertical stacking grades into lateral migration, resulting in the de-

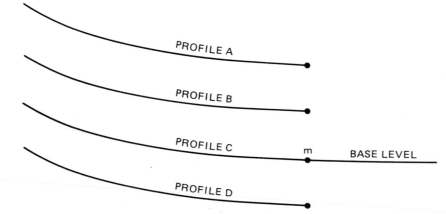

PROFILE A

PROFILE B

PROFILE C   m   BASE LEVEL

PROFILE D

FIG. 10.—Spatial dependence of equilibrium profile on base level; a group of possible profiles, all of which satisfy the conditions of grade, but only one, profile $\underline{C}$, that meets the actual base level at the point $\underline{m}$ (after Rubey, 1952).

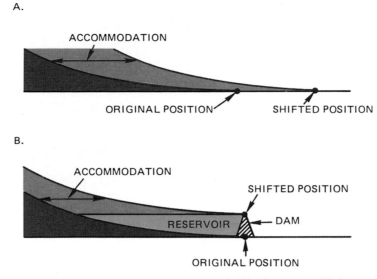

FIG. 11.—Subaerial accommodation as a result of shift of stream equilibrium profile.

velopment of sheet bedding geometry with relatively high lateral continuity at this time. This latter geometry is thus characterized by a low rate of aggradation as available accommodation is finally filled. Consequently, the model would suggest that, because of this stacking geometry, *correlation of latest highstand fluvial deposits may be less problematic because of their greater lateral continuity.*

Grain-size distribution and depositional mode (e.g., braided, meandering, and so on) within the fluvial depositional system is a complex function of many factors, such as varying sediment source, climate, vegetation, and tectonic activity, as well as accommodation. Locally, any one or a combination of these factors can be the dominant control. Here, only the effects of subaerial accommodation are considered. *Variations of accommodation as a function of sea-level change will be the only factor common to all subaerial settings at any given time.*

All else being constant, the gradual shift of the stream equilibrium profile during a late highstand should affect grain-size distribution in a predictable fashion. Intermediate positions of the stream profile are typically characterized by slope angles lower than the ultimate or "final" equilibrium profile (Fig. 15). In response to lower slope slope angles, stream velocity would decrease, resulting in decreased stream capacity and competence. Coarser sediment would be deposited preferentially at that time. When the rate of new space added decreases as the final equilibrium profile position is neared, sedimentation rate decreases as slopes gradually increase, and the stream regains the competency to carry the materials with which it is supplied from up-

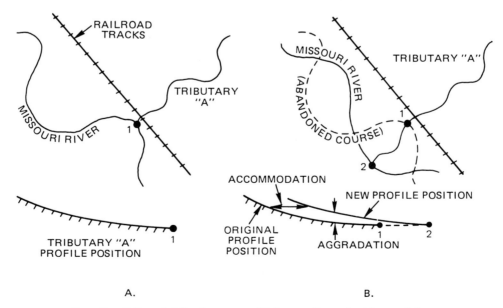

FIG. 12.—Effect of shift of stream equilibrium profile on fluvial aggradation.

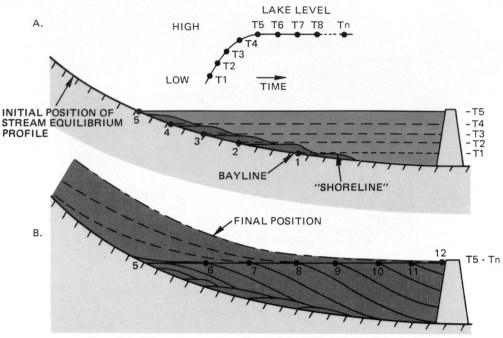

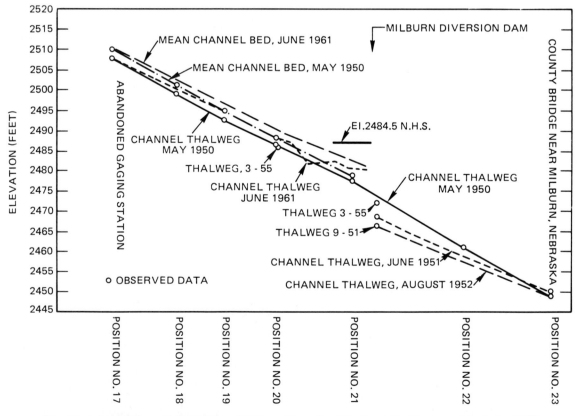

FIG. 13.—Effect of reservoir sedimentation on fluvial aggradation.

FIG. 14.—Aggradation and degradation at Milburn Dam site, Nebraska (after Simons and Senturk, 1977).

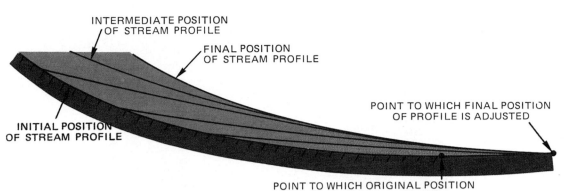

Fig. 15.—Fluvial aggradation due to shift of stream equilibrium profile.

stream. Thus, an overall upward-fining trend should be observed. It is important, however, to reemphasize that, locally, other factors such as tectonics may also exercise a major, if not a dominant, control on grain-size distribution and fluvial depositional mode.

Equilibrium profiles and eustatic change.—

The fundamental principle which governs sedimentation in fluvial environments (assuming constant sediment supply) is that a stream will aggrade if its equilibrium profile shifts basinward or upward and incise if its equilibrium profile shifts downward. In the same way that marine accommodation occurs by relative sea-level rise, subaerial accommodation occurs by basinward shifts of the stream equilibrium profile. The position in space of a stream equilibrium profile is determined by the position of the points to which the streams are graded. The way these points move in response to eustatic change provides the key to understanding deposition and erosion in the subaerial environment.

The point to which an equilibrium profile is adjusted lies at or near the regional bayline (except in the case of estuaries or embayments), as shown in Figure 9. Figure 16 illustrates a shelf margin experiencing beach progradation. At T1 (Fig. 16A), a slow eustatic rise coupled with subsidence results in a landward migration of the bayline. The equilibrium point lies somewhere landward of the hinge line at this time. In this example, sufficient sediment supply is brought in by longshore drift to allow the shoreline to prograde. (The sediment source is shown in Fig. 16A but not in Figs. 16B, C, and D.) Lagoonal sediment contemporaneously fills the new space being added between the bayline and the shoreline. As the bayline migrates landward, the points to which streams are graded gradually migrate landward along the equilibrium profiles of the streams. Consequently, the equilibrium profiles do not shift during this interval. Streams are at grade and no significant fluvial deposition occurs at this time.

The eustatic highstand occurs at T2 (Fig. 16B) and is characterized by a continuing landward shift of the bayline and regression of the shoreline. The points to which the streams are graded continue to shift landward, moving up the equilibrium profiles, as shown in Figure 13A. Again,

there are no shifts of the equilibrium profiles and no widespread fluvial deposition.

T3 (Fig. 16C) is characterized by a slow eustatic fall, resulting in a shift of the equilibrium point to a position basinward of the hinge line but still landward of the bayline. Consequently, the bayline still shifts landward at this time, and, as before, stream equilibrium profiles remain stationary.

Aggradation of lagoonal sediment continues within the new space being added between the bayline and the shoreline. As the addition of the new space slows, the beach tends to prograde more rapidly. Coastal onlap occurring at the bayline also continues its gradual landward shift at this time.

Finally, at time T4 (Fig. 16D), because of the increased rate of eustatic fall, the equilibrium point has reached the bayline and has shifted it basinward. In response to the seaward shift of the point to which the stream profile is graded, widespread alluvial deposition now begins. Coastal onlap continues shifting landward; however, with the addition of a fluvial component, onlap of fluvial rather than paralic sediment now occurs.

In a deltaic setting, the eustatic effects on fluvial deposition may be considered for two end members: (1) deltas that build beyond the regional bayline into open waters and are unconfined laterally (e.g., the Mississippi River), and (2) deltas that build into embayments, estuaries, or flooded valleys of tectonic origin whose lateral growth is constrained by valley walls (e.g., the Tigris-Euphrates River). In general, the model suggests that fluvial deposition associated with unconstrained deltas tends to occur after the equilibrium point reaches the regional bayline, whereas fluvial deposition associated with constrained deltas can occur well before this time. The latter situation generally describes the reservoir analogy discussed earlier. In addition, it is also analogous to the depositional setting of the proximal lowstand wedge of the lowstand systems tract.

Figure 17 illustrates a delta building into an open body of water. Deltaic deposition occurs when a stream encounters a standing body of water and flow velocity abruptly decreases. The mouth of the stream is gradually displaced basinward and eventually, because of the low gradient of the delta plain, channels become choked with sediment and flow is diverted, resulting in the formation of distinct lobes.

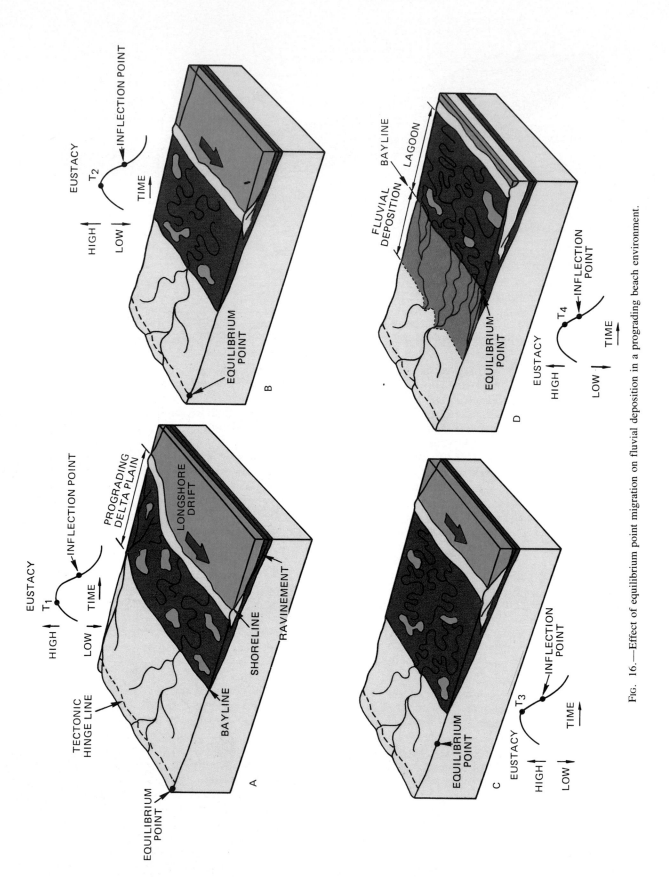

FIG. 16.—Effect of equilibrium point migration on fluvial deposition in a prograding beach environment.

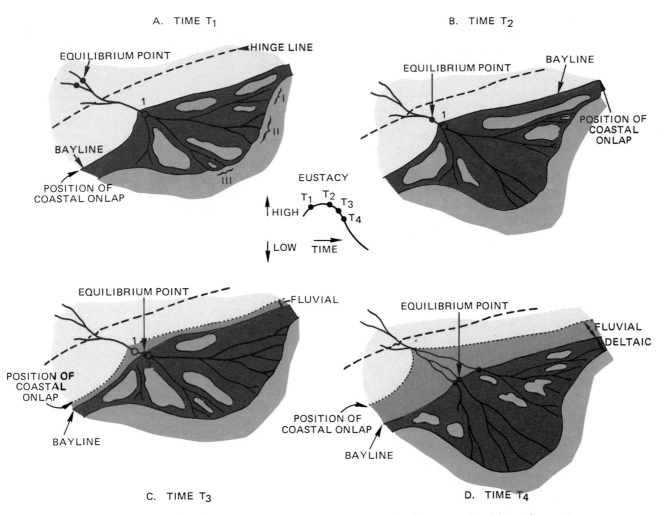

FIG. 17.—Effect of equilibrium point migration on fluvial deposition in a prograding delta environment.

Abandoned lobes are gradually submerged where subsidence, primarily due to compaction, exceeds eustatic fall. These lobes are sites of later deposition, when flow is once again diverted there. Coal of limited areal extent may form here at this time. At any given time, the stream mouth is at or near the delta front and is the point to which the stream profiles will be adjusted. Because of recurrent lobe abandonment, however, the position of this point varies considerably, moving slowly basinward and then shifting abruptly landward as lobe abandonment occurs. As a result of this recurrent lobe abandonment by streams seeking a shorter path to the sea, the net position of points to which stream profiles are adjusted tends to be near the head of the delta rather than at the delta front where the mouths of distributary streams are located.

From T1 to T2 (Figs. 17A, B), the distributary system, characterized by lobe II, is abandoned in favor of lobe III. The equilibrium point has shifted basinward but has not reached the bayline. Thus, the bayline continues to shift landward. A similar situation has been observed by Twidale (1976), who noted that, despite delta progradation, there is no overall advance of the land margin (bayline) of the Mis-

sissippi River system. The point to which the stream profile is adjusted is at or near location 1 near the head of the delta. This point is shifting landward but at a decreasing rate. In general, fluvial deposition does not occur landward of this point at this time, whereas coastal onlap is restricted primarily to deltaic/paralic sediment.

At some time between T2 and T3 (Figs. 17B, C), the equilibrium point reaches the bayline and location 1 and continues to migrate basinward. Although the process of lobe abandonment continues, the point to which the stream profile is adjusted has moved basinward from location 1 to location 2. This results in a basinward shift of the equilibrium profile and initiates fluvial deposition landward of this point. The strata at the point of onlap are no longer exclusively deltaic but are now primarily fluvial. The point of coastal onlap has now moved landward of the bayline. This situation persists through T4 (Fig. 17D). Throughout this interval (T2 through T4), fluvial coastal onlap shifts landward and the bayline and shoreline shift basinward. The zone of fluvial deposition is shown in this figure as being confined to a narrow belt because of display constraints. In the real world this belt would have significantly wider distribution since the tectonic hinge line is commonly much farther landward from the position shown here. Within the limits imposed by bathymetry, the rate of delta progradation gradually increases as the rate of addition of new shelf space decreases during this interval.

After the eustatic-fall inflection point, the equilibrium point and, consequently, the bayline reverse their migration direction and shift landward. The point to which the stream profile is adjusted also ceases shifting basinward and resumes its landward migration. Once the streams have adjusted to the basinward-most position of the profile, fluvial deposition ends. The model suggests that the stratal pattern of coastal onlap in the fluvial environment develops as shown in Figure 15. Completion of fluvial deposition in the upstream section of the graded profile at the end of a highstand probably does not occur instantaneously. Rather, it depends on the time lag between the $\underline{F}$ inflection point, when equilibrium profiles reach their maximum basinward position, and the time when streams complete aggradation to the new equilibrium profiles. The overall stratal pattern of this type 2 unconformity is shown in Figure 18.

A variation of this example would suggest that, with a sufficiently high rate of sediment supply, fluvial deposition can be initiated some time before the equilibrium point reaches the bayline. If there is a high rate of sediment supply, the points to which streams are graded may shift basinward despite lobe abandonment, thus triggering fluvial deposition; however, because of floodplain buildup in one stream system at a time when other stream systems in the area may be stable, stream piracy may eventually occur, possibly resulting in abandonment of the entire delta.

Figure 19 illustrates the case of a delta building into a narrow embayment of tectonic origin. Here, the delta of the Tigris-Euphrates system has built into an embayment where lateral shifting of delta lobes is severely constrained by valley walls. Consequently, as the delta advances, the river system is being extended basinward. The delta, and thus

the river system, has advanced over 300 km during the last 5,000 years (Twidale, 1976). In this way the points to which the stream profiles are adjusted shift basinward, resulting in a basinward shift of equilibrium profiles and concomitant fluvial deposition. Thus, fluvial deposition occurs despite the fact that the position of the equilibrium point is landward of the bayline the entire time.

A situation analogous to the Tigris-Euphrates example commonly occurs within lowstand wedge and early highstand depositional systems of type 1 sequences. Entrenched streams, established during rapid eustatic falls, become estuaries and embayments during subsequent relative sea-level rises. Constrained delta progradation will occur there and incised-valley fluvial deposition will occur until a rapid relative sea-level rise floods the valleys. The stratigraphic section filling an incised valley will consist of fluvial deposits deepening upward into estuarine and then marine deposits, which are possibly capped by a condensed section associated with the time of maximum flooding. If, by this time, the incised valley has not been filled, then slowed relative sea-level rise during the eustatic highstand will result in a shallowing-upward section grading from marine to deltaic to fluvial as constrained delta progradation leads to a basinward shift of the equilibrium profile and the filling of the incised valley.

The pattern of fluvial deposition within type 1 and type 2 sequences will differ. In type 1 sequences, incised-valley fills, composed of fluvial deposits with a linear or sinuous distribution pattern[4], are typical of the proximal lowstand wedge and early highstand systems tracts. In general, the conceptual model suggests that widespread unconfined fluvial deposition on broad floodplains will not occur until af-

---

[4]Some incised valleys can exceed 5 to 10 km in width and have a sediment distribution that may appear more sheetlike than linear and sinuous.

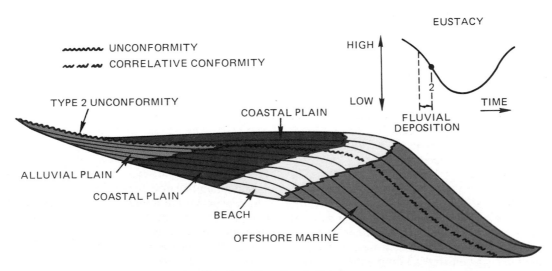

FIG. 18.—Type 2 unconformity.

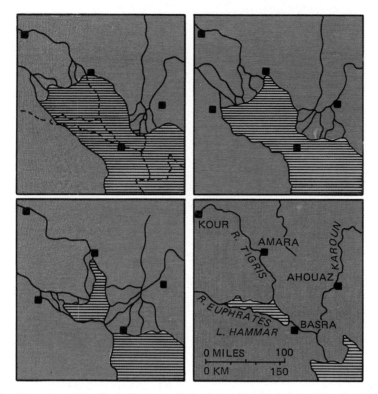

FIG. 19.—Extension of the Tigris-Euphrates Delta during historical times (after Twidale, 1976).

ter incised valleys are filled. On the other hand, fluvial deposition in type 2 sequences will be characterized only by widespread unconfined floodplain deposition during late highstand time since significant incised valleys are not associated with this sequence type.

Figure 20 summarizes the relationship between fluvial deposition and eustatic change. Note that widespread fluvial deposition is restricted to the interval T3 to T5. The equilibrium point reaches the bayline at T3 and reverses the bayline's migration direction until T5, when both resume a gradual landward shift. Following T5, rising sea level at the bayline results in landward shift of the bayline and gradual flooding of the river valleys. No fluvial deposition occurs after T5 because there is no concomitant landward shift of equilibrium profiles; therefore, no subaerial space is being added after this time. As a result, following T5, coastal onlap occurs at the bayline rather than farther landward in the fluvial environment, which characterized the interval prior to T5. This is observed as a basinward shift of coastal onlap occurring at T5.

In general, the occurrence of widespread fluvial deposition is dependent upon the position of the equilibrium point. Such deposition occurs during a relatively small portion of the time when the bayline and the equilibrium point migrate nearly horizontally basinward across a surface of low relief. Insofar as the equilibrium point position is a function of both eustacy and subsidence, the initiation of widespread fluvial deposition will vary from basin to basin. *The termination of widespread fluvial deposition preceding a type 2 unconformity, however, should be synchronous from ba-* *sin to basin, since it generally occurs at the eustatic-fall inflection point.*

LOWSTAND SYSTEMS TRACT

The lowstand systems tract is deposited during intervals characterized by relative sea-level fall (i.e., rate of eustatic fall exceeds rate of subsidence at the depositional shoreline break) and subsequent slow relative sea-level rise. If the lowstand systems tract is deposited in a basin with a discrete shelf edge, it can be divided commonly into two separate members which are not coeval: a *lowstand fan* (or *basin floor fan*) followed by a *lowstand wedge*. The lowstand fan is dominated by deposition of submarine fans at a time when sediments are bypassing the shelf through actively incising valleys. The subsequent lowstand wedge is dominated by finer grained, wedge-shaped slope deposits, which have been variously described as slope fan, slope front fill, wedge, cone, and submarine fan (e.g., see Mitchum, 1985), and contains contemporaneous shelf-depositional systems in the form of incised-valley fill.

If the lowstand systems tract is deposited on a ramp margin without a discrete shelf edge, then this systems tract consists of a two-part wedge. The first part is characterized by stream rejuvenation and sediment bypass of the coastal plain caused by a relative fall of sea level. Stream incision in this type of setting may not be great, as relative sea-level fall will expose a depositional profile not much different from the original stream equilibrium profile. In any event, sediment is delivered directly to the shoreline. This shoreline (and depocenter) steps basinward rapidly as a deposi-

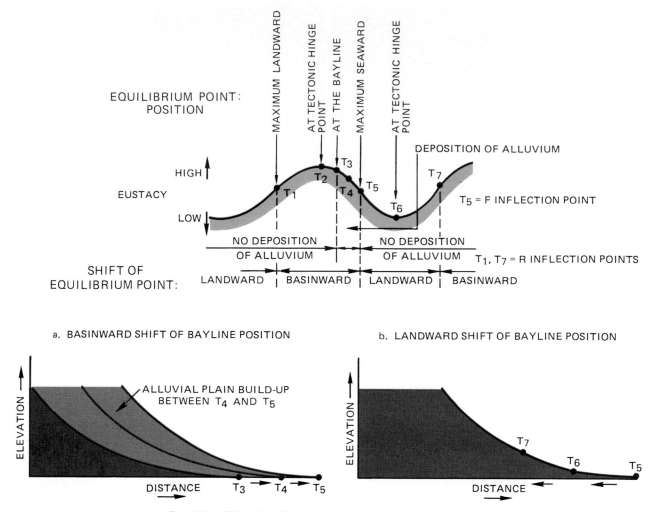

FIG. 20.—Effect of shifting equilibrium point on fluvial deposition.

tional profile of low relief is gradually exposed. This continues until relative sea level stabilizes and gradually begins to rise. This first part of the wedge will thus be characterized by a relatively coarse-grained basin-restricted wedge. The second part of the wedge is characterized by slow relative sea-level rise, resulting in the filling of incised valleys and slowed shoreline progradation coupled now with some aggradation. The fundamental similarity between deposition on the two types of basin margins is that deposition of the lowstand systems tract occurs as two separate non-coeval units in both cases.

Lowstand fan.—

Lowstand fans are deposited at basin margins that are characterized by a discrete physiographic shelf slope break. They are deposited when the rate of eustatic fall exceeds the rate of subsidence at the depositional shoreline break and when the depositional shoreline break is located at, or close to, the physiographic shelf slope break. At that time, the equilibrium point moves basinward of the depositional shoreline break and a relative fall of sea level occurs at this location (Fig. 21, after T1). Some or all of the shelf is exposed subaerially, and downcutting of streams commences. Sediment bypasses the shelf and is deposited directly on the slope and in the basin in the form of point-sourced submarine fans. It should be emphasized that although *submarine fans,* defined as "terrigenous, cone- or fan-shaped deposits located seaward of large rivers and submarine canyons" (Gary and others, 1974, p. 707), can be deposited at any time, they are most likely to be deposited and have the highest sand : mud ratio during lowstand fan time. The lowstand fan member derives its name from the fact that it is dominated by deposition of sand-prone submarine fans. Figure 6 illustrates geologic, geographic, and eustatic conditions conducive to the formation of these deposits. Note that during this interval, part of the slope as well as the shelf can be exposed, and active canyon cutting may occur. Figure 22 illustrates a modern analogue of lowstand fan time. In this case, "relative sea level" fell, resulting in stream rejuvenation and subsequent bypass of the "coastal plain." Fan formation generally ends when the rate of eustatic fall is again less than the rate of subsidence

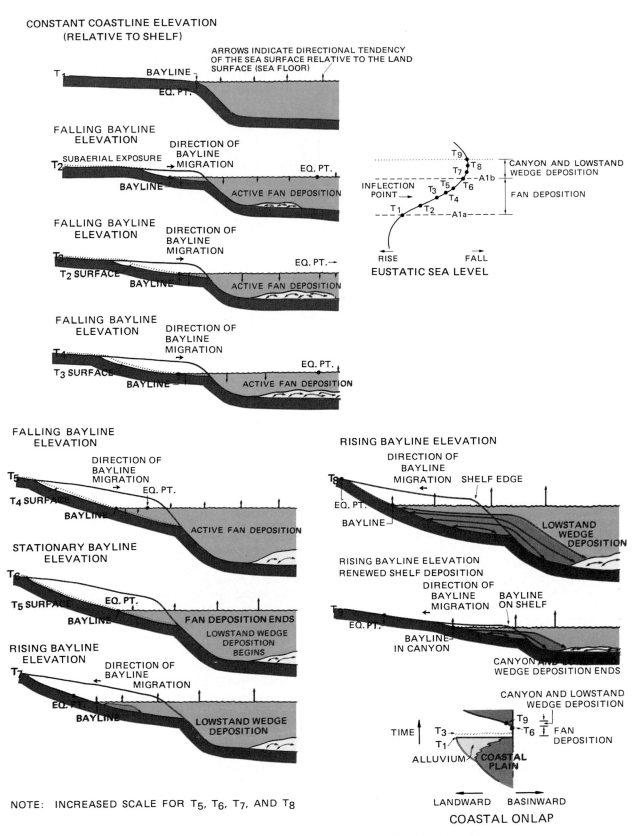

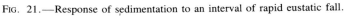

FIG. 21.—Response of sedimentation to an interval of rapid eustatic fall.

FIG. 22.—Modern analogue in the Pyrenees, Spain, of incised valley formed during lowstand fan time (note tire tracks in foreground for scale).

at the depositional shoreline break, and a relative rise of sea level occurs at this location. The lowstand fan is bounded below by a type 1 unconformity and above by a downlap surface. Internally, lowstand fans are characterized by mounded facies and may be extensively sand-prone.

The sequence of events resulting in lowstand deposition is shown in Figure 21. Immediately after T1, the bayline drops off the shelf onto the slope. This causes stream rejuvenation or incision as stream equilibrium profiles are lowered. The sediment load during this interval is greater and is characterized by a higher sand : mud ratio than during the preceding highstand systems tract time. This is because the stream sediment load during lowstand fan time consists of sediments being excavated by incising streams, in addition to sediments being supplied by the drainage basin; and, as no sediment is left behind on flood plains, all of this sediment load bypasses the shelf and is ultimately deposited at the shelf slope break. Subsequent instability at this location results in initiation of mass movement processes, and subsequent retrogressive failure there can result in formation of significant shelf edge notches or canyons. The sediment derived from streams and concurrent slumping of canyon walls is then transported into the basin by density currents, to be deposited where slopes become sufficiently gentle. The deposits are typically point-sourced submarine fans. As long as the bayline continues to drop in elevation down the slope, little or no deltaic or beach deposition will occur and fan building will continue as sediment is delivered directly into the basin. The equilibrium point reaches its maximum basinward position at the inflection point—T3—and thereafter shifts landward again. Active fan building continues until the equilibrium point reaches the bayline. As this time the bayline elevation stops falling and begins to rise as the equilibrium point continues

to shift landward. In response, lowstand wedge deposition is initiated and fan building ends. This continues until the bayline reaches the shelf surface and begins to migrate across the shelf—T9. Subsequent flooding of the shelf may eventually lead to a condensed section on the shelf.

Under certain conditions fan deposition may characterize one part of the basin margin, while deltaic deposition may characterize another part. This may occur along a basin margin, where the depositional shoreline break has reached the physiographic shelf edge in some areas but not in others. Where it has reached the shelf edge, submarine fans will form during a relative sea-level fall. Where it has not reached the shelf edge, however, deltas, rather than fans, may be deposited on the shelf margins. These prograding deltas would then be coeval with fans deposited elsewhere (see section on "Rapid Eustatic Falls of Short Duration"). Ultimately, with continued relative sea-level fall, even these areas may eventually be characterized by fan formation later during lowstand fan time.

Two types of discontinuity may be associated with rapid eustatic falls. One occurs at the base of deep-sea fan deposits—the type 1 unconformity; the other occurs at the top of deep-sea fan deposits and is referred to as a downlap surface or disconformity (Fig. 23). The type 1 unconformity occurs before the $\underline{F}$ inflection point at the time when the rate of subsidence equals the rate of eustatic fall at the shelf edge; the downlap surface or disconformity occurs after the $\underline{F}$ inflection point at the time when the rate of subsidence again equals the rate of eustatic fall at the bayline.

Lowstand wedge.—

The lowstand wedge is a regressive stratigraphic unit characterized by a progradational parasequence stacking pattern and is initiated during the latter part of a rapid eus-

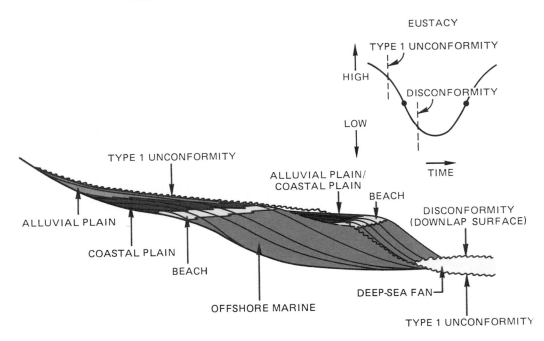

Fɪɢ. 23.—Type 1 unconformity.

tatic fall and subsequent eustatic lowstand, after the rate of seafloor subsidence at the depositional shoreline break once again exceeds the rate of eustatic fall (Fig. 3). This interval is characterized by the resumption of a slow rise of relative sea level at that location. As a result, stream incision ceases and incised valleys begin to fill with sediment (see earlier discussion). Sand : mud ratio, as well as total sediment load delivered to the depositional shoreline break, decreases at this time as coarse sediment is preferentially deposited within incised valleys. As downcutting of the canyon floor slows at this time and the frequency of density currents flushing out the canyon floor decreases, the oversteepened walls tend to collapse until a more stable profile is achieved. This results in leveed-channel and overbank-turbiditic deposits beyond the canyon mouth with a lower sand : mud ratio than that in the underlying lowstand fan. At this time, deltaic deposition is localized in the upper parts of canyons or embayments previously cut into the shelf edge. Lowstand wedge-deltaic deposits may eventually prograde onto the slope over the leveed-channel complex and onto submarine fans beyond the canyon mouths. The top of the lowstand wedge, deltaic depositional system lies at or below the level of the shelf, whereas the delta front will tend not to extend far beyond the shelf slope break. The slope gradient there is high, leading to instability of deltaic deposits. Consequently, mass-movement (debris flows) deposits frequently characterize the distal lowstand wedge. In addition, relatively high environmental energy at the shoreline (currents and waves) in this deeper water setting tends to enhance and focus wave energy, resulting in a wave-dominated environment, which further inhibits progradation. Thus, longshore drifting may tend to smear the lowstand wedge deposits along the outer shelf/upper slope (Fig. 24).

Figure 3 schematically illustrates the meandering pattern of the leveed channels associated with the leveed-channel complex of the lowstand wedge. They originate at the canyon mouth and may be downlapped subsequently by prograding deltaic deposits later during lowstand wedge time, as the canyon gradually fills. Consequently, lowstand wedge deposition may be divided into early and late parts (Fig. 25). The early lowstand wedge is characterized by active leveed-channel deposition with associated rhythmic turbidites. These leveed channels and overbank deposits are point-sourced from the canyon and are deposited in a more proximal position than the underlying lowstand or basin floor fan and may be referred to as a *slope fan*. Sediment is supplied by mass movement down the oversteepened canyon walls as the system tends toward equilibrium following the interval of active canyon downcutting during lowstand fan time. Deep-water sands of the slope fan unit of the lowstand wedge are usually restricted to the channels, although thin sands can occur as turbiditic overbank deposits and, in sandy systems, as sheet sands beyond the levees. The late lowstand wedge is characterized by progradation of the deltaic depositional system over the leveed-channel deposits of the early lowstand wedge and consequent filling of the canyon. Deep-water sands may be deposited as delta front rhythmic turbidites. This separation of early and late lowstand wedge would be most common in areas characterized by well-developed canyons. Both early and late lowstand wedge deposition is associated with deposition within incised valleys.

During the interval of lowstand wedge deposition, the rate of new shelf space added steadily increases as the rate of relative sea-level rise gradually increases. In response, topset beds will gradually increase in thickness (per unit time) as the rate of aggradation accelerates. Meanwhile, the rate of regression of the wedge gradually decreases until,

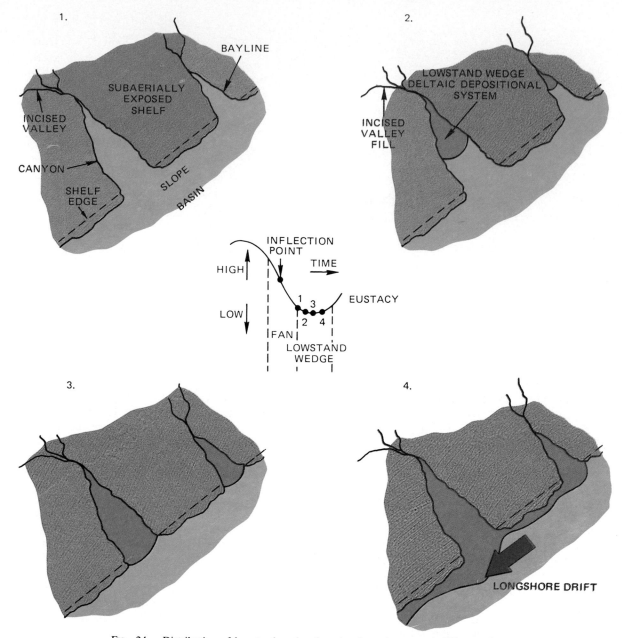

FIG. 24.—Distribution of lowstand wedge deposits along the outer shelf/upper slope.

eventually, sediment supply cannot keep pace with the rate at which new shelf space is added, and a transgression marks the end of lowstand wedge deposition.

A transgressive surface is typically observed at the upper boundary of the proximal lowstand wedge. Here, lowstand wedge deposition is restricted primarily to incised valleys. Whereas transgression can occur at any time during the filling of these valleys, it is observed that if incised valleys fill completely with sediment, a transgressive surface commonly develops atop the fill at the level of the surrounding shelf (Fig. 26D). This can be readily explained by an analysis of the rate of new space added. As sea level rises gradually, following a type 1 unconformity, the addition of new space is restricted to incised valleys (Figs. 26A, B), since the shelf remains subaerially exposed. Consequently, little new space is added until sea level rises above the pre-existing shelf surface (Fig. 26C). At that time, the addition of new space on the shelf, as well as over the incised valley, results in an abrupt increase of accommodation. Consequently, an abrupt landward shift of depocenter generates a transgressive surface atop the incised-valley fill as well as over the interfluves separating the incised valleys (Fig. 26D).

SHELF MARGIN SYSTEMS TRACT

The shelf margin systems tract is a regressive stratigraphic unit characterized by a decreasingly progradational,

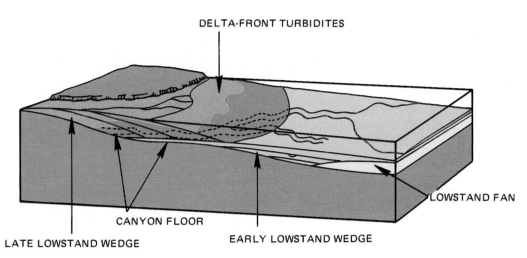

FIG. 25.—Separation of the lowstand wedge into early (slope fan) and late (lowstand delta) components.

followed by an aggradational, parasequence stacking pattern. It overlies a highstand systems tract and is usually deposited following the inflection point on a gentle eustatic fall (Fig. 6). This interval is characterized by a progressive increase in the rate of relative sea-level rise. The shelf margin systems tract is deposited on the outer part of the shelf and is bounded by an abrupt basinward shift of coastal onlap at its base. This basinward shift occurs because of the cessation of fluvial deposition at the F inflection point. The rate of regression during this interval gradually slows as topset beds thicken (per unit time) in response to the increasing rate of new shelf space added. Coal accumulation may be widespread near the top of this systems tract as the rate of addition of new shelf space increases and vast parts of the shelf may be temporarily inundated by brackish water. In general, the sediments are characterized by an increased tendency toward vertical stacking of facies with a gradual change from nonmarine to marine environment. In contrast with the highstand systems tract, the shelf margin systems tract is commonly not capped by widespread fluvial deposits.

The lower boundary of this systems tract is an erosional unconformity (or its correlative conformity) characterized by coastal plain or paralic/deltaic sediments overlying fluvial deposits. Where the lower boundary is the correlative conformity, the only expression of this basal contact is a change of parasequence stacking pattern from rapidly progradational to slowly progradational or aggradational. The upper boundary is marked by a transgressive surface separating the progradational-aggradational shelf margin systems tract from the overlying retrogradational transgressive-systems tract.

TRANSGRESSIVE-SYSTEMS TRACT

The transgressive-systems tract (Fig. 4) is composed of a succession of backstepping or retrogradational parasequences. It is initiated by the first significant flooding event (i.e., the transgressive surface) after the time of maximum regression of the lowstand wedge systems tract. It is characterized by a succession of flooding events and may be associated with concentrations of authogenic minerals and widespread coal deposits. The model suggests that wide-

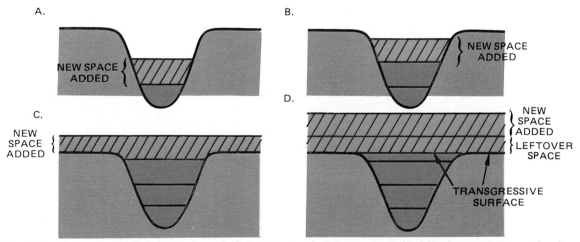

FIG. 26.—New space added during an interval of constant rate of relative sea-level rise following a type 1 unconformity.

spread transgressive deposits (outside of the incised valleys) should be significantly less common in association with type 1 unconformities relative to type 2 unconformities. In the former case, the early stages of the transgressive deposits are restricted to incised valleys, whereas in the latter case, where incised valleys do not occur, transgressive deposits will be more widespread. During the interval of gradual relative sea-level rise following a type 1 unconformity, sea level remains below the shelf. The eventual flooding of the shelf following a type 1 unconformity occurs when relative sea level is already rising quite rapidly. Consequently, depocenters shift abruptly and rapidly landward, resulting in little or no widespread transgressive deposition outside the incised valleys at that time. Flooding of the shelf following a type 2 unconformity is initiated more gradually, since incised valleys are absent and the shelf was never completely subaerially exposed. Because of this more gradual initiation of transgression, there is a greater possibility of an extensive deposition of transgressive sediments at that time.

VARIATIONS ON THE MODELS

The model described in this report initially incorporates certain assumptions regarding eustacy, subsidence, and sediment supply. The model is sufficiently flexible, however, to accommodate variations of these parameters. The following discussion considers some of the effects of varying these key parameters.

Rapid eustatic falls of short duration or *sequence deposition on a ramp margin.—*

When conditions favor rapid migration of the equilibrium point, a shift of this point beyond the depositional shoreline break would result in a type 1 unconformity, whereby the depositional shelf would be subaerially exposed and bypassed. Conditions conducive to rapid equilibrium point migration are a rapidly accelerating eustatic fall and/or a shelf with little differential subsidence. Lowstand deposition would then occur perched on a pre-existing submerged shelf if the depositional shoreline break had not reached the physiographic shelf edge.

At this time, the bayline drops down the shoreface or delta front en route to the physiographic shelf edge (Fig. 27). Stream incision is initiated and sediment loads are transported directly to the submerged shelf. This situation continues as long as the bayline continues to fall. During this interval, any delta formed at the stream mouth will be incised and eroded repeatedly as base level continues to drop. The net effect of this is to maintain the position of the depocenter at the delta front or shoreline and beyond without significant sediment deposition on the delta plain or coastal plain. Only some of the sediments deposited during lowstand fan time are preserved; much of the delta plain or coastal plain may have been removed by erosion. The preserved sediments are those delta front or lower shoreface/offshore sediments that were transported by density currents directly to the submerged shelf, bypassing the depositional shelf. Because of the shallow-water depth, gentle

shelf gradient, and short period of deposition, these lowstand deposits may initially be turbiditic but will not be fan-shaped and will be different from lowstand deposits occurring on the continental slope or in the basin. Coarse sediment will not be carried as far and may be more readily redistributed by the higher wave energy of the shallow water. On the submerged shelf, the result will be turbidite deposits that will appear less mounded and more sheetlike. In addition, because the period of rapid fall may have been short, comparatively little sediment may have been deposited during this interval. Thus, the rock record would show offshore fine-grained marine sediment, overlain by an abrupt introduction of coarser grained marine sediment (representing deposition at the delta front during lowstand fan time). This would be followed by finer grained delta front sediment characterized by a gradual coarsening-upward trend during subsequent lowstand wedge time. The left side of figure 27 illustrates how this would apply to a shallow intracratonic basin characterized by a ramp margin.

If a highstand sequence has not prograded to the physiographic shelf edge before rapid eustatic fall occurs, then it is possible that, locally, basin-deposited deep-sea fans may be initiated as shelf-perched lowstand deposits until continued relative sea-level fall at the bayline eventually exposes the entire shelf (T8, Fig. 27). On the other hand, true deep-sea fan deposition would be aborted if the equilibrium point reverses its migration direction (at $\underline{F}$ inflection point time) and reaches the bayline before the bayline reaches the physiographic shelf edge. This could occur with rapid eustatic falls of short duration, when the equilibrium point would shift well seaward of the bayline, initiating stream incision, but would then quickly reverse migration direction before the bayline reached the shelf edge. This condition would characterize a basin with a very broad shelf. The rising bayline would cause the gradual flooding of the incised valleys and the deposition of lowstand wedge deposits that are also perched on the submerged shelf, as shown in Figure 28.

In certain areas a eustatic fall may result in a type 1 unconformity in one location, while nearby a type 2 unconformity may develop. This occurs because the equilibrium point may not have passed the depositional shoreline break at one place (by inflection point time) yet may have done so nearby. This can occur in an area of a prograding delta where, because of rapid progradation and high rates of sea-floor subsidence due to compaction or salt withdrawal, the equilibrium point may not reach the delta front. This will result in a type 2 unconformity. Not far away, where beach progradation is occurring more slowly and subsidence is lower, the equilibrium point may travel well beyond the depositional shoreline break. This will result in a type 1 unconformity (characterized by. stream rejuvenation and sediment bypass of the concurrent shelf) with shelf-perched lowstand deposits. In areas where shelf width is narrow, the equilibrium point may pass the shelf edge despite low rates of eustatic fall. These areas may be characterized by a succession of type 1 unconformities, whereas areas with broad shelves, where the equilibrium point does not reach the shelf edge, would be simultaneously characterized by a succession of type 2 unconformities.

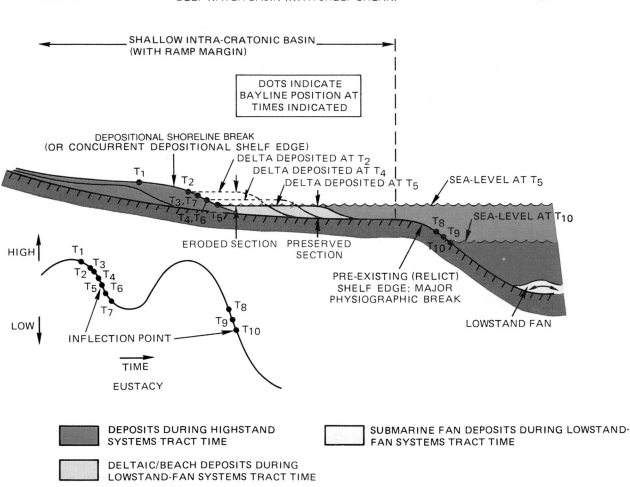

FIG. 27.—Bayline position and lowstand deposits during rapid short- and long-period eustatic falls.

Fluvial deposition—possible variations.—

Widespread alluvial-plain aggradation begins (provided there are no incised valleys present) when/if the equilibrium point reaches the bayline in the course of its basinward migration. Since the equilibrium point position is a function of the rates of eustatic fall and differential seafloor subsidence, both will affect the rate at which the equilibrium point travels basinward and thus the timing of its encounters with the bayline. Little differential subsidence and/or rapidly accelerating eustatic fall will cause rapid migration of the equilibrium point. Under these circumstances, alluvial-plain development would occur sooner than if differential subsidence were high and/or eustatic fall accelerated slowly. Thus, passive margins with little differential subsidence will be characterized by rapid migration of the equilibrium point across the shelf and beyond the depositional shoreline break. Since highstand fluvial deposition ends when the equilibrium point passes the depositional shoreline break, this interval of fluvial deposition will be short, resulting in thin highstand deposits on the shelf. One example of this is the apparent bypass of the west Texas Permian basin shelf by fluvial sediment, which may be a function of a rapidly moving equilibrium point caused by low differential subsidence.

Figure 29 compares two areas with different subsidence rates responding to the same eustatic change. Note that, in the case of low-subsidence rate (case 2), the equilibrium point moves more rapidly seaward after crossing the hingeline position. It reaches the bayline sooner, generating alluvium earlier and for a longer interval.

There is also a greater likelihood of developing a type 1 unconformity with concomitant lowstand deposits on a continental margin with a low-subsidence rate. For lowstand deposits to develop, sea level must drop below the depositional-shoreline break. This will occur when the equilibrium point passes the depositional-shoreline break en route basinward. In the two cases shown here (Fig. 29), the equilibrium point clearly travels farther into the basin in case 2 where rates of subsidence are lower and, all else being equal,

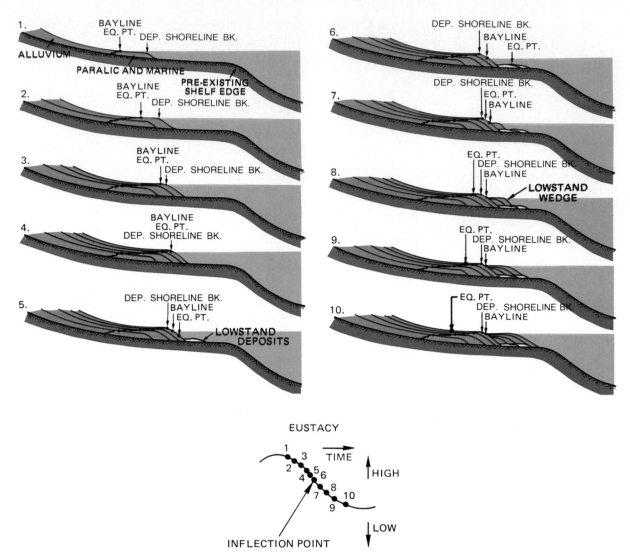

FIG. 28.—Shelf-perched lowstand deposits and eustatic change.

stands a greater chance of passing the depositional shoreline break, resulting in initiation of lowstand deposition. These effects are in response to differences in local subsidence conditions and are therefore not global.

Figure 30 shows a shelf margin responding to two rates of eustatic change. Although the equilibrium point passes the hinge line at the same time in both cases, in case 2, where there is a high rate of eustatic, it moves more rapidly seaward and reaches the bayline sooner. Fluvial deposition begins earlier and continues for a longer interval. Moreover, there is a greater chance in case 2 that the equilibrium point will reach and pass the depositional shoreline break and will produce a type 1 unconformity. Because this is a response to eustatic change, this effect will be global rather than local.

The symmetry of the eustatic curve can also play an important role in the occurrence of fluvial deposits. Figure 31 illustrates the effect of shifting the inflection point up or down the falling limb of the eustatic curve. If the inflection point occurs late in the eustatic fall, as it does in case 1, there will be a longer period of fluvial deposition than in those cases where the inflection point occurs early in the eustatic fall, as in case 2. If the slope of the eustatic curve or the rate of eustatic fall at the inflection point is the same in both cases, then the maximum seaward position of the equilibrium point will be the same, although the rate at which they have shifted will be greater for case 2.

The difference of the sediment response in these cases (Fig. 31) will be expressed in the timing and rate of fluvial deposition. In each case, the equilibrium point, bayline, and equilibrium profile shift laterally from position A to B, as shown in Figure 32. The A location corresponds to the point to which the equilibrium profile is adjusted just prior to initiation of its basinward shift. The B location corresponds

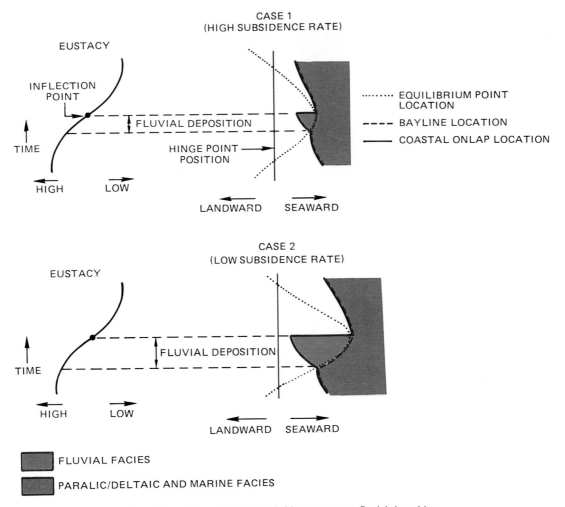

Fig. 29.—Effect of varying subsidence rates on fluvial deposition.

to the maximum basinward position of the equilibrium point and, hence, the maximum basinward position of the equilibrium profile. This situation occurs at inflection point time. In order to get to position B, however, the equilibrium point and profile must move faster across the shelf in case 1 than it must in case 2. From the time the equilibrium point reaches the bayline at time T0, two time increments are required to make the shift of the equilibrium profile to position B in case 1 and six are required in case 2. In case 1, assuming that sediment supply is constant, more sediment per time increment is required to shift the profile to its new equilibrium position. This implies that, after filling the space to the new equilibrium profile, less sediment will be left over to be deposited as paralic and marine facies in case 1 than in case 2. Thus, a lower rate of regression will characterize the former case. In general, the rate of regression is inversely related to the rate at which the equilibrium profile migrates basinward. Since these are responses to eustacy, these effects may be observed globally. Another ramification of a rapid shift of the equilibrium profile is that the

equilibrium point could shift beyond the depositional shoreline break and not the physiographic shelf edge, generating a type 1 unconformity with a deposition of lowstand units perched on a pre-existing shelf.

Implications of no fluvial deposition at an $\underline{F}$ *inflection point.—*

Under certain circumstances, the equilibrium point in its seaward migration may not reach the bayline; consequently, fluvial deposition may not occur. This can happen when the rate of subsidence is high and/or the rate of eustatic fall is low. Usually, the equilibrium point reaches the bayline sometime between the eustatic highstand and the $\underline{F}$ inflection point (Fig. 33), generating a type 1 or 2 unconformity. When the rate of eustatic fall is low, the equilibrium point, in the course of its basinward migration, may not reach the bayline at all before it starts shifting landward again (Fig. 34). Under these circumstances, the bayline would shift landward the entire time and fluvial deposition

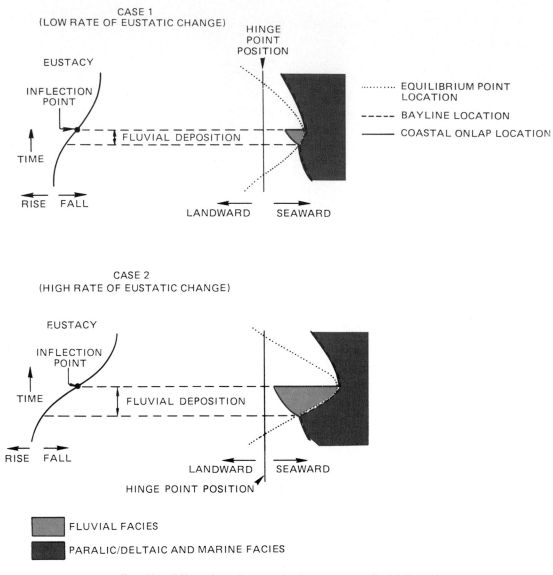

CASE 1
(LOW RATE OF EUSTATIC CHANGE)

HINGE
POINT
POSITION

EUSTACY

INFLECTION
POINT

FLUVIAL DEPOSITION

TIME

RISE   FALL

LANDWARD   SEAWARD

·········· EQUILIBRIUM POINT
           LOCATION

- - - - - BAYLINE LOCATION

———— COASTAL ONLAP LOCATION

CASE 2
(HIGH RATE OF EUSTATIC CHANGE)

EUSTACY

INFLECTION
POINT

TIME

FLUVIAL DEPOSITION

RISE   FALL

LANDWARD   SEAWARD

HINGE POINT POSITION

FLUVIAL FACIES

PARALIC/DELTAIC AND MARINE FACIES

FIG. 30.—Effect of varying eustatic-change rates on fluvial deposition.

would probably not occur. Consequently, no basinward shift of coastal onlap would characterize this inflection point time. Generation of a type 1 or 2 unconformity under these conditions (i.e., low rates of eustatic fall) would require low rates of subsidence, as this would have the effect of causing the equilibrium point to shift farther basinward. Thus, during eustatic cycles with low rates of fall, unconformities may not be generated in areas of moderate to high subsidence and may therefore be absent locally.

Fluvial deposits may also be absent in areas of rapid subsidence. High rates of subsidence will have the same effect of limiting the basinward migration of the equilibrium point (Fig. 35). Once again, no unconformity will develop under these circumstances. The only indication of an F inflection point, if no alluvium is deposited, is an increased rate of regression before and a decreased rate of regression after

this time. This occurs in response to the gradual thinning of topset beds per unit time prior to the inflection point and gradual thickening of topset beds afterward (Fig. 11 in Posamentier and others, this volume).

In areas of low sediment supply, it may be possible that little or no fluvial deposition is available to fill the subaerial accommodation. In these cases local coastal onlap would occur at the bayline, and the onlapping strata (in the absence of a fluvial component) would consist of paralic/deltaic sediments. Presumably, these sediments are transported into the area by longshore drift rather than directly by streams. The resulting stratal pattern is shown in Figure 36A; the chronostratigraphic distribution is shown in Figure 36B. Coastal onlap would shift landward until the equilibrium point reaches the bayline. Thereafter, the bayline, and therefore coastal onlap, would migrate basinward. At the F

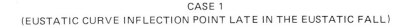

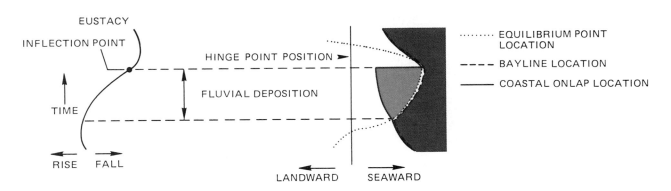

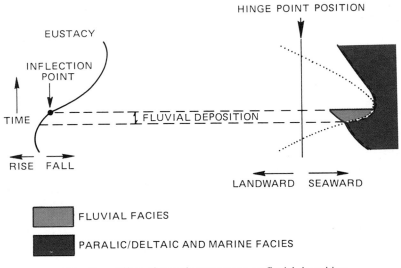

FIG. 31.—Effect of eustatic asymmetry on fluvial deposition.

inflection point, the bayline, as well as the equilibrium point, would resume its landward migration. Transgression probably would be more rapid during the ensuing relative sea-level rise, since the sea would be flooding a surface at base level (i.e., sea level rather than a stream equilibrium profile). This surface has a lower relief than the sloping surface that would have accompanied fluvial aggradation.

SUMMARY

The models discussed here provide a means for systematizing the geologic observations from a sedimentary basin. The general dependence of stratal patterns and facies on fluctuations of global sea level makes regional as well as global stratigraphic correlation possible. Global sea-level fluctuations result in a predictable succession of sequences bounded by unconformities associated with eustatic-fall inflection points. Each sequence is composed of a succession of systems tracts associated with specific segments of the sea-level curve. Finally, each systems tract is composed of one or more depositional systems within which lithofacies can be predicted when local subsidence, sediment supply, and physiography are incorporated into the model.

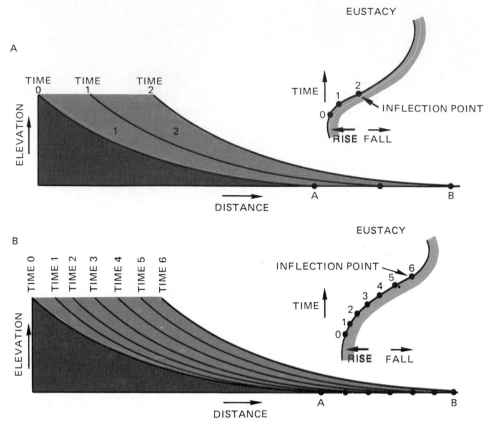

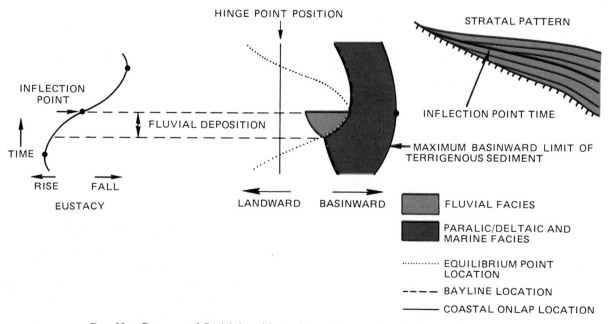

FIG. 32.—Effect of eustatic-fall asymmetry on shifting equilibrium profile.

FIG. 33.—Response of fluvial deposition and stratal pattern to equilibrium point position.

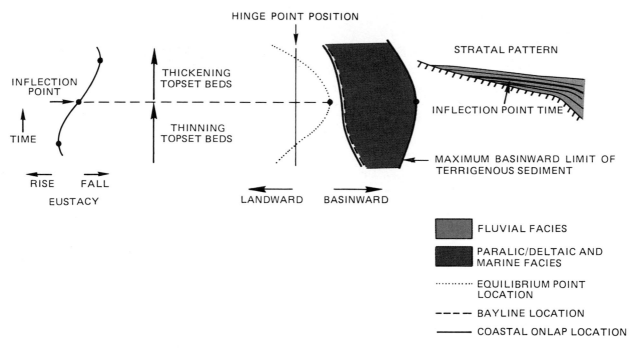

FIG. 34.—Effect of low rate of eustatic fall on stratal pattern.

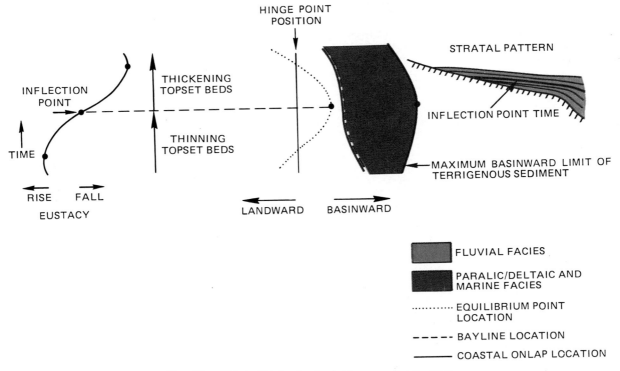

FIG. 35.—Effect of high rate of subsidence on stratal pattern.

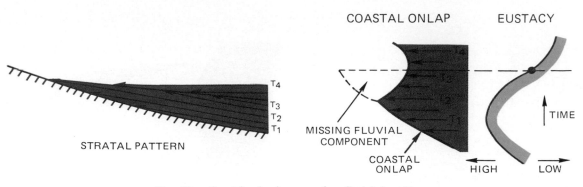

FIG. 36.—Coastal onlap in areas of no fluvial deposition.

ACKNOWLEDGMENTS

The author gratefully acknowledges all the support and feedback received during preparation of this report. A report of this nature relies heavily on the contributions of many co-workers over a period of time. It summarizes and draws from work done over many years by many individuals. Much early work on accommodation and sediment response to sea-level change was done through mathematical modeling by M. T. Jervey. This provided inspiration and motivation for much of our subsequent work. Thanks are also due to J. C. Van Wagoner, R. M. Mitchum, G. R. Baum, V. Kolla, J. E. McGovney, C. G. St. C. Kendall, K. M. Campion, J. M. Demarest, N. C. Lian, F. W. Schroeder, and S. R. Schutter for their patience and time in reviewing the manuscript. Other members of the Reservoir and Facies Division at Exxon Production Research Company, J. F. Sarg, G. J. Moir, W. A. Burgis, R. D. Erskine, R. A. Hoover, S. R. Morgan, G. Mirkin, T. R. Nardin, and G. R. Ramsayer, also contributed to this report with their helpful suggestions and comments. Ultimate responsibility, however, rests with the authors for the material herein.

REFERENCES

CHITCHOB, S., AND COWLEY, J. E., 1973, Bhumibol Reservoir—Sediment status after eight years of operation: International Association for Hydraulic Research, Proceedings, International Symposium on River Mechanics, 9–12 January, 1973, p. A4–1 to A4–11.
EAKIN, H. M., AND BROWN, C. B., 1939, Silting of reservoirs: U.S. Department of Agriculture Technical Bulletin 524, 168 p.
GARY, M., MCAFEE, R., JR., AND WOLF, C. L., 1974, Glossary of Geology, Washington, D.C., American Geological Institute, 805 p.
GILBERT, G. K., 1917, Hydraulic-mining debris in the Sierra Nevada:

U. S. Geological Survey Professional Paper 105, 154 p.
KUIPER, E., 1965, Water resources development; Planning, engineering and economics: Butterworths, Washington D.C. 471 p.
LANE, E. W., 1955, Design of stable channels: American Society of Civil Engineering, Proceedings, v. 81, paper 795, p. 1–17.
LEOPOLD, L. B., AND BULL, W. B., 1979, Base level, aggradation, and grade: American Philosophical Society, Proceedings, v. 123, no. 3, p. 168–202.
LEOPOLD, L. B., WOLMAN, M. G., AND MILLER, J. P., 1964, Fluvial processes in geomorphology: W. H. Freeman and Company, San Francisco, California, 522 p.
MACKIN, J. H., 1948, Concept of the graded river: Geological Society of American Bulletin, v. 59, p. 463–512.
MITCHUM, R. M., 1985, Seismic stratigraphic expression of submarine fans: American Association of Petroleum Geologists Memoir 39, Tulsa, Oklahoma, p. 117–136.
RUBEY, W. W., 1952, Geology and mineral resources of the Hardin and Brussels quadrangles (in Illinois): U.S. Geological Survey Professional Paper 218, 179 p.
SHULITS, S., 1941, Rational equation of riverbed profile: American Geophysical Union, Transactions, v. 22, p. 622–630.
SIMONS, D. B., AND SENTURK, F., 1977, Sediment transport technology: Water Resources Publications, Fort Collins, Colorado, p. 750–754.
STRAHLER, A. N., 1952, Dynamic basis for geomorphology: Geological Society of America Bulletin, v. 63, p. 923–928.
TWIDALE, C. R., 1976, Analysis of landforms: John Wiley and Sons, Sydney, Australasia Proprietary Ltd., p. 237–244.
VAIL, P. R., MITCHUM, R. M., JR., TODD, R. G., WIDMIER, J. M., THOMPSON, S., III, SANGREE, J. B., BUBB, J. N., AND HATLELID, W. G., 1977, Seismic stratigraphy and global changes of sea level, in Clayton, C. E., ed., Seismic stratigraphy—Applications to Hydrocarbon Exporation: American Association of Petroleum Geologists Memoir 26, Tulsa, Oklahoma p. 49–212.
———, AND TODD, R. G., 1981, North Sea Jurassic unconformities, chronostratigraphy and sea-level changes from seismic stratigraphy: Petroleum Geology of the Continental Shelf of Northwest Europe, Proceedings, p. 216–235.
YATSU, E., 1955, On the longitudinal profile of the graded river: American Geophysical Union, Transactions, v. 36, p. 655–663.

CARBONATE SEQUENCE STRATIGRAPHY

J. F. SARG

Exxon Production Research Company, P.O. Box 2189, Houston, Texas 77252-2189

ABSTRACT: The major controls on changes in carbonate productivity, as well as platform or bank growth and the resultant facies distribution, are interpreted here to be short-term eustatic changes superimposed on longer term tectonic changes (i.e., relative changes in sea level). Carbonate platforms associated with sea-level highstands are characterized by relatively thick aggradational-to-progradational geometry. They are bounded below by the top of a transgressive unit and above by a sequence boundary. Two types of highstand platform, *keep-up* and *catch-up*, are differentiated here. (1) A *keep-up* carbonate highstand platform is interpreted to represent a relatively rapid rate of accumulation that is able to keep pace with periodic rises in relative sea level. A *keep-up* carbonate is characterized at the platform margin by grain-rich, mud-poor lithofacies and nonpervasive submarine cementation. *Keep-up* platforms display a mounded/oblique stratal configuration at the platform/bank margin and in places on the platform. (2) A *catch-up* carbonate highstand platform is interpreted to represent a relatively slow rate of accumulation that is characterized by micrite-rich parasequences and pervasive early submarine cementation at the platform margin. A *catch-up* carbonate displays a sigmoid depositional profile at the platform/bank margin.

At the formation of a type 1 sequence boundary, where the rate of eustatic fall is interpreted to be greater than subsidence at the platform/bank margin, two major processes occur: (1) local-to-regional slope front erosion and (2) subaerial exposure of the shelf and major seaward movement of the regional meteoric lens. At a *large-scale type 1 sequence boundary*, sea level may fall from 75 to 100 m or more and for an extended period of time. When this occurs, the meteoric lens becomes established over the shelf for a long time, and its influence extends well into the subsurface. If there is sufficient rainfall and a permeable section with mineralogically unstable grains, significant solution will occur over the shelf in the shallow portion of the underlying highstand carbonate platform. Precipitation of phreatic cements will occur deeper or downdip in the section. At a *small-scale type 1 sequence boundary*, where sea level falls less than about 100 m and for a short period of time, the meteoric lens becomes less well established. It remains in a shallow position on the shelf, causing less extensive solution. Mixing and hypersaline dolomitization may be important processes during the late highstand and continuing through the formation of either a large- or small-scale type 1 sequence boundary. At a *type 2 sequence boundary*, in which the rate of eustatic fall is interpreted to be less than the rate of subsidence at the platform/bank margin, the inner-platform peritidal and outer-platform shoal areas will be exposed. The dominant meteoric effect will be in the inner-platform areas.

During sea-level lowstands, three types of carbonate deposits are recognized: (1) allochthonous material derived from erosion of the slope (i.e., debris sheets and allodapic carbonate sands); (2) autochthonous wedges deposited on the upper slope during type 1 sea-level lowstands; and (3) type 2 platform/bank margin wedges. In addition, given the appropriate climatic and hydrographic conditions (i.e., evaporation exceeds influx, and basin is restricted), evaporite lowstand wedges may occur associated with either type 1 or type 2 sequence boundaries. During evaporitic lowstands, hypersaline dolomitization, evaporite replacement, and solution may occur in associated carbonate highstand platforms. Siliciclastic lowstand deposition will occur in areas where an updip-source terrain is available.

INTRODUCTION AND BACKGROUND

Shallow-marine carbonate deposition occurs as relatively thick aggrading and prograding deposits in warm tropical areas, both around the rims of basins and as isolated platforms within basins (Wilson, 1975). The basin rim deposits may occur as broad regional platforms or ramps, or as relatively high angle (5°) prograded bank deposits. These features can usually be recognized on seismic sections. Facies can be predicted seismically where platform thickness is seismically resolvable. Where the carbonate platform deposits are thin and near seismic resolution, interpretation of well logs and core in combination with seismic stratigraphic interpretation and seismic modeling may allow prediction of facies. Steps in developing a carbonate facies and sequence interpretation include:

—recognizing regional basin settings and age relations in which carbonates occur;
—defining the sequence subdivisions and delineating their extent by mapping their external geometry (seismic-sequence analysis by using grids of seismic lines);
—outlining the lithofacies within sequences. Lithofacies distribution is predicted from reflection configuration, amplitude, and continuity (seismic facies) tied to log character, and core description.

In this paper criteria are presented for recognizing and delineating shallow-marine carbonate sequences and associated facies by using rock, seismic, and well-log data. A sequence is a relatively conformable succession of genetically related strata bounded by unconformities and their correlative conformities (Mitchum, 1977). An unconformity is a surface separating younger from older strata along which there is evidence of subaerial truncation (and, in some areas, correlative submarine erosion) or subaerial exposure, with a significant hiatus indicated. I propose a predictive depositional model supported by examples from different geologic ages ranging from Ordovician to Holocene, and from different depositional settings including regional platform/ramps with slopes of less than 5°, prograded banks with foreslopes of 5 to 35°, and isolated platforms. Some background information on platform geometry and facies characteristics of carbonate sequences is also presented. This paper does not discuss deep-marine chalk deposition.

SEQUENCE STRATIGRAPHY

There are four major variables that control the variations in stratal patterns and lithofacies distributions within carbonate rocks. They include (1) tectonic subsidence, which creates the space where the sediments are deposited; (2) eustatic change, which I believe is the major control over the stratal patterns and the distribution of lithofacies (Vail and Todd, 1981); (3) the volume of sediments, which controls paleo-water depth; and (4) climate. Climate is the major control over the type of sediments—rainfall and temperature are important to the distribution of carbonates and evaporites and to the type and amount of siliciclastic rocks deposited.

Sea-Level Changes—An Integrated Approach, SEPM Special Publication No. 42

The combination of eustasy and tectonic subsidence produces a relative change of sea level (Fig. 1). In Figure 1, tectonic subsidence is shown as a linear plot as it appears to change slowly with respect to eustasy. The relative change of sea level creates the available space for accommodation of the sediments. The thickness of sediments is primarily controlled by tectonic subsidence. The depositional stratal patterns and distribution of lithofacies, however, are controlled by the rate of relative change of sea level. This expresses itself by the change of slope of the relative sea-level curve, which is primarily controlled by eustasy (Fig. 1).

Sequences are composed of three parts, or systems tracts. A systems tract is a linkage of contemporaneous depositional systems (i.e., three-dimensional assemblage of lithofacies; Brown and Fisher, 1977). Systems tracts here are defined on the basis of types of bounding surfaces, stratal geometry, and position within the sequence. As shown in Figure 1, systems tracts are interpreted to be deposited during specific time intervals of the relative change of the sea-level curve. A sequence is interpreted to be deposited during a cycle of eustatic change of sea level starting and ending in the vicinity of the inflection points on the falling limbs of the sea-level curve.

Vail and Todd (1981) have recognized two types of sequences referred to as type 1 and type 2, in siliciclastic sequences (see also VanWagoner and others, Posamentier and Vail, this volume), and both types are recognized in carbonate sequences as well. A type 1 sequence boundary lies at the base of a type 1 sequence and is characterized by subaerial exposure and erosion of the platform, concurrent submarine erosion on the foreslope, onlap of overlying strata, and a downward shift in coastal onlap (Fig. 1). As a result of the basinward shift in coastal onlap, peritidal rocks commonly sharply overlie "deeper water" subtidal rocks. Because carbonate platforms tend to accrete to sea level over much of their areal extent, it is useful to define carbonate sequence boundaries in relation to the platform/bank margin. A type 1 sequence boundary is thus interpreted to form when the rate of eustatic fall exceeds the rate of basin subsidence at the platform/bank margin, producing a relative fall in sea level at that position. A type 2 sequence boundary (Fig. 1) is marked by subaerial exposure of inner-platform peritidal areas and platform shoal areas. As Figure 1 shows, a downward shift in coastal onlap occurs seaward of the underlying peritidal areas. If the platform top has accreted to sea level, this basinward shift may occur to a position at the preceding platform/bank margin. Onlap of overlying strata of peritidal origin occurs in platform lows that have not accreted to sea level and at the platform/bank margin. A type 2 sequence boundary is interpreted to form when the rate of eustatic fall is less than or equal to the rate of basin subsidence at the platform/bank margin.

The relation between relative changes of sea level and systems tracts is shown in depth and in geologic time on Figure 1. Each systems tract is identifiable on the basis of criteria observable in outcrops, on well logs, and on seismic data. There are four systems tracts: lowstand, shelf margin, transgressive, and highstand (VanWagoner and others, Posamentier and Vail, this volume). The lowstand systems tract is deposited basinward of the preceding platform/bank margin and overlies a type 1 sequence boundary. It laps out at or near the preceding platform margin, except where it fills incised valleys on the shelf. The shelf margin systems tract is a prograding and aggrading wedge that overlies a type 2 sequence boundary and laps out on the platform landward of the preceding platform/bank margin. Its lower boundary is a conformable sequence boundary, and its upper boundary is a transgressive surface. An unconformity exists landward of where it pinches out (Fig. 1).

Transgressive-systems tracts are composed of a set of backstepping or retrogradational units that thicken shelfward until they thin by onlap at the base. In general, the younger units are progressively thinner because of sediment starvation. The transgressive-systems tract thus thins basinward and upward, forming a condensed section at the top. The condensed section is a unit consisting of thin marine beds of hemipelagic or pelagic sediments deposited at very slow rates (see also Loutit and others, this volume). The surface at the base of the transgressive-systems tract is the transgressive surface, or the first flooding surface above the lowstand or shelf margin systems tract. Landward of where the lowstand systems tract pinches out, the lower boundary of the transgressive-systems tract coincides with the unconformable portion of the sequence boundary.

The highstand systems tract (Fig. 1) completes a sequence and is a sigmoid-to-oblique prograding unit that overlies the transgressive-systems tract. The boundary at the base of the highstand systems tract is a downlap surface that is associated with the condensed section. It is called the maximum-flooding surface and becomes conformable on the inner shelf. The upper surface of the highstand systems tract is a type 1 or type 2 sequence boundary.

DEPOSITIONAL PROFILES AND FACIES BELTS

Depositional Setting

Carbonate platform/bank margin profiles may be grouped into three classes, on the basis of both basin position (e.g., rimming basin margin, freestanding within basin) and on the slope of the stratigraphic profile. The classes are: regional platform/ramps attached to basin margin, with depositional slope of less than 5°; regional prograded banks/platforms rimming the basin margin, with foreslopes of 5 to 35°; and offshore or isolated platforms (Fig. 2). Each of these types of profile may be recognized on seismic sections (Fig. 3), and the internal seismic-facies character can aid in the prediction of growth history and enclosing geologic facies.

Regional platforms/ramps.—

Regional ramps vary widely in thickness from a few to many hundreds of meters, and their growth patterns range from aggradational to progradational. Carbonate ramps are built away from positive areas and down gentle regional paleoslopes. No striking break in slope exists, and facies patterns tend to be wide, irregular belts (Wilson, 1975). Seismically, ramps may display low-angle sigmoidal or

SEQUENCE STRATIGRAPHY DEPOSITIONAL MODEL
SHOWING SURFACES, SYSTEMS TRACTS AND LITHOFACIES

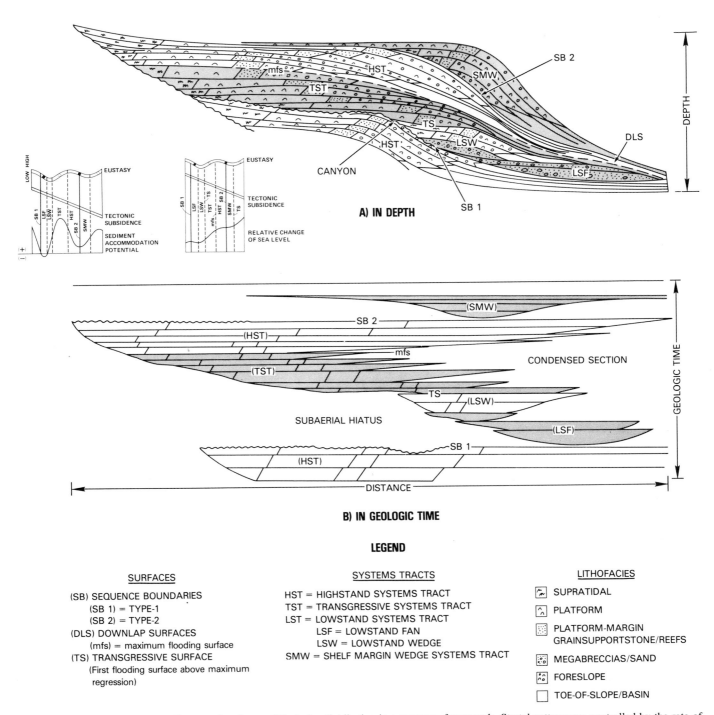

LEGEND

SURFACES

(SB) SEQUENCE BOUNDARIES
  (SB 1) = TYPE-1
  (SB 2) = TYPE-2
(DLS) DOWNLAP SURFACES
  (mfs) = maximum flooding surface
(TS) TRANSGRESSIVE SURFACE
  (First flooding surface above maximum
  regression)

SYSTEMS TRACTS

HST = HIGHSTAND SYSTEMS TRACT
TST = TRANSGRESSIVE SYSTEMS TRACT
LST = LOWSTAND SYSTEMS TRACT
  LSF = LOWSTAND FAN
  LSW = LOWSTAND WEDGE
SMW = SHELF MARGIN WEDGE SYSTEMS TRACT

LITHOFACIES

SUPRATIDAL

PLATFORM

PLATFORM-MARGIN
GRAINSUPPORTSTONE/REEFS

MEGABRECCIAS/SAND

FORESLOPE

TOE-OF-SLOPE/BASIN

FIG. 1.—Summary schematic diagram of carbonate lithofacies distribution in a sequence framework. Stratal patterns are controlled by the rate of relative change of sea level expressed as the change in slope of the eustatic curve. Systems tracts are deposited during specific time intervals of the sea-level curve. A relatively steep platform margin is predicted to have extensive slope front erosion during formation of a type 1 sequence boundary. Debris sheets and allodapic sands are deposited at the toe of slope (lowstand fan—LSF) and in submarine channels. A lowstand wedge (LSW) bank and transgressive-systems tracts (TST) onlap the slope and paleoplatform. The LSW and shelf margin wedge (SMW) are predicted to be more extensive and better developed on low-angle ramps or banks, where there will be abundant area for shallow-water carbonate production. A major transgressive surface (TS) separates the LSW from the TST. Following the time of maximum flooding, the highstand systems tract aggrades and progrades basinward. The maximum flooding surface (mfs) separates the highstand from the transgressive-systems tract.

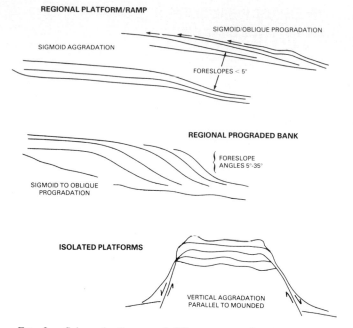

FIG. 2.—Schematic diagram of different types of carbonate depositional profiles, including (1) regional platform/ramp (foreslope 5° or less) displaying both sigmoid and oblique progradation; (2) regional prograded bank (foreslope 5–35°); and (3) isolated platforms displaying dominantly vertical aggradation with steep margins.

shingled progradation (Fig. 3A). Carbonate platforms are built up with a more or less horizontal top and in some cases possess abrupt margins. Platforms/ramps show very little expression of progradation, and where platform/ramps are seismically thin, identification of the platform/ramp margin is difficult. The integration of available well-log and core data into the sequence framework is therefore particularly important.

Regional prograded banks/platforms.—

Regional banks/platforms are characterized by a progradational pattern with foreslopes, which range upward from 5° to more than 35° (Fig. 3B). Bank thicknesses range from a few to several hundred meters, and progradation may be for many kilometers. These banks display sigmoid, sigmoid/oblique, and oblique progradational patterns. A common evolution within a sequence (see following sections) is a change from ramp or low-angle sigmoid progradation to oblique progradation. This is probably due to falling sea level at the end of a highstand.

Offshore isolated platforms.—

Isolated platforms occur as complex buildups of great size and thickness well offshore from regional basin margin ramps or platforms. The Bahama Late Neogene and the Miocene Terumbu (Fig. 3C) platforms are examples. Horst blocks within pull-apart basins commonly seed isolated platform development. These blocks may serve as sites of shallow-marine carbonate deposition, whereas muddy deep-water sediments are limited to the grabens. Platforms generally have steep margins, and there may be an open-ocean side

of the platform (Terumbu platform; Rudolph and Lehmann, 1987).

Facies Belts

A characteristic suite of facies occurs within each type of profile. Because most carbonate sediment originates in the basin of deposition and is largely of organic origin, the facies distribution is especially sensitive to changes in water depth, water chemistry, and water circulation. Figure 4 illustrates a representative carbonate profile from shelf to basin, with typical facies belts identified. The facies belts vary in width and uniformity. They are narrower and more regular where the shelf is narrow, and the shelf margin is steep. Where platform/bank margin is low and shelf areas wide, the belts are wider and more diffuse. From the coastal area to the basin, the following facies belts are recognized: *supratidal-intertidal flats*, *shallow-marine shelf*, *platform or bank margin*, *foreslope*, and *basin facies*.

Supratidal-intertidal flats.—

Tidal-flat facies commonly occur as small-scale shoaling-upward subtidal-to-supratidal cycles, or parasequences. A parasequence, as defined by Van Wagoner and others (this volume), is a conformable succession of genetically related beds or bed sets bounded by marine-flooding surfaces and their correlative surfaces. They may be several meters to more than 30 m thick and are zero to 1 million years or less in duration. They are the smallest recognizable allocyclic or autocyclic depositional sequence. Modern and ancient tidal-flat deposition has been described by many (see, for example, Shinn and others, 1969, on the Bahamas tidal flats; Evans and others, 1969, on the Persian Gulf; Wilson, 1975; Anderson and others, 1984; and Shinn, 1983 in a well-illustrated summary of tidal-flat characteristics).

Tidalites are composed of three basic depositional environments—supratidal, intertidal, and subtidal. The supratidal subfacies is characterized by mudcracks, storm-generated lamination composed of mud- or sand-size particles, laminites of algal origin, fenestral or birdseye fabric, and intraclast layers. The algal laminates may extend into the intertidal subfacies. The supratidal environment occurs above normal or mean high tide and is exposed to subaerial conditions most of the time. The intertidal subfacies is commonly mud-rich and contains tidal-channel complexes. The tidal channels generally contain a basal lag of intraclasts and lithoclasts overlain by burrowed skeletal packstone. The intertidal environment occurs between normal high tide and normal low tide. The adjacent subtidal subfacies of the tidal flat is commonly composed of pelleted carbonate mudstones and wackestones that lack primary sedimentary structures. In an evaporative climate, displacive nodular and rosette gypsum occur within the intertidal and supratidal subfacies.

Shallow-marine shelf.—

This facies belt commonly consists of parasequences shoaling from subtidal skeletal mudstones and wackestones to peloid or skeletal packstones and grainstones. If normal-marine water conditions existed, fauna and flora are abun-

CARBONATE FACIES BELTS
WITH REPRESENTATIVE TEXTURAL TYPES

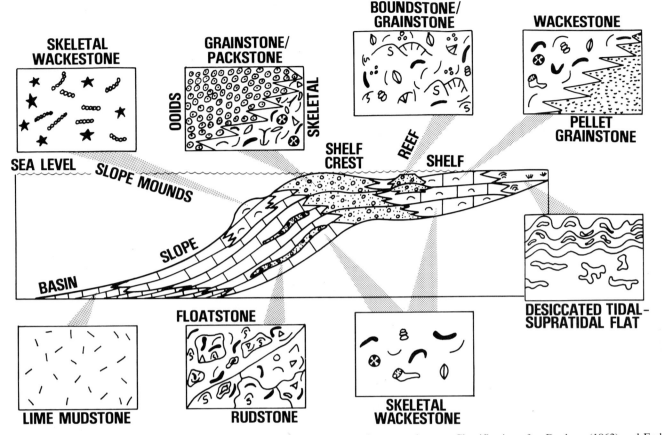

FIG. 4.—Schematic carbonate platform/bank with facies belts and representative textural types. Classification after Dunham (1962) and Embry and Klovan (1971).

dant and include corals, molluscs, brachiopods, sponges, arthropods, echinoderms, foraminifera, and algae. Bioturbation is common. This environment occurs seaward of tidal flats, and water depths are generally shallow, 10 to 20 m at most. Salinity varies from normal marine to mesosaline, and water circulation can range from low to moderately high, depending on the degree of platform margin restriction or dampening of tides and currents. When the shelf was restricted, broad evaporite lagoons may have formed and are characterized by brining-upward parasequences consisting of mudsupportstones capped by gypsum or anhydrite (e.g., Sarg, 1981; Bein and Land, 1983).

The seismic-facies expression of the shelf or platform interior and the tidal facies are generally a sheet or wedge-shaped unit containing concordant reflections and displaying onlap at the base. The continuity and amplitude of reflections range from low in facies tracts that are predominantly carbonate (Figs. 3A, B, 5) to high in a mixed clastic or evaporite/carbonate setting. The shelf may contain localized carbonate buildups with a mounded seismic geometry. The mounds display downlap at their base and concordance or truncation at their top. More subtle low-relief mounds may be identified by thickening of the seismic reflections into the buildup. Overlying units drape or onlap. Amplitude and continuity of reflections range from low in more massive-bedded or reefal buildups to high in well-stratified carbonate sandbank bodies (see Bubb and Hatlelid, 1977, and Figs. 5 and 6, this report, for examples).

Platform or bank margin.—

The platform/bank margin consists of a complex of lithofacies, depending upon available types of organics and water conditions. The complex can include shoaling skeletal or nonskeletal grainstones and packstones and organic/cement boundstone reefs. Bank margin parasequences are commonly capped by widespread correlatable exposure surfaces. In many cases, individual parasequences may be difficult to distinguish (see Wendte and Stoakes, 1982, for an example from the Devonian of Canada) because of vertical stacking of carbonate sand bodies that were deposited in an active high-energy wave-and-current regime. The bank margin facies commonly contain small- to medium-scale festoon cross-bedding and submarine hardgrounds. Bioherms contain masses and patches of organic/cement boundstone. Interstices are filled with lime mudstone and/or skeletal grainstones and packstones. This facies belt is deposited in water depths of from sea level to 50 m depth and in places may have built up into small supratidal islands that can be several kilometers across.

The seismic-facies expression of the platform or bank margin may be mounded (Figs. 3C, 5, 6) with a variable degree of break in slope (Figs. 3A, 5, 6), a series of toplapping reflection cycles (Figs. 3B, 6), or some combination of the two (Figs. 3C, 6). The platform/bank margin facies will grade shelfward into the shelf facies and basinward into foreslope facies.

Foreslope.—

The foreslope facies belt is located on the incline that formed seaward of the platform/bank margin break as the ramp or prograding bank built seaward. Depositional slopes may be as much as 35° or steeper and the water may be several hundred to over 1,000 m deep. Lithofacies consist of bedded lime mudstone with megaslumps and lens- or wedge-shaped strata consisting of lithoclastic or bioclastic lime sand, all deposited as debris shed off the adjacent bank or ramp. Siliciclastic material interbedded with the carbonate may be present (Fig. 3B).

Parasequence development is less obvious in this facies belt. It may be represented by carbonate (transgressive)/shale (regressive) couplets (see Stoakes, 1980) or by lime-mud/allodapic-sand couplets capped by submarine hardgrounds. Downslope buildups can occur and range from grain-rich to mud-rich; the Strawn buildups of the Permian basin are an example of the former (Fig. 3), and the Silurian pinnacle reefs, Michigan, and Mississippian Waulsortian mounds of New Mexico are examples of the latter.

The foreslope facies is characterized seismically by downlapping reflections, which vary from low angles (<5°; Figs. 3A, 3B) to moderate angles (5–12°; Figs. 3B, 6) to high angles (>12°; Figs. 3C, 5). Foreslope reflections are composed of interfingering foreslope debris and muddy carbonate. They have variable amplitude and continuity depending on the impedance contrasts between these two lithofacies.

Basin floor.—

This facies varies in composition according to the degree of circulation and water depth. Basinal environments as much as 100 m deep that were oxygenated and of normal-marine salinity with good current circulation are commonly characterized by burrowed skeletal wackestones with some packstones. Siliciclastic-rich beds occur interbedded with the limestones. Biota is diverse, may be abundant in places, and may include brachiopods, corals, cephalopods, and echinoderms. Deeper (several hundred meters) or more restricted basin areas are characterized by an oxygen-poor, quiet-water environment. Dark, thin-bedded, and commonly laminated lime mudstone is the dominant textural type. Chert is common. The biota contains sponge spicules and is dominantly nektonic-pelagic, including calpionellids, coccoliths, radiolarians, and diatoms. If the basin was sufficiently restricted, salinity stratification occurred, and the basinal carbonate sediment may contain penecontemporaneous gypsum or anhydrite.

Carbonate basin environments are commonly starved during highstands of sea level. The basinward-dipping foreslope units thin dramatically at the toe of slope. This is illustrated in Figures 3B and C, 5, and 6. Basin-restricted units interpreted to represent deposition during lowstands of sea level occur as onlapping units and may be composed of siliciclastic, evaporite, or carbonate sediment. Further discussion of carbonate lowstand units is included in following sections of this report.

CONTROLS ON CARBONATE PRODUCTIVITY AND DEPOSITION

The depositional geometry, facies distribution, and early diagenesis of a carbonate depositional sequence are controlled, most importantly, by *relative changes in sea level*, *depositional setting* (*basin architecture*), and *climate*.

Relative Changes in Sea Level

The first and primary control on carbonate productivity and platform or bank growth, and thus on the resultant facies distribution, is, I believe, relative sea-level change. This change is the sum of tectonic rates (subsidence or uplift) and rates of change in eustasy. The resultant *accommodation* (Posamentier and Vail, this volume) represents the accumulation potential of any carbonate sequence.

Carbonate sediment originates more or less *in situ*, within the depositional environment. Carbonate material is produced largely by organisms, and much is produced as a by-product of photosynthesis (Schlager, 1981). The process is thus dependent on light and decreases rapidly with water depth. Prolific production of carbonate is limited to the upper 50 to 100 m of the water column, which can sustain abundant growth of photosynthetic organisms (Schlager, 1981, summarized from others). Significantly, production is highest in water as much as 10 m in depth and drops off dramatically between the depths of 10 and 20 m (Fig. 7). These narrow depth limits of neritic carbonate production are an important factor in the ability of carbonate production to keep pace with changing sea level.

The history of carbonate reef deposition during the Holocene rise in sea level illustrates the effects of changing sea level on carbonate productivity. Although Holocene reef-building corals can exceed the rate of sea-level rise by an order of magnitude (10^4 to 10^5 versus 10^3), they have grown more slowly (Schlager, 1981). Their vertical growth is a function of total mass balance limited by the relative rise of sea level. The maximum growth rates of 12,000 to 15,000 μm/yr (Macintyre and others, 1977) have outpaced the fastest sea-level rise, namely, in the early Holocene at 8,000 μm/yr. Even so, a large number of reefs and platforms did not keep pace with the rising sea during the early Holocene and have been drowned (e.g., Campeche Bank) or have retreated along their seaward margins (e.g., Bahama Banks, Caribbean platforms; Schlager, 1981).

Reef growth and, probably, most carbonate production are easily disturbed by changes in the environment. Lower

1 KM

L − 5 X

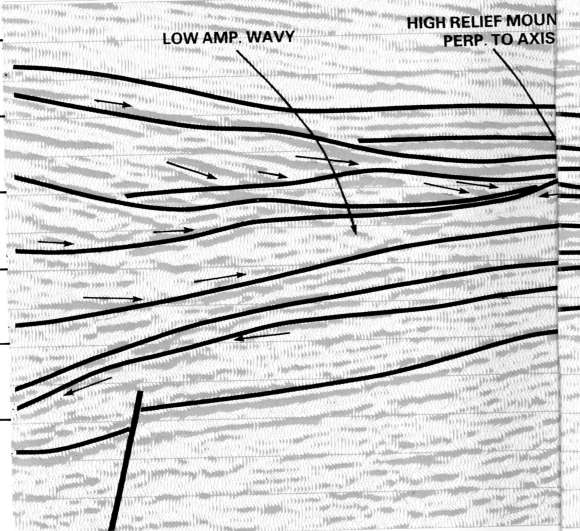

LOW AMP. WAVY

HIGH RELIEF MOUN
PERP. TO AXIS

HIGH AMP. OBLIQUE/PARALLEL

—Terumbu carbonate platform (Middle-Late Miocene), Natuna Field, offshore Borneo. Seismic line shows mounded highst
sive slopefront erosion on the 15.5-ma (SW end of line), 13.8-ma (NE end of line), and 10.5-ma sequence boundaries. A
nous carbonate lowstand wedge onlaps the 10.5-ma sequence boundary on the Southwest platform margin. A belt of high-r
platform.

DELAWARE BASIN SEISMIC LINE

LITHOFACIES INTERPRETATION
MIDDLE PERMIAN DEPOSITIONAL SEQUENCES

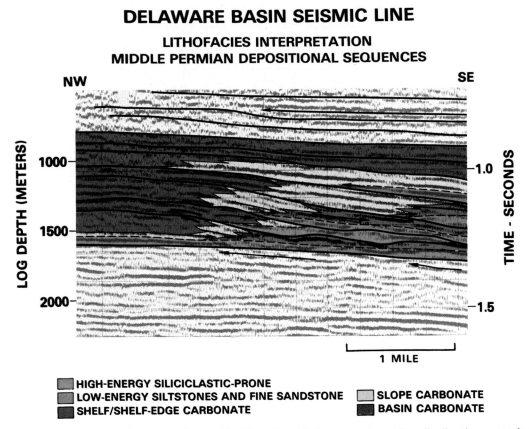

HIGH-ENERGY SILICICLASTIC-PRONE
LOW-ENERGY SILTSTONES AND FINE SANDSTONE   **SLOPE CARBONATE**
SHELF/SHELF-EDGE CARBONATE   **BASIN CARBONATE**

FIG. 6.—Portion of a CDP seismic line located northeast of Carlsbad, New Mexico, and oriented in a dip direction across the northwest margin of the Delaware basin. The lower two carbonate banks (Permian, Wolfcampian) display a sigmoid (lower) and mounded (upper) geometry at the bank margin. The upper mounded bank illustrates the changes in seismic facies in a basinward direction from continuous, higher amplitude reflections of the platform to discontinuous, mounded reflections at the bank margin, to more continuous and higher amplitude reflections of the foreslope. The next overlying bank shows sigmoid progradation with no mounding and, again, a change at the bank margin to a less continuous lower amplitude seismic facies.

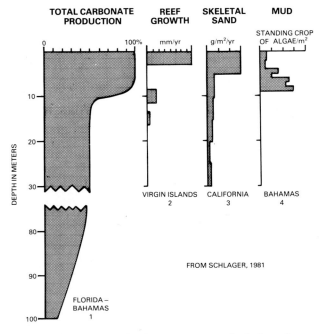

FIG. 7.—Carbonate production versus water depth from four areas, demonstrating high-carbonate productivity in shallow water and an abrupt, dramatic fall-off at a water depth of about 10 m.

accumulation rates can be attributed to the off-bank or off-platform flow of slightly hypersaline or nutrient-poor water from shallow platform top lagoons during the early Holocene transgression (Adey and others, 1977; Lighty and others, 1978), to the decrease of growth rates with water depth (Fig. 7), and to the slow growth during early carbonate production (Schlager, 1981). The actual long-term accumulation rate is thus likely a function of the following: changing water mass conditions during relative changes in sea level (i.e., salinity, nutrients, temperature, oxygen content); and the rate of change of accommodation (i.e., sea level plus subsidence) generated during any sequence interval.

The long-term accumulation rates of ancient carbonate platforms or banks are much less than Holocene rates (Fig. 8; Table I). Ancient rates range from 13 μm/yr for the Silurian of Michigan to 365 μm/yr for the lower Lower Clear Fork carbonate of the Midland basin, Texas. Holocene growth rates range from 500 to 1,100 μm/yr for oolite sands and tidal deposits during the late Holocene rise (500 μm/yr; Schlager, 1981, from others) to over 10,000 μm/yr for some reefs. Based on the maximum Holocene accumulation of 12 m along the Bahama Bank margin (Hine and others, 1981), the accumulation rate for the bank is 1,200 μm/yr. The Holocene rates are not anomalously high, however, when one considers that they are calculated over

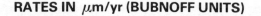

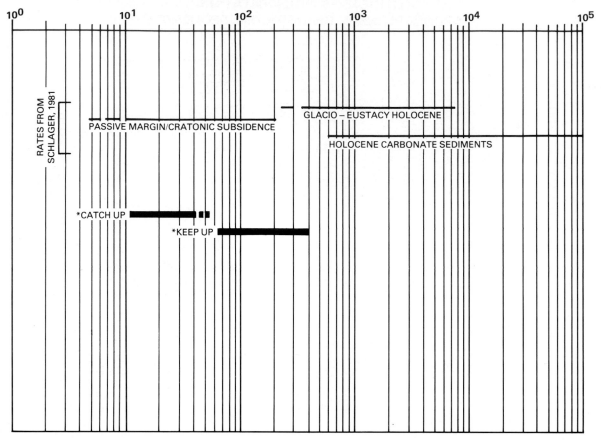

FIG. 8.—Comparison of *keep-up* and *catch-up* rates of deposition to tectonic and glacio-eustatic rates. Catch-up carbonate deposition appears significantly slower than keep up deposition. Both appear limited by the accommodation potential (i.e., subsidence plus eustasy), which, with the accompanying changes in water mass conditions, appears to be the dominant control on carbonate deposition. Potential carbonate depositional rates, as expressed by measured Holocene rates, are greater than subsidence rates and comparable to or greater than glacio-eustatic rates.

a much shorter time span (10,000 yr), they do not include burial compaction, parasequence hiatuses, or long-term periods of sea-level stillstand, and they are compared with the probable lower accommodation potential of many ancient sequences.

Depositional Settings

Another crucial effect on the overall potential growth geometry of a carbonate sequence is basin architecture. Unrestricted basins with normal, well-circulated marine waters will provide a favorable habitat for a more cosmopolitan biota with a growth potential different from that of a restricted basin. Basins with elevated salinities or deficient oxygen have a specialized or reduced biota. Abrupt breaks in seafloor slope, such as along rifted basin margins or at the edges of isolated horst blocks, may localize reef or carbonate-sand shoal development. Well-defined linear facies belts with abrupt, lateral facies changes will develop adjacent to shoal areas (Fig. 4).

Subsequent platform or bank margin growth will be by aggradation and progradation, and the resultant geometry will be a function of growth characteristics of the contributing organisms (e.g., wave-resistant reef; loose-sand shoal) and water depth. Progradation is common in shallow to moderately deep (100–600 m) basins with moderate-to-low subsidence rates. Deep-ocean–facing margins have dominantly aggradational geometries (i.e., Bahama Late Neogene and Terumbu platforms; Figs. 3C, 5). In contrast, settings in which the sea floor gradually deepens without any abrupt breaks may develop broader, less well-defined facies belts (i.e., craton settings).

Climate

A third especially important control on carbonate facies development is climate. When arid, it favors evaporite deposition. Evaporite deposition may occur in association with shelf carbonates, filling in shelf basins and lagoons, and within supratidal flats (i.e., sabkha deposition). During times

TABLE 1.—CARBONATE HIGHSTAND SYSTEMS TRACTS AND THEIR RATES OF ACCUMULATION AT THE SHELF MARGIN

Sequence	Average Thickness (m)	Age Approx. (ma)	Rate (μm/yr)
Ellenburger West Texas[a]	396 m	E. Ordovician 478–505	15
Pinnacle Reefs, Michigan[b]	180	Silurian 415–429	13
L. Keg River Reef Platform[c]	150	Devonian 378–389	14
Pillara Limestone, Canning Basin,[d] Australia		Devonian (Frasnian)	30
Wichita, West Texas[e]	550	Permian 268–279	50
Capitan Reef, New Mexico[e,f]	500	Permian (245)248–254	55–83
Grayburg Fm, West Texas[g]	160	Permian 255–256	160
Smackover Ls, Arkansas	335	Jurassic 144–148	83
Lower/Middle San Andres, New Mexico[g]	280	Permian 257–258	280
Upper San Andres, New Mexico[g]	180	Permian	180
Swan Hills, Canada[h]		Devonian 374–375	122
Shuaiba, Middle East[i]	155	Cretaceous 109–110	155
Haynesville, East Texas	190	Jurassic 136–138	95
Lower Clear Fork, Midland Basin	365	Permian 265–266	365
Terumbu Ls,[j]		U. Miocene	
5.5 ma	158	0.8	197
6.3 ma	442	5.2	85
10.5 ma	561	3.3	170
13.8 ma	134	1.7	80
15.5 ma	286	1.0	286
Bahama Bank[k]	12	Holocene 10,000 yr	1200

(a) Loucks and Anderson, 1980; (b) Mesolella and others, 1974; Sarg, 1982; (c) Schmidt and others, 1980; (d) Schlager, 1981, from Playford and Lowry, 1966; (e) Silver and Todd, 1969; (f) King, 1948; Silver and Todd, 1969; (g) Sarg and Lehmann, 1986; (h) Wendte, pers. commun. (i) Lehmann, pers. commun.; (j) Rudolph and Lehmann, 1987; (K) Hine and others, 1981.

of restriction, evaporites may fill basin areas. Climate is also an important control on the extent of early post-depositional diagenesis associated with exposure of the carbonate unit during times of falling sea level and during lowstands in sea level. The development of secondary karst porosity can vary tremendously in degree and areal extent. This variance relates to the time and length of exposure, and to whether the climate is arid because of low rainfall or humid because of abundant rainfall.

Assuming, then, the strong control that relative changes in sea level have on carbonate productivity and platform growth, the following sections present a depositional-sequence model that characterizes the deposition of carbonate platforms and banks. Use of this model and division of carbonate strata into their component depositional sequences suggests that the vertical stacking and areal distribution of carbonate facies are predictable.

CHARACTERISTICS OF HIGHSTAND SYSTEMS TRACTS

Carbonate depositional-systems tracts that occur during relative highstands in sea level are bounded below by the top of the transgressive-systems tract (a downlap surface in many cases) and above by a sequence boundary (Fig. 1). Highstand systems tracts are generally characterized by a relatively thick aggradational-to-progradational geometry. They form widespread platforms, ramps, or prograded banks and have offshore equivalents in isolated platforms. They are interpreted to be deposited during the late part of the eustatic rise, the eustatic stillstand, and during the early part of the eustatic fall (Jervey, this volume; Posamentier and Vail, this volume).

Carbonate highstand systems tracts are generally characterized by early and late stages that reflect different rates of accommodation and associated water mass conditions during the early and late highstand (Fig. 9). The early highstand is characterized by relatively greater increases in accommodation and by water conditions not necessarily conducive to high-carbonate productivity. The result is relatively slow deposition and aggradational accumulation in the shelf areas (Jervey, this volume) and is expressed as a sigmoidal pattern on the seismic section (Figs. 3, 6). Later, as global sea level begins to fall, the rate of increase of accommodation is reduced on the shelf (Figs. 1, 9). Shelf waters become better circulated and more stable, resulting in higher carbonate productivity.

Mounded aggradation to oblique progradation at the platform/bank margin characterizes the late highstand systems tract. This characteristic is well expressed in the Leonard (Middle Permian) strata of the Permian basin (Figs. 3, 10). An early highstand sigmoidal pattern composed primarily of carbonate is succeeded by oblique progradation composed of mixed carbonate and sandstone.

Highstand carbonate systems tracts appear to undergo two fundamentally different depositional histories. These histories are characterized by significant differences in micrite content and/or submarine cement at the platform/bank margin, which can be related to the rates of accumulation (Fig. 8). Modifying terms coined by Kendall and Schlager (1981), these two depositional styles are here termed *keep-up* and *catch-up* carbonate systems. Although they probably represent end members in a range of carbonate responses to sea-level rise, most highstand carbonate systems tracts can be characterized by either style. Each type of carbonate deposition creates a different depositional profile in outcrop, and where resolvable, on seismic profiles.

Keep-Up Carbonate Systems

A *keep-up* carbonate systems tract displays a relatively rapid rate of accumulation (see rates shown in Fig. 8) and is able to keep up with relative rises in sea level. A *keep-up* carbonate is characterized at the platform margin by relatively small amounts of early submarine cement and is generally dominated by grain-rich, mud-poor parasequences.

The *keep-up* carbonate systems tract displays a mounded/oblique geometry at the platform/bank margin and in places on the platform. The Abo and San Andres highstand systems tracts of the Delaware basin are expressed as mounded prograded banks (toplap at upper boundary) for most of their deposition (Figs. 6, 11). The lower San Andres contains

CARBONATE HIGHSTAND DEPOSITION

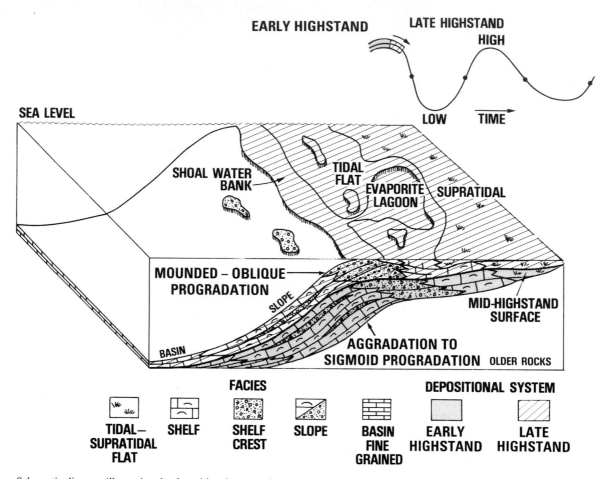

FIG. 9.—Schematic diagram illustrating the depositional geometries associated with early and late highstand deposition. Early highstand deposition shows aggradation to sigmoid progradation; late highstand deposition displays mounded-to-oblique progradation. The highstand systems tract may be dominated by either style of deposition.

only locally developed, early submarine cement in both the early and the late portions of the highstand (Sarg and Lehmann, 1986). A similar depositional system occurs in the upper San Andres Formation (Permian, Guadalupian; Fig. 11), where a mixed carbonate-clastic system displays a predominantly oblique progradational geometry for much of its history (Sarg and Lehmann, 1986). Bank edge pelloidal and ooid dolograinstones contain little early submarine cement (Figs. 12, 13).

Two other examples of interpreted *keep-up* deposition come from the platform areas of western Canada and offshore Indonesia. The Upper Devonian (Frasnian) highstand systems tracts of western Canada show a mounded-to-oblique geometry throughout their history. These well-studied reefs (Golden Spike by Walls and others, 1979; Judy Creek by Wendte and Stoakes, 1982) contain moderate amounts of early submarine cement in the buildup margin skeletal-grainstone lithofacies. Another example comes from the Middle and Upper Miocene sequences of the Terumbu carbonate platform, offshore Indonesia (Rudolph and Leh-

mann, 1987). There, the Miocene carbonates consist of four highstand systems tracts. They all display a mounded and/ or shingled geometry (Figs. 3C, 5) and contain abundant shallow-water, grain-rich facies. There was little early submarine cementation in the coral-red algae boundstone and coral-red algae-echinoderm packstones and grainstones that were deposited on the platform margin and crest (Rudolph and Lehmann, 1987).

Catch-Up Carbonate Systems

A *catch-up* carbonate system displays a relatively slow rate of accumulation (slow rates of Fig. 8). This response may result from the maintenance of water conditions throughout most of the highstand that are not conducive to rapid carbonate production—namely, oxygen-poor water, lack of nutrients, high salinity, or low water temperature. A *catch-up* carbonate is characterized at the platform margin by extensive, early submarine cementation, and it may contain abundant mud-rich parasequences. The extensive

DELAWARE BASIN SEISMIC LINE

LITHOFACIES INTERPRETATION
MIDDLE PERMIAN DEPOSITIONAL SEQUENCES

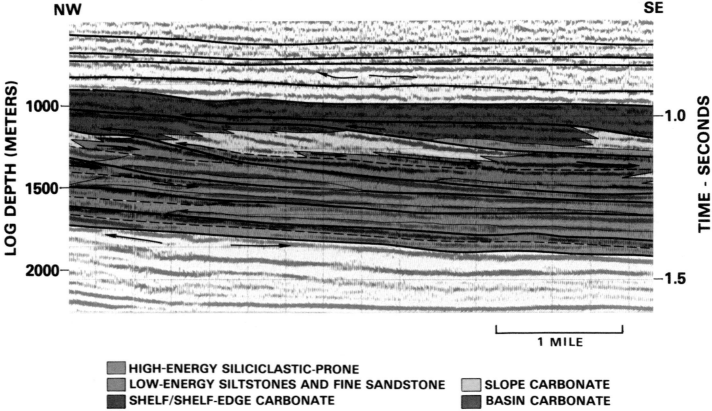

HIGH-ENERGY SILICICLASTIC-PRONE
LOW-ENERGY SILTSTONES AND FINE SANDSTONE
SHELF/SHELF-EDGE CARBONATE
SLOPE CARBONATE
BASIN CARBONATE

FIG. 10.—Portion of Delaware basin CDP seismic line immediately basinward of Figure 6. Prograded bank (Permian, Leonardian) in center of section displays early highstand aggradation in the lower part and late highstand progradation in its upper basinward part (lithofacies as in Fig. 6).

early cementation may be the result of a relatively longer time available during deposition for active pore fluid migration and cement precipitation. Only during the latest portion of the highstand, when accommodation is reduced because of falling sea level, will a *catch-up* system display *keep-up* characteristics.

The *catch-up* carbonate system displays a sigmoidal depositional profile at the bank or platform margin. The outcrop profile of the Capitan reef (Permian, Upper Guadalupian) on the margin of the Delaware basin, West Texas/ New Mexico, displays a sigmoidal progradational pattern (Fig. 14). Concordant topset beds change basinward to gently basinward-dipping shelf crest beds. These pass farther basinward into the massive bank margin Capitan reef. The Capitan reef, in turn, passes farther basinward into steeply inclined (approximately 35°) foreset beds. Submarine cement is pervasive throughout the Capitan massive, shelf crest, and foreslope beds (Babcock, 1977; Yurewicz, 1977; Hurley, 1979).

Two other examples of Paleozoic *catch-up* highstand de-

position come from the Silurian of the United States and the Devonian of Australia. The Middle Silurian reef sequences of the Michigan basin contain two highstand systems tracts. Each systems tract shows dominantly aggradational growth on a regional scale (Fig. 15) and little evidence of shoal water deposition, except in the latest portion of the younger highstand. Both systems tracts are interpreted to represent *catch-up* carbonate systems. Early submarine cement is pervasive in each systems tract in the platform margin and slope deposits (Lehmann, 1978; McGovney, 1978; Sarg, 1982). The Upper Devonian (Frasnian) Pillara Limestone of the Canning basin, Australia, also displays a sigmoid pattern in outcrop (Playford and Lowry, 1966) and contains abundant early submarine cement.

TYPES OF UNCONFORMITY AND ASSOCIATED PROCESSES

Type 1 Sequence Boundary

Two major processes occur during the formation of a type 1 sequence boundary, when sea level falls at a rate rapid

DELAWARE BASIN SEISMIC LINE
SHELF EDGE SEGMENT

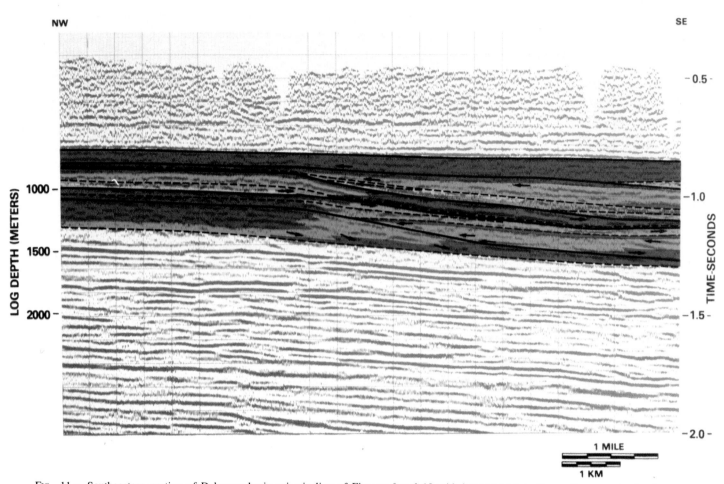

FIG. 11.—Southeastern portion of Delaware basin seismic line of Figures 6 and 10 with interpreted late Leonardian-to-early Guadalupian sequences. (A) Major shelf margin present in center of interpreted seismic section occurs within the San Andres Formation, The shelf edge has a mounded aggradational geometry. The immediately overlying sequence represents the upper part of the San Andres Formation and displays an oblique progradational geometry (lithofacies as in Fig. 6). Both San Andres banks are interpreted to represent *keep-up* deposition and contain little early submarine cement in their outcrop equivalents (see Fig. 13).

enough to drop below the preceding platform/bank margin. These processes are (1) slope front erosion and (2) seaward movement of the regional freshwater meteoric lens (Fig. 16).

Slope front erosion.—

During the formation of a type 1 sequence boundary, significant slope front erosion can occur and results in substantial loss of platform/bank margin and upper-slope material. The result is downslope deposition of carbonate megabreccias and traction or density current deposition of carbonate sands. The erosion can be local or regional in extent.

Three Middle to Upper Permian type 1 sequence boundaries, exposed on the western escarpment of the Guadalupe Mountains at the edge of the Delaware basin, display significant slope front erosion (Figs. 17, 18). The sequence boundary at the base of the Cutoff sequence has eroded into the underlying Victoria Peak Formation and has removed as much as 250 m of bank front material (L. C. Pray, pers. commun., 1980). The sequence boundary at the top of the Cutoff sequence also displays significant slope front erosion and, in addition, contains several well-developed erosional cuts (Fig. 17), interpreted here to represent portions of Permian submarine canyons. In places, the entire Cutoff has been removed by erosion.

DELAWARE BASIN SEISMIC LINE
BASIN MARGIN SEGMENT

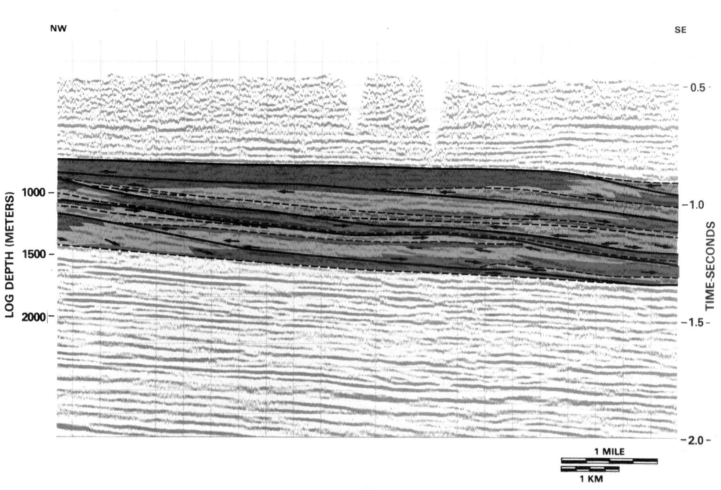

FIG. 11.—(B) Identified lowstand wedges on this portion of the seismic line occur onlapping the lowermost interpreted highstand bank (upper Leonardian) and the middle and upper San Andres banks. Well data indicate the lower LSW is composed of carbonate bank lithofacies, and the upper two LSWs are composed of mixed carbonates and siliciclastics.

The third sequence boundary showing slope front erosion on the western escarpment of the Guadalupe Mountains occurs in the upper portion of the Grayburg Formation (Upper Permian), where submarine erosion has removed 70 m or more of the bank edge and has eroded back into the shelf strata of the Grayburg Formation (Fig. 18). Massive bedded, skeletal floatstone and rudstone beds were deposited directly in front of the erosional surface (Fekete and others, 1986). Seismically, this type of erosion is identified by abrupt truncation at the platform/bank margin of the shelf and bank edge reflections (Fig. 11).

Other platform/bank margins display significant slope front erosion: the Upper Jurassic bank margin in offshore Morocco (Mitchum and others, 1977; their figure 4, p. 61), the Terumbu Platform sequences (Fig. 5) and the Middle Permian Clear Fork sequences of the northern Midland basin (Fig. 3B).

Seaward movement of freshwater lens.—

The second major process that is interpreted to occur during formation of a type 1 sequence boundary is regional migration of the freshwater lens in a basinward or seaward direction (Fig. 16). Regional diagenetic events affecting large portions of the carbonate highstand facies tracts will be associated with this meteoric lens. The degree to which the lens extends down into the carbonate section is related to the magnitude and rate of sea-level fall as well as to the length of time that sea level remains below the platform/bank margin. This will affect the degree of meteoric and mixing diagenesis in each carbonate sequence.

I propose that during the formation of a *large-scale type 1 sequence boundary*—that is, when sea level falls 75 to 100 m or more and for an extended time—the freshwater lens may be established for a long time over the shelf, and its influence may extend well into the subsurface and perhaps into underlying sequences. With sufficient rainfall, significant secondary solution and solution compaction may occur over the shelf in the shallow portion of the section. Precipitation of abundant meteoric cement will occur deeper in the phreatic zone. Unstable aragonite and high-Mg calcite grains may be dissolved and reprecipitated as low-Mg calcite cement. The Vail eustatic-cycle charts indicate that major type 1 falls in global sea level are rare (Haq and others, this volume). Generally, sea-level falls are of much smaller magnitude. During formation of *small-scale type 1 sequence boundaries*—sea level falls less than 75 to 100 m and for a short time—the meteoric lens is less well established and remains in a shallow position on the shelf, resulting in less extensive solution and precipitation of phreatic cement.

Mixing and hypersaline dolomitization may be important processes during the late highstands and may continue through the formation of either large-scale or small-scale type 1 sequence boundaries. During the formation of a small-scale type 1 sequence boundary, the dolomitization process may affect only the shallow portions of a carbonate sequence.

Examples of carbonate highstand systems tracts that have undergone extensive secondary solution associated with one or more large-scale type 1 sequence boundaries include:

—Pleistocene limestones of the Caribbean (Land, 1973a, b);
—Upper Miocene reefs of Spain (Armstrong and others, 1980) and, possibly, of the Terumbu Platform, offshore Borneo (Rudolph and Lehmann, 1987), where the highstands are capped by the 5.5-, 6.3-, and 10.5-ma sequence boundaries (see Haq and others, this volume for global-cycle chart and ages of sequence boundaries);
—Middle Cretaceous Golden Lane platform of Mexico, affected by the 94-ma unconformity, contains porosities greater than 30 percent in places and is dominated by solution-enlarged inter- and intra-particle, moldic, and vug porosity (see Wilson, 1975, for summary);
—Cretaceous (Aptian) Shuaiba shelf and reef bank-edge facies of the Middle East area capped by the 109-ma sequence boundary (Litsey and others, 1983; Frost and others, 1983);
—Upper Mississippian limestones of the United States (e.g., New Mexico; see Meyers, 1974, 1978, 1980) display solution over the shelf and phreatic cementation in outer-shelf and slope positions (Meyers, 1978) associated with the pre-Pennsylvanian (top Chester) sequence boundary;
—Middle Devonian Sulphur Point–Keg River carbonates of the northern Alberta basin, Canada (Bebout and Maik-

lem, 1973), affected by a major Middle Devonian unconformity;
—Middle Ordovician Ellenberger Dolomite of the Permian basin, West Texas, displays significant solution porosity (Loucks and Anderson, 1980).

In contrast to the examples above, most type 1 sequence boundaries are smaller in scale, and the underlying carbonate highstand systems tracts generally appear to have only locally developed solution porosity and phreatic cementation. Dissolution would be most prevalent in the inner shelf and on topographic highs of shelf and platform/bank margin buildups. Meteoric cementation would occur away from these areas, deeper within buildups, and in the outer-platform and slope positions of the sequence. The upper Jurassic (Tithonian) carbonate platforms of the Arab A–C cycles are affected by the small-scale type 1 sequence boundaries at 134, 135, and 136 ma, and display minor solution porosity and phreatic-spar cementation (P. J. Lehmann, pers. commun., 1984). The highstand shelf limestones of the midcontinental United States Pennsylvanian cyclothems commonly display dissolution of unstable grains, solution compaction, and local phreatic cementation associated with periods of subaerial exposure (Heckel, 1983; Wilson, 1975) during formation of small-scale type 1 sequence boundaries. The Albian Edwards and Stuart City limestones of the Gulf Coast possess locally preserved inter- and intraparticle porosity in platform margin reefal areas (Bebout and Loucks, 1974). Significant early fibrous-to-bladed submarine cement (14 percent bulk volume; i.e., *catch-up* deposition) and near-surface, meteoric, equant-spar cement (16 percent bulk volume; Prezbindowski, 1983), here interpreted to be associated with the 97.5- and 98.5-ma type 1 sequence boundaries, have occluded most of the depositional porosity in the Stuart City platform margin. Dolomitized and recrystallized lime mud contains secondary intercrystalline and moldic porosity of as much as 15 percent in the inner-shelf area (Griffith and others, 1969; Rose, 1972).

Type 2 Sequence Boundary

The processes and deposition associated with a type 2 sequence boundary differ somewhat from those of the type 1 sequence boundary. During the formation of a type 2 sequence boundary, sea level is interpreted to fall to a position at or just below the bank margin, and the inner-platform area is exposed (Fig. 19). The outer-platform and platform margin areas may experience brief subaerial exposure. In general, the dominant meteoric effect will be in the inner platform. There, meteoric diagenetic processes will be similar to the effects interpreted to occur during a small-scale type 1 eustatic fall. These effects include dissolution of grains, especially unstable aragonite and high-Mg calcite; precipitation of minor amounts of vadose and phreatic cement; and mixing-zone dolomitization. Hypersaline dolomitization may be initiated during formation of a type 2 sequence boundary. In contrast to the type 1 sequence boundary, sea level is interpreted to begin rising within a relatively short time and to flood back over the outer-plat-

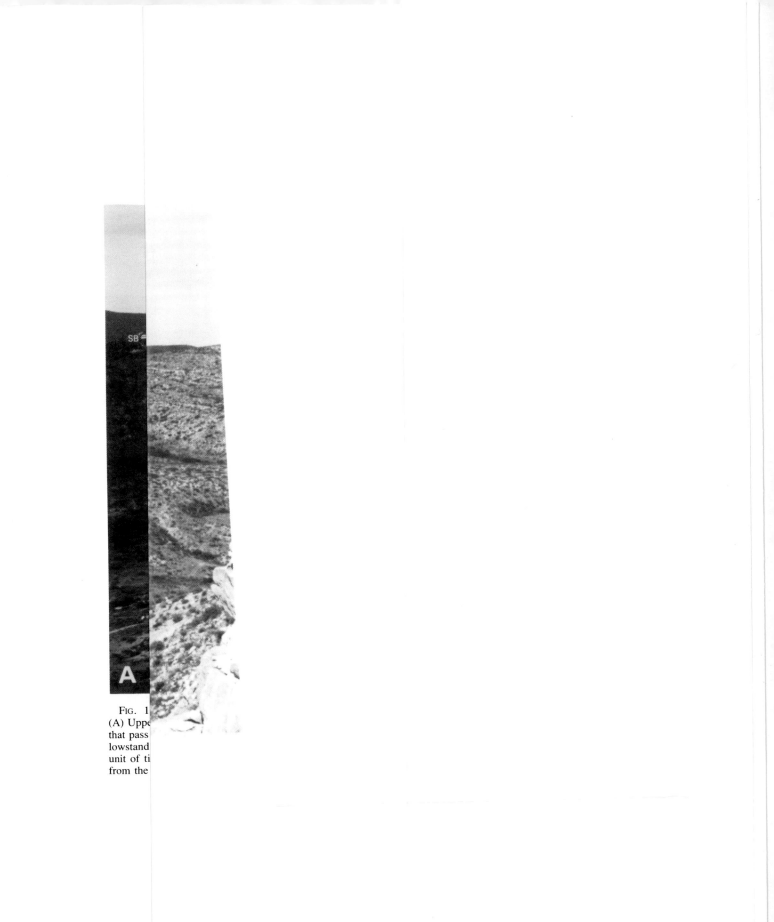

Fig. 1
(A) Uppe
that pass
lowstand
unit of ti
from the

FIG. 14.—Outcrop photo of Capitan Limestone (Permian, Guadalupian) of North McKittrick Canyon, Guadalupe Mountains, West Texas. The massive Capitan displays a *catch-up* sigmoidal progradational geometry. Concordant topset beds change basinward to thicker bedded, gently dipping, shelf crest beds. These, in turn, pass into the massive bank margin Capitan reef. Early submarine cement is pervasive throughout the Capitan massive, shelf crest, and foreslope facies.

form area. Deposition of a platform/bank margin wedge will commence (Fig. 19) at or below the underlying platform margin and onlap in a landward direction. In contrast to the occurrence of erosion, in type 1 sequence boundaries, slope front erosion does not appear to be a major process associated with type 2 sequence boundaries.

The Smackover Limestone (Oxfordian, Jurassic) of Arkansas and north Louisiana is a well-studied carbonate platform that culminates in a type 2 sequence boundary (Fig. 20). It is composed of two shoaling-upward highstand systems tracts. The Smackover displays *keep-up* deposition in the later parts of each highstand, and reservoirs are in thick oolitic grainstones. Marine cementation is not abundant but occurs in thin zones where it may essentially fill interparticle porosity (Moore and Druckman, 1981; Wagner and Matthews, 1982). The upper highstand systems tract is

overlain by the Buckner anhydrite and redbeds in central Arkansas and Buckner redbeds in southernmost Arkansas and northern Louisiana (Moore and Druckman, 1981). The Buckner in southern Arkansas and northern Louisiana is interpreted to represent the abrupt downward shift of the inner platform to nearshore clastic/evaporite facies over the Smackover Limestone during formation of the 144-ma type 2 sequence boundary. The inner-shelf area of central Arkansas is dominated by the dissolution of ooids, with intergranular porosity entirely filled with very fine, clear, equant, sparry calcite cement. Reservoir-quality porosity and permeability occur where the oomoldic grainstones have been dolomitized (Midway Field). The diagenesis predates burial and compaction and is interpreted to represent the effects of subaerial exposure and meteoric diagenesis (Moore and Druckman, 1981). The degree of dissolution dimin-

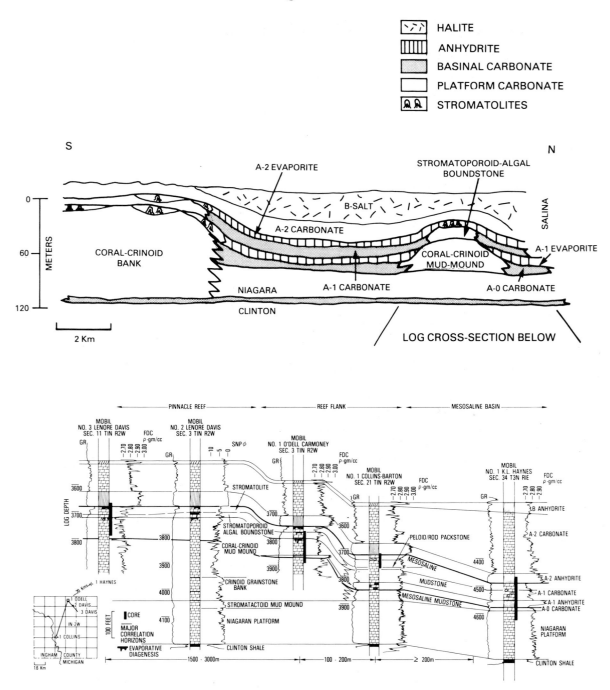

FIG. 15.—Regional schematic and well-log cross sections of Silurian reef interval, Michigan basin. *Catch-up* aggradation characterizes the basin margin pinnacle refs. Submarine cementation is pervasive throughout two highstand systems tracts. Each reef interval is punctuated by interpreted lowstand evaporite deposition (A-1 and A-2 evaporites) that onlaps and drapes the reefs.

ishes in a basinward direction. In southernmost Arkansas, porosity has been reduced by (1) minor fibrous marine cement, (2) solution compaction, and (3) precipitation of coarse poikilitic calcspar. Marine cementation is not abundant but occurs in thin zones where it may essentially fill interparticle porosity (Moore and Druckman, 1981; Wagner and Matthews, 1982).

CHARACTERISTICS OF LOWSTAND AND TRANSGRESSIVE SYSTEMS TRACTS

Lowstand and transgressive carbonate systems tracts represent an important part of carbonate sequence stratigraphy. Lowstand systems tracts are divided into three types: type 1 carbonate lowstand deposits (Figs. 16, 21); type 2 shelf

TYPE 1 CARBONATE SEQUENCE

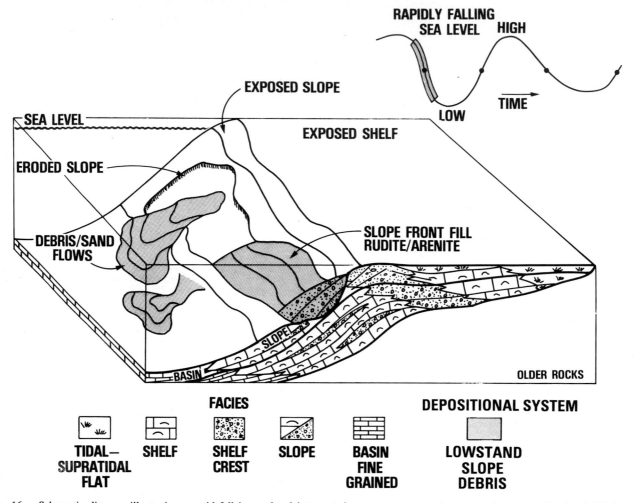

FIG. 16.—Schematic diagram illustrating a rapid fall in sea level interpreted to occur at a type 1 sequence boundary. See level falls below the platform/bank margin, subaerially exposing the platform and resulting in significant slope front erosion. Major slope front carbonate debris sheets and turbidite sands occur as localized onlapping wedges and channel fills. Erosion may continue through the lowstand in sea level.

or platform/bank margin carbonate wedges (Fig. 19); and basinally restricted and onlapping evaporite wedges (Fig. 22).

Type 1 Lowstand Deposits and the Transgressive-Systems Tract

Type 1 sequence lowstand deposits can be divided into *allochthonous debris*, derived from erosion of the slope front (Fig. 16), and *autochthonous carbonate wedges*, deposited on the upper slope during the interpreted eustatic lowstand (Fig. 21). The allochthonous deposits form wedges of carbonate sediment composed of carbonate debris flows and carbonate sands deposited at the base of and against the eroded slope. Allochthonous debris is also shed during highstand progradation (the slope and base-of-slope aprons of Cook, 1983), but, unlike the lowstand debris, these can be traced back up clinoforms to equivalent-age platform

material, and they are not associated with extensive erosion of the slope. As the interpreted eustatic lowstand is reached and the rate of sea-level fall slows, *in situ* carbonate growth may occur in the shallowed slope areas (Fig. 21). During this time, a slow relative rise in sea level will generate accommodation in the upper-slope and outer-platform areas. The lowstand wedge will then onlap back across the slope and outer platform.

The development of this wedge is affected both by the basin water conditions (i.e., salinity, circulation) and the angle of the underlying highstand foreslope (i.e., steep, gentle). If the basin retains normal-marine water conditions and is well circulated, and the underlying depositional slope is gentle, there will be a large area for abundant shallow-water carbonate deposition. A significant lowstand wedge may develop. More restricted basin conditions or steep depositional slopes may preclude development of a lowstand wedge.

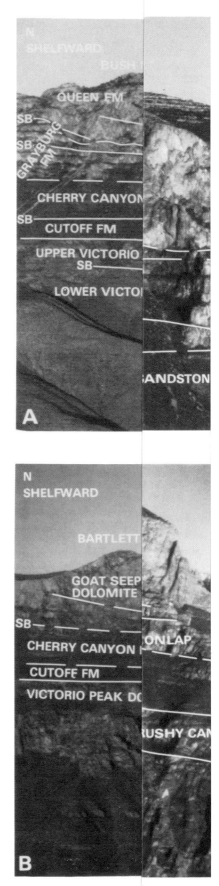

A

N
SHELFWARD
BUSH
QUEEN FM.
SB
SB
GRAYBURG FM
CHERRY CANYON
SB
CUTOFF FM
UPPER VICTORIO
SB
LOWER VICTOR
SANDSTON

B

N
SHELFWARD
BARTLETT
GOAT SEEP
DOLOMITE
SB
CHERRY CANYON   ONLAP
CUTOFF FM
VICTORIO PEAK DC
RUSHY CAN

FIG. 17.—Outcrop photos of U
the Delaware basin showing sequ
are composed of quartz sandstone
erosion cut and erosional channels

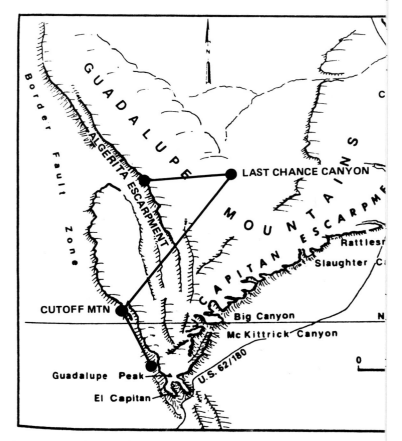

FIG. 18.—Regional cross section showing depositional-sequence interpretation and lithofaci
strata. Outcrop photos of Figure 17 keyed to cross section.

SMACKOVER–HAYNESVILLE LOG

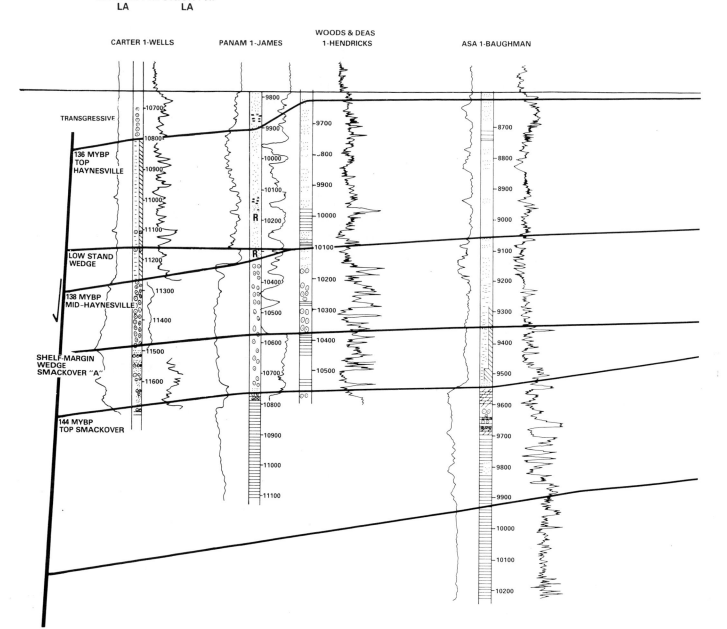

FIG. 20.—Regional Northeast-Southwest log cross section of Gulf basin Jurassic carbonates showing log character and lithofacies from cutting samples. Lowstand systems tracts of the 144-ma and 138-ma sequences are characterized by abrupt basinward or downward shifts in the redbed and

CROSS SECTION NE LOUISIANA

NE

UNION PH.
LA

HOME 1-GREEN

OLIN B-1   OLIN FEE

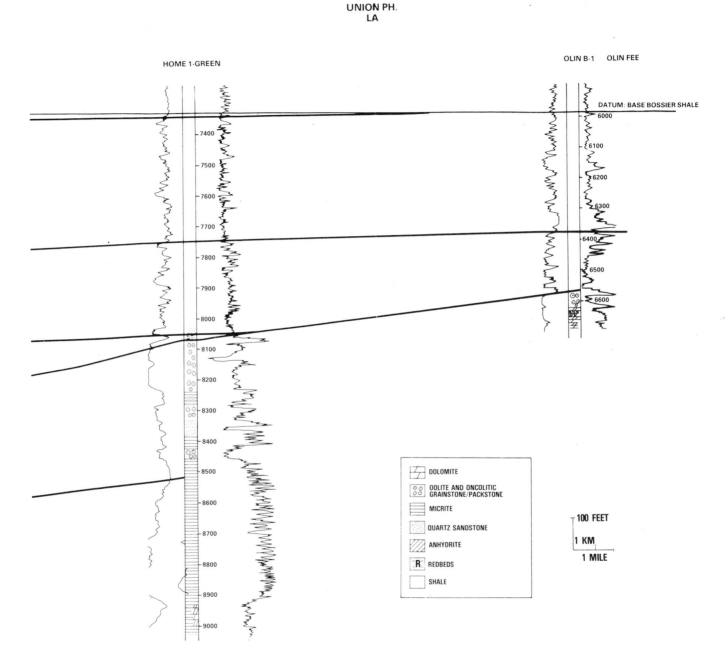

evaporite facies to, at, or below preceding platform margins. Highstand Smackover Limestone is composed of a shoaling-upward limestone ramp capped by ooid grainstones. During deposition of the shelf margin wedge, this ooid facies occurs basinward of the Buckner redbeds and evaporites.

TYPE 1 CARBONATE SEQUENCE

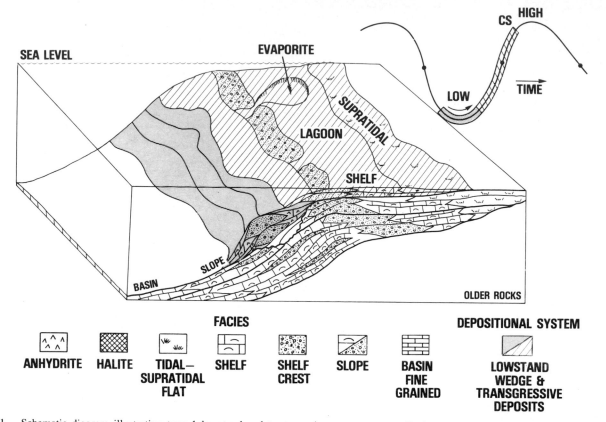

FIG. 21.—Schematic diagram illustrating type 1 lowstand and transgressive-systems tracts. During the eustatic lowstand, as relative sea level begins to rise, an autochthonous carbonate bank may be deposited in the upper slope or in a down-ramp position. As the paleoshelf becomes flooded, this lowstand bank is drowned, and a retrogradational transgressive-systems tract is deposited over the shelf. Offshore areas become starved, and a condensed section is deposited. The subsequent highstand systems tract downlaps over these underlying systems tracts.

gin exhibits widespread failure of carbonate platform margins. Massive slope failure occurred, resulting in submarine slides, debris flows, and turbidity current flows. Slope failure was coeval with siliciclastic deposition in the slope and basin areas and karsting of the platforms (Cook and others, pers. commun., 1987). A third example comes from the Triassic of Italy, where extensive earliest Carnian debris sheets occur at the toe of slope of Ladinian carbonate highstand platforms. These debris sheets are interpreted as having been deposited during a type 1 fall in sea level at the end of the Ladinian (Biddle, 1984; Bosellini, 1984; Fig. 23).

Autochthonous lowstand wedges.—

Depending on its configuration, a basin may have restricted or open-marine conditions during lowstands in sea level. Autochthonous lowstand deposits in a closed carbonate basin are predicted to be composed of *catch-up* carbonates dominated by micrite-rich slope front fill units. If arid climatic conditions prevail, evaporites will form in basinal areas. Basins that maintain open-marine conditions

during lowstands in sea level are predicted to develop *keep-up* carbonate wedges.

The Bahama and St. Croix bank margins display examples of *in situ* lowstand wedges deposited during the early stages of the Holocene rise in sea level. A series of Holocene reefs, now drowned in more than 20 m of water, were deposited at the bank margins. They onlap the upper slope bank margin, and, in the St. Croix example, the upper portion of submarine canyons (see Hine and Neumann, 1977, their figures 7 and 20; Hine and others, 1981, their figure 6; Hine, 1983, his figure 12; and Hubbard and others, 1986) and are now being covered by prograding Holocene bank margin carbonates.

Other examples of ancient, autochthonous, carbonate-lowstand deposits include: the Triassic of the Dolomites region of the Southern Alps (Bosellini, 1984; Fig. 23); the Miocene Natuna Field area in offshore Indonesia (Rudolph and Lehmann, 1987); and the lower part of the Permian basin Grayburg Formation.

In the Dolomites of northern Italy, the work of Bosellini (1984) suggests two lowstand wedges in the Triassic, one at the base of the Carnian associated with the 231-Ma se-

SEQUENCE STRATIGRAPHY DEPOSITIONAL MODEL SHOWING CARBONATE AND EVAPORITE LITHOFACIES

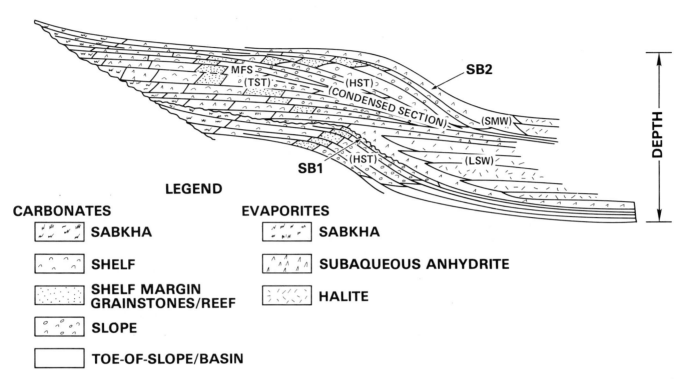

FIG. 22.—Summary schematic diagram of evaporite-carbonate lithofacies distribution in a sequence framework. Evaporites occur both as onlapping lowstand and shelf margin wedges and as lagoonal/sabkha facies in interior or back-shelf positions in carbonate banks.

TRIASSIC OF THE DOLOMITE MOUNTAINS ITALY

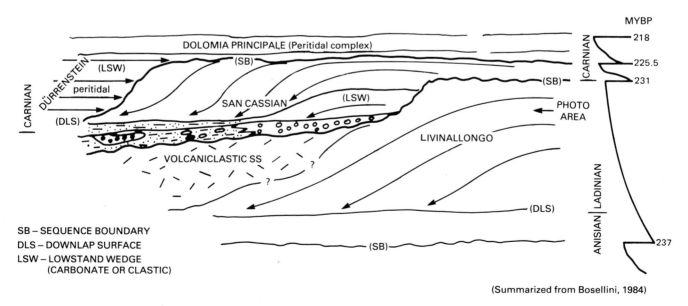

(Summarized from Bosellini, 1984)

FIG. 23.—Schematic summary of Triassic carbonate sequences, Dolomite Mountains, northern Italy (interpreted from Bosellini, 1984).

quence boundary, and one in the Upper Carnian associated with the 225.5-Ma sequence boundary (Fig. 23). In the lower sequence, the basal Carnian, small reef and platform bodies are scattered in the Late Ladinian basin and are onlapped by associated nonfossiliferous, thin-bedded shales and by turbidite sandstones (Bosellini, 1984, his fig. 17). A later Carnian highstand systems tract prograded over these lowstand wedges. The second lowstand wedge occurs on a sequence boundary of late Carnian age. During deposition of the wedge, the Carnian platforms were subaerially exposed. Sea level dropped and peritidal dolomites with stromatolites, vadose pisolites, and tepee structures were deposited on the upper slope (Bosellini, 1984; his fig. 19). This Carnian tidal-flat unit onlaps over the Carnian highstand slopes (Fig. 23).

In the Natuna Field area, the upper part of the Miocene Terumbu Formation contains two well-developed autochthonous lowstand wedges associated with the 10.5-Ma (Fig. 5) and the 6.3-Ma (Fig. 3C) sequence boundaries (Rudolph and Lehmann, 1987). Each lowstand wedge occurs as a belt of high-relief mounds interpreted as a fringing-reef complex deposited on the underlying platform slope. The mounds occur within an oblique progradational seismic facies interpreted to have been deposited under moderate- to high-energy conditions (Fig. 5). A core of the mounded lithofacies from one of the highstand systems (well L-2X, Fig. 3C) contains a coral-red algae grainstone interpreted to represent shoal water deposition. High-amplitude oblique/parallel seismic facies containing low-relief mounds occurs platformward of the high-relief mounds and is interpreted to represent a low- to moderate-energy lagoon with patch banks and reefs.

The Grayburg Formation (Guadalupian, Permian) of the Guadalupe Mountains, New Mexico, provides an excellent outcrop example of a bank margin lowstand wedge (Figs. 12, 13; Sarg and Lehmann, 1986). This wedge is equivalent to the lowstand wedge that onlaps the upper San Andres bank on the seismic section of Figure 11B. The wedge thins from a maximum to 40 m at the seaward margin of the outcrop to 12 m at the shelfward end (Figs. 12B, 13). The wedge is dominated by a series of sandstone-carbonate couplets (Naiman, 1982) that onlap the bank margin grainstones and packstones of the underlying San Andres highstand systems tract (Figs. 12, 13). The basal 10 to 12 m of this wedge at the San Andres bank margin is composed of well-sorted fine- to medium-grained quartz sandstone containing reactivation ripple sets. This basal unit is overlain by carbonate-sandstone couplets that are 1 to 6 m thick. The sandstones are similar to the basal sandstones and grade upward into sandy dolomite and dolomite. The dolomite ranges from stromatolitic and pelloidal-lump-oncolite dolopackstones with sparse fenestral fabric to peloid-coated grain-intraclast dolopackstones (Fig. 13). The reactivated ripple sets, stromatolites, and fenestral fabric suggest a tidal-flat origin for the updip portion of this autochthonous lowstand wedge.

Type 2 Platform/Bank Margin Wedge Deposits

An example of a type 2 sequence platform/bank margin wedge includes the Buckner wedge in southern Arkansas and Louisiana. The Buckner is a thin, onlapping, platform margin wedge deposited basinward of the Smackover highstand platform margin in southernmost Arkansas and northeastern Louisiana (Fig. 20). The wedge is composed of anhydrite and red shale, north of a line that corresponds to the position of McKamie Patton Field. Red shale with minor anhydrite occurs south of that line (Moore and Druckman, 1981). The Buckner anhydrite/shale is generally interpreted as tidal flat (sabkha) in origin (most recently, Harris and Dodman, 1982). It changes facies in southernmost Arkansas and northeast Louisiana to shallow-water Smackover-type limestone (Smackover A).

In places in northern Louisiana, the Buckner Limestone platform margin wedge is composed of shoal water ooid grainstone and algal boundstone lithofacies (Oaks Field, Claiborne Parish: Erwin and others, 1979; North Haynesville Field, Claiborne Parish: Bishop, 1968, and Baria and others, 1982; Hico Knowles Field, Lincoln Parish: Baria and others, 1982).

Basinally Restricted and Onlapping Evaporite Wedges

The third type of lowstand systems tract is the *basinally restricted and onlapping evaporite wedge* associated with either type 1 or type 2 carbonate sequence boundaries (Fig. 22). Evaporites can occur in each of the systems tracts: (1) as onlapping lowstand or shelf margin wedges; (2) as onlapping and retrogradational units of the transgressive-systems tract; and (3) as lagoonal/sabkha facies in platform interior settings of the highstand systems tract. The transgressive evaporites are predicted to occur during times of slowly rising sea level, when the platform or bank top waters remain hypersaline. As the rate of rise of sea level increases, the platform becomes more normal marine in character, and evaporite deposition is replaced by carbonate sedimentation. Two examples illustrating these effects come from the Silurian basin of Michigan and the Middle Devonian basin of western Canada.

The Silurian reefs of the Michigan basin were deposited on a basinward-dipping carbonate ramp during two Middle Silurian (Wenlock and Ludlow) eustatic cycles (Fig. 19; Sarg, 1982, 1983; Lehmann and others, 1982). Highstand deposition is characterized by reef development in a stratified basin laterally adjacent to thinner, laminated, anhydritic mudstones deposited in the basin (Fig. 19). Lowstand deposition occurred during two type 2 eustatic falls. The basin became restricted, reef growth ended, and the A-1 and A-2 evaporites were deposited as onlapping and draping-basin wedges.

In the southern Michigan reef trend, both the A-1 and A-2 evaporites are anhydritic and are interpreted to have been deposited subaqueously, except over reef crests. They are characterized by bedded massive/mosaic anhydrite and contain, in reef-proximal positions, allochthonous dolomitic anhydrite breccias, and anhydritic pelloidal grainstones. Reef crests show a basal stromatolite facies with associated nodular mosaic anhydrite, suggesting a supratidal sabkha origin (Sarg, 1982). The sabkha deposits pass abruptly upward to bedded mosaic anhydrite of probable subtidal origin.

Pervasive dolomitization followed initial reef lithification

(McGovney, 1978; Sarg, 1982) and may have formed during the deposition of the two lowstand evaporites. Interparticle, intercrystalline, and microvug porosity is characteristic of the reef crest stages. Micritization, fracturing, and vug formation have also occurred in the upper parts of the reefs, enhancing porosity and permeability. This alteration may have occurred as the result of exposure of the reefs during the second lowstand and/or from the onlap of hypersaline brines over the reefs, causing micritization and evaporite replacement of reef material. Subsequent solution of the evaporites has left a porous reservoir rock (Sarg, 1982).

Two major Middle Devonian type 1 sequence boundaries resulted in subaerial exposure of pinnacle-reef and platform areas (Keg River and Sulphur Point carbonates). The formation of the first sequence boundary resulted in deposition of the Muskeg–Cold Lake Salt in the basinal areas (Bebout and Maiklem, 1973; Maiklem, 1971). Bebout and Maiklem (1973) suggest that a major drawdown occurred in the basin areas (i.e., Elk Point basin), and their interpretation is that sea level fell as much as 60 m (Maiklem, 1971). The fall at the end of Sulphur Point deposition was smaller.

Pervasive early marine cementation significantly reduced depositional porosity of these reefs (Schmidt and others, 1980)—the *catch-up* deposition of this report. Fracturing and leaching in the vadose zone developed vuggy porosity, and anhydrite replacement of carbonate occurred in the reef and platform flank areas. As sea level is interpreted to have slowly risen during the lowstand, evaporites onlapped and eventually overlapped the reefs (Schmidt and others, 1980). Dolomitization appears to be associated with evaporite deposition. A downward decrease in intensity of dolomitization, a lack of dolomite in the immediately overlying limestone, and dolomite-mudstone clasts contained within the overlying limestone all suggest early dolomitization (Schmidt and others, 1980).

CONCLUSIONS

(1) Depositional stratal patterns, facies distribution, and productivity of carbonate platforms are controlled most importantly by the rates of relative changes in sea level (i.e., sum of rate of change of eustasy and subsidence). Depositional setting and climate also exert strong controls on basin water chemistry and carbonate productivity.

(2) Carbonate highstand systems tracts are characterized by aggradational-to-progradational geometry. A *keep-up* systems tract is interpreted to represent a relatively rapid rate of accumulation and displays a mounded/oblique stratal configuration at the platform/bank margin. A *catch-up* systems tract is interpreted to represent a relatively slow rate of accumulation and displays a sigmoid depositional profile at the platform/bank margin. *Catch-up* systems tracts appear to have significantly greater amounts of early submarine cement.

(3) During formation of a type 1 sequence boundary, two major processes may occur: (1) local-to-regional slope front erosion and (2) subaerial exposure of the platform. For a large-scale type 1 sequence boundary and given the appropriate climatic conditions, the meteoric lens may remain established for a long time over the platform. During formation of a small-scale type 1 or a type 2 sequence boundary, the meteoric lens is less well established, and the dominant effects will be in the inner-platform areas.

(4) Two types of carbonate lowstand systems tracts are recognized: (1) allochthonous deposits derived from erosion of the bank margin and slope and characterized by channelled megabreccia deposition, and (2) autochthonous wedges deposited on the upper slope during either type 1 or type 2 sea-level lowstands. Gentle depositional slopes and well-circulated basins favor lowstand bank growth. With the appropriate climatic, hydrographic, and provenance conditions, evaporite or siliciclastic-lowstand deposition will occur.

ACKNOWLEDGMENTS

I gratefully acknowledge the support of Exxon Production Research Company for this research and thank them for permission to publish this work. The concepts and interpretations reported herein depend in part on the work of other geologists, both within and outside of Exxon, who have also been fascinated with stratigraphic problems and who have shared their perspectives with me. In addition, this manuscript was greatly improved by the reviews of John Van Wagoner, Christopher Kendall, and especially Charles Ross.

REFERENCES

ADEY, W. H., MACINTYRE, I. G., STUCKENRATH, R., AND DILL, R. F., 1977, Relict barrier reef system off St. Croix: Its implications with respect to late Cenozoic coral reef development in the western Atlantic: Proceedings, Third Coral Reef Symposium, University of Miami, Miami, Florida, v. 2, p. 15–21.

ANDERSON, E. J., GOODWIN, P. W., AND SOBIESKI, T. H., 1984, Episodic accumulation and the origin of formation boundaries in the Helderberg Group of New York State: Geology, v. 12, p. 120–123.

ARMSTRONG, A. K., SNAVELY, P. D., AND ADDICOTT, W. O., 1980, Porosity evolution of Upper Miocene reefs, Almera Province, southern Spain: American Association of Petroleum Geologists Bulletin, v. 64, p. 188–208.

BABCOCK, J. A., 1977, Calcareous algae, organic boundstones, and the genesis of the Upper Capitan Limestone (Permian, Guadalupian), Guadalupe Mountains, West Texas and New Mexico, in Hileman, M. E., and Mazzullo, S. J., eds., Upper Guadalupian Facies, Permian Reef Complex, Guadalupe Mountains, New Mexico and West Texas: Permian Basin Section, Society of Economic Paleontologists and Mineralogists Special Publication 77-16, p. 3–44.

BARIA, L. R., STOUDT, D. L., HARRIS, P. M., AND CREVELLO, P. D., 1982, Upper Jurassic reefs of Smackover Formation, United States Gulf Coast: American Association of Petroleum Geologists Bulletin, v. 66, p. 1449–1482.

BEBOUT, D. G., AND LOUCKS, R. G., 1974, Stuart City Trend Lower Cretaceous—A carbonate shelf-margin model for hydrocarbon exploration: University of Texas, Austin, Bureau Economic Geology, Report of Investigations 78, 80 p.

———, AND MAIKLEM, W. R., 1973, Ancient anhydrite facies and environments, Middle Devonian Elk Point basin, Alberta: Bulletin of Canadian Petroleum Geology, v. 21, p. 287–343.

BEIN, A., AND LAND, L. S., 1983, Carbonate sedimentation and diagenesis associated with Mg-Ca-chloride brines: The Permian San Andres Formation in the Texas Panhandle: Journal of Sedimentary Petrology, v. 53, p. 243–260.

BIDDLE, K., 1984, Triassic sea level change and the Ladinian-Carnian stage boundary: Nature, v. 308, no. 5960, p. 631–633.

BISHOP, W. F., 1968, Petrology of upper Smackover Limestone in North Haynesville Field, Claiborne Parish, Louisiana: American Association of Petroleum Geologists Bulletin, v. 52, p. 92–128.

BOSELLINI, A., 1984, Progradation geometries of carbonate platforms: Examples from the Triassic of the Dolomites, northern Italy: Sedimentology, v. 31, p. 1–24.

BROWN, L. F., JR., AND FISHER, W. L., 1977, Seismic-stratigraphic interpretation of depositional systems: Examples from Brazilian rift and pull-apart basins, in Payton C. E., ed., Seismic Stratigraphy—Applications to Hydrocarbon Exploration: American Association of Petroleum Geologists Memoir 26, p. 213–248.

BUBB, J. N., AND HATLELID, W. G., 1977, Seismic stratigraphy and global changes of sea level, part 10: Seismic recognition of carbonate build-ups, in Payton C. E., ed., Seismic Stratigraphy—Applications to Hydrocarbon Exploration; American Association of Petroleum Geologists Memoir 26, p. 185–204.

COOK, H. E., 1983, Ancient carbonate platform margins, slopes, and basins, in Cook, H. E., Hine, A. C., and Mullins, H. T., eds., Platform Margin and Deep Water Carbonates: Society of Economic Paleontologists and Mineralogists Short Course Notes No. 12, p. 5.1–5.189.

DUNHAM, R. J., 1962, Classification of carbonate rocks according to depositional texture, in Ham, W. E., ed., Classification of Carbonate Rocks: American Association of Petroleum Geologists Memoir 1, p. 108–121.

EMBRY, A. F., AND KLOVAN, J. E., 1971, A Late Devonian reef tract on northeastern Banks island, Northwest Territories: Bulletin of Canadian Petroleum Geology, v. 19, p. 730–781.

ERWIN, C. R., EBY, D. E., AND WHITESIDES, V. S., 1979, Clasticity index: A key to correlating depositional and diagenetic environments of Smackover reservoirs, Oaks Field, Claiborne Parish, Louisiana: Transactions, Gulf Coast Association of Geological Societies, v. 29, p. 52–62.

EVANS, G. V., SCHMIDT, V., BUSH, P., AND NELSON, H. W., 1969, Stratigraphy and geologic history of the Sabkha, Abu Dhabi, Persian Gulf: Sedimentology, v. 12, p. 145–159.

FEKETE, T. E., FRANSEEN, E. K., AND PRAY, L. C., 1986, Deposition and erosion of the Grayburg Formation (Guadalupian, Permian) at the shelf-to-basin margin, Western Escarpment, Guadalupe Mountains, Texas, in Moore, G. E., and Wilde G. L., eds., Lower and Middle Guadalupian facies, Stratigraphy, and Reservoir Geometries, San Andres/Grayburg Formations, Guadalupe Mountains, New Mexico and Texas: Field Trip Guidebook, Permian Basin Section, Society of Economic Paleontologists and Mineralogists Publication 86-25, p. 69-81.

FROST, S. H., BLIEFNICK, D. M., AND HARRIS, P. M., 1983, Deposition and porosity evolution of a lower Cretaceous rudist buildup, Shuaiba Formation of eastern Arabian Peninsula, in Harris, P. M., ed., Carbonate Buildups: Society of Economic Paleontologists and Mineralogists Core Workshop No. 4, Dallas, p. 381–410.

GRIFFITH, L. S., PITCHER, M. G., AND RICE, G. W., 1969, Quantitative environmental analysis of a Lower Cretaceous reef complex, in Friedman, G. M., ed., Depositional Environments in Carbonate Rocks: Society of Economic Paleontologists and Mineralogists Special Publication 14, p. 120–138.

HARRIS, M. T., 1982, Sedimentology of the Cutoff Formation (Permian), Western Guadalupe Mountains, West Texas and New Mexico: Unpubl. M.S. Thesis, University of Wisconsin, Madison, 186 p.

HARRIS, P. M., AND DODMAN, C. A., 1982, Jurassic evaporites of the U.S. Gulf Coast: The Smackover-Buckner contact, in Handford, C. R., ed., Depositional and Diagenetic Spectra of Evaporites: Society of Economic Paleontologists and Mineralogists Core Workshop No. 3, p. 174–192.

HECKEL, P. H., 1983, Diagenetic model for carbonate rocks in Midcontinent Pennsylvanian eustatic cyclothems: Journal of Sedimentary Petrology, v. 53, p. 733–759.

HINE, A. C., 1983, Relict sand bodies and bedforms of the northern Bahamas: Evidence of extensive early Holocene sand transport, in Peryt, T. M., ed., Coated grains: Springer-Verlag, Heidelberg, p. 116–131.

———, AND NEUMANN, A. C., 1977, Shallow carbonate bank margin growth and structure, Little Bahama Bank: American Association of Geologists Bulletin, v. 61, p. 376–406.

———, WEBER, R. J., AND NEUMANN, A. C., 1981, Carbonate sand bodies along contrasting shallow bank margins facing open seaways in northern Bahamas: American Association of Petroleum Geologists Bulletin, v. 65, p. 261–290.

HUBBARD, D. K., BURKE, R. B., AND GILL, I. P., 1986, Styles of reef accretion along a steep, shelf-edge reef, St. Croix, U.S. Virgin Islands: Journal of Sedimentary Petrology, v. 56, p. 848–861.

HURLEY, N. F., 1979, Seaward primary dip of fall-in beds, Lower Seven Rivers Formation (Permian), Guadalupe Mountains, New Mexico (abst.): American Association of Petroleum Geologists Bulletin, v. 63, p. 471.

KENDALL, C. G. ST. C., AND SCHLAGER, W., 1981, Carbonates and relative changes in sea level: Marine Geology, v. 44, p. 181–212.

KING, P. B., 1948, Geology of the southern Guadalupe Mountains, Texas: U.S. Geological Survey Professional Paper 215, 183 p.

LAND, L. S., 1973a, Contemporaneous dolomitization of middle Pleistocene reefs by meteoric water, North Jamaica: Bulletin of Marine Science, v. 23, p. 64–92.

———, 1973b, Holocene meteoric dolomitization of Pleistocene limestones, North Jamaica: Sedimentology, v. 20, p. 411–422.

LEHMANN, P. J., 1978, Deposition, porosity evolution, and diagenesis of the Pipe Creek Jr. reef (Silurian), Grant County, Indiana: Unpub. M.S. Thesis, University of Wisconsin, Madison, 234 p.

———, MCGOVNEY, J. E., AND SARG, J. F., 1983, Middle Silurian (Niagaran) sea level cycles—North American and Europe: Program with Abstracts, Geological Society of America National Meeting, Indianapolis, p. 626.

LIGHTY, R. G., MACINTYRE, I. G., AND STRICKENRATH, R., 1978, Submerged early Holocene barrier reef, southeast Florida shelf: Nature, v. 275, p. 59–60.

LITSEY, L. R., MACBRIDE, W. L., AL-HINAI, K. M., AND DISMUKES, N. B., 1983, Shuaiba reservoir geological study, Yibal Field, Oman, in Third Middle East Oil Show, Proceedings, Bahrain: Society of Petroleum Engineers of American Institute of Mining Engineers, p. 131–142.

LOUCKS, R. G., AND ANDERSON, J. H., 1980, Depositional facies and porosity development in lower Ordovician Ellenburger dolomite, Puckett Field, Pecos County, Texas, in Halley, R. B., and Loucks, R. G., eds., Carbonate Reservoir Rocks: Society of Economic Paleontologists and Mineralogists Core Workshop No. 1, p. 1–31.

MACINTYRE, I. G., BURKE, R., AND STRICKENRATH, R., 1977, Thickest recorded Holocene reef section, Isla Perez core hole, Alacran Reef, Mexico: Geology, v. 5, p. 749–754.

MAIKLEM, W. R., 1971, Evaporative drawdown—A mechanism for water-level lowering and diagenesis in the Elk Point basin: Bulletin of Canadian Petroleum Geology, v. 17, p. 194–233.

MCGOVNEY, J. E., 1978, Deposition, porosity evolution and diagenesis of the Thornton Reef (Silurian), Northeastern Illinois: Unpubl. Ph.D. Dissertation, University of Wisconsin, Madison, 447 p.

MESOLELLA, J. J., ROBINSON, J. D., MCCORMICK, L. M., AND ORMISTON, A. R., 1974, Cyclic deposition of Silurian carbonates and evaporites in Michigan Basin: American Association of Petroleum Geologists Bulletin, v. 58, p. 34–62.

MEYERS, W. J., 1974, Carbonate cement stratigraphy of the Mississippian Lake Valley Formation, Sacramento Mountains, New Mexico: Journal of Sedimentary Petrology, v. 44, p. 837–861.

———, 1978, Regional cementation patterns in Mississippian limestones of southwestern New Mexico: Sedimentology, v. 25, p. 371–400.

———, 1980, Compaction in Mississippian skeletal limestones, southwestern New Mexico: Journal of Sedimentary Petrology, v. 50, p. 457–474.

MITCHUM, JR., R. M., 1977, part eleven, Glossary of terms used in Seismic Stratigraphy, in Payton, C. E., ed., Seismic Stratigraphy–Applications to Hydrocarbon Exploration: American Association of Petroleum Geologists Memoir 26, p. 205–212.

———, VAIL, P. R., AND THOMPSON, S., III, 1977, Part II: The depositional sequence as a basic unit for stratigraphic analysis, in Payton, C. E., ed., Seismic Stratigraphy—Applications to Hydrocarbon Exploration: American Association of Petroleum Geologists Memoir 26, p. 53–62.

MOORE, C. H., AND DRUCKMAN, Y., 1981, Burial diagenesis and porosity evolution, Upper Jurassic Smackover, Arkansas and Louisiana: American Association of Petroleum Geologists Bulletin, v. 65, p. 597–628.

NAIMAN, E. R., 1982, Sedimentation and diagenesis of a shallow marine carbonate and siliciclastic shelf sequence: The Permian (Guadalupian) Grayburg Formation, southeastern New Mexico: Unpubl. M.S. Thesis, University of Texas, Austin, 197 p.

PLAYFORD, P. E., AND LOWRY, D. C., 1966, Devonian reef complexes of the Canning Basin, Western Australia: Geological Survey of Western Australia Bulletin, v. 118, 50 p.

PREZBINDOWSKI, D., 1983, Burial cementation—Is it important? A case study—Stuart City trend, south-central Texas (abst.): American Association of Petroleum Geologists Bulletin, v. 67, p. 536–537.

ROSE, P. R., 1972, Edwards Group, surface and subsurface, central Texas: University of Texas, Austin, Bureau of Economic Geology, Report of Investigations 74, 198 p.

ROSSEN, C., 1985, Sedimentology of the Brushy Canyon Formation (Permian, Early Guadalupian) in the onlap area, Guadalupe Mountains, West Texas: Unpubl. M.S. Thesis, University of Wisconsin, Madison, 314 p.

RUDOLPH, K. W., AND LEHMANN, P. J., 1987, Platform evolution and sequence stratigraphy of the Natuna L-Structure, South China Sea: Abstracts with Program, American Association of Petroleum Geologists National Meeting, Los Angeles, p. 608.

SARG, J. F., 1981, Petrology of the carbonate-evaporite facies transition of the Seven Rivers Formation (Guadalupian, Permian), southeast New Mexico: Journal of Sedimentary Petrology, v. 51, p. 73–96.

————, 1982, Off-reef Salina deposition (Silurian), southern Michigan Basin: Implications for reef genesis, in Handford, C. R. ed., Depositional and Diagenetic Spectra of Evaporites: Society of Economic Paleontologists and Mineralogists Core Workshop No. 3, p. 354–384.

————, 1983, Eustatic control of cyclic carbonate-evaporite deposition (Silurian), southern Michigan Basin: Program with Abstracts, Geological Society of America national Meeting, Indianapolis, p. 678.

————, AND LEHMANN, P. J., 1986, Lower-Middle Guadalupian facies and stratigraphy San Andres/Grayburg formations, Permian Basin, Guadalupe Mountains, New Mexico, in Moore, G. E., and Wilde, G. L., eds., Field Trip Guidebook, San Andres/Grayburg Formations, Guadalupe Mountains, New Mexico and Texas: Permian Basin Section, Society of Economic Paleontologists and Mineralogists Publication 86–25, p. 000.

SCHLAGER, W., 1981, The paradox of drowned reefs and carbonate platforms: Geological Society America Bulletin, v. 92, p. 197–211.

SCHMIDT, V., McDONALD, D. A., AND McLLREATH, I. A., 1980, Growth and diagenesis of Middle Devonian Keg River cementation reefs, Rainbow Field, Alberta, in Halley, R. B., and Loucks, R. G., eds., Carbonate Reservoir Rocks: Society of Economic Paleontologists and Mineralogists Core Workshop No. 1, p. 43–63.

SHINN, E. A., 1983, Tidal flat environment, in Scholle, P. A., Bebout, D. G., and Moore, C. H., eds., Carbonate Depositional Environments: American Association of Petroleum Geologists Memoir 33, p. 171–210.

————, LLOYD, R. M., AND GINSBURG, R. N., 1969, Anatomy of a modern carbonate tidal flat, Andros Island, Bahamas: Journal of Sedimentary Petrology, v. 39, p. 1202–1228.

SILVER, B. A., AND TODD, R. G., 1969, Permian cyclic strata, northern Midland and Delaware basins, West Texas and southeastern New Mexico: American Association of Petroleum Geologists Bulletin, v. 53, p. 2223–2251.

STOAKES, F. A., 1980, Nature and control of shale basin fill and its effect on reef growth and termination: Upper Devonian Duvernay and Ireton formations of Alberta, Canada: Bulletin of Canadian Petroleum Geology, v. 28, p. 345–410.

VAIL, P. R., AND TODD, R. G., 1981, North Sea Jurassic unconformities, chronostratigraphy, and sea-level changes from seismic stratigraphy, in Illing, L. V., and Hobson, G. D., eds., Proceedings, Petroleum Geology of the Continental Shelf, Northwest Europe Conference, London: Heydon and Sons, p. 216–235.

WAGNER, P. D., AND MATTHEWS, R. K., 1982, Porosity preservation in the Upper Smackover (Jurassic) carbonate grainstone, Walker Creek Field, Arkansas: Response of paleophreatic lenses to burial processes: Journal of Sedimentary Petrology, v. 52, p. 3–18.

WALLS, R. A. MOUNTJOY, E. W., AND FRITZ, P., 1979, Isotopic composition and diagenetic history of carbonate cements in Devonian Golden Spike Reef, Alberta: Geological Society of America Bulletin, v. 90, p. 963–982.

WENDTE, J. C., AND STOAKES, F. A., 1982, Evolution and corresponding porosity distribution of the Judy Creek reef complex, Upper Devonian, central Alberta, in Cutter, W. G., ed., Core Workshop Manual, Canada's Giant Hydrocarbon Reservoirs: Canadian Society of Petroleum Geology, p. 63–81.

WILSON, J. L., 1975, Carbonate Facies in Geologic History: Springer-Verlag, New York, 471 p.

YUREWICZ, D. A., 1977, The origin of the massive facies of the Lower and Middle Capitan Limestone (Permian), Guadalupe Mountains, New Mexico and West Texas, in Hileman, M. E., and Mazzullo, S. J., eds., Upper Guadalupian Facies, Permian Reef Complex, Guadalupe Mountains, New Mexico and West Texas: Permian Basin Section, Society of Economic Paleontologists and Mineralogists Publication 77-16, p. 45–92.

CONDENSED SECTIONS: THE KEY TO AGE DETERMINATION AND CORRELATION OF CONTINENTAL MARGIN SEQUENCES

TOM S. LOUTIT, JAN HARDENBOL, PETER R. VAIL[1]
Exxon Production Research Company, P. O. Box 2189, Houston, Texas 77252-2189
AND
GERALD R. BAUM
Arco Oil and Gas Company, 2300 West Plano Parkway, Plano, Texas 75075

ABSTRACT: Condensed sections play a fundamental role in stratigraphic correlation, both regionally and globally. Condensed sections are thin marine stratigraphic units consisting of pelagic to hemipelagic sediments characterized by very low-sedimentation rates. Areally, they are most extensive at the time of maximum regional transgression of the shoreline. Condensed sections are associated commonly with apparent marine hiatuses and often occur as thin, but continuous, zones of burrowed, slightly lithified beds (omission surfaces) or as marine hardgrounds. In addition, condensed sections may be characterized by abundant and diverse planktonic and benthic microfossil assemblages, authigenic minerals (such as glauconite, phosphorite, and siderite), organic matter, and bentonites and may possess greater concentrations of platinum elements such as iridium.

Condensed sections are important because they tie the temporal stratigraphic framework provided by open-ocean microfossil zonations to the physical stratigraphy provided by depositional sequences in shallower, more landward sections. Condensed sections represent a physical stratigraphic link between shallow- and deep-water sections and are recognized by the analysis of seismic, well-log, and outcrop data. Within each depositional sequence, condensed sections are best recognized and utilized within an area from the shelf/slope break landward to the distal edge of inner neritic-sand deposition. Where sedimentation rates are generally low, as in the deep ocean, a number of condensed sections may coalesce to form a composite condensed section.

Data from detailed analyses of continental-margin condensed sections are presented to illustrate the nature and importance of condensed sections for dating and correlating continental-margin sequences and reconstructing ancient depositional environments.

INTRODUCTION

Developments in biostratigraphy during the past 70 years, particularly during the last 20 years, have played a dominant role in the evolution of stratigraphy during the twentieth century (Fig. 1). The recognition that benthic and planktonic microfossil evolutionary events could be used to divide the rock record into biozones has resulted in a shift toward a biostratigraphically dominated global-correlation system. The product of this shift has been the formation of biostratigraphic definitions for the classic European stages (see Hardenbol and Berggren, 1978). Many of the European stages were bounded originally by unconformities and disconformities or recognized by major changes in macrofossil genera in shallow-marine to marginal-marine sections. The recognition of evolutionary-based planktonic and benthic microfossil biozonation schemes resulted in the correlation of the classic stages, defined in incomplete sections, to more complete sections represented by deeper water facies containing good planktonic microfossil populations. This trend toward biostratigraphic definitions of classic stages and more local stages is exemplified by the evolution of the stage system in New Zealand (see discussion in Loutit and Kennett, 1981). By about 1970, biostratigraphic zonation schemes provided the key to the global correlation of sedimentary rocks.

Dramatic developments in stratigraphy during the past 20 years have further strengthened the role of biostratigraphy in the regional and international correlation of sedimentary rocks. One of these events was the development of seismic stratigraphy in the 1970s and, more recently, the development of sequence stratigraphy (Figs. 1, 2). The development of seismic stratigraphy resulted in the recognition and definition of fundamental stratal units, called depositional sequences, that are bounded by unconformities and their correlative surfaces (Mitchum and others, 1977). Sequence stratigraphy involves the recognition and correlation of depositional sequences by using a variety of stratigraphic disciplines in both outcrop and subsurface sections (Vail and Todd, 1981; Hardenbol and others, 1986; Baum and Vail, this volume). The importance of this new stratigraphic discipline to biostratigraphy is that it has provided a physical stratigraphic framework (based on the recognition of surfaces in the rock record) within which sample-based disciplines, such as biostratigraphy and geochemistry, can be evaluated.

Three other events have resulted in the development of a system to define and date biochronozones (Fig. 2): (1) the development of paleomagnetic stratigraphy on the sea floor (Cox, 1969, 1973) and in sediments (Opdyke, 1972); (2) the initiation of the Deep Sea Drilling Project (DSDP) in 1968 and the Ocean Drilling Project (ODP) in 1984; and (3) the invention of the Hydraulic Piston Corer (HPC) and its derivatives. Paleomagnetic stratigraphy provided a semi-independent stratigraphy within which biostratigraphic and chemostratigraphic events could be calibrated. The major breakthrough in stratigraphy was the invention of the HPC in the late 1970s, which finally allowed the direct correlation between magnetic reversals and microfossil evolutionary datum planes. The HPC resulted in the recovery of undisturbed cores that could be analyzed for magnetic reversals, biostratigraphic datum planes, and chemostratigraphic events. Prior to 1979, core quality was often poor because of drilling disturbance and thus unsuitable for magnetostratigraphy. The large number of deep-ocean cores obtained with the HPC since 1979 formed the basis for the rigorous microfossil biochronozone compilations of Berggren and others (1985a,b) and Haq and others (1987). Thus, these few events have had a major effect on the develop-

[1]Present address: Department of Geology and Geophysics, Rice University, Houston, Texas 77251

1820-1920   **LITHOSTRATIGRAPHY**
            MACROFOSSILS
            (ABUNDANCE, GENERA)   OUTCROP

1920-1960'S **BENTHIC FORAMINIFERA** (SPECIES)   OUTCROP
            LITHOSTRATIGRAPHY   WELL LOGS

1960-1970'S **PLANKTONIC MICROFOSSILS** (SPECIES)
            BENTHIC FORAMINIFERA   OUT CROP
            LITHOSTRATIGRAPHY   WELL LOGS
            SEISMIC STRATIGRAPHY   SEISMIC

1980'S   **SEQUENCE STRATIGRAPHY**
            BIOSTRATIGRAPHY   OUTCROP
            LITHOSTRATIGRAPHY   WELL LOGS
            SEISMIC STRATIGRAPHY   SEISMIC

FIG. 1.—Major periods in the development of stratigraphy, stratigraphic disciplines, and data sources that have played a major role within each period.

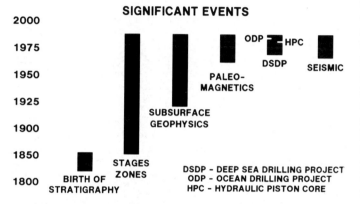

FIG. 2.—Schematic outline of the timing of major events in the development of stratigraphy.

ment of stratigraphy by dramatically improving the reliability of the biostratigraphic-age framework.

Deep-ocean stratigraphy, based primarily on biostratigraphy and magnetostratigraphy, and more recently on chemostratigraphy, provides a relatively well-tested, but still evolving, biochronozonal framework for the global correlation of deep-ocean and continental-margin sedimentary sections. The biochronozone is the basic unit of this deep-ocean age framework and has only become well established during the past few years as the number of HPC cores with magnetostratigraphic data has increased. The distinction between a biochronozone and a biozone (Fig. 3) is important in both deep-ocean and continental-margin stratigraphy. A biochronozone represents all the sediments deposited globally between specific faunal or floral evolutionary events, whereas a biozone represents only those sediments in any given place that actually contain the taxa used for definition of the biozone. The time elapsed during deposition of a biozone differs from place to place. Water mass chemistry and temperature generally define the latitudinal distribution of each taxon, resulting in a restricted area over which biozones and biochronozones are identical. Away from these optimum areas, the range of a particular age-diagnostic taxon may vary considerably because of environmental factors. In such cases, the chronozone cannot be defined and the term biozone is used (Fig. 3). In open-ocean settings, this means that correlations between DSDP holes drilled in sediments below discrete water masses, such as those in the tropical regions, are generally reliable. Correlations between tropical and temperate regions are more difficult, however, because of the fewer number of organisms common to both areas and the environmental control on the occurrences of zonal indicator taxa (Fig. 4). Because of (1) the availability of undisturbed sections; (2) relatively

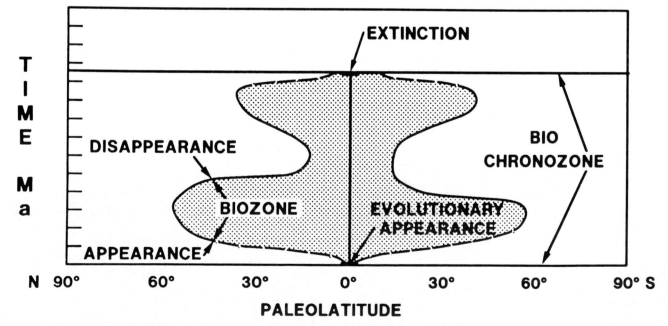

FIG. 3.—Distribution of planktonic microfossils with respect to time and paleolatitude. See text for definition of biochronozone and biozone.

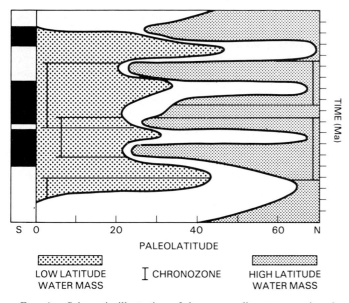

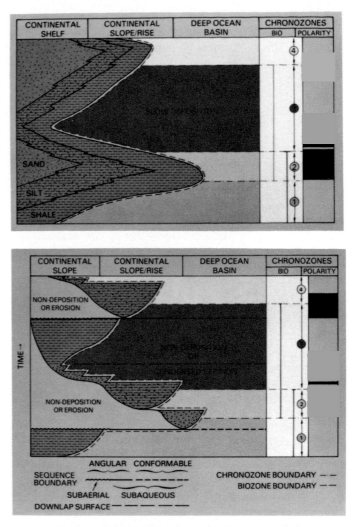

FIG. 4.—Schematic illustration of deep-sea sedimentary stratigraphy using biostratigraphy and magnetostratigraphy. Biochronozones, defined in high- and low-latitude water masses, can only be correlated within their respective water masses. Each water mass migrated latitudinally through time, thereby controlling area over which biochronozones can be recognized.

FIG. 5.—(A) Schematic chronostratigraphic diagram showing deep-ocean biochronozones (labelled 1 through 4) that generally cannot be recognized in continental-margin sections because of the influence of terrigenous sedimentation, controlled by transgressions and regressions. (B) Chronostratigraphic diagram illustrating importance of combining physical stratigraphic framework provided by stratal surfaces with age framework provided by faunal and floral zonation schemes. The combination produces a higher resolution stratigraphy than is shown in A.

constant and low-sedimentation rates; (3) the abundance of biostratigraphic data; and (4) more recently, the availability of magnetostratigraphic and chemical stratigraphic data, correlation in deep-sea sections appears to be more straightforward than correlation in continental-margin sections.

This paper is concerned with demonstrating that the integration of biostratigraphy and sequence stratigraphy provides a powerful tool for dating and correlating continental-margin sequences. Correlation within continental-margin sections is not simple, either in outcrop or subsurface, because of abrupt changes in terrigenous sedimentation rates caused by a number of processes, including eustasy, subsidence, and climate. Appearances and disappearances of age-diagnostic taxa in continental-margin sections rarely represent biochronozone boundaries as defined in open-ocean sections (Figs. 5A,B). Despite our inability to define biochronozones in continental-margin sections, stratigraphic correlation has traditionally been guided by biostratigraphy. Biostratigraphic first- and last-appearance data are recorded with respect to depth or thickness in continental-margin sections or to transgressions and regressions. Transgressions and regressions strongly affect the distribution of the depositional environments in which the organisms live (Fig. 5A). Transgressions, in particular, are critical to biostratigraphy because they transport and concentrate age-diagnostic microfossils in areas that are generally devoid of biostratigraphic control (Fig. 5A). Thus, an important aspect of dating continental-margin sections is an understanding of the controls on the distribution of age-diagnostic microfossils within the rock record. The combination of water mass chemistry (oxygen concentration, nutrient concentration, salinity, and so on), temperature, and sedimentation rate changes is probably the dominant control on the distribution

of planktonic microfossils. The distribution of benthic microfossils may be controlled by other factors, in addition to the above, which include type of substrate and water turbidity. Benthic microfossils, particularly benthic foraminifera, are used commonly for regional correlation of continental-margin sequences, but they are becoming less important for dating. In addition, because of the restricted distribution of most benthic taxa, they are unsuitable for interregional correlation. The task then is to apply the planktonic-microfossil biochronozonal scheme established in the deep ocean to the continental margins, where the biostratigraphy is anything but complete.

Because the distribution of all organisms in sedimentary sections is environmentally controlled, it is important to be able to observe their distribution within an independent stratigraphic framework in order to establish synchronous biostratigraphic correlations. In the deep ocean this framework is provided routinely by magnetostratigraphy and chemostratigraphy. On continental margins, however, this framework, if available, is fragmented and incomplete because of the frequent occurrence of unconformities and related erosion. Fortunately, a number of physical surfaces (Fig. 5B) can provide a semi-independent stratigraphic framework within which biostratigraphic observations can be evaluated. Three major surfaces are critical to continental-margin stratigraphy—sequence boundaries, transgressive surfaces, and downlap surfaces (Fig. 5B). The sequence boundary is the most widespread. It is recognizable from non-marine to deep-ocean environments and is defined by stratal onlap or truncation. The transgressive surface is formed during a transgression as the high-energy (wave-dominated), nearshore facies transgress the underlying strata, causing minor erosion and sediment starvation basinward of the transgressing beach (marine-flooding surface). The downlap surface represents a starvation surface produced during a time of transgression that subsequently forms a surface onto which prograding delta clinoforms downlap. These three types of surfaces, which are associated with depositional sequences and condensed sections, occur predictably in the sedimentary rock record and provide the framework for assessing the effect of transgressions and regressions on the distribution of age-diagnostic microfossils in continental-margin sections.

This paper attempts to demonstrate that the key to successful regional continental-margin correlation and dating is to project the excellent age control provided by open-ocean microfossils (biochronozones) into continental-margin sections with higher sedimentation rate by using a combination of depositional-sequence boundaries (physical stratigraphy) and biostratigraphic data concentrated in condensed sections (Fig. 6). The stratigraphic occurrence of condensed sections is predictable because these sections form during periods of rapid relative sea-level rise (Fig. 7; see also Posamentier and others, this volume). It is the predictability of condensed sections in space and time that makes them so important to stratigraphy (1) as a distinct unit that may be correlated physically from the deep-ocean to classic

continental-margin outcrop sections (Fig. 8), and (2) because of the concentration of age-diagnostic microfossils and authigenic minerals necessary for age dating. The events that produce condensed sections are also conducive to producing, concentrating, and sometimes preserving organic matter.

The paper defines condensed sections and discusses their position within a depositional sequence and the changes in water depth associated with the development of a condensed section. A series of examples is provided to illustrate the nature of condensed sections in outcrop and in the subsurface and the importance of condensed sections to stratigraphy.

RECOGNITION OF CONDENSED SECTIONS

Definition.—

Condensed sections are thin marine stratigraphic units consisting of pelagic to hemipelagic sediments characterized by very low-sedimentation rates. They are areally most extensive at the time of maximum regional transgression of the shoreline.

Condensed sections are associated commonly with apparent marine hiatuses and often occur either as thin but continuous zones of burrowed, slightly lithified beds (omission surfaces) or marine hardgrounds (for examples, see Bromley, 1974; Kennedy and Garrison, 1975; Vail and others, 1984; Baum and others 1984). Condensed sections may also be characterized by abundant and diverse planktonic and benthic microfossil assemblages, authigenic minerals (glauconite, phosphorite, and siderite), organic matter, and bentonites and may possess greater concentration of platinum elements such as iridium. Most ocean basin sediments are within the definition of condensed sections, because sedimentation rates in the deep ocean are generally low (<1 cm/1,000 years). This paper is concerned, however, mainly with condensed sections that form on continental margins during transgression of the shoreline in response to relatively rapid rises in relative sea level.

Depositional sequences and the condensed section.—

A condensed section extends, as a thin sedimentary unit, from the basin to the shelf in the middle of a depositional sequence (Fig. 9A). In a time-distance plot (Fig. 9B), however, the condensed section assumes far greater significance because it is produced during a period of extremely low-sedimentation rates as a result of a relative sea-level rise and abrupt transgression of the shoreline. As the shoreline transgressed, the loci of terrigenous deposition associated with shallow-water deposits moved landward, effectively starving pre-existing shelf areas and the deeper parts of the basin of terrigenous material.

Depositional sequences are bounded by unconformities and their correlative surfaces (Figs. 9A,B). The sequence boundary is characterized by subaerial exposure along part of its length and onlap of the overlying coastal strata onto it. Truncation of the underlying strata is also a relatively common feature of depositional-sequence boundaries. Within a depositional sequence, two other major surfaces exist that

SEQUENCE STRATIGRAPHY

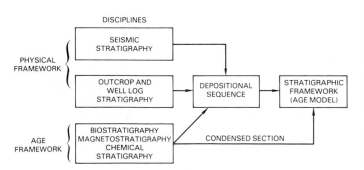

FIG. 6.—Diagram illustrating the role that condensed sections play in integrating the major disciplines of sequence stratigraphy to form an age model.

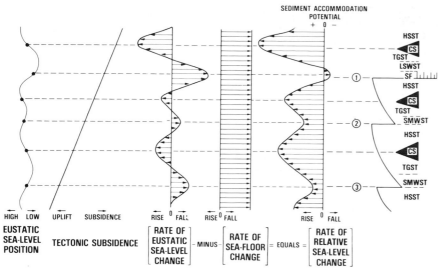

FIG. 7.—One-dimensional model depicting global sea level, subsidence, rate of eustatic change, subsidence rate, rate of change of accommodation, and estimated coastal onlap chart (modified from Posamentier and Vail, this volume). Negative accommodation potential at sequence-boundary 1 represents a period when sediment is transported into the basin, thus forming a submarine fan. During the formation of sequence boundaries 2 and 3, sediment accommodation potential remains positive and space is available for sedimentation. Condensed sections form at peaks in accommodation potential associated with the period of maximum rates of eustatic rise.

divide the depositional sequence into genetic packages called systems tracts (Van Wagoner and others, this volume). The two surfaces are the first transgressive surface and the downlap surface (Figs. 9A,B). The transgressive surface is produced during a transgression of high-energy nearshore environments across the underlying strata. The first transgressive surface and its correlative marine-flooding surface form a boundary between the lowstand systems tract and the overlying transgressive-systems tract. The downlap surface is a starvation surface that also begins to form during a time of transgression as the loci of terrigenous sedimentation move landward. The downlap surface is then buried by prograding clinoforms that downlap onto the top of the underlying systems tracts during regression of the shoreline.

The transgressive surface and the downlap surface are the bounding surfaces between the series of systems tracts that constitute a depositional sequence. Each systems tract, the lowstand, transgressive, highstand, and shelf margin, can be defined uniquely by the nature of the bounding surfaces and the geometry of the parasequence stacking patterns within systems tracts (Van Wagoner and others, this volume).

The systems tracts are interpreted as forming in response to changes in relative sea level that are primarily controlled by eustatic changes. During the formation of a type 1 depositional sequence in clastic sediments (see Vail and Todd, 1981), three systems tracts are produced during one eustatic cycle.

When the rate of sea level fall exceeds the rate of subsidence at the depositional-shoreline break, sediment accommodation on the shelf decreases significantly. Sediment bypasses the shelf and a lowstand fan may be formed (Fig. 9).

As the rate of relative sea-level fall decreases to a min-

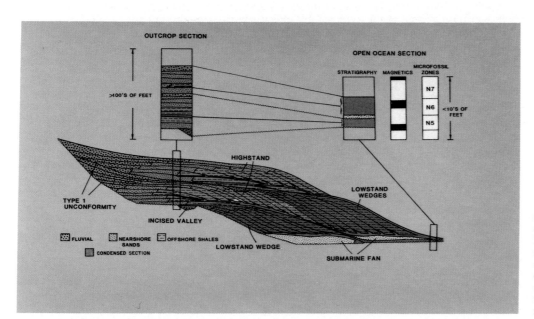

FIG. 8.—Depth-distance cross section showing distribution of condensed section that provides physical link between deep-sea sedimentary sections and outcrop sections. Classic time-rock units, defined in outcrop section, generally represent neritic to non-marine environments that cover only a portion of the rock record.

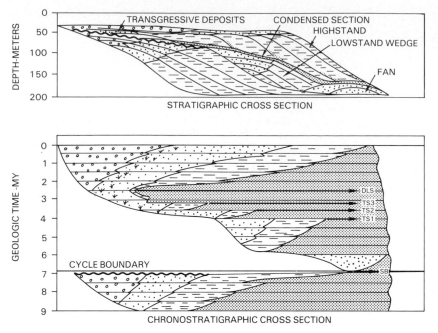

FIG. 9.—(A) Depth-distance diagram of a type 1 depositional sequence deposited on top of a highstand system tract. Condensed section (thin, solid black unit) is in the middle of the sequence and incorporates distal portion of the transgressive- and highstand systems tracts. (B) Time-distance diagram of the same stratigraphic succession shown in A. Condensed section (solid black area) represents period of slower sedimentation as shoreline transgressed across the shelf. Its duration increases basinward. Note three major surfaces that provide physical stratigraphic framework: the sequence boundary (SB), transgressive surface (TS1), and downlap surface (DLS).

imum and then begins to rise, sediment accommodation on the uppermost slope increases and a restricted onlapping unit, the lowstand wedge, forms. The lowstand fan and lowstand wedge are included within the lowstand systems tract. As the rate of sea-level rise increases toward a maximum, the loci of deposition of terrigenous sediments transgress rapidly across the shelf, depositing a succession of laterally extensive nearshore deposits, termed the transgressive-systems tract. The transgressive-systems tract is characterized generally in a vertical sequence by an overall upward-deepening trend and an overall upward decrease in terrigenous sediment. Maximum transgression is associated with, or after, the point of maximum rate of sea-level rise, and, during this time, terrigenous sedimentation is increasingly restricted to more landward regions. Hemipelagic and pelagic sediments may be deposited over a large area of the shelf, initiating the formation of a condensed section. As the rate of relative sea-level rise begins to decrease, sediment accommodation updip decreases. The loci of deposition then prograde basinward, forming the highstand systems tract that is characterized in a vertical succession by a coarsening- or shoaling-upward trend and an upward increase in terrigenous sediment content. Sedimentation at the distal toes of prograding clinoforms is initially very slow, representing the continuation of the condensed section. Hence, as presently defined, the condensed section incorporates slowly accumulating sediments laid down in the distal parts of both the transgressive and the highstand deposits. For more detailed discussions of depositional sequence sedimentation models, see Posamentier and Vail, this volume.

Water depth and condensed sections.—

In general, condensed sections are associated with maximum water depths during a depositional sequence. In the absence of sediments, maximum water depth occurs at the time when the sum of the rate of eustatic rise and the rate of subsidence is at a maximum.

Water depth is a function of relative sea level. Water depth is also a function of sediment yield. The rate and amount of sediment yield affect water depth directly at any point on a margin. Small sediment yield moves the point of maximum water depth later in time (or farther basinward) than the point of maximum rate of eustatic rise (Fig. 10, Well B). Conversely, large sediment yield means that maximum water depth may occur closer to the point of maximum rate of eustatic rise. Generally, maximum water depths will center around the interval between the point of maximum rate of eustatic rise and the highest point of eustasy (Fig. 10, Well A). Thus, paleobathymetric estimates may be used to record transgressions and regressions produced by eustatic cycles. Sediment yield rates can have a major effect on the timing of water depth changes, however, and each water depth change is therefore not directly linked to eustasy.

SEISMIC EXPRESSION OF A CONDENSED SECTION

On seismic sections, condensed sections are generally identified at the base of prograding clinoforms of the highstand systems tract. Each clinoform downlaps onto the underlying transgressive- and lowstand systems tracts. Hence, the downlap surface is often a good indication of the presence of a condensed section. In outcrop sections and on well logs, the term downlap surface is also used to define a surface associated with condensed sections that formed during a period of nondeposition or extremely slow sedimentation.

In many outcrop examples, the downlap surface is represented by an omission surface. It is generally not possible to observe the downlap of prograding highstand clinoforms in outcrop sections. We believe, however, that the term

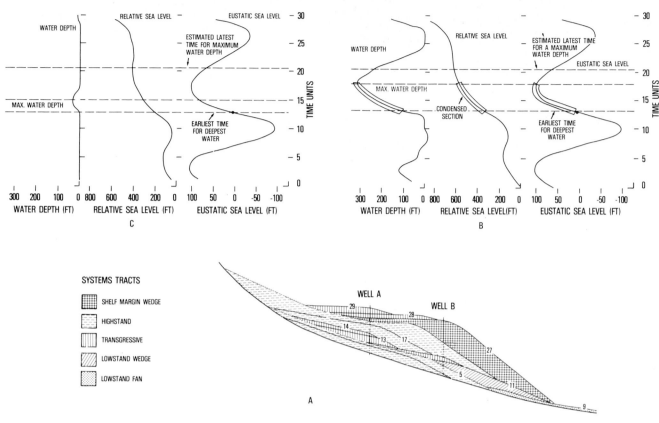

FIG. 10.—(A) Stratigraphic cross section through type 1 (time lines 5–21) and type 2 (time lines 21–29) depositional sequences, showing position of Wells A and B. The cross section was constructed by T. R. Nardin and H. W. Posamentier (pers. commun., 1984) by using three variables– global sea level, subsidence, and sediment supply (an equal amount of sediment was supplied per unit time). (B) Plot of global sea level, relative sea level, and water depth versus time at Well B. Maximum water depth occurs at time 18 at the end of a period of slow-sedimentation rates in the condensed section. (C) Plot of global sea level, relative sea level, and water depth versus time at Well A. Maximum water depth occurs at time 15. No condensed section formed at well A because terrigenous sedimentation was continuous during the period when a condensed section formed at Well B.

downlap surface accurately illustrates the nature of sedimentation associated with the cessation of condensed-section conditions by the progradation of highstand sediments. The downlap surface is often burrowed and bioturbated or sometimes lithified and bored. A surface of nondeposition or downlap surface will only form under certain oceanographic conditions, for example, when oceanic currents prevent any pelagic sedimentation. In general, some form of sedimentation occurs, whether it is pelagic or authigenic.

Last Chance Canyon, on the eastern side of the Guadalupe Mountains, cuts through the Middle Guadalupian Cherry Canyon Sandstone Tongue and the upper San Andres and Grayburg formations. The canyon cut exposes a good example of stratal geometry associated with condensed sections, at a scale that is resolvable by using seismic reflection methods (Sarg and Lehmann, 1986; see also Fig. 11A this paper). The downlap surface occurs on top of the Cherry Canyon Sandstone Tongue that onlaps the San Andres Formation. Prograded clinoforms of the upper San Andres For-

mation downlap onto lower members of the Cherry Canyon Sandstone Tongue (Sarg and Lehmann, 1986). The top of the Cherry Canyon Sandstone Tongue is intensely burrowed and the toes of the upper San Andres Formation clinoforms are represented by a dolomitic mudstone deposited in a few hundred feet of water (Fig. 11B). The environment of deposition in the San Andres Formation highstand systems tract changes from relatively quiet, deeper water conditions above the downlap surface to higher energy, shallower water conditions represented by fusilinid banks and sandstones just below the Grayburg Formation.

Greenlee and others (this volume) recognize a condensed section at the base of Tertiary section in the Baltimore Canyon area on the east coast of the United States and in the northwestern Gulf of Mexico. The condensed section occurs at the base of major progradational clinoforms, which show a characteristic downlap pattern onto the underlying Cretaceous sediments. Prior to deposition of the prograding Tertiary package, the top of the Cretaceous section was

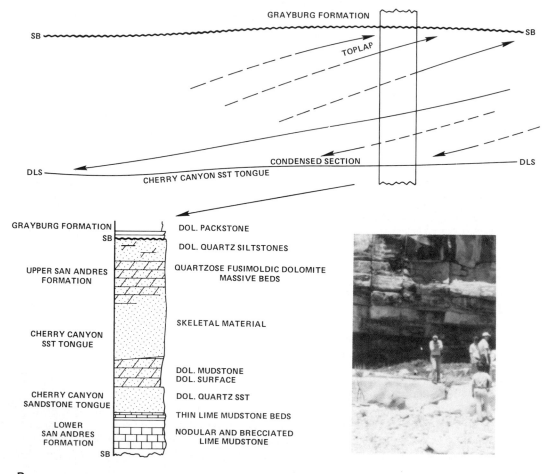

B

FIG. 11.—(A) Photograph and schematic diagram of stratal geometry exposed in Last Chance Canyon, Guadalupe Mountains, New Mexico. Downlap surface is defined at the base of prograded clinoforms of upper San Andres Formation and Cherry Canyon Sandstone Tongue onto the top of a lower member of the Cherry Canyon Sandstone Tongue. Sequence boundaries occur at base of Cherry Canyon Sandstone Tongue and at the top of the upper San Andres Formation. (B) Schematic lithologic section through middle Guadalupian depositional sequence (shown in A) illustrating changes that occur within a condensed section and the highstand systems tract. Deepest water and quietest energy are associated with condensed section. Depositional environment shoaled and energy increased at top of prograded highstand systems tract.

starved for a considerable period of time. Analysis of the stratal geometry suggests that a rise in sea level at the end of the Cretaceous resulted in a major transgression of the shoreline and subsequent starvation of the Cretaceous relict shelf area. The rise in sea level, coupled with continued thermal subsidence of the Cretaceous section during the transgression and loading-related subsidence (a response to the progradation of the Tertiary sedimentary wedge), combined to produce relatively deep water on top of the Cretaceous section.

The second example of the seismic expression of condensed sections is from the east coast of the South Island of New Zealand. A major, extensive composite condensed section formed during the Oligocene to early Miocene in response to a rise in relative sea level. Sedimentation on the New Zealand margin during the Mesozoic and Tertiary is characterized by a major onlap-offlap cycle related to plate-tectonic stresses involving the Pacific and Australian plates. Cretaceous-to-Oligocene sedimentation is represented by a major transgressive package, composed of many depositional sequences, that exhibits a characteristic onlap pattern onto metamorphosed basement sediments (Fig. 12A). The Oligocene-to-Recent section is a major regressive package, also composed of many depositional sequences, that downlaps sequentially basinward onto Eocene-early Oligocene sediments. The Oligocene generally represents an interval starved of terrigenous sedimentation that formed at the end of the Cretaceous-to-Oligocene transgression. Thus, during the Oligocene, and prior to the late Oligocene-Miocene regression, a condensed section formed that

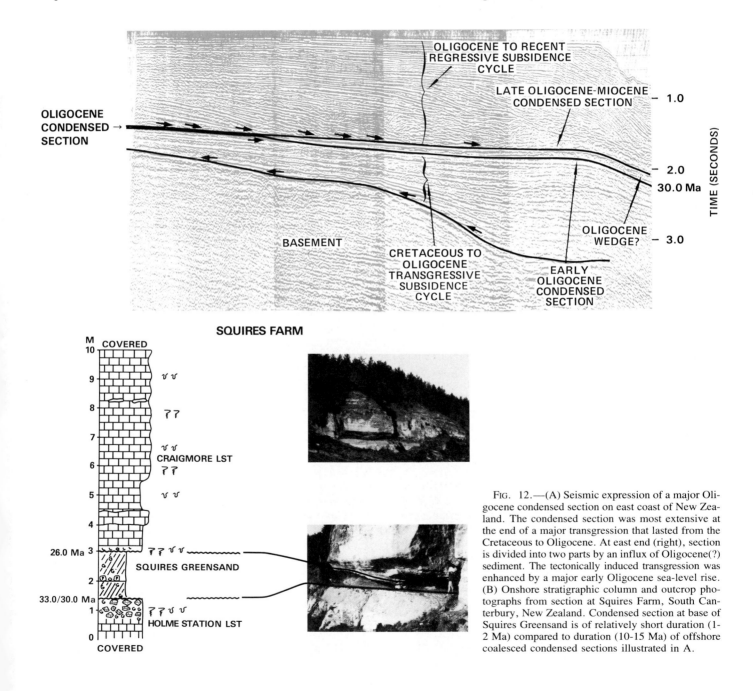

FIG. 12.—(A) Seismic expression of a major Oligocene condensed section on east coast of New Zealand. The condensed section was most extensive at the end of a major transgression that lasted from the Cretaceous to Oligocene. At east end (right), section is divided into two parts by an influx of Oligocene(?) sediment. The tectonically induced transgression was enhanced by a major early Oligocene sea-level rise. (B) Onshore stratigraphic column and outcrop photographs from section at Squires Farm, South Canterbury, New Zealand. Condensed section at base of Squires Greensand is of relatively short duration (1-2 Ma) compared to duration (10-15 Ma) of offshore coalesced condensed sections illustrated in A.

covered most of the New Zealand area. The condensed section actually represents the coalescing of a number of condensed sections associated with a number of eustatic cycles during the Cenozoic. In this example, the interaction of a major tectonic cycle and many eustatic cycles has resulted in the formation of a major composite condensed section formed over millions of years. Sediments deposited during the formation of the condensed section are largely glauconitic limestones and marls that are often burrowed and bioturbated. Exploration wells that penetrate the Oligocene condensed section provide evidence for as much as 15 Ma of nondeposition or slow sedimentation rates in far offshore wells to as little as 1 to 2 Ma, or less, in outcrop sections in the North Otago–South Canterbury region (Fig. 12B). The Oligocene condensed section in New Zealand formed for two reasons: (1) in response to long-term thermotectonic subsidence and loading of the margin from the Cretaceous to the Oligocene, and (2) in response to high global sea level in the early to middle Oligocene, which enhanced the starvation of the Oligocene shelf area of New Zealand. The condensed section terminated in response to the increase in compression between the Pacific and Indian plates during and following the late Oligocene (Walcott, 1978), and a major drop in sea level at 30.0 Ma (Haq and others, 1987). Uplift of the New Zealand region during the early Miocene, in response to the compression, provided a source for terrigenous sedimentation that gradually prograded across the Oligocene condensed section.

The New Zealand example provides an excellent illustration of the seismic criteria used to recognize condensed sections. The combination of sea level and subsidence worked in concert to produce a major, areally extensive condensed section. The long period of subsidence from the Cretaceous to the Oligocene, followed by the Oligocene-to-Recent regressive cycle, provides an example of the long-term effects of subsidence. Sea-level oscillations superimposed onto the major subsidence cycle in the New Zealand region are responsible for the more local distribution of facies within each depositional sequence.

Early to middle Oligocene sections are often condensed on continental margins of the world, providing good evidence for a major transgression during the Oligocene prior to the major fall in sea level at 30 Ma. The New Zealand Oligocene provides a spectacular example of the effects of subsidence and sea-level movements on stratal geometry.

EXAMPLES OF CONDENSED SECTIONS

The examples presented in this section illustrate the nature and importance of condensed sections. The first example provides an easily recognizable and verifiable data set that illustrates the characteristics of a condensed section. The Wisconsinan-to-Holocene transition in the Gulf of Mexico provides an excellent counterpart for ancient condensed sections. A series of examples (Fig. 13) from the coastal plain of the Gulf of Mexico provides examples of ancient condensed sections and their recognition in outcrop and subsurface sections by using a variety of tools, including seismic and well-log analysis. Examples 4 and 5 illustrate the role that condensed sections have played in the

development of stratigraphy in the Gulf coastal plain. The last example illustrates the correlation of the Cenomanian-Turonian condensed sections and the importance of detailed analyses of condensed sections for stratigraphy, paleoceanography, and source-rock prediction.

1. Gulf of Mexico—Wisconsinan-to-Holocene transition.—

Today, sea level, which reached its present level at approximately 6 ka, is high relative to its position during the past 50,000 years (Fig. 14). The most rapid and highest magnitude of eustatic rise in the latest Pleistocene began at about 14 ka and ended in the Holocene at about 7 ka (Mix and Ruddiman, 1985). During this time (14 to 7 ka), a condensed section began to form on continental shelves around the world. In fact, relative sea level is still rising, although very slowly, and in many areas transgression is still occurring on the inner shelf. In high-sedimentation rate areas, such as the mouth of the Mississippi River, highstand deposition, represented by the Mississippi Delta, is occurring farther basinward on the relict outer shelf and slope. Today, the middle and outer continental shelves are starved of terrigenous material and provide a good analogy for illustrating the nature of a condensed section. The transition from the Wisconsinan to the Holocene provides an insight into the formation of ancient condensed sections. The enormous latitudinal variation in sediment types on the present continental shelves, both in shallow and deep waters, demonstrates that no single generalization adequately describes the geology of condensed sections.

A brief and very generalized description of the geologic evolution of the Mississippi River system over the past 30,000 years is as follows. As a result of a eustatic fall at about 25 ka, the shoreline moved abruptly basinward to a position approximately coincident with the present shelf slope break (Fig. 15). Between 25 and 14 ka, sediment deposition was restricted to the Gulf of Mexico basin. River systems, such as the Brazos, Rio Grande, Neches, Sabine, Calcasieu, Mermentau, and Mississippi, all deposited sediment directly into the basin or in depocenters along the present shelf slope breeak and upper slope (Suter and Berryhill, 1985; Fisk and McFarlan, 1955). Most of the present shelf area was subaerially exposed (Fisk and McFarlan, 1955; Frazier, 1974).

Following the rise in sea level that began approximately 16 to 14 ka, the site of major sediment distribution moved abruptly landward from the Gulf of Mexico basin to a shoreline position landward of the present shoreline. This rapid movement of the site of sediment deposition across the present shelf was produced by about a 300-ft eustatic rise (Fig. 14). The net result of the shoreline transgression between 14 ka and 7 ka was to shut off the bulk of terrigenous sediment supply to the Gulf of Mexico basin and present shelf area. A marine disconformity (Frazier, 1974), or condensed section, that formed on the present shelf is characterized by nondeposition or minor hemipelagic-to-pelagic deposition.

The marine disconformity recognized by Frazier (1974) in the area of the Mississippi Delta actually represents two separate surfaces that may be coincident with one another.

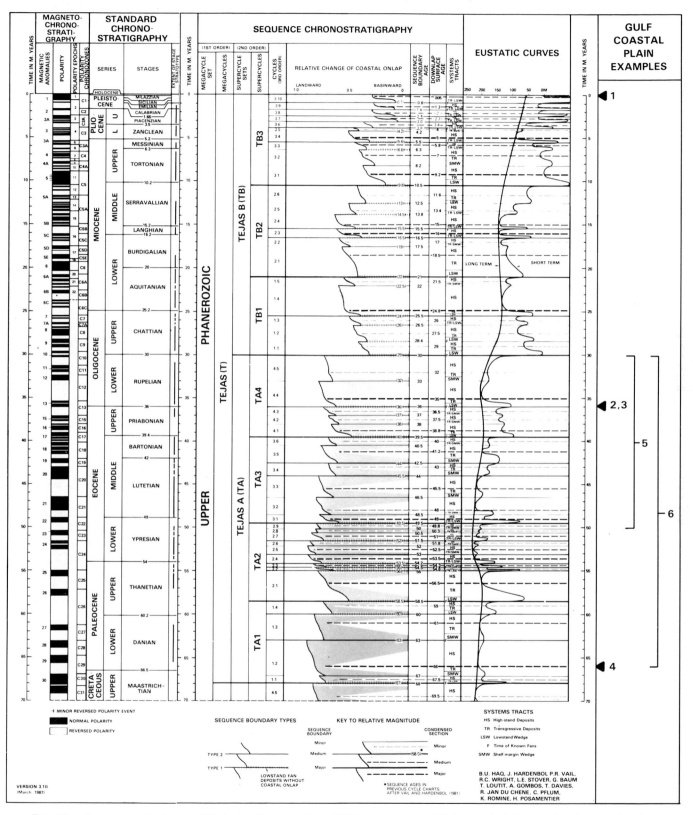

Fig. 13.—Cenozoic cycle chart modified after Haq *and others* (1987) showing age of Gulf coastal plain examples discussed herein.

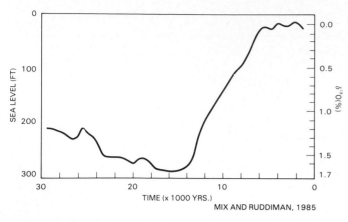

FIG. 14.—Stacked oxygen isotope record for the last 30 ka. Record represents approximate sea-level index and is calibrated to a sea-level change of about 300 ft (91 m) for the last glacial-to-interglacial transition.

The marine disconformity may represent (1) a transgressive surface, an irregular, sometimes erosional surface, formed by the transgression of high-energy nearshore facies or (2) a surface of nondeposition or extremely slow sedimentation, known as a downlap surface. In many areas the transgressive surface and the downlap surface may be coincident. In some areas they are separated by sediments of the transgressive-systems tract (Fig. 16).

During the past 6 ka, the rate of sea-level rise has dropped to negligible amounts, and eustasy has remained almost constant during this time. In the vicinity of the Mississippi River, sediment yield was large enough to fill drowned estuaries and river valleys and to produce a major progradation of the highstand systems tract. The shoreline of the Mississippi Delta has prograded almost all the way to the previous lowstand shoreline (Fig. 17). The delta has prograded across the top of the condensed section formed during the transgression and early phases of progradation (Fig. 16). Prior to the progradation of the Mississippi Delta across the shelf, the present shelf area beneath the delta was starved of terrigenous material. Water depths increased from zero to about 600 ft (near the present physiographic shelf slope

break) in response to a rise of about 300 ft in sea level between 14 ka and 6 ka. About 50% of the water depth increase occurred as a result of sediment (Fisk and MacFarlan, 1955) and water loading on the underlying sediment pile. The abundance and diversity of microfossils are high within the condensed section below the Mississippi Delta because of the decrease in sedimentation caused by the transgression of the shore line and the increase in water depth (Fig. 18). The increase in abundance is primarily due to a decrease in sedimentation rate. The increase in diversity is partially due to the deposition of a variety of benthic and planktonic microfossil assemblages associated with the increase in water depth and eustatic rise between 14 ka and 6 ka. This concentration of microfossils resulting from sedimentation rate decreases and water depth increases is an important characteristic of ancient condensed sections.

Away from the Mississippi Delta, the condensed section is still exposed (Fig. 17). Nearly all of the river systems that drain into the Gulf of Mexico have such small drainage basins that the sediment supplied to these rivers today has been insufficient to produce any significant progradation of deltas onto the shelf (Gould and Stewart, 1955). The thin veneer of sediment overlying the late Wisconsinan subaerial exposure surface consists of reworked sediment that was deposited during lower levels of eustasy, hemipelagic sediment deposited from suspension and probably derived from the Mississippi River, and pelagic material produced in the surface waters of the Gulf of Mexico (Shepard and others, 1960). The characteristics of the Wisconsinan-Holocene condensed section in the Gulf of Mexico vary considerably from east to west. In the east the shelf is characterized by carbonate environments, whereas in the west the shelf is predominantly clastic, with isolated carbonate banks. Thus, it is obvious that the criteria used to recognize a condensed section can change dramatically, even within a single basin.

In addition to lithofacies data, modern biofacies distribution data in the Gulf of Mexico provide an independent line of evidence to characterize condensed sections. Figure 19 illustrates the present distribution of benthic foraminiferal biofacies in the Gulf of Mexico. There is a clear as-

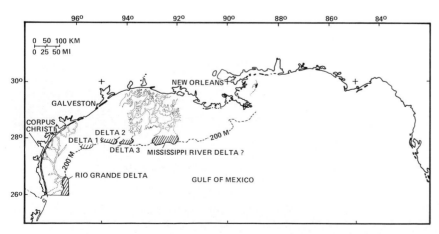

MODIFIED FROM SUTER & BERRYHILL, 1984

FIG. 15.—A series of deltas perched on present physiographic shelf slope break approximately defines position of last major (25 ka) lowstand shoreline as being almost coincident with shelf slope break. Area represented by present shelf was incised by a number of river systems that supplied clastic sediments to the deep Gulf of Mexico (Fisk and McFarlan, 1955).

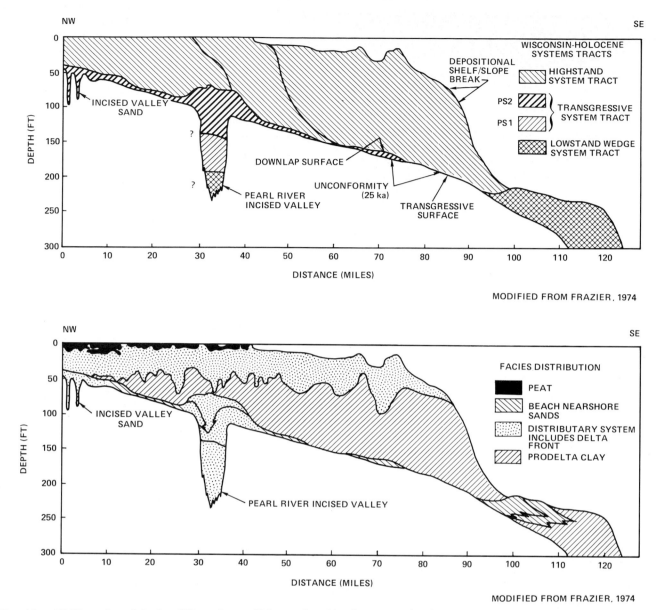

FIG. 16.—(A) Illustration of the late Wisconsinan-to-Holocene depositional sequence showing the late lowstand wedge, two transgressive parasequences, and a highstand system tract. Location of cross section runs southeast from New Orleans through a number of delta lobes within Saint Bernard Delta complex. (B) The late Wisconsinan-to-Holocene depositional sequence showing distribution of major lithofacies. Deepest water depth and most abundant and diverse faunas are associated with the lowermost portion of prodelta clays of highstand and transgressive-systems tract.

sociation between benthic foraminiferal biofacies and depositional regimes in the Gulf (Poag, 1981). The active sedimentation regime on the shelf is restricted to areas of major river discharge. The Mississippi River is the dominant sediment source in the Gulf today. Large areas of the shelf exist, however, that are receiving little or no sediment at present and represent a condensed section. These areas have been classified as relict and/or palimpsest sediment by Shepard and others (1960) and are characterized by the *Elphidium, Elphidium-Hanzawaia,* and *Bigenerina* biofacies on the Texas and Louisiana shelf (Poag, 1981). Other relict deposits off Florida are characterized by the *Planulina* and *Cibicidoides* biofacies.

The distribution of sediment types and biofacies in the Gulf today is controlled, primarily, by the major rise in sea level since the last glacial and, secondarily, by (1) the amount and geographic position of sediment discharge by rivers; (2) climate (glacial versus interglacial); and (3) oceanic circulation. The sea-level rise since 14 ka has moved the loci of terrigenous deposition in a landward direction, but the other factors listed above are now modifying sediment patterns to conform to the present, higher sea level.

The complex sediment-biofacies relations (Fig. 19) seen in the Gulf of Mexico today are assumed to have remained fairly constant during the past 15 to 20 Ma, and their geographic distribution through time provides important infor-

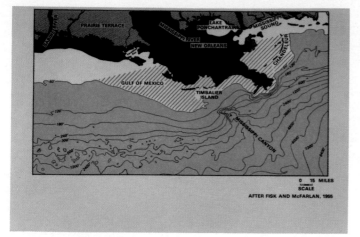

FIG. 17.—Position of prograded highstand systems tract shoreline (present shoreline). Mississippi Delta highstand has prograded almost to position of lowstand shoreline, defined as nearly 300 ft (91 m) from the delta and closer to 600 ft (183 m) to the vicinity of sediment-loading effects by the delta. Condensed section can be recognized beneath Mississippi Delta (diagonal pattern) and over much of the modern continental shelf (shaded area) away from influence of Mississippi River sediment supply.

mation on ancient condensed sections. For example, the foraminiferal *Bulimina* biofacies is associated with the present slope and approximately outlines the position of the oxygen minimum zone (Poag, 1981). If present in sufficient numbers, buliminids, bolivinids, and uvigerinids may be used to identify areas of lower oxygen and of slower terrigenous sedimentation. Thus, their distribution in the Gulf of Mexico subsurface may be used to recognize ancient condensed sections.

2. Eocene-Oligocene boundary condensed sections in Alabama—outcrop and biotic criteria.—

The sequence of lithological and biological changes recorded across the paleontologically defined Eocene-Oligocene boundary in two southwestern Alabama outcrop sections (Little Stave Creek and St. Stephens Quarry) provides a good example of an ancient condensed section in an overall setting of low-sedimentation rate. The Paleogene in western Alabama is represented by a thickness of over 2,000 ft and in eastern Alabama by less than 1,000 ft. During the Paleogene, sedimentation rates in the Alabama area were considerably lower than those in Mississippi, Louisiana, and Texas, where the Paleogene may be thicker than

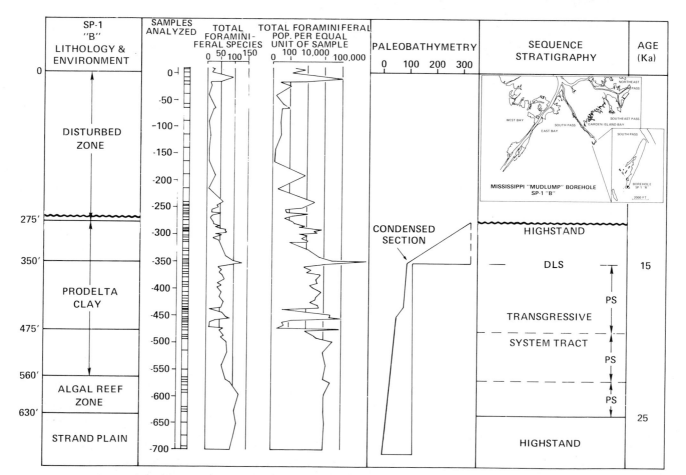

FIG. 18.—Location of "mudlump" borehole at mouth of Mississippi River (inset) and interpreted lithology, foraminiferal abundance, water depth, and depositional sequence reported from borehole. Note that abrupt increase in water depth at 350 ft (107 m) within prodelta clays is probably associated with period of nondeposition in condensed section at about 10 ka. Increase in foraminiferal abundance at 350 ft (107 m) is probably related to decrease in sedimentation rate as shoreline transgressed in response to a sea-level rise of about 300 ft (91 m). Decrease in diversity and abundance in upper part of core are related to progradation of prodelta muds into the area and dilution of faunal populations by terreginous sediments.

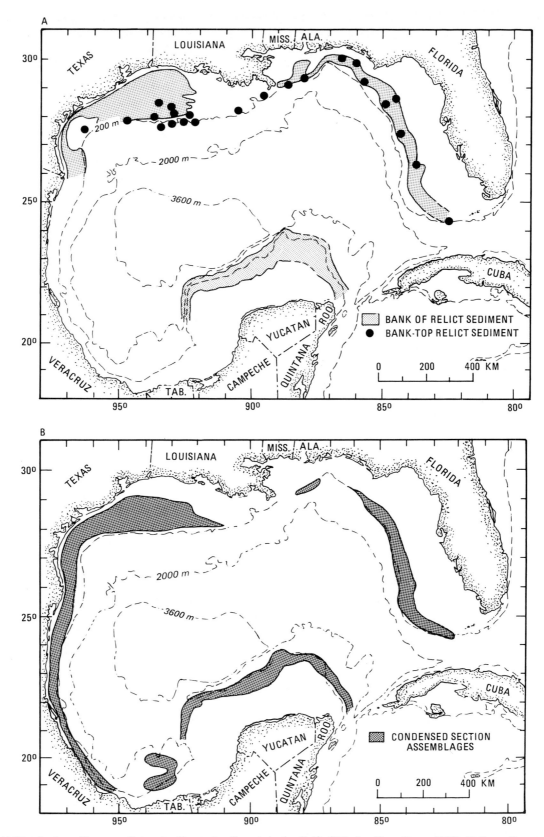

FIG. 19.—(A) Distribution of known relict and palimpsest sediments in the Gulf of Mexico (from Poag, 1981). Relict sediments were deposited during a sea-level rise from 15 to 6 ka following the glacial maximum at 18 ka. (B) Distribution of benthic foraminifera associated with relict sediments in the Gulf of Mexico (from Poag, 1981). The *Elphidium, Elphidium-Hanzawaia, Bigenerina, Planulina,* and *Cibicidoides* assemblages may be used to recognize ancient condensed sections.

10,000 ft. Even within the lower sedimentation rate setting in Alabama, however, the effects of sea-level oscillations can be recognized.

The sections at Little Stave Creek and St. Stephens Quarry consist of fossiliferous, glauconitic marls (Yazoo Clay), a glauconitic clay (unnamed blue clay), glauconitic clays and marls interbedded with a silty limestone (Bumpnose Limestone), and carbonaceous clays (Red Bluff Clay). The condensed section incorporates some of the Shubuta Clay, the unnamed blue clay, and the lowermost portion of the Bumpnose Limestone (Fig. 20). The unnamed clay consists predominantly of pelagic microfossils and represents the period of deposition when the rate of sea-level rise and starvation of the shelf and basin, with respect to terrigenous sediments, reached a maximum (Loutit and others, 1983). An overall upward increase occurs in gamma ray counts in the Yazoo Clay, reaching a maximum in the unnamed clay (Fig. 20A). The upward increase reflects an increasing abundance of uranium and potassium probably associated with organic-matter deposition (and oxidation) in the unnamed clay.

The last occurrence of the planktonic foraminifer *Globorotalia cerroazulensis cocoaensis* occurs within the unnamed clay (Mancini, 1979), and the last occurrence of the Eocene calcareous nannofossil *Discoaster saipanensis* occurs at the top of the Pachuta Marl (Bybell, 1982). In addition, there is a maximum in the relative abundance of foraminifera such as *Bolivina* spp. and *Uvigerina* spp. (indicating lower oxygen conditions) throughout the condensed section, particularly in the Shubuta Clay, the unnamed clay, and the Bumpnose Limestone (Fig. 20B).

The last occurrence of *Globorotalia cerroazulensis cocoaensis* occurs in a pelagic unit with a low-sedimentation rate. This unit exhibits a high gamma ray count and represents the deepest and lowest oxygen concentration waters recorded in the middle of the 37-Ma depositional sequence at St. Stephens Quarry. The association between important microfossil events, peaks in water depth, high gamma ray counts, and low terrigenous sedimentation rates is nearly always associated with transgressions of the shoreline and formation of a condensed section in continental-margin sections.

The association of age-diagnostic microfossils with condensed sections is common in continental-margin sections. Several microfossil zones may be juxtaposed in long-duration condensed sections.

3. Interregional correlation of the Eocene-Oligocene boundary.—

Planktonic microfossils generally provide the age control for correlating depositional sequence boundaries between continents. The example presented in this section illustrates the sequence-stratigraphic approach to interregional correlation of the Eocene-Oligocene boundary. The top of the Priabonian Stage in Italy is marked by the 36-Ma depositional-sequence boundary (Haq and others, 1987). The Priabonian-Rupelian boundary is approximated and correlated generally by using the last appearance of planktonic microfossils such as *Discoaster saipanensis* and *Globorotalia cerroazulensis cerroazulensis*. In the Alabama outcrop

sections that span the Eocene-Oligocene boundary, the last appearance of *G. cerroazulensis cocoaensis* Oligocene boundary in the Gulf Coastal Plain occurs at the top of the Yazoo Clay in a condensed section (see previous section). The condensed section occurs below the 36-Ma sequence boundary at the base of the Mint Springs that correlates with the Eocene-Oligocene boundary at the top of the Priabonian (Fig. 21A).

The Eocene-Oligocene boundary in New Zealand is defined paleontologically by the last appearance of *Globigerapsis index*. The paleontological approximation of the boundary occurs in a condensed section within the Kaiata Mudstone at Cape Foulwind on the west coast of the South Island of New Zealand. The 36-Ma sequence boundary is interpreted to be at the base of the Little Totara Sand overlying the Kaiata Mudstone (Fig. 21B). The synchroneity of the biostratigraphic correlation was confirmed by oxygen-isotopic studies of foraminifera in both outcrop and DSDP sites (Keigwin, 1980; Keigwin and Corliss, 1986; Burns and Nelson, 1981). A dramatic positive shift in $\delta^{18}O$ in benthic and planktonic foraminifera, first reported by Shackleton and Kennett (1975), records the formation of cold deep-ocean–bottom water coincident with the last appearance of *Globorotalia cerroazulensis* and *Globigerapsis index* (Keigwin, 1980; Fig. 22). The oxygen-isotopic shift, recorded by foraminifera, represents a synchronous change in ocean stratification and circulation that permits a calibration of biostratigraphic events within the world oceans. The last appearance of *Globorotalia cerroazulensis cerroazulensis* occurs at the $\delta^{18}O$ shift in the low-latitude DSDP Site 292. The last appearance of *Globigerapsis index* is also associated with a major $\delta^{18}O$ shift in the high southern latitude DSDP Site 277 (Fig. 22). The last appearance of *Globorotalia cerroazulensis cocoaensis* occurs in a condensed section in Alabama (Fig. 21A), and the last appearance of *Globigerapsis index* occurs within a poorly developed condensed section at Cape Foulwind in New Zealand (Fig. 21B). The downward shift in coastal onlap (produced by a eustatic fall) that occurs at the top of the Priabonian can be recognized at the base of the Mint Springs Formation in Alabama and at the base of the Little Totara Sand in New Zealand.

Interregional correlation of key stage boundaries can be accomplished by using a combination of physical stratigraphy (depositional sequences) and biostratigraphic correlation of marine planktonic microfossils associated with condensed sections in continental-margin sections. Very little work was done on the recognition of the physical or chemical expression of depositional-sequence boundaries in deep-sea sections, and most correlations are achieved by using biostratigraphy, magnetostratigraphy, and chemical stratigraphy. If depositional-sequence boundaries are recognizable in the ocean basins, however, then interregional physical correlation of continental-margin sections will become even more accurate.

4. Cretaceous-Tertiary boundary section in Alabama: outcrop, biotic, and chemical criteria.—

The Cretaceous-Tertiary boundary section at Braggs, Alabama, is one of a number of intensively studied classic

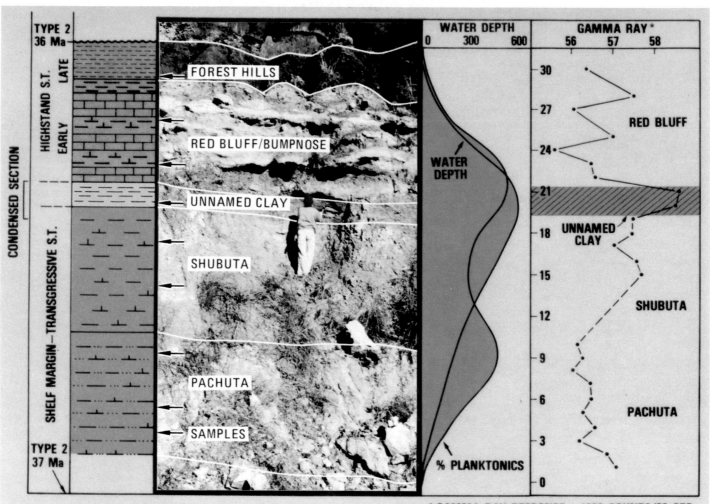

* GAMMA RAY RESPONSE x 1000 COUNTS/30 SEC.

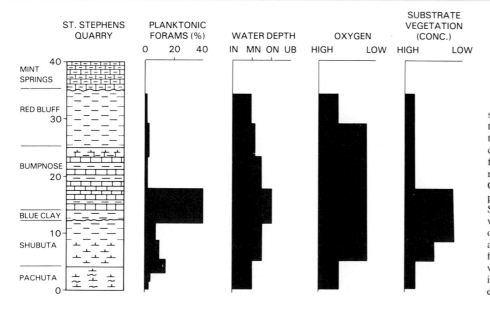

FIG. 20.—(A) Eocene-Oligocene boundary section in St. Stephens Quarry. Condensed section is associated with the unnamed clay (center of section), a pelagic unit consisting of predominantly calcareous nannofossil and foraminiferal tests. Total gamma ray counts were recorded on a Scintrex GAD-6 recorder and GSP-3 sensor. (B) Paleobathymetric and depositional environmental estimates for the St. Stephens Quarry section. Deepest water occurs within unnamed blue clay and lowermost part of Bumpnose Limestone. Lower oxygen values and lower concentrations of attached benthic foraminifera are also found in this interval, which supports the interpretation of deeper water in the condensed section (from Loutit and others, 1983).

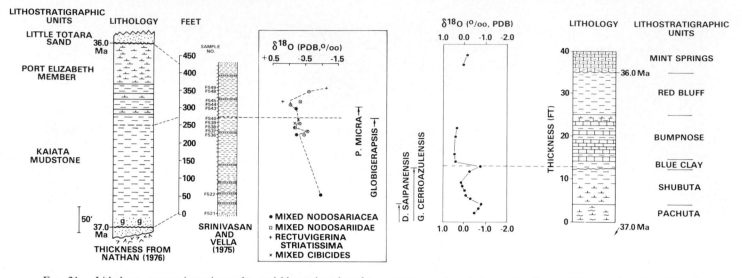

FIG. 21.—Lithology, oxygen-isotopic results, and biostratigraphy of two outcrop sections that span the Eocene-Oligocene boundary. The Cape Foulwind section (A) is on the west coast of South Island, New Zealand, and the St. Stephens Quarry section (B) is in Alabama in the Gulf of Mexico coastal plain. The last appearance of Eocene planktonic foraminifera in both sections occurs in an interval of slow sedimentation and decreased energy in the depositional environment. Isotopic data from St. Stephens Quarry provided by L. D. Keigwin (written commun., 1983). Dashed line represents paleontological approximation of Eocene-Oligocene boundary.

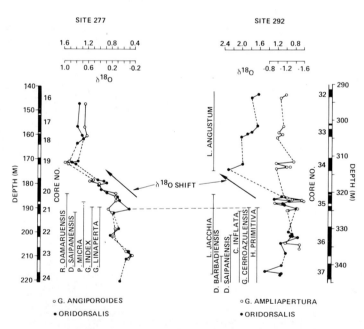

FIG. 22.—Oxygen-isotopic results from DSDP Site 277 (52.S) and DSDP Site 292 (16.N) in the Pacific Ocean (Keigwin, 1980). Note a 1-per-mil shift in δ18O that records a synchronous global cooling of bottom waters and provides a time-synchronous event that can be used to correlate last appearance of age-diagnostic planktonic taxa (near the Eocene-Oligocene boundary) in continental sections in New Zealand and Alabama. Dashed line represents paleontological approximation of the Eocene-Oligocene boundary.

boundary sections. Detailed studies based on intense sampling of these boundary sections over the past few years have showed a considerable amount of information on condensed sections in a variety of depositional environments. Marine continental-margin sections across the Cretaceous-Tertiary boundary are characterized by low-sedimentation rates, both terrigenous (Alvarez and others, 1980; Baum and others, 1984) and pelagic (Arthur and others, 1987). Many open-ocean Cretaceous-Tertiary boundary sections also exhibit slow pelagic-sedimentation rates (Arthur and others, 1986).

The Braggs section in Alabama was interpreted as a condensed section (Baum and others, 1984; Donovan and others, this volume). Based on paleontology, the boundary occurs between 3 and 6 ft in the Clayton Limestone Member within a transgressive-systems tract above the 67-Ma sequence boundary (Fig. 23). The lowermost Pine Barren Member (above 14 ft) of the Clayton Limestone Formation is interpreted as being within the lower part of a highstand systems tract. The condensed section spans the upper part of the micrite-rich lower Clayton Limestone and lowermost Pine Barren Member.

Increased micrite concentration toward the top of the Clayton Limestone Member (3 to 14 ft in Fig. 23) suggests that water depths also increased during deposition of this unit, effectively shutting off the terrigenous sediment yield. The top of the Clayton Limestone Member is an intensely highly burrowed/bored(?) surface that represents a time of nondeposition (Baum and others, 1984). Sedimentation resumed with the deposition of the relatively unfossiliferous, glauconitic, lime mud (Pine Barren Member). The terrigenous content increases toward the top of the Pine Barren Member, suggesting that water depths may have been shoaling during deposition of the Pine Barren.

A surface of nondeposition characterized by burrowing or boring often is found associated with condensed sections. The term "downlap surface," often used to describe this surface of nondeposition, is derived from the seismic expression of a condensed section (downlap over concordant; see examples in preceding sections). Highstand systems tract deposits prograde or downlap onto transgressive- and lowstand systems tracts, producing a downlap surface recognizable on seismic lines.

The downlap surface is areally most extensive at the time of maximum regional transgression of the shoreline and maximum rates of relative sea-level rise and is associated with maximum water depths at a given geographic locality within a depositional sequence. Condensed sections and downlap surfaces can often be recognized in outcrop, and the former may also be readily discernible on electric logs (Fig. 24) and on gamma ray logs. Gamma ray logs are especially useful because of the radioactive elements associated with organic matter (Meyer and Nederlof, 1984) and potassium-rich glauconite that are often concentrated in condensed sections. The downlap surface may also be recognizable on electric logs, notably when a hardground or a unit more rich in calcite is present. The downlap surface is generally detectable on electric and gamma ray logs as the point where progradation of highstand sediments begins above the retrogradational stacking of the transgressive-systems tracts (Fig. 24).

The decrease in terrigenous sedimentation rates across the Cretaceous-Tertiary boundary, produced by a sea-level rise, has not always been discussed adequately by various workers studying boundary sections. Low terrigenous sedimentation rates associated with the boundary are partially responsible for (1) the condensed nature of biozones associated with the boundary; (2) the apparent abrupt nature of the changes in environments; (3) the concentration of authigenic minerals; and (4) the enhancement and variation in shape of geochemical events, such as $\delta^{13}C$ shifts, or spikes (recorded in marine carbonates), and the platinum group element spikes. The decrease in pelagic sedimentation across the boundary in open-marine sections, reported by Arthur and others (1987) produced similar effects on the character of biostratigraphic and geochemical events to those observed in the continental-margin sections.

The Cretaceous-Tertiary boundary at the Braggs section occurs within a transgressive-systems tract that represents part of the condensed section within the 67-Ma (TA1.2 Fig. 13) depositional sequence. Sequence-stratigraphic analysis of Cretaceous-Tertiary boundary in continental-margin sections provides additional insights into the nature of boundary events, both biostratigraphic and geochemical. A eustatic rise in the latest Maastrichtian caused a significant effect on sedimentation patterns on continental margins and must be included in any scenario of events across the Cretaceous-Tertiary boundary. The Braggs example makes it clear that sample-based observations should be made within a depositional framework and particular attention paid to sedimentation-rate changes.

5. Claiborne-to-Vicksburg condensed sections—coastal plain of the Gulf of Mexico: outcrop and subsurface criteria.—

The fifth example illustrates the correlation of condensed sections from traditional outcrop sections to their subsurface equivalents. Murray (1961), in a summary of Gulf Coast stratigraphy, illustrated three thin, fossiliferous, deeper water units (condensed sections) within the sediments deposited between the top of the Wilcox Group to the base of the Vicksburg Stage (Fig. 25). The base of the Cane River, the Crockett and Jackson formations, and their lateral equivalents in the Gulf Coast all represent periods of maximum transgression of the shoreline and greatest water depths. Each transgression disconformably overlies regressive, progradational formations such as the Wilcox Group, the Sparta Sand, and the Yegua Formation. The cyclical nature of these transgressive-regressive packages within the Claiborne-Vicksburg section was long used as a basis for regional correlation within the Gulf basin (Murray, 1961). Each cycle is traditionally bounded by marine disconformities at the base of the transgressive portion of the cycle. Higher orders of cyclicity have also been recognized within the Sparta Sand and Yegua Formation (Murray, 1961), but have not been correlated as widely. Stenzel (1952) recognized a number of additional marine disconformities within a higher order of transgressive-regressive cycles in the Claiborne Stage and was able to correlate them through much of the Gulf Coast Claiborne outcrop belt.

The Cane River and the Crockett and Jackson formations are recognized easily as major shale packages within the subsurface (Fig. 25). In the outcrop and shallow-subsurface example shown in Figure 25, the lower portion of the Cane River and the Crockett and Jackson formations are represented by middle- to outer-neritic environments. In the deeper subsurface example in Figure 25, outer-neritic to upper-bathyal environments are associated with the condensed sections. The sand content of the Sparta and Yegua formations has also decreased considerably. Key age-diagnostic microfossils are associated with each condensed section and provide the information necessary to date the major depositional sequences. The same three transgressive-regressive cycles observed in the outcrop sections are easily recognized in the subsurface example, even though the depositional environment has changed considerably. The condensed sections constitute only a minor portion of the sedimentary section shown in Figure 25; however, they are critical to age determination in the Claiborne section. The efforts of stratigraphers during the past 100 years have demonstrated the synchroneity of these three thin transgressive units across the Gulf Coast from Mexico to Florida (see references in Toulmin, 1977, and Murray, 1961). The three major depositional-sequence boundaries associated with the major Claiborne-Jackson condensed sections are the 49.5-Ma boundary at the base of the Carizzo Sand, the 44.0-Ma boundary within the Sparta Sand, and the 39.5-Ma boundary within the Yegua Formation (Fig. 25).

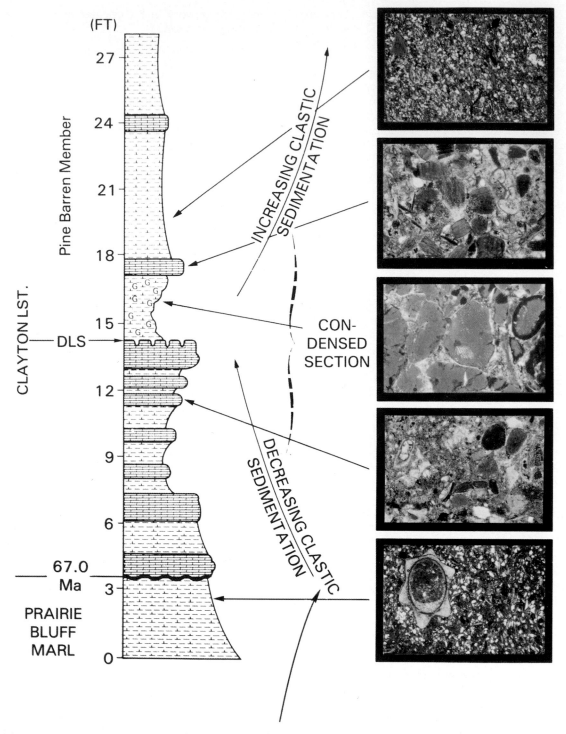

FIG. 23.—Braggs section in Lowndes County, Alabama. Condensed section between the Navarro and Midway stages is characterized by low terrigenous content and is richer in glauconite and organic matter than the rocks above or below (modified from Baum and others, 1984).

The three major condensed sections in the Claiborne-Jackson section provide a physical link between the classic outcrop and subsurface exploration sections. Key plank-tonic microfossils found in these condensed sections pro-vide age information necessary to date the 49.5-, 44-, and 39.5-Ma sequence boundaries. Each dated sequence bound-ary can be correlated into shallow- or marginal-marine en-vironments that characterize many of these classic outcrop sections, thus providing additional age information. The combination of open-ocean planktonic foraminifera found

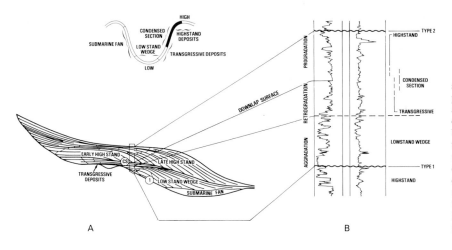

FIG. 24.—(A) Composite depth-distance diagram showing systems tracts produced within a type 1 depositional sequence. (B) SP and resistivity logs through a depositional sequence. Condensed section is recognized most easily in region shown as shaly interval between a fining- or deepening-upward (retrogradational) stacking pattern and a shoaling- or coarsening-upward (progradational) unit. Section generally occurs at or before time of maximum water depth within each depositional sequence. Downlap surface occurs at point in section where parasequence stacking pattern changes from retrogradation to progradation. Downlap surface occurs below downlapping clinoforms of prograding highstand systems tract.

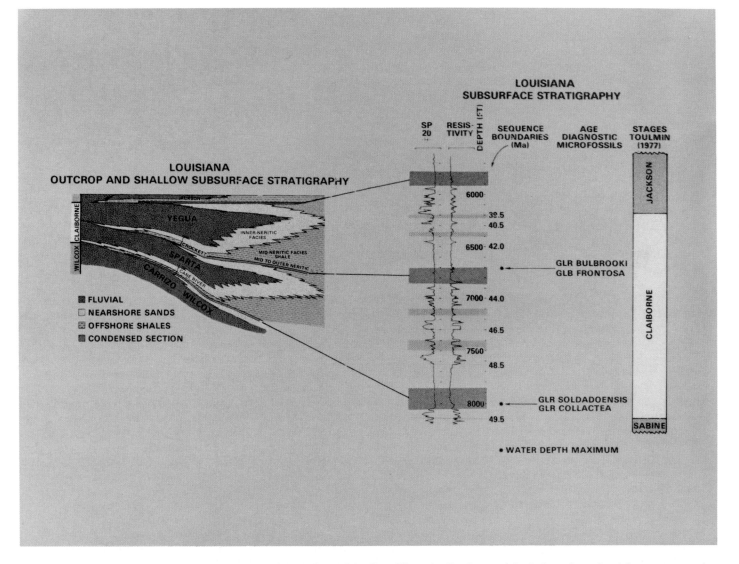

FIG. 25.—Correlation of three major condensed sections (at base of the Cane River, the Crockett, and the Jackson formations) from outcrop and shallow subsurface to deep subsurface. Each condensed section represents a maximum in water depth and contains age-diagnostic planktonic microfossils.

in condensed sections and depositional-sequence boundaries provides a powerful integrated correlation technique that is applicable to a wide variety of environments.

At least seven depositional sequences and associated condensed sections have been recognized in the Claiborne-Jackson section (Fig. 25). A number of these depositional sequences correspond to transgressive-regressive correlation units defined by Fisk (1940), Stenzel (1952), Murray (1961, and references therein), and Toulmin (1977). With the exception of Toulmin (1977) and, to a lesser extent, Stenzel (1952), most workers have not recognized the presence of subaerial unconformities in the Tertiary of the Gulf basin.

6. Paleogene condensed sections of the Gulf Coast: stratigraphic correlation philosophy.—

Transgressive marine-shale tongues containing age-diagnostic fauna, both macrofauna and microfossils, have proven to be the key to stratigraphic correlation within the Gulf Coast region. Despite approximately 100 years of discussion concerning the exact placement of Gulf Coast stage and major formation boundaries, considerable confusion still exists. Most workers agree, however, that a well-developed marine shale occurs above each of the Paleogene stage boundaries, particularly at the base of the Midway, Wilcox, Claiborne, and Jackson stages. Many workers, including Fisk (1940), MacNeil (1946), Stenzel (1952), Murray (1953, 1955, 1961), and Toulmin (1977), have discussed the definitions of the Paleogene stages and the approaches used to define them. Murray (1961) and Toulmin (1977) provide good examples of these differing stratigraphic philosophies. Murray (1961) redefined the Gulf Coast stages by utilizing transgressive-regressive cycles and the evolutionary appearance of age-diagnostic marine organisms in outcrop and shallow-subsurface sections. Murray's (1961) stage boundaries correspond primarily to a transgressive-flooding surface (produced by transgression of nearshore environments) or, to a lesser extent, a condensed section (Fig. 26). In contrast, Toulmin (1977) defined stage boundaries based on disconformities or unconformities defined by subaerial erosion or marine transgression (transgressive surface; Fig. 27). The significance of these two approaches will be discussed in a later section. The main point to be made here is that the lower part of the Midway, Wilcox, Claiborne, and Jackson stages represent transgressive units of variable lithology and thickness that are overlain by marine shales or carbonates deposited during a period of increasing water depth. The depositional environment of these marine shales, or condensed sections, varies from shallow-marine in the outcrop belt to bathyal depths and greater in the subsurface. Each of these thin marine shales has been buried beneath a major progradational package containing a series of higher frequency transgressive-regressive cycles. Age-diagnostic microfossils that consistently occur in these thin marine shales, or condensed sections, have provided confirmation of physical correlations from Florida to Texas made by both Murray (1961) and Toulmin (1977). Classic age-diagnostic macrofossils are, in general, associated with thin, trans-

gressive, somewhat coarser grained, units below the condensed section. The *Ostrea thirsae* beds are a good example of macrofossil-rich units that are restricted to a transgressive-systems tract and provide age control over a large area of the Gulf Coast.

Both Murray (1961) and Toulmin (1977) agree on the placement of the Jackson-Vicksburg stage boundary. The Jackson-Vicksburg boundary is represented by a downlap surface within a condensed section (Figs. 26, 27). The base of the Vicksburg Stage is defined by the top of the Yazoo Clay in Mississippi. The top of the Yazoo Clay represents the deepest water within the 37-Ma to 36-Ma (TA 4.3) sequence and contains a diverse suite of planktonic microfossils and nannofossils (Figs. 20A,B; Loutit and others 1983). In Alabama, the Yazoo Clay is overlain by prograding Bumpnose Limestone/Red Bluff Clay and Forest Hill Sand highstand sediments. The downlap surface defined at the top of the Yazoo Clay is an easily recognized surface that is relatively well defined paleontologically. The combination of good paleontological control and stratigraphic break at the top of the Yazoo Clay is one of the reasons why there is little confusion about the placement of the Jackson-Vicksburg stage boundary in Gulf Coast outcrop and subsurface sections.

The condensed sections at the base of each of the Gulf Coast Paleogene stages provide both physical and temporal control for stratigraphic correlation of each stage. Thus, the condensed sections have provided the key to regional correlations in the Gulf Coast region.

7. Condensed-section geochemistry.—

The geochemistry of condensed sections is as complex as the oceanographic conditions in the overlying water column and is directly related to these conditions. It is not possible to illustrate this complexity adequately by one or two examples, but the examples presented do illustrate the interplay of slow sedimentation or nondeposition (induced by sea-level rise), local oceanographic conditions, global-ocean chemistry, and climate. The geochemical complexity of condensed sections can only be unraveled by a wide variety of analyses of closely spaced samples.

Cenomanian-Turonian boundary sections in continental-margin settings illustrate some of the geochemical variabilities associated with condensed sections. Many workers have recognized that high sea level during the middle Cretaceous, coupled with a Late Cenomanian transgression, often resulted in the formation of a distinct organic-rich layer coincident with the Cenomanian-Turonian boundary. Most sections which span the Cenomanian-Turonian boundary provide good examples of condensed sections that can be well correlated regionally and between continents by using biostratigraphic and chemical stratigraphic techniques.

The South Ferriby section in Humberside, England, has been studied in detail by Wood and Smith (1978, lithology, stratigraphy); Hart and Bigg (1981, micropaleontology) and more recently by Schlanger and others (1986, geochemistry). The environmental interpretation of the South Ferriby section is consistent with an interpretation of the "Black

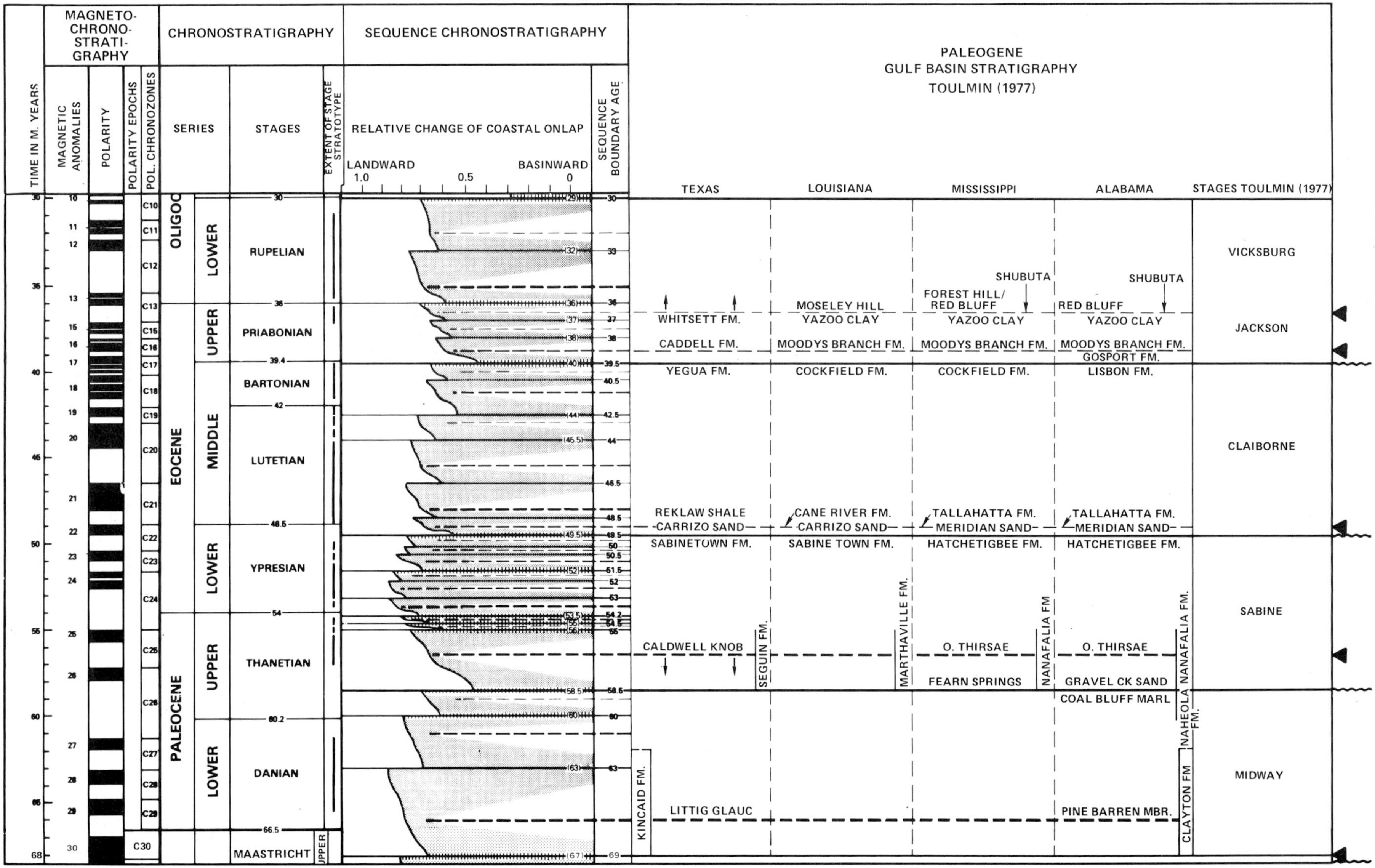

FIG. 26.—Stage classification correlated to coastal onlap chart of Haq and others (1987). Each stage boundary corresponds to either a transgressive surface or a condensed section. Black triangles mark position of condensed sections.

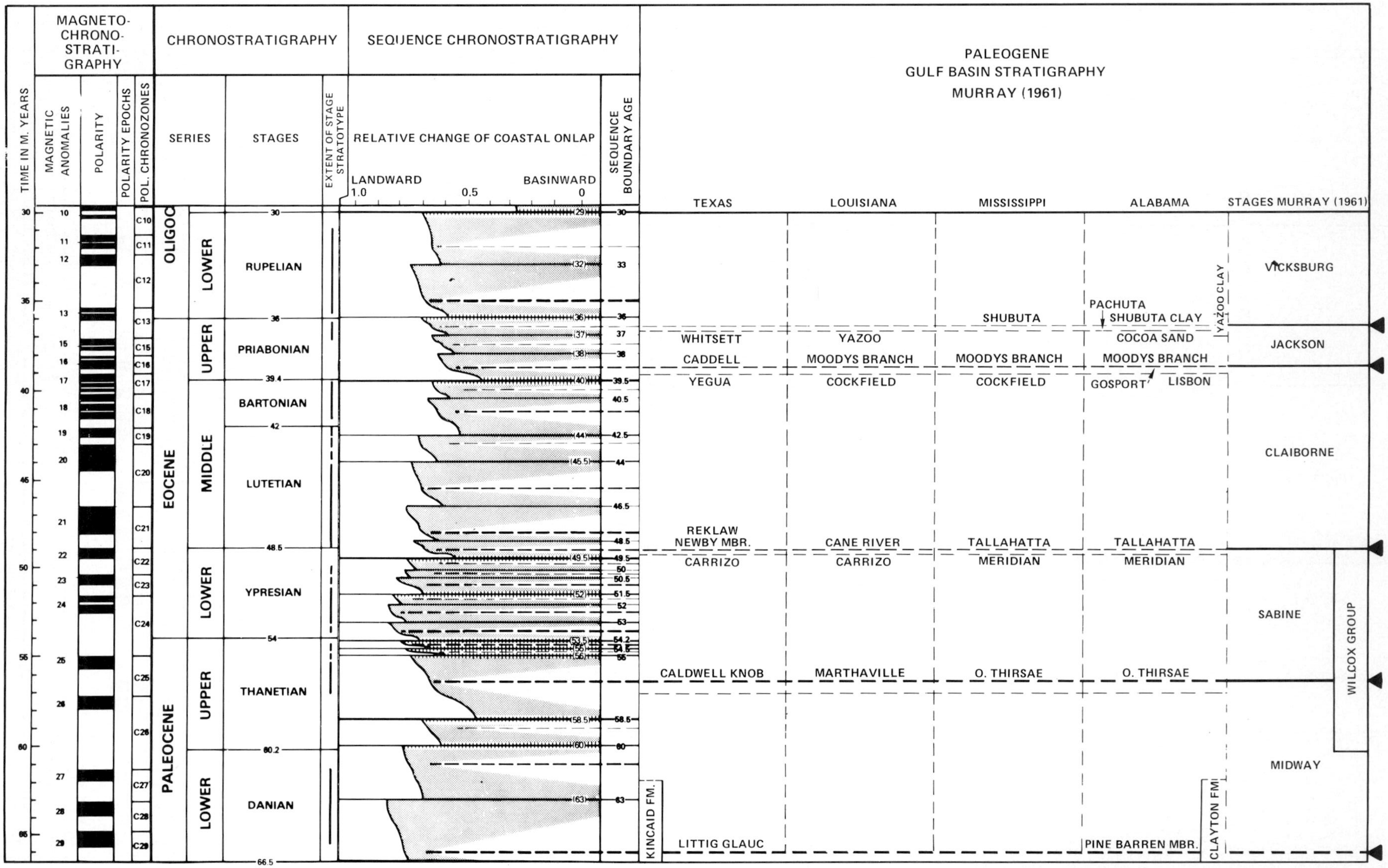

FIG. 27.—Stage classification correlated to coastal onlap chart of Haq and others (1987). Except for the Jackson-Vicksburg boundary, each stage boundary generally corresponds to a depositional sequence boundary. Black triangles mark position of condensed section.

Band" as part of a condensed section that formed within the 93-Ma depositional sequence. The 93-Ma sequence boundary may be represented by an erosional surface overlain by a pebble lag, less than a meter below the "Black Band" (Fig. 28). Water depths probably reached a maximum either within or just above the "Black Band." Planktonic foraminiferal concentrations are greatest just above the band and are not found within it. The overlying Turonian section represents a return to "normal" sedimentation rates within a highstand systems tract. Schlanger and others (1986) postulated that the "Black Band" was deposited near the upper limit of an overall extensive and expanded oxygen-minimum zone that formed during the late Cenomanian transgression.

The "Black Band" and the sequence of events above and below it provide some interesting insights into the nature of condensed sections. The *Whiteinella archeocretacea* interval biozone is represented by the "Black Band" and is less than 50 cm thick. The time scale of Haq and others, (1987) suggests that the duration of the *W. archeocretacea* interval chronozone is about 1.5 Ma. If the *W. archeocretacea* interval biozone in the South Ferriby section approximated the *W. archeocretacea* chronozone, then sedimentation rates associated with the "Black Band" in the South Ferriby section are about 0.03 cm/103 yrs. It is likely that the *W. archeocretacea* biozone at South Ferriby represents only a portion of the *W. archeocretacea* interval chronozone, and thus sedimentation rates may be slightly higher.

Another interesting feature of the South Ferriby section is the sharp-based positive $\delta^{13}C$ excursion, recorded in marine carbonates, just below and within the "Black Band" (Schlanger and others, 1986). The $\delta^{13}C$ values within the "Black Band" are approximately 1.0 to 1.5‰ heavier than the $\delta^{13}C$ values in either the Ferriby Chalk below or the Welton Chalk above (Fig. 28). The positive $\delta^{13}C$ values in the South Ferriby section are thought to be the result of an Oceanic Anoxic Event (OAE; Schlanger and Jenkyns, 1976),

in which a large quantity of isotopically light organic carbon was removed from the oceanic-carbon reservoir. The removal of the "light" organic carbon produced a positive $\delta^{13}C$ spike, or event. The relatively flat-based spike suggests that either the excursion occurred rapidly or that sedimentation rates decreased dramatically, producing an apparently rapid transition from light to heavy $\delta^{13}C$ values.

Scholle and Arthur (1980) suggested that the $\delta^{13}C$ event at the Cenomanian-Turonian boundary provided a chemical stratigraphic marker for international correlation (Fig. 29). Studies since then have documented the Cenomanian-Turonian $\delta^{13}C$ event in many different sections, particularly in the Northern Hemisphere. The carbon isotope event provides an ideal marker for global correlation; however, the characteristic "shape" of the $\delta^{13}C$ event varies considerably from section to section. The primary control on the overall shape of the curve is sedimentation rate. Nondeposition or very slow sedimentation below the "Black Band," associated with the 93-Ma sequence transgression, resulted in the formation of a flat-based spike. Slightly higher sedimentation rates associated with the transgression in other areas may produce a more gradual event. Thus, the character and stratigraphic position of the $\delta^{13}C$ spike may vary considerably between continental-margin sections depending on sedimentation rate variations. In general, the peak $\delta^{13}C$ values associated with the OAE are recorded within the condensed section in a continental-margin setting and their equivalents in deeper ocean sections.

The combination of micropaleontology and chemical stratigraphy has proven powerful for correlating the OAE into other sections where no "Black Band" was deposited (Fig. 28). Because of local oxidizing conditions, no organic matter was deposited in the correlative section at the Cap D'Antifer locality north of Le Havre, France. The oxidizing environments, however, associated with a reduction in sedimentation rates in the condensed section, produced a distinct suite of minerals and elements that is often asso-

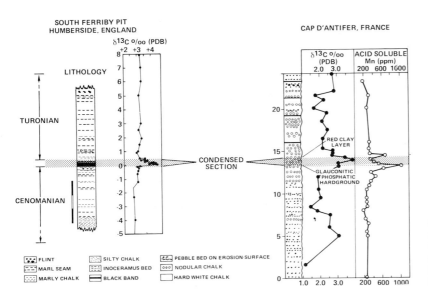

FIG. 28.—Lithology, carbon-isotopic results, and Mn concentration from the South Ferriby section in Humberside, England, and the Cap d'Antifer section in France. Shape of carbon spike at the Cenomanian-Turonian boundary is controlled by sedimentation rate changes associated with a transgression during deposition of the 93-Ma depositional sequence.

ciated with condensed sections. In the Cap D'Antifer section, the Cenomanian-Turonian boundary is associated with a glauconitic-phosphatic hardground. Both glauconite and phosphorite are authigenic minerals that form in distinct oceanographic environments away from the direct influence of terrigenous sedimentation. Radiometric dating of glauconites from outcrop sections (Odin and others, 1978), now recognized as condensed sections, is an important technique for calibrating biozones (Haq and others, 1987). In addition, Schlanger and others (1986) reported greater concentrations of manganese associated with the $\delta^{13}C$ event. The manganese increase is thought to be related to the increased solubility of manganese in anoxic waters and its precipitation in oxygenated waters above the oxygen-minimum zone (Force and others, 1983; Cannon and Force, 1983).

The consistent association of slow-sedimentation rate zones, a $\delta^{13}C$ "event" in marine carbonates, the concentration of authigenic minerals and elements, condensed microfossil zones, relatively deep paleobathymetry estimates in each section, and the paleontologic pick for the Cenomanian-Turonian boundary are consistent with the interpretation that sections that span the Cenomanian-Turonian boundary are condensed. The condensed section occurred as the result of a major sea-level rise during the Late Cenomanian sea-level rise.

DISCUSSION

The condensed section examples presented herein illustrate only a portion of the variability of facies that can be found within condensed sections. Despite this variability, condensed sections are relatively easy to recognize in outcrop, core, well logs, and seismic reflection profiles, particularly in continental-margin settings where their recognition is so critical to stratigraphy.

The Wisconsinan-to-Holocene example from the northern Gulf of Mexico provides a good illustration of the nature of condensed sections through both carbonate and clastic provinces. Even in areas with larger sedimentation rates,

such as the Mississippi Delta, a distinct starvation surface or marine disconformity (Frazier, 1974) can be recognized immediately above the subaerial exposure surface that formed during the last glacial period. Large numbers and diversity of microfossils are concentrated in the condensed section below the prograding Mississippi Delta. The condensed section can be recognized by the abrupt change from nonmarine or marginal-marine sediments below (transgressive- or lowstand systems tract) to relatively deep-water marine shales above. Away from the Mississippi Delta, the condensed section is still exposed because no rivers other than the Mississippi have prograded very far onto the shelf. In fact, the condensed section and the subaerial exposure surface (25-ka sequence boundary) may be coincident, except where the sequence boundary is beneath incised valleys (Fig. 30).

The important points to recognize concerning the Wisconsinan-to-Holocene condensed-section example are that most of the modern shelf area represents a condensed section that has been starved of major terrigenous sedimentation for at least the past 10 ka; and that the depositional environments that may be represented in a condensed section vary greatly. An additional point to emphasize is that the sequence boundary formed during the last major fall of sea level can be recognized either as a subaerial-exposure surface or as an erosional surface at the base of an incised valley. Over most areas of the modern shelf, the sequence boundary and the condensed section are almost coincident. The exception is in areas where river systems eroded deeply into the modern shelf area when it was subaerially exposed during previous lowstands. In the vicinity of these incised valleys, the condensed section and sequence boundary will be separated by incised-valley fill (as much as several hundred meters in deeply incised valleys).

The three surfaces discussed in the Wisconsinan-Holocene example, the sequence boundary, the transgressive surface, and the downlap surface (in the condensed section), are the same surfaces that formed the basis for a number of stratigraphic-correlation philosophies that are discussed in examples 5 and 6. Examples 2, 3, and 4 are

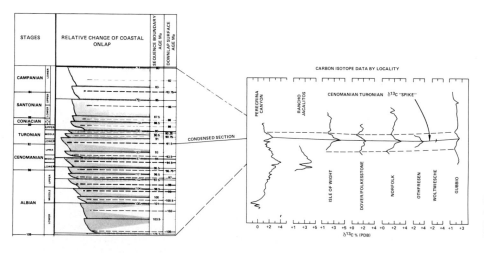

FIG. 29.—Correlation of the $\delta^{13}C$ spike of Scholle and Arthur (1980) to coastal onlap chart of Haq and others (1987). The $\delta^{13}C$ spike is generally associated with condensed section in the 93-Ma depositional sequence.

primarily presented to illustrate the nature of condensed sections and techniques for their recognition. Examples 5 and 6 illustrate the importance of mapping surfaces in combination with biostratigraphic information and the importance of condensed sections to the stratigraphic correlation of continental-margin sections.

The stratigraphy of the upper Wilcox-to-Vicksburg outcrop sections in the Gulf Coast was developed and relatively well understood early in the twentieth century because of the abundance of marine rocks in the outcrop belt. This is especially true for major transgressive units, such as the Claiborne, Jackson, and Vicksburg stages. The stratigraphy in the regressive Wilcox and Yegua sections was initially less well understood because of the non-marine to marginal-marine nature of these units in the outcrop belt. Condensed sections were instrumental in the stratigraphic correlation of the major transgressive units (Fig. 31) and still provide the key to subsurface correlation. Condensed sections also play a role in the correlation within the Wilcox and Yegua, but only in subsurface sections. Toulmin (1977) and Murray (1961) defined and correlated the Gulf Coast stages in outcrop and shallow-subsurface sections by using two different approaches. In these landward sections, stratigraphic correlations using these two approaches are not that different because of the vertical proximity of the sequence boundary, transgressive surface, and downlap surface within each depositional sequence. A fundamental difference exists, however, between Toulmin's and Murray's approaches to stratigraphy that is not obvious in correlations along strike of the Paleogene stages in the Gulf Coast outcrop and shallow-subsurface sections. The difference only becomes obvious when each stage boundary is correlated basinward into the subsurface (Fig. 32). Under Murray's classification the stage boundaries that are defined by the first appearance of marine organisms often coincide with transgressive surfaces (e.g., the top of the Sabine Stage is defined by "the initial invasion of the Claibornian sea" (Murray, 1961, p. 374); Fig. 26). The stage boundaries under Toulmin's classification are generally defined by unconformities or surfaces of subaerial exposure (e.g., the top of the Sabine Stage is below the Meridian or Carizzo sands; Fig. 27).

Landward of the physiographic shelf slope break, the transgressive surface and sequence boundary within each depositional sequence are generally close or coincident with each other (Fig. 32). The exception is in the vicinity of incised valleys, where the sequence boundary and the transgressive surface may be separated by a considerable thickness of sediment (Figs. 32A,B). Beyond the shelf slope break, the sequence boundary and transgressive surface diverge significantly where the sequence boundary is under a lowstand or shelf margin wedge (Figs. 29A,B). The transgressive surface marks the top of the landward portion of the lowstand wedge or shelf margin wedge. Because some of the wedges may be thousands of meters thick, the difference between the two approaches can be significant.

Which of these two approaches is more practical? In this paper we suggest that depositional sequence boundaries are an ideal way of defining a stage boundary because sequence boundaries (1) represent boundaries separating older from younger rocks (all rocks above a sequence boundary are younger than all rocks below it); (2) can be identified physically in a stage type section; (3) can be correlated from non-marine to deep-basin environments; and (4) can be dated by using the techniques outlined. Furthermore, Vail and others (1977), Vail and Hardenbol (1979), Vail and Todd (1981), and Haq and others (1987) demonstrated that depositional sequences provide the best way to divide the rock record into time-rock units. The return to a physical division of the rock record and a reassessment of the classic stages of Europe in terms of depositional sequences suggests that stratigraphy has traveled full circle since its early days. Much of the early stratigraphy was based on division of the rock record into either physically defined units, usually by lithology or unconformities and disconformities, or evolutionary changes in macrofossils. On the other hand, the evolution of biostratigraphy over the past 80 years has resulted in the redefinition of some of these physically de-

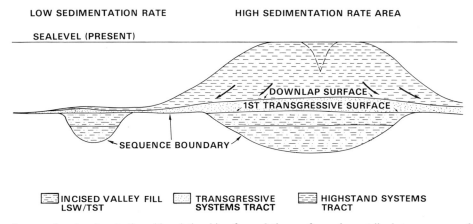

LOW SEDIMENTATION RATE   HIGH SEDIMENTATION RATE AREA

SEALEVEL (PRESENT)

DOWNLAP SURFACE

1ST TRANSGRESSIVE SURFACE

SEQUENCE BOUNDARY

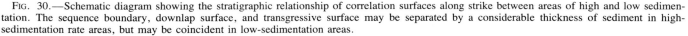

INCISED VALLEY FILL LSW/TST   TRANSGRESSIVE SYSTEMS TRACT   HIGHSTAND SYSTEMS TRACT

FIG. 30.—Schematic diagram showing the stratigraphic relationship of correlation surfaces along strike between areas of high and low sedimentation. The sequence boundary, downlap surface, and transgressive surface may be separated by a considerable thickness of sediment in high-sedimentation rate areas, but may be coincident in low-sedimentation areas.

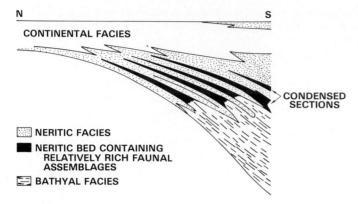

FIG. 31.—Thin, richly fossiliferous shale intervals (or condensed sections) have played a significant role in the development of stratigraphy in the Gulf Coast. Figure modified from Lowman (1949).

fined units into biostratigraphically defined units. Each biostratigraphically defined package closely approximates the original physical unit in the updip outcrop sections, where the condensed section, transgressive surface, and sequence boundary are close together. The biostratigraphically based system also works very well in open-ocean sections for the reasons mentioned earlier; but, the main problem with a biostratigraphically based correlation system is most apparent in subsurface continental-margin stratigraphy, where stratal geometry can be imaged by using seismic-reflection data and portrayed on well-log cross sections. Because the distribution of age-diagnostic microfossils is controlled strongly by the environment of deposition (and sedimen-

tation rate), it is virtually impossible to correlate chrono-zone boundaries in the subsurface of continental margins. Planktonic microfossil biozone boundaries may approximate chronozone boundaries in more basinal settings, but rarely in settings influenced by rapid terriginous sedimentation. Regional correlations of biozone boundaries in continental-margin dip sections are generally controlled by low-sedimentation rates associated with condensed sections. The environmental, or sedimentation, effect on biozone boundaries is usually recognized when a particular age-diagnostic microfossil is "depressed" in a landward direction or "climbs" in a basinward direction, relative to a physical correlation surface.

The alternative approach for stratigraphic correlation in a continental-margin setting is to divide the rock record into depositional sequences, bounded by unconformities or their correlative conformities, and to correlate these sequences with those recognized and dated in stage type sections. A depositional sequence can be correlated between continents by using biostratigraphy as long as the difference between biochronozones and biozones is taken into account. The combination of depositional sequences (surfaces defined on seismic, well logs, and outcrop) and condensed sections (biostratigraphic control) provides a powerful tool for subsurface correlation that cannot be achieved by using either approach alone.

The Paleogene stage boundaries defined by Toulmin (1977) and correlated by using planktonic microfossils are the most widely applicable in the subsurface of the Gulf of Mexico. The base of the Midway corresponds to the 67-Ma (TA 1.2) sequence boundary, the base of the Sabine corresponds to the 58.5-Ma (TA 2.1) sequence boundary, the base of the

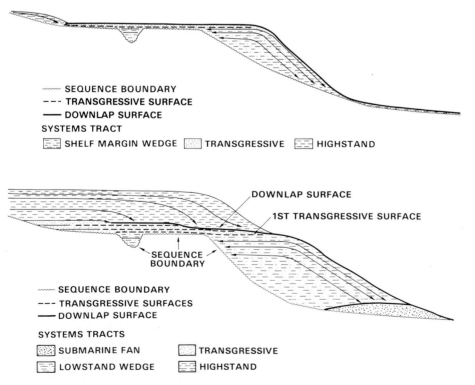

FIG. 32.—(A) Schematic diagram showing stratigraphic relations of major physical correlation surfaces on a dip section in a low-sedimentation-rate type 2 depositional sequence. Downlap surface, transgressive surface, and sequence boundary are generally close to each other landward of relict shelf slope break but may be separated by the shelf margin wedge in a basinward direction. (B) Schematic diagram showing stratigraphic relations of major physical correlation surfaces on a dip section in a high-sedimentation-rate type 1 depositional sequence. Downlap surface, transgressive surface, and sequence boundary are generally close together landward of relict shelf slope break but are separated by a lowstand wedge in a basinward direction. Transgressive surface is restricted to area landward of most regressive nearshore-sand deposits in the lowstand wedge.

Claiborne to the 49.5 Ma (TA 3.1) sequence boundary, and the base of the Jackson to the 39.5-Ma (TA 4.1) sequence boundary (see Fig. 27). The base of the Vicksburg is defined by both Toulmin and Murray at the condensed section at the top of the Yazoo Clay. This is an easily recognizable stage boundary over most of the Gulf Coast, except where sedimentation rates were high and significant lowstand deposition occurred.

Continental-margin stratigraphy requires the combination of a physical framework (depositional sequences) and a temporal framework consisting of biostratigraphy (and other age indicators). In open-ocean settings, where depositional-sequence boundaries become conformable, the primary tool for dividing the rock record is biostratigraphy, with support from both magnetostratigraphy and chemical stratigraphy.

This discussion of the stratigraphic philosophy concerning the definition of Gulf Coast stage boundaries illustrates that (1) both approaches to stage-boundary definition are physically similar in shelf environments, where the sequence boundary, transgressive surface, and condensed section are close together. (2) The classification of Toulmin (1977), when coupled with biostratigraphic data (concentrated in condensed sections), is the most widely applicable. This is particularly true for dip-section correlation in continental-margin sections. (3) In open-ocean sections, a biostratigraphically based classification supported by magnetostratigraphy and chemical stratigraphy is the most widely applicable at present. Depositional sequence boundaries, however, are sometimes recognizable in open-ocean sections, thus providing a physical stratigraphic framework for correlation in the deep ocean.

The Cenomanian-Turonian example (example 7) is important because it illustrates the importance of (1) condensed sections to global stratigraphy; (2) detailed geochemical analyses of condensed sections for chemical stratigraphy and paleoceanography; (3) chemical stratigraphy (using a variety of elements and analyses) for global stratigraphy; and (4) shows that condensed sections in a shelf setting provide a window or glimpse of the oceanographic conditions and hydrocarbon-source potential that may exist in the adjacent basin beyond the shelf slope break.

An additional point, which was mentioned but not illustrated, is the importance of radiometric dating of stratigraphically well-defined glauconite occurrences (Haq and others, 1987). Glauconite is often found in transgressive-systems tracts and condensed sections. Glauconite with greater concentrations of potassium seems to be common in condensed sections. Thus, radiometric dating of glauconites in condensed sections in individual depositional sequences has the potential to provide additional information for the calibration of biozone occurrences in continental-margin sections and age-dating of depositional sequences.

The Cenomanian-Turonian boundary example also provides considerable insight into the interaction of sea level, hypsometry, climate, and ocean chemistry. Detailed analyses of condensed sections suggest that they represent the best places to examine the interaction of physical processes, such as rising sea level, and chemical processes, and the oceanographic, biologic, and sedimentologic response to these processes.

CONCLUSIONS

Condensed sections provide a physical link between the highly refined and well-calibrated biostratigraphic chronozonal schemes of the deep-ocean basins and the classic stage classification scheme defined in shallow-marine to marginal-marine outcrop sections. Age-diagnostic microfossils are preferentially concentrated in condensed sections and provide the key to the age determination and correlation of continental-margin sections. The distribution of condensed sections, and therefore age-diagnostic microfossils in the rock record, is predictable since they are related to eustatic rises.

A combination of depositional-sequence analysis (physical framework) and biostratigraphic analysis (temporal framework) is necessary for the regional and interregional correlation of continental-margin sections. The combination is termed sequence stratigraphy. Biostratigraphy, combined with magnetostratigraphy and chemical stratigraphy, is presently the best tool for correlation of deep-ocean sections; but, biostratigraphy, and the other tools mentioned above, must be combined with an analysis of stratal geometry and facies distribution for correlation in continental-margin sediments where sedimentation rates control the distribution of age-diagnostic organisms. Biostratigraphic analysis alone is not sufficient for the stratigraphic correlation of continental-margin sequences.

Condensed sections can be recognized in outcrop, core, and subsurface by using a variety of tools, including facies analyses of outcrop and well logs, biostratigraphic analyses, and seismic stratigraphy. Continental-margin condensed sections provide a window or glimpse of the oceanographic conditions in the adjacent basin. Because of the decreased sedimentation rates, condensed sections may represent sites for accumulation of organic matter, but the preservation of organic matter will depend on the oceanographic conditions in the overlying water column.

ACKNOWLEDGMENTS

The development of this paper has benefited significantly from discussions with a number of colleagues, especially S. M. Greenlee, T. R. Nardin, V. D. Rahmanian, and J. F. Sarg. Detailed reviews by L. D. Keigwin, Jr., G. de Vries Klein, A. D. Partridge, and K. K. Romine significantly improved the manuscript.

REFERENCES

ALVAREZ, L., ALVAREZ, W. W., ASARO, F., AND MICHEL, M., 1980, Extraterrestrial cause for the Cretaceous-Tertiary extinctions: Science, v. 208, p. 1095–1108.
ARTHUR, M. A., ZACHOS, J. C., AND JONES, D. S., 1987, Primary productivity and the Cretaceous/Tertiary boundary event in the oceans: Cretaceous Research, v. 8, p. 43–54.
BAUM, G. R., LOUTIT, T. S., BLECHSCHMIDT, G. L., WRIGHT, R. C., AND SMITH, T., 1984, The Maastrichtian/Danian Boundary in Alabama: A stratigraphically condensed section: Geological Society of America, Abstracts with Programs, v. 16, p. 6.
BERGGREN, W. A., KENT, D. V., AND FLYNN, J. J., 1985, Paleogene geochronology and chronostratigraphy: Geological Society of London, Special Paper, p. 141–186.

————, ————, AND VAN COUVERING, J. A., 1985, Neogene geochronology and chronostratigraphy: Geological Society of London, Special Paper, p. 211–250.

BROMLEY, R. G., 1974, Trace fossils at omission surfaces, in Frey, R. W., ed., The Study of Trace Fossils: Springer-Verlag, New York Inc., p. 399–428.

BURNS, D. A., AND NELSON, C. S., 1981, Oxygen isotopic paleotemperatures across the Runangan-Whaingaroan (Eocene-Oligocene) boundary in New Zealand shelf sequence: New Zealand Journal of Geology and Geophysics, v. 23, p. 529–538.

BYBELL, L. M., 1982. Late Eocene to early Oligocene calcareous nannofossils in Alabama and Mississippi: Gulf Coast Association of Geological Societies, Transactions, v. 32, p. 295–302.

CANNON, W. F., AND FORCE, E. R., 1983, Potential for high grade shallow marine manganese deposits in North America, in Shanks, W. P., ed., Cameron Volume on Unconventional Mineral Deposits: American Institute of Mining, Metallurgical and Petroleum Engineers, Inc., New York, p. 175–189.

COX, A., 1969, Geomagnetic reversals: Science, v. 163, p. 237–245.

————, 1973, Plate Tectonics and Geomagnetic Reversals: W. H. Freeman, San Francisco, 702 p.

FISK, H. N., 1940, Geology of Avoyelles and Rapides Parishes, Louisiana: Louisiana Geological Survey Bulletin 18, 240 p.

————, AND MCFARLAN, E., JR., 1955, Late Quaternary deltaic deposits of the Mississippi River: Geological Society of America Special Paper 62, p. 279–302.

FORCE, E. R., CANNON, W. F., KOSKI, R. A., PASSMORE, K. T., AND DOE, B. R., 1983, Influences of ocean anoxic events on manganese deposition and ophiolite-hosted sulphide preservation, in Paleoclimate and Mineral Deposits: U.S. Geological Survey Circular 822, p. 26–29.

FRAZIER, D. E., 1974, Depositional episodes: Their relationship to the stratigraphic framework in the Gulf Basin: Texas University Bureau of Economic Geology Circular 74–1, 28 p.

GOULD, H. R., AND STEWART, R. H., JR., 1955, Continental terrace sediments in the northeastern Gulf of Mexico, in Hough, J. L., and Menard, H. W., eds., Finding Ancient Shorelines: Society of Economic Paleontologists and Mineralogists Special Publication 3, p. 2–20.

HAQ, B. U., HARDENBOL, J., AND VAIL, P. R., 1987, Chronology of fluctuating sea levels since the Triassic (250 million years ago to present): Science, v. 235, p. 1156–1167.

HARDENBOL, J., AND BERGGREN, W. A., 1978, A new Paleogene numerical time scale, in Cohee, G. V., Glaessner, M. F., and Hedberg, H. D. eds., Contributions To The Geologic Time Scale: American Association of Petroleum Geologists Studies in Geology No. 6, p. 213–234.

————, VAIL, P. R., AND LOUTIT, T. S., 1986, Sequence stratigraphy: An integrated approach to global stratigraphic correlation (abstract): Second International Conference on Paleoceanography, Woods Hole, Massachusetts (not paginated).

HART, M. B., AND BIGG, P. J., 1981, Anoxic events in the chalk seas of northwest Europe, in Neale J. W., and Brasier, M. D., eds., Microfossils from Recent and Fossil Seas: The British Micropaleontological Society, p. 177–185.

KEIGWIN, L. D., JR., 1980, Paleoceanographic change in the Pacific at the Eocene-Oligocene boundary: Nature, v. 287, p. 722–725.

————, AND CORLISS, B. H., 1986, Stable isotopes in late middle Eocene to Oligocene foraminifera: Geological Society of America Bulletin, v. 97, p. 335–347.

KENNEDY, W. J., AND GARRISON, R. E., 1975, Morphology and genesis of nodular chalks and hardgrounds in the Upper Cretaceous of southern England: Sedimentology, v. 22, p. 311–386.

LOUTIT, T. S., BAUM, G. R., AND WRIGHT, R. C., 1983, Eocene-Oligocene sea-level changes as reflected in Alabama outcrop sections: American Association of Petroleum Geologists Bulletin, v. 67, p. 506.

————, AND KENNETT, J. P., 1981, New Zealand and Australian Cenozoic sedimentary cycles and global sea level changes: American Association of Petroleum Geologists Bulletin, v. 65, p. 1586–1601.

LOWMAN, S. W., 1949, Sedimentary facies in the Gulf Coast: American Association of Petroleum Geologists Bulletin, v. 33, p. 1939–1947.

MACNEIL, F. S., 1946, Summary of the Midway and Wilcox stratigraphy of Alabama and Mississippi: U.S. Geological Survey, Strategic Minerals Investigations, Preliminary Report 3–195, 29 p.

MANCINI, E. A., 1979, Eocene-Oligocene boundary in southwest Alabama: Gulf Coast Association of Geological Societies, Transactions, v. 29, p. 282–289.

MEYER, B. L., AND NEDERLOF, M. H., 1984, Identification of source rocks on wireline logs by density/resistivity and sonic transit time-resistivity cross plots: American Association of Petroleum Geologists Bulletin, v. 68, p. 121–129.

MITCHUM, R. M., JR., VAIL, P. R., AND THOMPSON, S., III, 1977, The depositional sequence as a basic unit for stratigraphic analysis, in Peyton, C. E., ed., Seismic Stratigraphy—Applications to Hydrocarbon Exploration: American Association of Petroleum Geologists Memoir 26, p. 53–62.

MIX, A. C., AND RUDDIMAN, W. F., 1985, Structure and timing of the last deglaciation: Oxygen-isotope evidence: Quaternary Science Reviews, v. 42, p. 59–108.

MURRAY, G. E., 1953, History and development of Paleocene-Lower Eocene nomenclature, central Gulf coastal plain: Guidebook, 10th Field Trip, Mississippi Geological Society, p. 48–54.

————, 1955, Midway Stage, Sabine Stage, and Wilcox Group: American Association of Petroleum Geologists Bulletin, v. 39, p. 671–696.

————, 1961, Geology of the Atlantic and Gulf Coastal Province of North America: Harper and Brothers, New York, 692 p.

NATHAN, S., 1976, Sheets S23/9 and S24/7 Foulwind and Westpart (first edition). "Geological Map of New Zealand 1:25,000." Map (1 sheet) and notes (16 p.). New Zealand Department of Scientific and Industrial Research, Wellington.

ODIN, G. S., CURRY, D., AND MUNZIKER, J. C., 1978, Radiometric dates from North West European glauconites and the Paleogene time scale: Journal of the Geological Society of London, v. 135, p. 481–497.

OPDYKE, N., 1972, Paleomagnetism of deep sea cores: Reviews of Geophysics and Space Physics, v. 10, 213–249.

POAG, C. W., 1981, Ecologic Atlas of Benthic Foraminifera of Gulf of Mexico: Hutchinson Ross Publishing Company, 174 p.

SARG, J. R., AND LEHMANN, P. J., 1986, Lower-Middle Guadalupian facies and stratigraphy, San Andres/Grayburg Formations, Permian Basin, Guadalupe Mountains, New Mexico, in Moore, G. E., and Wilde, G. L., eds., Field Trip Guide Book, (San Andres/Grayburg Formations, Guadalupe Mountains, New Mexico and Texas): Permian Basin Section, Society of Economic Paleontologists and mineralogists, Publication 86-25, p. 1–94.

SCHLANGER, S. O., ARTHUR, M. A., JENKYNS, H. C., AND SCHOLLE, P. A., 1986, The Cenomanian-Turonian oceanic anoxic event. Stratigraphy and distribution of organic carbon-rich beds and the marine $\delta^{13}C$, in Brooks, J., and Fleet, A., eds., Marine Petroleum Source Rocks: Geological Society of London, Special Publication No. 26, p. 371–400.

————, AND JENKYNS, H. C., 1976, Cretaceous oceanic anoxic events: Causes and consequences: Geologist Mijnsonn, v. 55, p. 179–184.

SCHOLLE, P. A., AND ARTHUR, M. A., 1980, Carbon isotope fluctuations in Cretaceous pelagic limestones: Potential stratigraphic and petroleum exploration tool: American Association of Petroleum Geologists Bulletin, v. 64, p. 67–87.

SHACKLETON, N. J., AND KENNETT, J. P., 1975, Paleotemperature history of the Cenozoic and the initiation of Antarctic glaciation: Oxygen and carbon isotope analyses in DSDP Sites 277, 279, and 281: Initial Reports of the Deep Sea Drilling Project, v. 29, U.S. Government Printing Office, Washington, D.C., p. 743–755.

SHEPARD, F. P., PHLEGER, F. B., AND VAN ANDEL, T. H., eds., 1960, Recent sediments, northwest Gulf of Mexico: American Association of Petroleum Geologists, Tulsa, Oklahoma, 394 p.

SRINIVASAN, M. S., AND VELLA, P., 1975, Upper Eocene—Lower Oligocene benthonic foraminifera, Port Elizabeth and Cape Foulwind, New Zealand: New Zealand Journal of Geology and Geophysics, v. 18, p. 21–37.

STENZEL, H. B., 1952, Boundary problems: Guidebook, 9th Field Trip, Mississippi Geological Society p. 11–33.

SUTER, J. R., AND BERRYHILL, H. L., 1985, Late Quaternary shelf margin deltas, northwest Gulf of Mexico: American Association of Petroleum Geologists Bulletin, v. 69, p. 77–91.

TOULMIN, L. O., 1977, Stratigraphic distribution of Paleocene and Eocene fossils in the Eastern Gulf Coast Regions: Alabama Geological Survey Monograph 13, v. 1, 602 p.

VAIL, P. R., AND HARDENBOL, J., 1979, Sea level changes during the Tertiary: Oceanus, v. 22, p. 71–79.

————, AND ————, 1984, Jurassic unconformities, chronostratigraphy, and sea-level changes from seismic and biostratigraphy: American Association of Petroleum Geologists Memoir 36, p. 129–144.

————, MITCHUM, R. M., JR., THOMPSON, S., III, TODD, R. G., SANGREE, J. B., WIDMIER, J. M., BUBB, N. N., AND NATELID, W. G., 1977, Seismic stratigraphy and global sea-level changes: American Association of Petroleum Geologists Memoir 26, p. 49–212.

————, AND TODD, R. G., 1981, Northern North Sea Jurassic unconformities, chronostratigraphy and sea-level changes from seismic stratigraphy *in* Illing, L. V., and Hobson, G. D., eds., Proceedings of the Petroleum Geology of the Continental Shelf of Northwest Europe Conference: Heyden and Son, Ltd., London, England, p. 216–235.

WALCOTT, R. I., 1978, Present tectonics and Late Cenozoic evolution of New Zealand: Geophysical Journal of the Royal Astronomical Society, v. 52, p. 137–164.

————, 1984, Reconstructions of the New Zealand region for the Neogene: Palaeoeography, Palaeoclimatology, Palaeoecology, v. 46, p. 217–231.

WOOD, C. J., AND SMITH, E. G., 1978, Lithostratigraphical classification of the chalk in North Yorkshire, Humberside and Lincolnshire: Proceedings, Yorkshire Geological Society, v. 42, p. 263–287.

Part III
SEA-LEVEL CHANGES THROUGH GEOLOGIC TIME

MEASUREMENT OF SEA-LEVEL CHANGE IN EPEIRIC SEAS: THE MIDDLE ORDOVICIAN TRANSGRESSION IN THE NORTH AMERICAN MIDCONTINENT

JOHN L. CISNE AND RAYMOND F. GILDNER

Department of Geological Sciences, Institute for the Study of Continents, and Division of Biological Sciences, Cornell University, Ithaca, New York 14853; Department of Geological Sciences and Institute for the Study of Continents, Cornell University, Ithaca, New York 14853

ABSTRACT: Carbonate sediments of tectonically quiescent continental interiors are nearly ideal for precisely tracing eustatic sea-level change during major transgressions. Over roughly 10 million years during the later Middle Ordovician (Rocklandian through the middle Denmarkian stages), sea levels measured in the American Midwest rose about 10 m relative to the continent. Because the sediment accumulation rate in the epeiric sea was proportional to water depth, the time trend of sea level can be reconstructed from cumulative sediment thickness and from measurements on water depth throughout a stratigraphic section. Sea level is reconstructed as a function of time for six sections in the midwestern United States, and the reconstructed time trends are compared for common eustatic components based on the section time correlations by geochemically fingerprinted volcanic-ash layers. Relative water depth is measured through gradient analysis of fossil assemblages by reciprocal averaging ordination. Sample ordination scores are calibrated as a measure of absolute depth by use of the offshore depth estimated from stratigraphic expressions of the shoreline edge effect in lithospheric flexure. Sea-level change during the Middle Ordovician transgression had at least two components: (1) a steady rise at a slowly varying rate around 1 m per million years (2) pulses no more than 0.1 to 1 million years long during which sea level fell roughly 1 m and then rose about the same amount. The long-term trend is attributable to steady decrease in the mean age of oceanic lithosphere. The pulse correlations from section to section and the remarkably small sea-level changes involved testify to the tectonic quiescence, spatial homogeneity, and essential tidelessness of the epeiric sea, and to the precision of its stratigraphic record. Pulses were probably related to climatic fluctuations, and their association with particularly frequent volcanic-ash deposition is suggestive that climatic effects of increased explosive volcanism may have been controlling factors.

INTRODUCTION

Coastal onlap curves from seismic-profile studies of continental margins have pointed the way toward a quantitative global sea-level history, a relative time scale based on eustatic sea-level fluctuations, and a new understanding of mechanisms of sea-level change. Thus far, studies have been concerned primarily with relatively large-scale, long-term patterns, such as the eustatic sea-level history of eras or periods (Vail and others, 1977; Vail and Hardenbol, 1979) or the subsidence history of an entire continental margin (Hardenbol and others, 1981). Much remains to be done on finer-scale features. Of particular importance are the shortest term fluctuations, for these are the ones that may give the most precise time correlations.

The central parts of seas in tectonically quiescent continental interiors are nearly ideal for studying the fine details of sea-level change during major marine transgressions (for discussion, see Cisne and others, 1984). Carbonate sediments, in general, are faithful recorders of sea-level history (for review, see Kendall and Schlager, 1981), and the loose carbonate sediments typical of the middle portions of epicratonic seas should nearly ideal recorders. The record of a given transgression may be much thinner on a continental platform than on a thermally subsiding continental margin, but several factors make the sea-level history more precise and easy to read in epeiric carbonates. The first factor is the absence of tectonic effects that complicate interpretation of the record on continental margins (Pitman, 1978; 1979; Steckler and Watts, 1978; Watts, 1982). Next is the natural limitation of carbonate sedimentation to areas below some elevation in the high intertidal or supertidal setting. The sensitivity of carbonate-secreting organisms to depth-related environmental factors is significant. Finally, depth-dependent dispersion of loose sediment by wave-and-current action occurs such that depth gradients recorded in individual stratigraphic sections tend to reflect sea-level change rather then temporally shifting spatial gradients related to localized carbonate buildups or to lithospheric flexure resulting from spatially uneven sediment and water loading.

In this paper, we develop and apply a new approach to reconstruction of sea-level history that takes advantage of a special feature of epeiric carbonates, the depth dependence of the sediment accumulation rate (Fig. 1). As explained, the relative time scale for measuring sea-level change from a stratigraphic section can be reconstructed from measurements on relative water depth as a function of position in that section. Data on water depth as a function of stratigraphic position and on sediment thickness can then be used to reconstruct sea-level change as a function of time for individual sections.

After explaining the way in which sea-level change is measured, reconstruct sea-level histories for stratigraphic sections in the midwestern United States (Fig. 2) that record a part of the general Middle Ordovician transgression (Rocklandian through Denmarkian stages). These histories (Figs. 3, 4) are checked for common eustatic components by comparing the reconstructed time trends for sea level against the section time correlations by layers of geochemically fingerprinted volcanic ash.

Our use of certain general terms is tailored to describing sea-level history from individual stratigraphic sections in areas generally far from shore. As used in describing the sea-level history of one section, transgression and regression refer, respectively, to rise and fall in sea level relative to local continental basement. Transgression and regression in this sense do not necessarily entail landward or seaward movement of generally distant shorelines. Sea level is measured relative to the continent, as detailed by Cisne and others (1984). The zero point for measuring sea-level change over a stratigraphic section is taken to be a specified point in that section or in another with which the section has been

Sea-Level Changes—An Integrated Approach, SEPM Special Publication No. 42

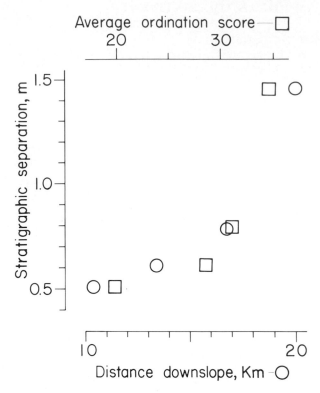

time correlated. Eustatic sea-level change refers to global sea-level change. Transgressive or regressive change measured in a particular section may represent local subsidence or uplift, not necessarily eustatic transgression or regression. Absolute water depth refers to the depth in meters as averaged over the tidal cycle or some longer period (the tide was evidently negligible in epeiric seas such as that in question in the Middle Ordovician; see Hallam, 1981). Relative water depth refers to depth in relative units (e.g., distance downslope along time surfaces or dimensionless ordination scores for fossil assemblages). Sea-level history and sea-level curve refer to sea level as a function of time.

Fig. 1.—Sediment accumulation rate as a function of relative water depth—stratigraphic separation between two volcanic-ash layers (M2 and M3 of Cisne and Rabe, 1978; Cisne and others, 1982) as a function of two measures of relative water depth: (1) distance downslope from the Trenton Falls section along transect parallel to the vector mean paleocurrent direction of gravity flows; and (2) first-axis reciprocal averaging ordination score for fossil assemblages of benthic macroinvertebrates (values from 20 to 35 on normalized 0–100 scale, increasing with increasing depth; each value indicated is the mean of scores for four fossil assemblages, one above and one below each ash layer at each section). The trend of reciprocal averaging scores and the distance downslope show the depth dependence discussed in the text.

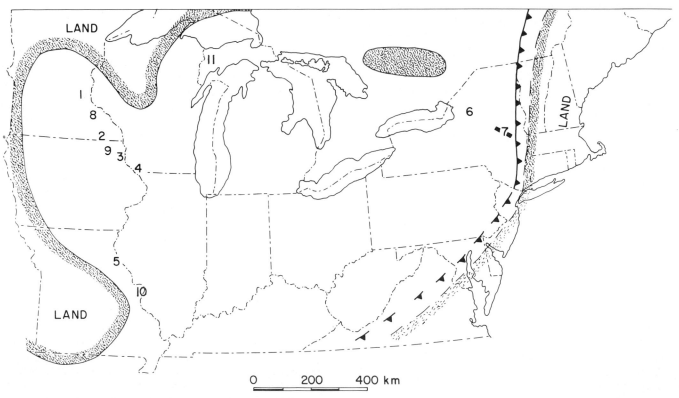

Fig. 2.—Sections studied, indicated by number on generalized map of later Middle Ordovician paleogeography (based on Schuchert, 1955): (1) St. Paul, MN (Templeton and Willman, 1963). (2) Greenleafton, MN. (3) Guttenberg, IA (Templeton and Willman, 1963; Levorson and Gerk, 1972, 1975). (4) Dickeyville, WI (Templeton and Willman, 1963; Willman and Kolata, 1978; Kolata and others, 1983). (5) New London, MO, including (5a) New London (Templeton and Willman, 1963) and (5b) south of New London (Templeton and Willman, 1963; Willman and Kolata, 1978). (6) Black River Valley, NY, including (6a) Martinsburg and (6b) Lowville (Kay, 1937; Titus and Cameron, 1976; Titus, 1982). (7) Mohawk River Valley, NY (Cisne and others, 1982; samples from over 20 sections used only for comparative purposes in this study). (8) Rochester, MN (Templeton and Willman, 1963). (9) Decorah, IA (Templeton and Willman, 1963; Levorson and Gerk, 1972, 1975). (10) Kimmswick, MO (Templeton and Willman, 1963; Kolata and others, 1983). Escanaba, MI (Templeton and Willman, 1963).

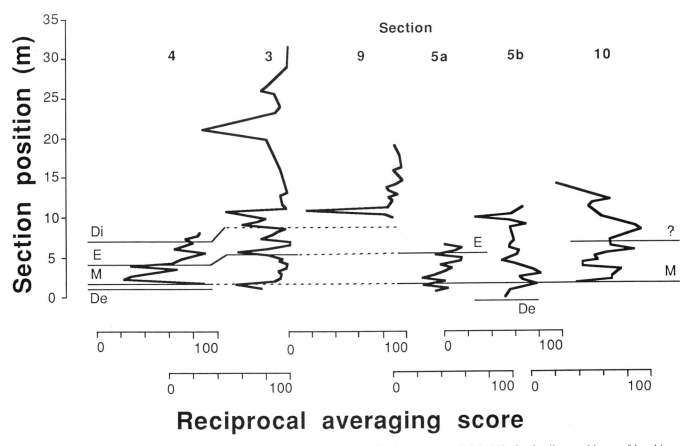

FIG. 3.—Coenocorrelation curves showing first-axis reciprocal averaging ordination score (scaled 0–100) for fossil assemblages of benthic macroinvertebrates as a function of stratigraphic position for certain sections indicated in Figure 2. The Deicke (De), Millbrig (M), Elkport (E), and Dickeyville (Di) K-bentonites are indicated with solid lines where observed, dashed lines where inferred from lithostratigraphy. These volcanic-ash layers, and lithostratigraphic units keyed in with them, are the basis for the time-correlation of sections (Kolata and others, 1983). The unnamed K-bentonite in section 10 is at nearly the position expected for the Dickeyville K-bentonite (D. R. Kolata, pers. commun., 1986).

MEASUREMENT OF SEA-LEVEL CHANGE IN EPICRATONIC CARBONATE SEAS

Sea-Level Change and Sedimentation

Sea-level change, sediment and water loading, and subsidence.—

As sea level rises relative to a continent and as sediment accumulates in an epeiric sea, the continent will subside under the sediment and water loads. For rates of sea-level rise less than about 10^{-4} m/yr, and for comparably slow-sedimentation rates, the sediment and water loads in the central portion of the sea can be assumed to remain in instantaneous isostatic equilibrium (Cisne and others, 1984). The edges of the sediment and water loads at the onshore and offshore margins of the sea create an edge effect in lithospheric flexure. This effect dies out within about 100–200 km of those margins (Cisne, 1985) and so does not substantially affect subsidence in the middle of the sea.

Under isostatic equilibrium, the rates of change with time t in sea level $L(t)$, sediment thickness $S(t)$, and water depth $H(t)$ are related as follows (Cisne and others, 1984):

$$\frac{dL}{dt} - \left(\frac{\rho_m - \rho_s}{\rho_m}\right)\frac{dS}{dt} - \left(\frac{\rho_m - \rho_w}{\rho_m}\right)\frac{dH}{dt} = 0 \quad (1)$$

where $\rho_m = 3300$ kg/m^3, $\rho_s = 2100$ kg/m^3, and $\rho_w = 1000$ kg/m^3 are the densities, respectively, of the displaced mantle, carbonate sediment, and sea water. Long-term compaction is assumed to be negligible, and sediment density is assumed constant with time over the deposition of a transgressive sequence. These should be reasonable assumptions for a relatively thin (<100 m) sequence dominated by grain-supported calcarenites, such as the Middle Ordovician sediments analyzed here. The sediment density we use here is based on a porosity of 35% as appropriate for surficial sediments (Enos and Sawatsky, 1981).

Depth-dependent sedimentation.—

Accumulation of carbonate sediments depends strongly on water depth and depth-dependent environmental conditions (for review, see Kendall and Schlager, 1981). The carbonate accumulation rate will be zero above some elevation very near sea level. In much deeper water, the rate will again approach zero as water depth approaches the cal-

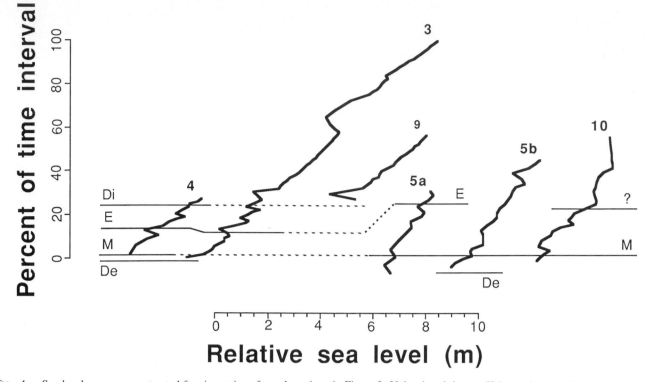

FIG. 4.—Sea-level curves reconstructed for six sections from data given in Figure 3. Volcanic-ash layers (K-bentonites) are indicated as in Figure 3. The curves show a common, evidently eustatic sea-level rise. Despite variations owing to differences in the depth dependence of sediment accumulation and syndepositional tectonism (sec. 10 only), the curves show small-scale, evidently eustatic fluctuations that correspond in time relative to ash layers.

cium carbonate compensation depth. In carbonate environments, the rate will reach a maximum at some depth between the two extremes. Judging from modern carbonate environments (Ginsburg and James, 1974), the depth of maximum accumulation rate would probably be not much more than 10 m. On the assumption that maximum depths in the particular epeiric sea were not much greater than the depth of maximum accumulation rate, the following linear approximation is used for the more complicated (and unknown) dependence of sedimentation rate on water depth (Cisne and others, 1984):

$$\frac{dS}{dt} = cH \qquad (2)$$

where c is the *depth-dependence frequency* for the particular depositional setting. In the following discussion, c is assumed constant with time in individual stratigraphic sections. Temporal variation in c will be discussed later in connection with tests of reconstructed sea-level histories.

When applied to synthesizing stratigraphies (equation 3), equation 2 predicts that water should asymptotically approach a depth directly proportional to the ambient transgression rate (Cisne and others, 1984). The apparent bathymetric constancy of many epeiric "keep up" carbonate sequences (in the sense of Kendall and Schlager, 1984), deposited during long, seemingly steady transgressions, tends to confirm the prediction in principle.

Because of the commonness of widespread volcanic-ash layers, Middle Ordovician strata in eastern and midwestern North America provide some of the best information (and almost the only exact information) on depth dependence in carbonate accumulation rates (Cisne and others, 1984). Figure 1 is an example of depth dependence according to equation 2. This example involves shallow-water carbonates on the very edge of the eastern North American Middle Ordovician carbonate platform in the Mohawk Valley, New York (Cisne and others, 1982). The strata are lithologically and biostratigraphically very much like correlative strata in the midcontinent. The spacing of two ash layers (i.e., net sediment accumulation over a precisely defined time interval) increases with distance downslope along a transect parallel to the vector mean of paleocurrent directions for gravity flows (i.e., relative water depth). Figure 1 also shows mean reciprocal averaging ordination scores for four fossil assemblages at each location, one just above and one just below each ash layer. As determined from studies on downslope transects along individual ash layers that extend farther down the slope, this ordination score is an approximately linear measure of relative depth (Cisne and others, 1982). The curvature of both trends plotted in Figure 1 probably reflects bottom topography, the change from the level bottom of the platform to the slope leading down into the Taconic Trench (Cisne and others, 1982). Near coincidence of the trend for the spacing of the ash layers and the trend for ordination score indicates a basically linear relationship between rock accumulation rate and wider depth, as in equation 2.

Cisne and others (1984) reported quantitative evidence of depth-dependent carbonate accumulation according to equation 2 from comparison of the bathymetric histories of time-correlative Middle Ordovician sections in regions then undergoing differential subsidence. A further test of depth dependence according to equation 2 concerns the characteristic signature it can generate in stratigraphic expressions of the shoreline edge effect in lithospheric flexure (Cisne, 1985). Middle Ordovician strata that lap onto the Wisconsin Arch show this signature, and further analysis gives $c \approx 3$ per million years and an asymptotic offshore water depth of about 2 m, which is consistent with sedimentological evidence on bathymetry (Cisne, 1985).

The record of sea level history.—

Combining equations 1 and 2 gives a differential equation for water depth H in terms of transgression rate dL/dt (Cisne and others, 1984)

$$\frac{dH}{dt} + c\left(\frac{\rho_m - \rho_s}{\rho_m - \rho_w}\right)H = \left(\frac{\rho_m}{\rho_m - \rho_w}\right)\frac{dL}{dt} \quad (3)$$

from which sediment thickness S can be determined from equation 2 once the equation is solved for H. This equation applies only for $H > 0$, and does not apply in cases of exposure (i.e., $H < 0$).

Cisne and others (1984) synthesized stratigraphies for various sea-level histories $L(t)$. A fundamental implication of depth-dependent sedimentation according to equation 3 is that sea-level history should be recorded in depth-dependent variation in the rate of rock accumulation as well as in variation in water depth.

Reconstruction of Sea Level History

Reconstruction for a single stratigraphic section.—

Reconstructing sea level as a function of time involves two basic steps: (1) determining the relationship between stratigraphic position and time; and (2) measuring change in sea level and relating sea level to the time scale.

Consider a stratigraphic interval of thickness S and duration T which is sampled at $n + 1$ evenly spaced stations numbered $i = 0, 1, 2, \ldots, n$. Choosing the sampling interval $\Delta S = S/n$ amounts to choosing a generally unknown average time interval $\Delta T = T/n$ between samples. Rewriting equation 2 in terms of discrete intervals gives the relationship between ΔS and $\Delta t(i)$ for a given sampling interval i:

$$\Delta S = c\,h(i)\Delta t(i), \quad 1 \leq i \leq n \quad (4)$$

where $h(i)$ is the average water depth during the interval, and $\Delta t(i) = t(i) - t(i - 1)$. Assuming a constant (possibly zero) acceleration in sea-level change over the interval, then

$$h(i) = \frac{H(i) - H(i - 1)}{2} \quad (5)$$

where $H(i)$ and $H(i - 1)$ are the water depths at the end and beginning of the interval, respectively (see Cisne and others, 1984, equations 23–28). Equation 2 also gives

$$\Delta S = c\Delta hT \quad (6)$$

$$\frac{1}{h} = \frac{1}{n}\sum_{i=1}^{n}\frac{1}{h(i)} \quad (7)$$

where h is the harmonic mean of all $h(i)$.

Combining equations 4 and 6 gives the time duration of an interval $\Delta t(i)/\Delta T$ in terms of relative water depth:

$$\frac{\Delta t(i)}{\Delta T} = \frac{h}{h(i)}. \quad (8)$$

Summing equation 8 to the position of the jth sample gives relative age $t(j)/T$ of the ith sample:

$$\frac{t(j)}{T} = \frac{1}{n}\sum_{i=1}^{j}\frac{\Delta t(i)}{\Delta T} = \frac{1}{n}\sum_{i=1}^{j}\frac{h}{h(i)}, \quad 1 \leq j \leq n. \quad (9)$$

Applying equation 6 to equation 9 gives $t(j)$ in absolute terms:

$$t(j) = \frac{S}{c}\sum_{i=1}^{j}\frac{1}{h(i)}, \quad 1 \leq j \leq n. \quad (10)$$

Equation 9 can be applied to measuring relative time when only relative depth is measurable, and equation 10 can be applied to measuring absolute time when c is known and absolute depth is measurable.

Rewriting equation 1 in terms of discrete intervals gives

$$\Delta L(j) = \left(\frac{\rho_m - \rho_s}{\rho_m}\right)\Delta S$$
$$+ \left(\frac{\rho_m - \rho_w}{\rho_m}\right)\Delta H(j), \quad 1 \leq j \leq n \quad (11)$$

where $\Delta L(j) = L(j) - L(j - 1)$ and $\Delta H(j) = H(j) - H(j - 1)$, with $L(0) \equiv 0$ defined as the arbitrary zero point for measuring sea level in the particular section. Summing equation 11 over the stratigraphic interval to sample j gives relative sea level $L(j)$ at time $t(j)$:

$$L(j) = \sum_{i=1}^{j}\Delta L(i), \quad 1 \leq j \leq n \quad (12a)$$

$$= j\left(\frac{\rho_m - \rho_s}{\rho_m}\right)\Delta S$$
$$+ \left(\frac{\rho_m - \rho_w}{\rho_m}\right)[H(j) - H(0)] \quad (12b)$$

$$= j\left(\frac{\rho_m - \rho_s}{\rho_m}\right)\Delta S$$
$$+ \bar{H}\left(\frac{\rho_m - \rho_w}{\rho_m}\right)\left(\frac{H(j) - H(0)}{\bar{H}}\right) \quad (12c)$$

$$\bar{H} \equiv \frac{1}{n + 1}\sum_{i=0}^{n}H(i) \quad (13)$$

where $\bar{H}$ is the arithmetic mean water depth for the section.

Suppose that some measure of relative depth R is directly proportional to absolute depth:

$$R = CH \qquad (14)$$

where C is the proportionality constant. Substituting equation 14 into equations 9 and 12c puts relative time and sea-level change in terms of relative depth R and average absolute depth $\bar{H}$:

$$\frac{t(j)}{T} = \frac{1}{n} \sum_{i=1}^{j} \frac{r}{r(i)}, \quad 1 \le j \le n. \qquad (15)$$

$$L(j) = j\left(\frac{\rho_m - \rho_s}{\rho_m}\right)\Delta S + \bar{H}\left(\frac{\rho_m - \rho_w}{\rho_m}\right)\left(\frac{R(j) - R(0)}{\bar{R}}\right) \qquad (16)$$

where r, $r(i)$, $\bar{R}$, and $R(j)$ are analogous with h, $h(i)$, $\bar{H}$, and $H(j)$. Substituting equation 9 into equation 10 gives absolute time in terms of absolute depth:

$$t(j) = \frac{CS}{c} \sum_{i=1}^{j} \frac{1}{r(i)}, \quad 1 \le j \le n. \qquad (17)$$

Note that C does not appear in expressions for relative time and sea level in equations 15 and 16.

Comparison of reconstructions among sections.—

Consider N sections of lengths S_k with varying durations T_k, depth-dependence frequencies c_k, numbers of sampling intervals n_k, and harmonic and arithmetic mean depths of deposition h_k and $\bar{H}_k$, $1 \le k \le N$. Suppose that the sections are time correlative and that the same measure R of relative depth applies to all of them.

A first step in attempting to compare their respective sea-level histories is to find their time durations relative to that of a standard section. Let us take the one of length S_1 as the standard section and find T_k/T_1. Multiplying equation 6 by n_k gives

$$S_k = c_k h_k T_k. \qquad (18)$$

Dividing equation 18 for the kth section by the corresponding expression for the standard section and rearranging gives

$$\frac{T_k}{T_1} = \frac{S_k}{S_1} \frac{c_1}{c_k} \frac{h_1}{h_k} \qquad (19a)$$

$$= \frac{S_k}{S_1} \frac{c_1}{c_k} \frac{r_1}{r_k} \qquad (19b)$$

where r_k is the kth section's harmonic mean in the relative depth measure R.

Suppose that the arithmetic mean depth of deposition is known only for the standard section. What is the mean depth $\bar{H}_k$ needed for reconstructing the kth section's sea-level history? Applying equation 14 to both sections, dividing one expression by the other, as above, and rearranging, gives

$$\bar{H}_k = \frac{\bar{R}_k}{\bar{R}_1} \bar{H}_1. \qquad (20)$$

The relative depth-dependence frequency c_k/c_1 can be es-

timated for the interval between corresponding time horizons separated by distances S_1' and S_k'. Adapting equation 19b with $T_1 = T_k$ gives

$$\frac{c_k}{c_1} = \frac{S_k'}{S_1'} \frac{r_1'}{r_k'}. \qquad (21)$$

where r_k' is the harmonic mean depth measure over that interval of the kth section.

Tests of reconstructed sea-level histories.—

When sea-level history is known only from the reconstructions to be tested, as it is in cases considered here, tests must be ones of consistency among reconstructions. In tectonically quiescent areas, where eustatic change would be expected to dominate over local change related to vertical tectonic movement, stratigraphic sections should record the same sea-level history. A general test of reconstructions is whether the histories of such sections accord with one another both in time scale and in magnitude of sea-level change. Correspondences among histories are corroborative evidence for the corresponding features' eustatic nature. The more widely distributed the sections are, the stronger will be the evidence for eustasy.

The reconstructed time scale can be tested more rigorously by comparing the time correlation of sections by their reconstructed sea-level histories with the time correlation of the sections by independent evidence. If eustatic sea level fluctuated, it should be possible to time-correlate the sections: (1) by matching corresponding fluctuations in their sea-level curves; or (2) more rigorously, by statistically cross-correlating once-differentiated sea-level curves or, alternatively, bathymetric curves (i.e., plots of death H or R against time) that in this case should approximate once-differentiated relative sea-level curves (see earlier discussion; in addition, see Rabe and Cisne, 1980; Cisne and Chandlee, 1982). As long as tectonic "noise" did not drown out eustatic fluctuations, this test should apply to sections undergoing differential vertical movement at the time of deposition.

Because the reconstructed sea-level history of a section depends sensitively on depth-dependence frequency c, the degree of correspondence between reconstructed sea-level curves for independently time-correlated sections provides a critical test of the assumed constancy of c within a section. Variation in c within a section would show up as variation in water depth and rate of rock accumulation unrelated to eustatic history, and as corresponding distortion of the reconstructed time scale. Substantial and independent variation in c within different sections would cause eustatic sea-level changes to appear different in both magnitude and timing in the reconstructed sea-level curves of the sections. The many factors and uncertainties involved in the reconstruction process make the ranges of error in magnitude and timing seem practically impossible to assess *a priori*. Comparison of curves among a number of time-correlated sections gives some idea of variation in c within sections as compared with variation between sections, which is discussed further in the next section.

THE MIDDLE ORDOVICIAN TRANSGRESSION IN THE NORTH AMERICAN MIDCONTINENT

Deposits of the later Middle Ordovician (Rocklandian, Kirkfieldian, and Shermanian stages) epeiric sea in the midwestern United States (Fig. 2) are predominantly carbonate strata. These deposits represent the continuation of the general Middle Ordovician transgression after a minor interruption that is marked throughout the area by an unconformity between Rocklandian and underlying Blackriveran strata (Templeton and Willman, 1963; Ross and others, 1982). References cited in the caption for Figure 2 describe the midwestern sections bed by bed.

The uninterrupted Rocklandian-Shermanian sequence especially lends itself to reconstructing and testing sea-level histories. Except in areas surrounding the Ozark Uplift and the Taconic arc-continent collision zone, the midcontinent was tectonically quiescent during the period considered here (Templeton and Willman, 1963; Willman and Kolata, 1978; Cisne and others, 1982). It was not yet affected by warping associated with later stages of the Taconic Orogeny (Quinlan and Beaumont, 1984). Although the Midwest was a tectonically stable area roughly 2,000 km from the nearest subduction zone, strata in the Midwest contain numerous volcanic-ash (K-bentonite) layers (Templeton and Willman, 1963). Age relationships of midwestern strata have been determined in relation to geochemically fingerprinted ash layers (Kolata and others, 1983; W. D. Huff and D. R. Kolata, pers. commun., 1986).

Materials and Methods

Sampling and sample processing.—

Sections for study were chosen to represent not only the tectonically quiescent, carbonate-dominated interior of the epeiric sea but also the clastic-dominated inner margins (secs. 1, 8 in Fig. 2) and areas affected by gentle, syndepositional tectonism (secs. 5a and 5b on the Lincoln Fold; secs. 6a and 6b in an area then undergoing tectonic subsidence related to the Taconic Orogeny; sec. 10 flanking the Ozark Uplift). Some sections were collected solely to broaden the representation of fossil assemblages by facies and geographic area (an important consideration for the paleontologically-based approach to paleobathymetry) and not for reconstruction of sea-level history in the particular areas. Data on the bathymetric distribution of fossil assemblages on downslope transects along volcanic-ash layers in area 7 (Fig. 2), the very edge of the carbonate platform and the outer slope of the Taconic Trench, are not included in present analyses but have been used for comparison.

This study applies the same procedures used by Cisne and others (1984) to a more extensive data set. Small bulk samples (1–2 kg) were collected from single beds at regularly spaced intervals through sections indicated in Figure 2. All identifiable macrofossils found in microscopic examination of ultrasonically cleaned rock pieces were identified to the level of genus or group of similar genera. The 250 samples, containing at least 10 identifiable specimens each (and averaging about 90 specimens per sample, the total exceeding 22,000 identified specimens), were ordinated based on the relative abundances of taxa by reciprocal averaging (Hill, 1973; 1974). This eigenvector ordination method has generally proven most powerful in resolving the principal, environmentally related compositional trend in modern communities (Gauch, 1982; Austin, 1985) and in fossil assemblages (Cisne and Rabe, 1978; Gauch, 1982). Only first-axis scores for fossil assemblages are used here (i.e., the analog of the first principal component in principal components analysis, a related method).

Paleobathymetry.—

Inspection of vertical sequences and onshore-offshore sequences along volcanic-ash layers (e.g., among secs. 1, 8, 2, 9, and 3, from onshore to offshore) shows a bathymetrically related onshore-offshore gradient in the composition of fossil assemblages (Cisne and others, 1984; for a synthetic stratigraphic reconstruction of the onshore-offshore profile, see Cisne, 1985). Like the similar gradient developed over a larger range of depth in coeval assemblages in the Mohawk Valley (Cisne and others, 1982), the gradient in midcontinent assemblages appears to reflect the depth gradient. To be sure, the gradient may not reflect water depth itself so much as it reflects gradients in many biologically and taphonomically important environmental factors related both to water depth and to distance from shore.

First-axis reciprocal averaging scores for fossil assemblages (scaled 0–376, increasing with depth) reflect this gradient quite strongly (Cisne and others, 1984) and are used as the measure of relative depth. Scores for Mohawk Valley assemblages are linearly related to relative depth, as measured by distance downslope along volcanic-ash layers (Cisne and others, 1982). Because low scores for what seem to be the shallowest water midcontinent assemblages are consistent with a zero ordination score corresponding to a depth of practically zero, we assume that ordination score is directly proportional to water depth. A test of this working assumption is the degree of consistency among sea-level histories reconstructed for different sections.

Reconstruction of sea-level history.—

Sea-level histories are reconstructed for six stratigraphic sections (secs. 3, 4, 5a, 5b, 9, 10 in Fig. 2) as outlined previously. The reconstructions are based on two working assumptions that are evaluated through comparison of results for different sections: (1) first-axis reciprocal averaging score is directly proportional to water depth as per equation 14, (2) depth-dependence frequency c is constant within and among sections. Coencorrelation curves (Cisne and Rabe, 1978) in Figure 3 give the basic data on ordination score as a function of stratigraphic position for the six sections. Section 3, the longest Upper Middle Ordovician section in the Midwest, is taken as the standard. The arithmetic mean depth of deposition in this section is taken to be $\bar{H}_3 = 2$ m, the asymptotic offshore water depth estimated from analysis of stratigraphic expressions of shoreline edge effects in the area (Cisne, 1985). The duration of the section is $T_3 \simeq 10$ million years (Ross and others, 1982). Mean depths $\bar{H}_k$ and durations T_k of other sections are measured relative to these.

Curves in Figure 4 show absolute sea level relative to the base of each section as a function of reconstructed relative time. The curves are located along the time axis in Figure 4 according to the time correlation of sections by geochemically fingerprinted ash layers, and particularly in the case of section 9, by lithostratigraphic units that have been found to key in with the sequence of ash layers (Templeton and Willman, 1963; Willman and Kolata, 1978; Kolata and others, 1983).

Tests of reconstructions.—

As expected, all six sea-level curves conform to the same pattern in the overall transgressive time trend and in fine details of apparently eustatic fluctuations (Fig. 4). The spacing of ash layers on the reconstructed time scales shows that depth-dependence frequencies c were very nearly the same in sections 3, 4, and 10 (for reference, the unnamed K-bentonite in sec. 10 is nearly in the location expected for the Dickeyville K-bentonite; D. R. Kolata, pers. commun., 1986) but were somewhat higher in sections 5a and 5b (Fig. 4). The relative depth-dependence frequency (equation 21) for section 4 with respect to section 3 is $c_4/c_3 = 1.03$ between the Deicke and Dickeyville K-bentonites, the longest interval for comparison. For section 5a, in contrast, the relative frequency is $c_{5a}/c_3 = 2.08$ between the Millbrig and Elkport K-bentonites.

The steeper slopes of the curves for sections 5a and 5b (Fig. 4) reflect the greater sediment accumulation rates and the correspondingly overestimated time durations, not a substantially different net rate of transgression. The higher sedimentation rate could be due to clastic input from surrounding lands affected by the Ozark Uplift, and perhaps to a higher carbonate production rate made possible by higher input of land-derived nutrients. The upturn in the curve for section 10, signifying a change to near stillstand, may be an expression of the Ozark Uplift, which affected that area as well.

Perhaps most obvious among the small-scale correspondences is the regressive-transgressive pulse immediately around the Elkport K-bentonite (Fig. 4). A similar pulse is associated with this ash layer at sections 3, 4, and 5a; it also occurs close to the expected position near the top of section 5b and in section 10. Another obvious correspondence is the regressive-transgressive pulse followed by the long, steady rise in sections 3 and 9 (Fig. 4).

Non-constancy in depth-dependence frequency c between sections raises the question of temporal within-section variation in c. Close correspondence in sea-level curves for different sections indicates that independent temporal variation in c within sections was relatively unimportant and, in general, probably much smaller than variation in c between sections (see previous sections).

Sea-Level History

We take the sea-level curve reconstructed for section 3 (Fig. 4), the longest section, to represent best the eustatic history. Comparison of this curve with others (see previous discussions) shows that the curve for section 3 is representative of the group and that even minor fluctuations correlate in time over a large area (Fig. 2). This suggests that the reconstructed time trends record eustatic sea-level change.

Sea-level change during the Rocklandian through Denmarkian transgression had at least two components. The first was a steady rise at a very slowly varying rate around 1 per million years relative to the continent. The second component was a group of regressive-transgressive pulses no more than 0.1 to 1 million years long, during which sea level fell and then rose roughly 1 m. The remarkably small magnitude of the pulses testifies to the tectonic quiescence, spatial homogeneity, and essential tidelessness (Hallam, 1981) of the epeiric sea, and to the precision of its stratigraphic record.

Overall, the curve (Fig. 4, sec. 3) shows a basically steady rise in sea level of about 10 m relative to the continent over about 10 million years. For the 30 to 60% and 85 to 100% intervals on the time scale (Fig. 4), the rise is remarkably steady. Interestingly, the transgression rates appear to differ somewhat between these intervals of a few million years. Steady rise in sea level at this rate is consistent with a transgression due to decrease in the mean age of oceanic lithosphere (Donovan and Jones, 1979). This is the expected consequence of the origin and growth of new spreading ridges during continental dispersion in the latest Proterozoic and earlier Paleozoic (Heller and Angevine, 1985).

The regressive-transgressive pulses may represent minor glacio-eustatic fluctuations such as those associated with climatically controlled fluctuations in small glaciers (Meier, 1984). Although durations of pulses cannot be accurately estimated from available evidence, their small 0.1-to-1-million-year time scale is consistent with climatic control. Caradocian diamictites (Hambrey and Harland, 1981) show that glaciation was broadly contemporaneous with the strata analyzed, and Ashgillian continental glaciation in Gondwanaland followed not long after. As expected for an episode of glacial advance and retreat, each pulse begins with a regressive departure from the overall trend of the sea-level curve and returns to very nearly the same sea stand at its end. The sea-level changes are small enough that climatically controlled, temperature-dependent changes in the specific volume of ocean water could have been involved.

The falls in sea level do not appear to be large enough to result in exposure and development of unconformities. The stratigraphic sequence is evidently continuous from the Blackriveran-Rocklandian contact through the Denmarkian. The Blackriveran-Rocklandian contact itself, however, a continent-wide unconformity (Ross and others, 1982), may represent the regressive phase of an especially large pulse that did lead to exposure.

Interestingly, regressive-transgressive pulses are often, although not always, associated with volcanic-ash layers (Fig. 4), suggesting that climatic effects of explosive volcanism may have been a controlling factor. The pulses are not simply artifacts of the effects of volcanic ash on the benthic fauna. As Figure 4 shows, pulses often begin before the ash is deposited, and they occur in sections where the particular ash layers do not. As Cisne and others (1984, fig. 8) show, faunal composition changes continuously from well below to well above an individual ash layer.

ACKNOWLEDGMENTS

This contribution benefited from our discussions with other participants in the symposium. We thank reviewers Dennis R. Kolata of the Illinois State Geological Survey (who also kindly provided unpublished information on K-bentonites); Ian Lerche of the University of South Carolina, and Henry Posamentier of Exxon Production Research Company for constructive criticism; Cheryl K. Wilgus of Everest Geotech for editorial advice; and Carl Bass of Cornell University for help with computer graphics. The work was supported by National Science Foundation Grants DEB-8021158, EAR-8117987, EAR-8121052, and EAR-8407723.

REFERENCES

AUSTIN, M. P., 1985, Continuum concept, ordination methods, and niche theory: Annual Review of Ecology and Systematics, v. 16, p. 39–61.

CISNE, J. L., 1985, Depth-dependent sedimentation and the flexural edge effect in epeiric seas: measuring water depth relative to the lithosphere's flexural wavelength: Journal of Geology, v. 93, p. 567–576.

———, AND CHANDLEE, G. O., 1982, Taconic Foreland Basin graptolites: Age zonation, depth zonation, and use in ecostratigraphic correlation: Lethaia, v. 15, p. 343–363.

———, GILDNER, R. F., AND RABE, B. D., 1984, Epeiric sedimentation and sea level: Synthetic ecostratigraphy: Lethaia, v. 17, p. 267–288.

———, KARIG, D. E., RABE, B. D., AND HAY, B. J., 1982, Topography and tectonics of the Taconic Foreland Basin as revealed through gradient analysis of fossil assemblages: Lethaia, v. 15, p. 229–246.

———, AND RABE, B. D., 1978, Coenocorrelation: Gradient analysis of fossil assemblages and its applications in stratigraphy: Lethaia, v. 11, p. 341–363.

DONOVAN, D. T., AND JONES, E. J. W., 1979, Causes of world-wide changes in sea level: Journal of the Geological Society of London, v. 136, p. 187–192.

ENOS, P., AND SAWATSKY, L. H., 1981, Pore space in Holocene carbonate sediments: Journal of Sedimentary Petrology, v. 51, p. 961–985.

GAUCH, H. G., 1982, Multivariate Analysis in Community Ecology: Cambridge University Press, Cambridge, 298 p.

GINSBURG, R. N., AND JAMES, N. P., 1974, Holocene carbonate sediments of continental shelves *in* Burke, C. A., and Drake, C. L., eds., The Geology of Continental Margins: Springer-Verlag, New York, p. 137–155.

HALLAM, A., 1981, Facies Interpretation and the Stratigraphic Record: Freeman, Oxford and San Francisco, 291 p.

HAMBREY, M. J., AND HARLAND, W. B., eds., 1981, Earth's Pre-Pleistocene Glacial Record: Cambridge University Press, Cambridge, 1004 p.

HARDENBOL, J., VAIL, P. R., AND FERRER, J., 1981, Interpreting paleoenvironments, subsidence history, and sea level changes of passive margins from seismic and biostratigraphy: Oceanologica Acta, v. 3, Supplement, p. 334–344.

HELLER, P. L., AND ANGEVINE, C. L., 1985, Sea level cycles during the growth of Atlantic-type oceans: Earth and Planetary Science Letters, v. 75, p. 417–426.

HILL, M. O., 1973, Reciprocal averaging and other ordination techniques: Journal of Ecology, v. 61, p. 237–249.

———, 1974, Correspondence analysis, a neglected multivariate method: Journal of the Royal Statistical Society, Series C., v. 23, p. 340–354.

KAY, G. M., 1937, Stratigraphy of the Trenton Group: Geological Society of America Bulletin, v. 48, p. 233–302.

KENDALL, C. G. St. C., AND SCHLAGER, W., 1981, Carbonates and relative sea level change: Marine Geology, v. 44, p. 181–212.

KOLATA, D. R., HUFF, W. D., AND FROST, J. K., 1983, Correlation of K-bentonites in the Decorah Subgroup of the Mississippi Valley by chemical fingerprinting, *in* Delgado, D. J., ed., Ordovician Galena Group of the Upper Mississippi Valley—Deposition, diagenesis, and paleoecology: Guidebook, 13th Annual Field Conference, Society of Economic Paleontologists and Mineralogists, Great Lakes Section, p. F1–F15.

LEVORSON, C. O., AND GERK, A. J., 1972, A preliminary stratigraphic study of the Galena Group of Winneshiek County, Iowa: Iowa Academy of Science, Proceedings, v. 79, p. 111–122.

———, AND ———, 1975, Field recognition of the subdivisions of the Galena Group in Winneshiek County, Iowa: Field Trip Guide, Iowa, Minnesota, and Wisconsin Academies of Science (unpaginated).

MEIER, M. F., 1984, Contribution of small glaciers to global sea level: Science, v. 226, p. 1418–1421.

PITMAN, W. C., III, 1978, Relationship between eustasy and stratigraphic sequences of passive margins: Geological Society of America Bulletin, v. 89, p. 1389–1403.

———, 1979, The effect of eustatic sea level changes on stratigraphic sequences at Atlantic margins: American Association of Petroleum Geologists Memoir 29, p. 453–460.

QUINLAN, G. M., AND BEAUMONT, C., 1984, Appalachian thrusting, lithospheric flexure, and the Paleozoic stratigraphy of the Eastern Interior of North America: Canadian Journal of Earth Sciences, v. 21, p. 973–996.

RABE, B. D., AND CISNE, J. L., 1980, Chronostratigraphic accuracy of Ordovician ecostratigraphic correlation: Lethaia, v. 13, p. 109–118.

ROSS, R. J., AND 27 OTHERS, 1982, The Ordovician System in the United States: International Union of Geological Sciences Publication 12, 73 p.

SCHUCHERT, C., 1955, Atlas of Paleogeographic Maps of North America: John Wiley, New York, 177 p.

STECKLER, M. S., AND WATTS, A. B., 1978, Subsidence of the Atlantic-type continental margin off New York: Earth and Planetary Science Letters, v, 41, p. 1–13.

TEMPLETON, J. S., AND WILLMAN, H. B., 1963, Champlainian Series (Middle Ordovician) in Illinois: Illinois State Geological Survey Bulletin, v. 89, 260 p.

TITUS, R., 1982, Fossil communities of the middle Trenton Group (Ordovician) of New York State: Journal of Paleontology, v. 56, p. 477–485.

———, AND CAMERON, B., 1976, Fossil communities of the lower Trenton Group (Middle Ordovician) of central and northwestern New York State: Journal of Paleontology, v. 50, p. 1209–1225.

VAIL, P. R., AND HARDENBOL, J., 1979, Sea-level changes during the Tertiary: Oceanus, v. 22, p. 71–79.

———, MITCHUM, R. M., JR., AND THOMPSON, S., III, 1977, Seismic stratigraphy and global changes of sea level, part 4: Global cycles of relative sea level *in* Payton, C. E., ed., Seismic Stratigraphy Applications to Hydrocarbon Explanation: American Association of Petroleum Geologists Memoir 26, p. 83–97.

WATTS, A. B., 1982, Tectonic subsidence, flexure and global changes of sea level: Nature, v. 297, p. 469–474.

WILLMAN, H. B., AND KOLATA, D. R., 1978, The Platteville and Galena Groups in northern Illinois: Illinois State Geological Survey Circular 502, 75 p.

LATE PALEOZOIC TRANSGRESSIVE-REGRESSIVE DEPOSITION

CHARLES A. ROSS AND JUNE R. P. ROSS

Chevron U.S.A., Inc., P.O. Box 1635, Houston, Texas 77251; Department of Biology, Western Washington University,
Bellingham, Washington 98225

ABSTRACT: Approximately sixty transgressive-regressive depositional sequences are present in Carboniferous and Permian shallow-marine successions on the world's stable cratonic shelves. These sequences were synchronous depositional events that resulted from eustatic sea-level changes. Based on currently available age correlations of rapidly evolved late Paleozoic tropical, subtropical, and temperate shelf faunas, the sequences on different cratonic shelves were time equivalent. These transgressive-regressive sequences averaged about 2 million years and ranged from 1.2 to 4.0 million years in duration.

Local depositional conditions are important in controlling sedimentary patterns on different cratonic shelves. These conditions are affected by changes in sea level, strandline position, and drainage base level and are reflected in the sedimentary record. Because mid-size sea-level fluctuations are usually widely identifiable in the stratigraphic record, they are useful aids in correlation. They are particularly helpful between regions that have contrasting depositional conditions, such as between a carbonate shelf starved of clastic sediments and a clastic-dominated shelf on which carbonates are rare.

The appearance of new species and genera generally occurs above unconformities that signal new marine transgressive events and new depositional sequences. The durations of the hiatuses between these transgressive-regressive sequences are difficult to estimate. The hiatuses may represent cumulatively as much time, if not more, than the rock record. The numerous worldwide synchronous unconformities marking hiatuses of considerable duration within late Paleozoic shelf strata suggest that the fossil record may be very incomplete and preserves mostly biota that were extant during times of high sea level. Such an incomplete fossil record could easily be misinterpreted as a punctuated evolution having a highly irregular mutation rate.

INTRODUCTION

Late Paleozoic sea-level changes are recognized mainly in relatively thin deposits left by marine transgressions and regressions on the stable cratonic shelves of the Paleozoic world (Fig. 1). This stratigraphic record contrasts with that of the Cenozoic and Mesozoic, which preserves not only shallow deposits of the cratonic shelves, but also thick wedges of sediment along the margins of the continents and adjacent portions of the ocean basins. These younger, more complete records have been extensively studied as "depositional sequences." Most of these younger depositional sequences are sufficiently thick to be identified by seismic stratigraphic methods.

Similar thick sedimentary wedges were deposited adjacent to the Paleozoic cratons. Seafloor spreading and plate-tectonic events during the Mesozoic and Cenozoic, however, deformed these Paleozoic ocean and ocean margin deposits. As a result, nearly all of them have been either consumed in subduction zones or transported, strongly deformed, and accreted as linear belts, or terranes, to cratonic margins as parts of orogenic belts. The Ouachita-Marathon fold belt along the southern margin and the Antler and Sonoma orogenic belt along the western margin of Paleozoic North America are good examples of these deformed sedimentary wedges. To date, individual depositional sequences within these Paleozoic terranes have not been identifiable because they are generally too extensively deformed, too poorly studied, and contain long-ranging deeper water fossils that are difficult to correlate with the shallow-shelf faunas. Many of these terranes are allochthonous and far traveled so that they are no longer adjacent to the craton from which they were originally derived. In fact, we often do not know which cratonic shelves were originally adjacent to which of the accreted terranes.

The thin, updip transgressive edges of Paleozoic depositional sequences are widely distributed on the cratons (Figs. 2–5). The lower and lower middle Carboniferous record of

these sea-level changes has been extensively studied in Great Britain (Ramsbottom, 1973, 1977, 1979, 1981). Analysis of the facies and depositional history of Tournaisian and Visean strata in Great Britain (Ramsbottom, 1973) indicate that these strata can be divided into eleven transgressive-regressive depositional cycles (Fig. 6A). The overlying Namurian Stage (Figs. 6A, B), which has a detailed and precisely defined set of cephalopod (goniatite) zones, consistently shows eleven transgressions and regressions throughout Great Britain (Ramsbottom, 1977). These could be reliably dated as being synchronous. Although such deposits could result from relative local vertical motion of the crust, the consistency and large number of repeated episodes more strongly suggest repeated changes in world sea level, that is, changes in eustasy. The Westphalian coal measures above the Namurian show evidence of 10 additional depositional cycles (Ramsbottom and others, 1978). For these sedimentary cycles, Ramsbottom (1977) applied the name "mesothem," because he considered them larger than cyclothems (Wanless and Shepard, 1936) and smaller than "sequences" (as used by Sloss, 1963). For the same magnitude of cycle, Vail and Mitchum (1977) used the term depositional sequence, which we shall also use for consistency.

In North America, Ross (1973) was able to trace 20 Pennsylvanian and Early Permian regional transgressive-regressive depositional units and unconformities across large areas of the southern part of the North American craton (Figs. 6B, C). Ross (1979), Ross (1978, 1981a,b, 1984), and Ross and Ross (1979, 1981, 1985a,b) examined the stratigraphic details of the worldwide distribution of Carboniferous and Permian fusulinaceans, bryozoans, and other faunas. They concluded that at least 50 of these middle level transgressions and regressions are clearly identifiable in Carboniferous and Lower Permian strata deposited in the marine shallow-water cratonic shelves of the late Paleozoic tropical and subtropical regions. More recently, Ross and Ross (1987a,b) have systematized the late Paleozoic sea-level cycle terminology and have brought together the fossil

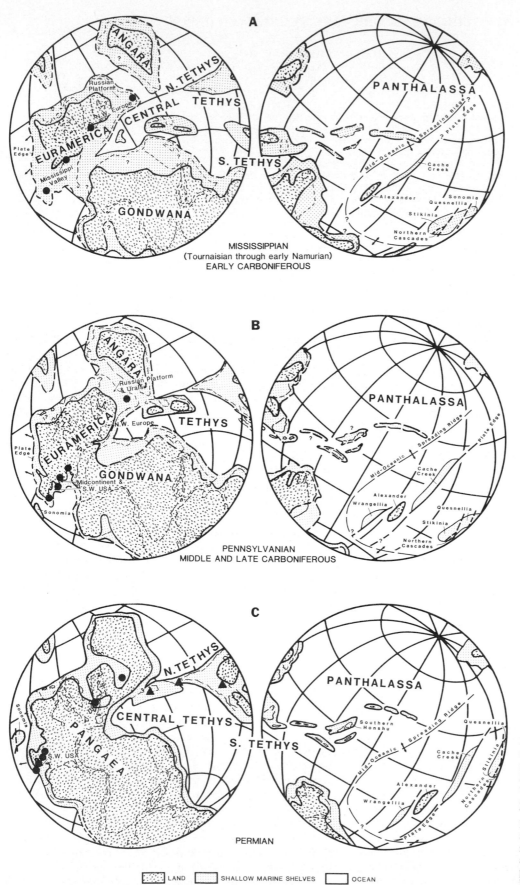

FIG. 1.—Maps show location of late Paleozoic cratonic shelf areas and stratigraphic sections. Dots in A represent stratigraphic sections shown in Figure 2; in B sections shown in Figure 3; in C Lower Permian and Guadalupian sections shown in Figure 4. Triangles in C represent uppermost Permian sections shown in Figure 5. Stratigraphic sections are located on relatively stable portions of cratonic areas and are near the late Paleozoic paleoequator (modified from Ross and Ross, 1985a).

zonations used to date the various transgressions. To date, the difficulties of separating the effects of unstable tectonic depositional settings and later orogenic deformation from sea-level changes has hindered division of the strata of the late Paleozoic orogenic belts, such as the Carboniferous Hercynian-Appalachian-Ouachita-Marathon belt, the Antler-Sonoma belt, or the Permian Tethyan belt, into individual depositional sequences that can be correlated directly to the shelf sequences.

PHYSICAL FEATURES

Identification and recognition of transgressive-regressive depositional sequences on the craton rely on interpreting depositional environments and facies changes, finding and tracing unconformities, and establishing the age relationships of the strata with as much precision as possible. These transgressions and regressions are comparable to the megacyclothems of Kansas and are of considerably longer duration than most Illinois basin-type cyclothems. A Middle and Upper Carboniferous transgressive-regressive depositional sequence may have four or five cyclothems or partial cyclothems within it (Heckel, 1986), particularly in areas rich in clastic-sediment sources.

A typical succession may start with an unconformity on which fluvial conglomerates and other non-marine clastics rest. These may pass upward into finer clastics and coals and then usually into shallow-water marine limestones as sea level rose. Higher, the limestones may pass into deeper water shelf limestone and/or dark shales during high stands of sea level. Areas starved of clastic sediments are common on these cratonic shelves because of low relief, low stream gradients, and extensive areas of shallow-water carbonate platform buildups that acted as sediment traps (Lane and De Keyser, 1980). Regressive phases as sea level fell also commonly are clastic-sediment poor. In some areas, a regressive limestone facies may be important in many of the cycles.

The unconformity between depositional sequences may involve channels and significant erosion so that the succeeding cycle may start on a deeply eroded or weathered surface. Karst surfaces and regoliths, particularly red residuum on weathered limestone, commonly are readily traceable for long distances. Fossils found in the weathered zones are often recrystallized or replaced by chert. Some fossils are reworked as clastic grains in the sediments just above an unconformity. These fossils usually show evidence of having been in a zone of weathering. They may be recrystallized (indistinct wall structure), may be abraided, and may have a different internal pore filling than the cement of the rock in which they are now located.

REGIONAL EXTENT

Within a geologic region or province, the upper Paleozoic depositional sequences may be traced laterally for considerable distances on the basis of their physical features and stratigraphic position. In southwestern Arizona, Ross (1973) was able to use two reddish clastics intervals in the Black Prince and Horquilla limestones to identify distinctive post-early Atokan and post-Desmoinesian transgressive clastics. Fossil faunas were used to confirm the position of the transgressive-regressive sequences between these reddish intervals.

In the Illinois basin and midcontinent areas, the correlation of Middle and Upper Pennsylvanian transgressive-regressive sequences and their subdivision, cyclothems, over distances of 1,000 km or more has been well established for more than 50 years (Moore and others, 1944). Similarly, Middle and Upper Carboniferous strata of the Russian platform, China, and adjacent areas also have well-established intraregional correlations.

During the Late Paleozoic, these various shelf regions were in the tropics, subtropics, or warm temperate regions (Fig. 1) and shared some depositional features in common. Higher stands of sea level usually resulted in considerable carbonate deposition across much of the shelves in these regions. Shales, sandstones, and, locally, thin conglomerates are common and represent a range of deposits from non-marine to deeper marine shelf environments. In contrast, cooler water environments on stable shelves, such as in western Alberta and eastern British Columbia (McGugan, 1984), are mainly sandstone and cherty-sandstone facies with only minor amounts of limestone and relatively little shale. McGugan's detailed studies show that the succession also was deposited as a series of transgressive-regressive depositional units. Although these lack an abundant calcareous fauna, there is an indication that these units represent much of the Middle and Late Carboniferous and Early Permian.

In each area, many regional and local features and events contribute to the details of the depositional patterns. For example, the progression of events in the Appalachian orogeny provided increasingly larger amounts of clastics to the Illinois and midcontinent basins and finally resulted in uplift of the Illinois basin before the end of the Carboniferous. Also, the horst-and-graben structures of the Ancestral Rockies and areas to the south in West Texas, southern New Mexico, and southeastern Arizona greatly modified the location of depocenters and the availability of clastics between the Middle Carboniferous, the Late Carboniferous, and the Early Permian.

DEPOSITIONAL ENVIRONMENTS

In general, the stratigraphic succession of Carboniferous and Early Permian transgressive-regressive depositional sequences commonly shows very similar types of lithologies and has similar lithologic changes vertically through this portion of the stratigraphic column (Figs. 2–5). What should be borne in mind is that these shallow-water, carbonate-rich shelf deposits were, for the most part, in similar latitudes at the time of their deposition. Thus, the systematic changes in general lithologies, particularly in the abundance of limestone, reflect, in large part, changes in low-latitude ocean conditions, especially water temperatures. These changes are well displayed in the amount of carbonate in the depositional sequence and in the fluctuations of faunal and floral composition and diversity of these carbonate-producing communities.

The ratio of calcium to magnesium in calcium carbonate

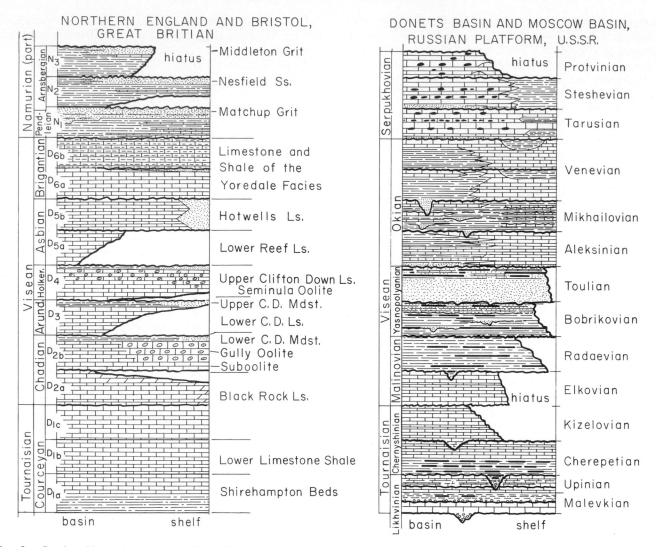

FIG. 2.—Stratigraphic sections of Lower Carboniferous rocks show correlation of unconformities and major rock units from New Mexico and Illinois basin (right) and Great Britain and the Russian Platform (left). Evaporites are common in restricted basins during several of the Visean

fossils gives an indication of temperature trends of the water in which the organisms lived (Fig. 7; Chave, 1954). Yasamanov (1981) gives Ca/Mg data and interpretations of Carboniferous and Permian temperatures based on studies of coral, brachiopod, and fusulinid data (Fig. 8). Most of his data are from Transcaucasia. In Figure 8, the heavy line showing the temperature trend follows the data of the Transcaucasian brachiopods. Note the lack of values for the Middle Carboniferous. Dodd (1967) and Popp and others (1986) found that in well-preserved Devonian brachiopods, ^{18}O showed little alteration from that of modern brachiopods. This suggests that Ca/Mg ratios may also show little alteration in well-preserved brachiopod shells and that Yasamonov's temperature trends, and possibly even the magnitude of these trends, are real.

The four Lower Carboniferous successions from different parts of Euramerica shown in Figure 2 show many similarities in depositional sequences and in their general lithologies. These successions lie close to the Carboniferous paleoequator. Tournaisian deposits are predominantly shallow-water carbonates in contrast to Upper Devonian (Fras-

nian and Famennian) carbonate-poor siltstones and shales. By mid-Tournaisian time, the carbonate-producing organisms dominated extensive areas of the shelves and had divided into many diverse communities. Carbonate mudstone mound communities are the most striking and typically separated the shallow-water carbonate platform communities from slightly deeper, more shaly parts of the shelf (Lane and De Keyser, 1980).

Ecological diversity continued almost to the end of Visean time. The broad, shallow-water carbonate platforms during the early and middle Visean commonly were sites of extensive evaporite deposits. These evaporites accumulated in shallow topographic depressions on the shelves between carbonate platform buildups and formed a prominent facies within some Visean depositional sequences. Oolites also are common in middle and upper Visean depositional sequences, particularly along the platform margins. These evaporite and oolite facies suggest global climatic warming, because they are widespread in much of Euramerica during the Visean.

Upper Visean depositional sequences show the begin-

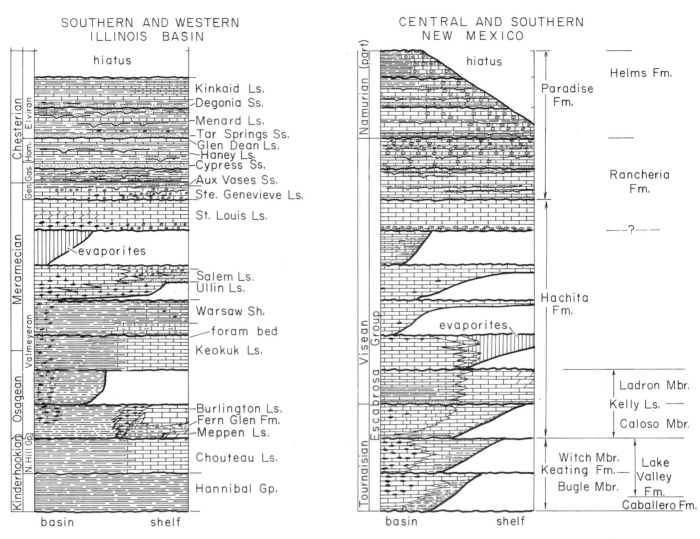

SOUTHERN AND WESTERN
ILLINOIS BASIN

hiatus

— Kinkaid Ls.
— Degonia Ss.
— Menard Ls.
— Tar Springs Ss.
— Glen Dean Ls.
— Haney Ls.
— Cypress Ss.
— Aux Vases Ss.
— Ste. Genevieve Ls.

St. Louis Ls.

evaporites

— Salem Ls.
— Ullin Ls.

Warsaw Sh.

— foram bed

Keokuk Ls.

— Burlington Ls.
— Fern Glen Fm.
— Meppen Ls.

Chouteau Ls.

Hannibal Gp.

basin   shelf

CENTRAL AND SOUTHERN
NEW MEXICO

hiatus

Helms Fm.

Paradise
Fm.

Rancheria
Fm.

—?—

Hachita
Fm.

evaporites

Ladron Mbr.
Kelly Ls.
Caloso Mbr.

Witch Mbr.
Keating Fm.
Bugle Mbr.

Lake
Valley
Fm.

Caballero Fm.

basin   shelf

cycles. The upper Visean and lower Namurian cycles are complex and contain numerous subcycles. Data from Ramsbottom, 1973, 1979; Willman and others, 1975; Yablokov and others, 1975; George and others, 1976; Armstrong and others, 1979, 1980.

nings of two trends. Within each depositional sequence, smaller subcycles appear (or at least become more evident), and these subcycles continue to be present through the remainder of the Carboniferous and into Lower Permian sequences. The other trend is a marked decrease in carbonate-producing communities and an accompanying decrease in faunal diversity to the point that the trophic levels were apparently disrupted. By the beginning of Namurian time, most of the shallow-water carbonate-producing communities had become extremely patchy in their distributions. The common lower Namurian carbonate-producing communities that did survive were deeper water or outer-shelf communities, presumably adapted to cooler water, which appear to have been displaced to shallower water at this time. Along with these changes in communities, shelly faunas and calcareous algae during the early and middle Namurian continued to become progressively less diverse at the species and generic level because of large numbers of extinctions and are less widely distributed geographically. This seems to characterize a gradual change to cooler ocean surface water starting near the end of the Visean and pro-

ceeding until about the middle of the Namurian. On most shelves (Fig. 9), the Lower-Middle Carboniferous boundary falls within a long hiatus, and the Chokierian and Alportian stages of the Namurian are missing. This was a time when relative sea level remained generally low and only occasional and brief transgressions reached farther than the edges of the shelves.

Middle and Upper Carboniferous depositional sequences (Fig. 3) are marked by many major, rapid shoreline transgressions and regressions. Some of these are major transgressions and regressions that have minor shoreline retreats and advances within them. Others are single events that have a low sea-level shoreline position near the shelf margin and a high sea-level shoreline position very high on the craton, with no intermediate sea-level stillstands. Interspersed between these larger single-event depositional sequences are usually one or more cycles having considerably less shift in the position of the shoreline (Heckel, 1986).

The Atokan-Desmoinesian part of the Middle Carboniferous appears to represent a time of warmer temperatures

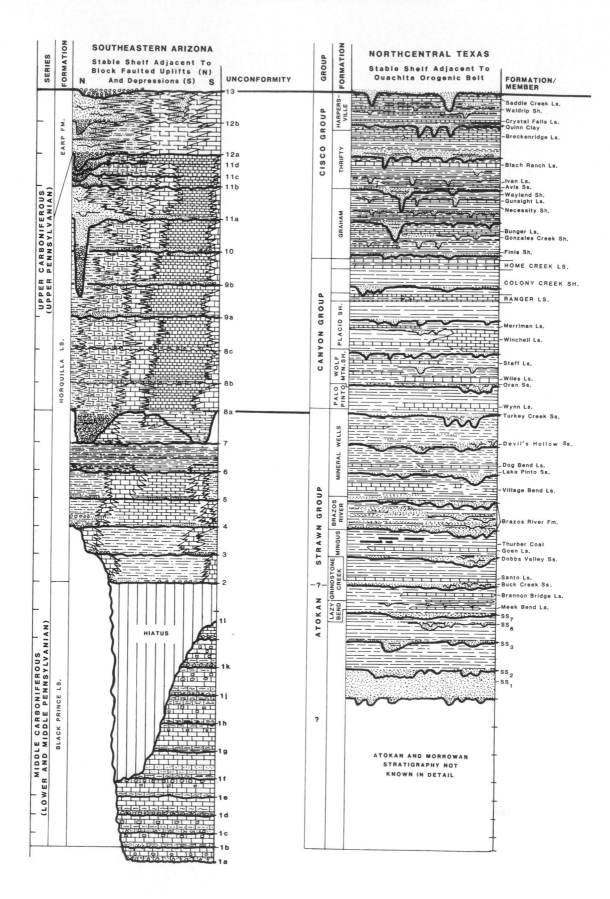

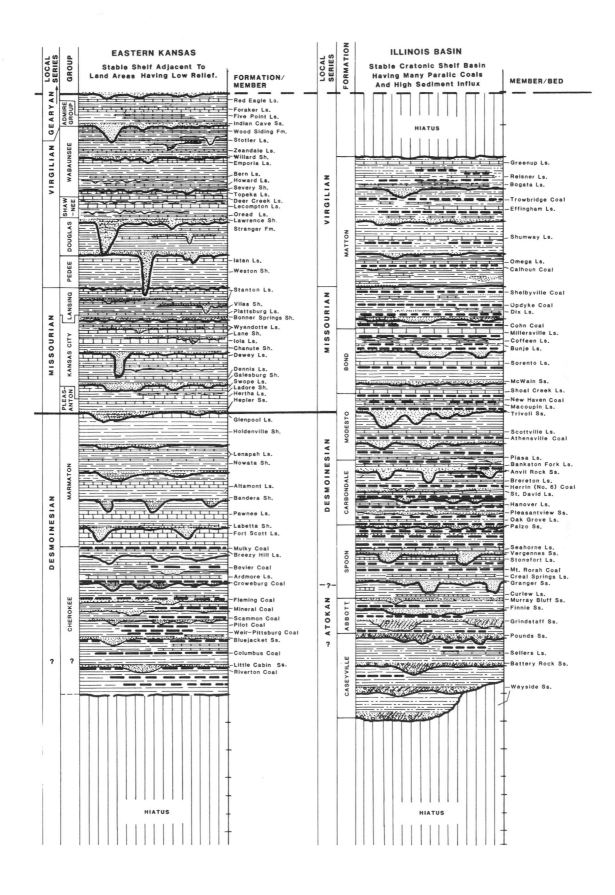

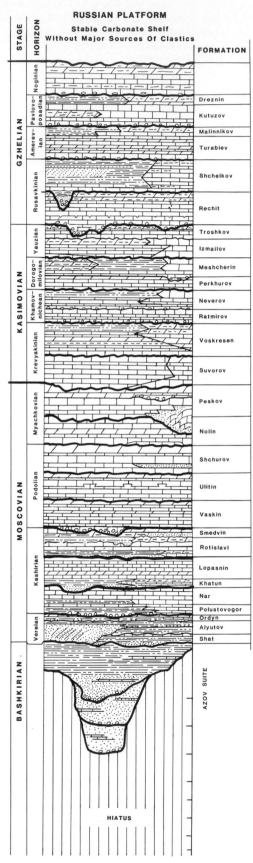

RUSSIAN PLATFORM

Stable Carbonate Shelf
Without Major Sources Of Clastics

STAGE	HORIZON		FORMATION
GZHELIAN	Noginian		
	Pavlovo-posadian		Dreznin
			Kutuzov
	Amerev-ian		Malinnikov
			Turabiev
	Rusavkinian		Shchelkov
			Rechit
KASIMOVIAN	Yauzian		Troshkov
			Izmailov
	Dorogo-milovian		Meshcherin
	Khamov-nichean		Perkhurov
			Neverov
			Ratmirov
	Krevyakinian		Voskresen
			Suvorov
MOSCOVIAN	Myachkovian		Peskov
			Nolin
	Podolian		Shchurov
			Ulitin
			Vaskin
	Kashirian		Smedvin
			Rotislavl
			Lopasnin
			Khatun
			Nar
	Verelan		Polustovogor
			Ordyn
			Alyutov
			Shat
BASHKIRIAN		AZOV SUITE	

HIATUS

than those of the middle Namurian, but not so warm as those during the middle Visean. Faunal and algal diversity increases at the species and generic level, and carbonate-producing communities show increased diversification, but not to the same degree as in the Visean. Oolites locally are common. Carbonates dominate many of the shallow-water shelves. These carbonates tend to be darker gray than those in the middle Visean or those in the overlying Upper Carboniferous. Possibly they were adapted to slightly deeper or cooler water environments than their Visean or Late Carboniferous counterparts.

Within the later part of the Middle Carboniferous, paralic coal is a widely distributed facies in paleoequatorial successions (Fig. 3). Extensive coal preservation implies low bacterial action (cool climates) and high rates of carbon production. The world climates were probably repeatedly cool within a portion of these depositional patterns. In the Upper Carboniferous, paralic coals give way to liminic coals in importance. This suggests global climatic warming, as has also been suggested by Yasamanov's data (Fig. 8), and that average sea surface temperatures rose after the beginning of the Late Carboniferous.

The end of the Middle Carboniferous and beginning of the Late Carboniferous were other times when the extinctions of species and genera were relatively numerous. The causes for this are not immediately evident; however, the carbonate-producing communities of the late Desmoinesian were strongly affected. In fact, many of the shallow-marine Visean lineages that were able to survive the Namurian became extinct before the beginning of the Late Carboniferous. Rapid climatic cooling is a possible cause.

Upper Carboniferous shelf deposits typically begin with a large proportion of clastics. Limestones became increasingly important only gradually, as new carbonate-producing communities evolved. These began as relatively simple communities in comparison with Visean ones. Most were very shallow-water communities in which green calcareous algae contributed much of the carbonate material. Carbonate mudbanks formed the platform edges and separated algal meadows and channels from slightly deeper parts of the shelves that tended to receive fine sands and silts that had bypassed the platform. The limestones are generally light colored and have a large amount of micritic matrix of probable algal origin. It is possible that these shallow-water carbonate-producing communities had adapted to broad areas of shallow water because these were much warmer than the open ocean. Deeper water shelf carbonates are rare. In the

FIG. 3.—Stratigraphic sections of Middle and Upper Carboniferous rocks showing correlation of unconformities and major rock units from (A) southeastern Arizona and northcentral Texas; (B) eastern Kansas and Illinois basin; and (C) the Russian Platform, U.S.S.R. Heavy, short dashed lines indicate coal seams. On cratonic shelves, the lower part (Bashkirian) of the Middle Carboniferous is usually preserved as fluvial clastics deposited in river channels. On slopes of the cratonic margins, the lower Middle Carboniferous includes interbedded carbonates and clastics, as in southeastern Arizona. Data from Moore and others, 1951; O'Connor, 1963; Brown, 1969, 1979; Ross, 1973; Sinitsyn, 1975; Willman and others, 1975; Yablokov and others, 1975; Armstrong and others, 1979; Kier and others, 1979.

middle part of the Upper Carboniferous, at least two successive times of deep-stream erosion cut across these shelf sediments.

The Late Carboniferous communities set the stage for the greatly diversified communities of the Early Permian. Few lineages became extinct between the Late Carboniferous and the Permian, and many well-established stocks provided the diversity from which Permian faunas evolved.

In the Permian, transgressive-regressive sequences (Figs. 4, 5) illustrate the gradual change from Lower Permian marine deposition to predominantly Upper Permian non-marine deposition on the cratonic shelves. The general pattern for Lower Permian carbonate deposition is the gradual development of wave-resistant, carbonate-producing reefal communities. These were mostly cemented together by specialized brachiopods. Mound communities supported by bryozoans and sponges also are common. An ecological succession gradually evolved so that more complex calcareous-producing reefal communities succeeded earlier ones to form reefs of considerable topographic relief. This trend of reefal development continued into the Late Permian Guadalupian reefs and the reefs in the Tethyan areas. Faunas and algae that formed the Early Permian reefal communities evolved and were frequently replaced by newly introduced species and genera. Thus, as communities, they also exhibited considerable change. The marked increase in species and generic diversity and the wider geographical distribution of these communities suggest continued warming of the oceans.

At the same time as these reefal communities were evolving, much of Pangaea was still undergoing tectonic adjustments in the Early Permian. Originating from local sources, clastics covered progressively more and more of the shelves and modified many of the sedimentary patterns. Carbonate platforms behind the reef systems commonly were buried by clastics and evaporites as in Kansas, Oklahoma, southeastern New Mexico, and the panhandle of Texas.

The uppermost Permian sequences on the cratons (Fig. 4) are non-marine, such as the Tatarian of the Russian platform, or evaporites, such as the Ochoan of West Texas. Correlations and identification of depositional sequences in these shelf deposits have relied on tracing unconformities and establishing relative stratigraphic positions, both of which are fraught with difficulties. Uppermost Permian marine sections are rare. Three are shown in Figure 5 and lie off the edges of the cratons along the Tethyan orogenic belt.

FOSSIL ZONATION

The length of time involved in each of the transgressive-regressive depositional sequences appears to range from about 1.2 to 4.0 million years. This apparently provided sufficient time for many shallow-water organisms to evolve and to differentiate within each transgressive event. The two benthic groups on which we have relied are the calcareous foraminifera (including the fusulinaceans) and bryozoans. Both evolved abundant, numerous, and diverse sets of species during the later Paleozoic, many of which were widely distributed. Of these two groups, the calcareous foraminifera were more restricted in their environmental requirements

and usually inhabited shallower water (10 m or less), normal marine, warm tropical, and subtropical environments. Bryozoans also inhabited normal marine, warm tropical, and subtropical habitats; however, they usually preferred slightly deeper water (3 to 50 m). In addition, bryozoans were more widely distributed in temperate and possibly even cooler waters where, however, their generic and species diversity was considerably less.

Conodonts and cephalopods also are used for correlation of parts of this stratigraphic interval, particularly the upper part of the Lower Carboniferous and the lower part of the Middle Carboniferous. Both of these fossil groups were apparently nektonic and, where common, serve as good stratigraphic guides.

During the Carboniferous, pieces of continental crust were assembled to form Pangaea and resulted in the closing of an equatorial east-west tropical seaway between western North America and eastern Europe (Ross and Ross, 1985a). This joining together of Gondwana and Euramerica during the Namurian (Fig. 1) and the subsequent northward displacement of the northern coast of Pangaea (i.e., north coast of Euramerica) into progressively cooler latitudes greatly disrupted the dispersal patterns of shallow-water, tropical marine faunas. The eastern and western coasts of Pangaea developed strongly endemic shallow-marine faunas, and changes in ocean circulation patterns also affected the distribution of some of the nektonic organisms; for example, some of the families of Lower Permian ammonoids have restricted distributions.

Worldwide correlation of Middle Carboniferous through Early Permian (and also Late Permian) strata requires a paleogeographical perspective that recognizes strongly endemic faunas adapted to regional or provincial community structure. Rare, and commonly short-lived, dispersals between these provinces were not assured and usually were the result of chance events. Because of these limitations, almost no fusulinacean species occur in both the western and eastern tropical provinces, and the few common species that do also occur along the northern, warm temperate coast of Euramerica (now the Canadian Arctic Islands, Greenland, and Spitzbergen). Even the generic ranges differ between the faunal provinces, as do the relative importances of genera in the endemic community structure. The shallower water bryozoans appear to have similar distributional limitations; however, the limitations are ameliorated to some degree for bryozoans in deeper water and cooler water communities.

The irregular faunal dispersals between these provinces introduce new species into the other province or provinces that have no apparent endemic ancestors. By comparing the morphological features of the introduced species with those of the other provinces, it is possible to work out the ancestry and point of dispersal of the species from the evolutionary tree of the genus in the other province. This is a refinement on the "stage of evolution" concept and appears to work well as a correlation tool as long as sufficient lineages are considered. Using this method, a detailed zonation of these strata was constructed for the shallow, warm-water benthic foraminifera (Fig. 2).

The dispersals that can be worked out in some detail for

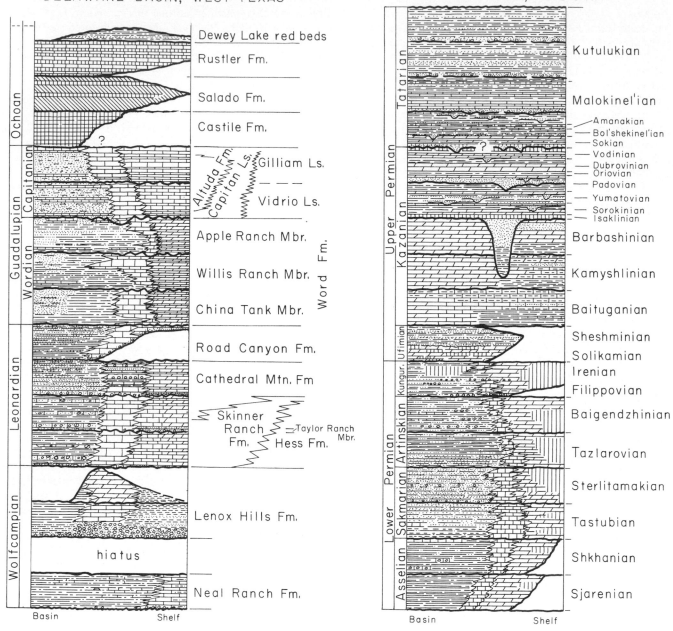

Fig. 4.—Stratigraphic sections of Permian rocks show correlation of unconformities and major rock units from (A) West Texas, and the Russian Platform, U.S.S.R.; (B) north central Texas and Kansas; and (C) northern England and southeastern Arizona. Data from King, 1930, 1948; Sellards and others, 1932; Dunbar and Skinner, 1937; Moore and others, 1951; Likharev, 1966; Ross, 1973.

fusulinaceans (Ross and Ross, 1985a) suggest that they occur during the transgressive phases of transgressive-regressive deposition cycles and are therefore related to rises in sea level. Although this may be an artifact of the preserved stratigraphic record, our interpretation is that higher sea level flooded more of the cratonic shelves and made dispersals for these organisms more likely. If these higher sea levels also meant warmer climates and ocean surface temperatures, then dispersal of these warm-water-adapted faunas would have been easier.

Lower Carboniferous Tournaisian and Visean faunas are remarkably worldwide in their distribution, and latitudinal differences are more strongly developed than are longitudinal differences. This feature is well shown in shallow-water calcareous foraminifera, corals, and brachiopods, and in conodonts and cephalopods, which have nearly worldwide distribution and consistent successions of fossil zones. The subdivisions of the Tournaisian and Visean of northwestern Europe can be readily correlated by fossil zonation with most parts of the world, including the Russian Platform, western North America; northern Canada; Australia; and southern, southeastern, and eastern Asia. The corre-

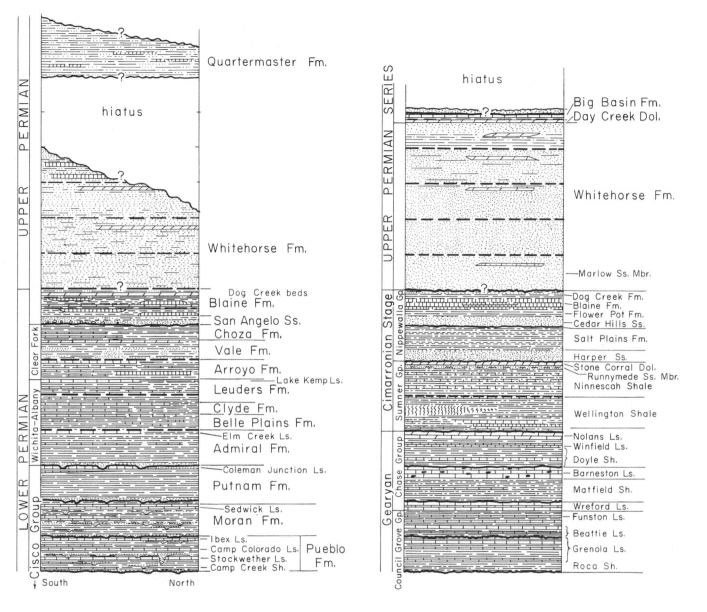

NORTH CENTRAL TEXAS

KANSAS

lations within the Lower Carboniferous are particularly good worldwide, except within the type area of the North American Mississippian Subsystem, where the faunal assemblage was strongly provincial or ecologically restricted.

The Namurian benthic faunas are less widely distributed and are generally impoverished compared to those of the Visean. Successful zonation with benthic faunas is closely related to the occurrence of suitable limestone facies. Conodonts and cephalopods have proven useful in the clastic-dominated marine parts of the shelf edges, as shown by Saunders and others (1979) for the southern Ozark shelf, where, using goniatite cephalopods, they were able to correlate individual transgressions with those of Great Britain.

The Westphalian deposits of northwestern Europe are mainly paralic continental beds interrupted by thin tongues of brackish-marine bands that demonstrate 10 transgressions and regressions. A similar number of depositional se-

quences occurs in normal marine deposits on the Russian Platform and on the southern shelves of North America. The marine deposits can be readily correlated even though the foraminiferal assemblages contain a great percentage of endemic species and genera.

Faunas near the Middle-Upper Carboniferous boundary show many features in common with those near the Early-Middle Carboniferous boundary. Most underwent major changes through evolution and extinction.

Rare fusulinacean faunas from the lower part of the Upper Carboniferous of Kansas include two genera that are best known from strata of this age on the Russian Platform, indicating that dispersals were possible even at times of generally low sea levels. Upper Carboniferous fusulinid zones have evolutionary counterparts in each faunal province; however, most species were relatively simple and morphologically conservative.

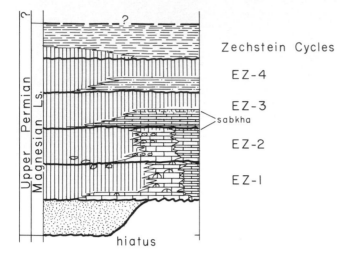

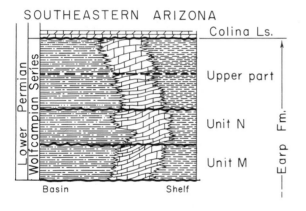

Correlation of Lower Permian zones is well established through the Irenian horizon of the Artinskian. The Wolfcampian zones include the development of inflated fusulinacean shells in several different lineages, some of which probably were planktonic. Bryozoans also had a rapid deployment of many new species and genera into an increasingly diverse set of complex communities. By about Middle Leonardian time, dispersals stopped between the two tropical benthic faunal provinces, and zonations start to be increasingly difficult to correlate between the east and west coasts of Pangaea.

Upper Permian shelf deposits (Fig. 4) are generally difficult to correlate because of either a lack of fossils, strongly developed faunal provinciality, or dissimilar types of fossils from one shelf area to another. For example, the Guadalupian in West Texas has a normal marine-reef fauna, the Kazanian of the Russian platform and Zechstein of northwestern Europe have restricted hypersaline/hyposaline faunas, and the Whitehorse Group of Kansas, Oklahoma, and north-central Texas generally lacks a fauna. Upper Permian marine strata are difficult to correlate between faunal provinces and latitudes because of strongly developed endemic faunas. In some areas, such as on the Russian Platform or in northwestern Europe, the shelf sections are brackish to supersaline at different times and may become filled by non-marine clastics. Species of advanced *Parafusulina* and of *Polydiexodina* mark zones within the Guadalupian. The most diverse of the Upper Permian marine faunas, and the most useful for zonation, however, are in the Tethyan biogeographic province. This was an active tectonic belt, or belts, during this time and transgressive-regressive sequences, as identified on stable shelves, are difficult to recognize.

From stratigraphic evidence, the Lower Guadalupian (Wordian) may contain four transgressive-regressive depositional sequences, and the Upper Guadalupian (Capitanian) may contain four more. Whether these sequences are the result of eustasy or are from other causes, and whether they represent worldwide or local events is at present not known. The Zechstein of northwestern Europe and the Kazanian of the Russian Platform also show evidence of three or four or more transgressive-regressive events, but again faunal evidence is too poor to establish these in a worldwide zonation.

Although the uppermost Permian deposits on the cratons are poorly fossiliferous, those from the Tethyan region (Fig. 5) appear to contain several depositional sequences. Of these, the Lopingian of South China has the most diverse marine fauna, of which mainly fusulinids and ammonoids are well known. The lower, or Wuchiapingian, part has the zone of *Reichelina* spp. and may include two depositional sequences. The upper, or Changshsingian, part is divided into the zone of *Paleofusulina simplex* (below) and the zone of *P. sinenus* (above), each of which is probably a depositional sequence.

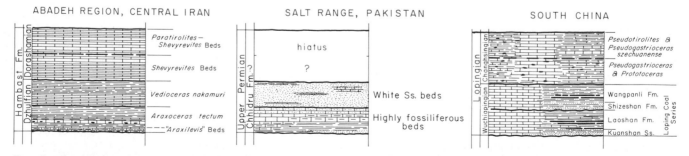

FIG. 5.—Stratigraphic sections of Upper Permian rocks in central Iran, Salt Range of Pakistan, and South China. Data from Sheng, 1963; Chao, 1965; Kummel and Teichert, 1970; Taraz and others, 1981.

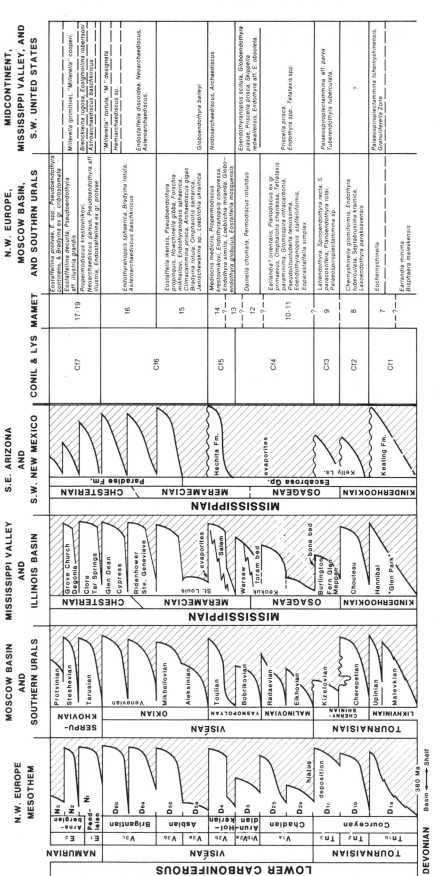
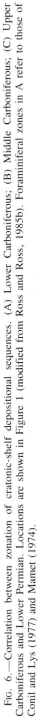

Fig. 6.—Correlation between zonation of cratonic-shelf depositional sequences. (A) Lower Carboniferous; (B) Middle Carboniferous; (C) Upper Carboniferous and Lower Permian. Locations are shown in Figure 1 (modified from Ross and Ross, 1985b). Foraminiferal zones in A refer to those of Conil and Lys (1977) and Mamet (1974).

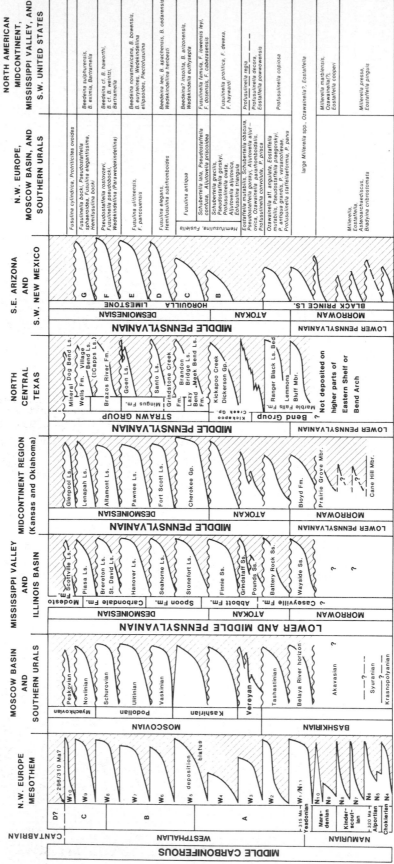

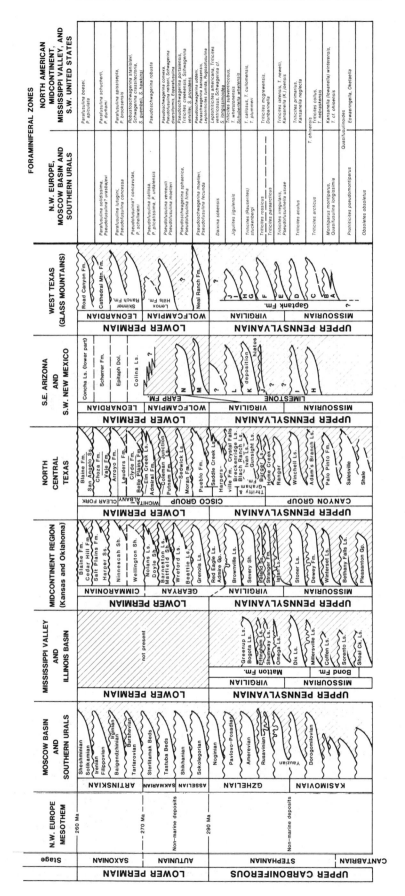

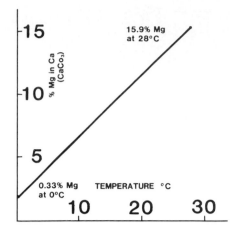

FIG. 7.—General relationship of the percent of magnesium in calcium carbonate as a function of temperature (after Chave, 1954).

COMPARISON WITH EARLIER PALEOZOIC TRANSGRESSIVE-REGRESSIVE SEQUENCES

Several lines of evidence suggest that transgressive-regressive depositional events of similar magnitude and duration also occur in Cambrian to Devonian strata (Fig. 10). The best documented is evidence from the Devonian (Conkin and Conkin, 1984 a,b; J. G. Johnson and others, 1985; Dennison, 1985). Conkin and Conkin (1984a,b) were able to correlate surfaces of minor erosion and nondeposition (they termed these paracontinuities) and volcanic-ash beds

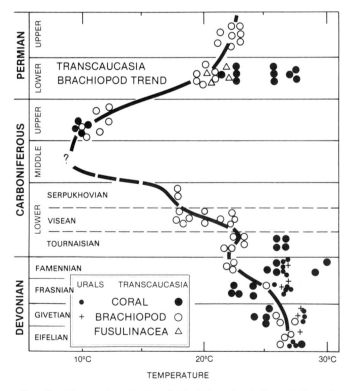

FIG. 8.—Temperature trends of late Paleozoic shallow-shelf marine waters based on Ca/Mg in the southern Ural region and in the Transcaucasia region. Data and interpretation after Yasamanov (1981).

over large areas of the midwestern United States (Kentucky, Ohio, Indiana, Tennessee, and Illinois), J. G. Johnson and others (1985) and Dennison (1985) correlated changes in depositional facies and what they considered to be minor local unconformities with a sea-level curve. Although the actual magnitude of Devonian sea-level fluctuations may not be so great as in the Carboniferous, J. G. Johnson and others (1985) recognize 12 (14 if Dennison's two are included, 1985) significant transgressive events. This averages to one event per 3.5 to 4 million years for the Devonian transgressive-regressive depositional sequences. This average is similar to the longer depositional sequences of the Carboniferous.

Silurian transgressive-regressive deposition is less well documented. M. E. Johnson and others (1985) compared depositional environments of Lower Silurian Llandovery beds on the eastern North American cratonic shelf and on the Yangtze shelf of China with the data presented by McKerrow (1979) and arrived at the tentative conclusion that the Llandovery included four eustatic transgressive-regressive events, or about one per 2 to 3 million years. For the Middle and Late Silurian Wenlock, Ludlovian, and Pridolian, J. G. Johnson and Murphy (1984) indicate that these are represented by only one shallowing-upward cycle on the shelf edge in eastern Nevada; however, three clearly dated Silurian pulses of turbidites were transported from the shelf into the adjacent basin. Even if the three turbidite pulses are related to only slight drops in sea level, this is still a period of very stable sea level because the drops average one per 7 million years.

Ordovician sea level fell near the Cambrian-Ordovician boundary and twice during the Early Ordovician Ibexian Epoch (Tremadocian and Arenigian) (Miller, 1984), during the Llanvirnian-Llandeilian, and again during the early part of the Ashgillian (Spjeldnaes, 1961; Seslavinski, 1979; Ross, 1985). A minor drop is recorded also for the later part of the Caradocian. These Ordovician sea-level events suggest long periods between each as they average 15 million years. A detailed stratigraphic study by Templeton and Willman (1963) on Middle Ordovician strata in Illinois and across much of the craton south of the North American transcontinental arch suggests that "hard grounds," paracontinuities, unconformities, widespread transgressive marine sequences, and similar abrupt facies changes are common within the Middle Ordovician part of the succession. There is, therefore, good reason to believe that this part of the Ordovician is at least as complicated by marine transgressions and regressions as middle parts of the Devonian and that our figure is only an indication of major sea-level trends. The three Early Ordovician events average about 8 million years; the single, large Late Ordovician event was nearly 35 million years.

In the upper part of the Middle Cambrian and in the Upper Cambrian of the western half of North America, there appear to be three or possibly four thick, shallowing-upward depositional sequences (when one analyzes the depositional record rather than the theory of the zonation, evolution, and biomerization of trilobites; Lockman-Balk, 1970; Palmer, 1984; Ludvigsen and Westrop, 1985). The time interval for these depositional sequences is about 25

COASTAL ONLAP CURVE AS AN INDICATION
OF EUSTATIC SEA LEVEL CHANGES

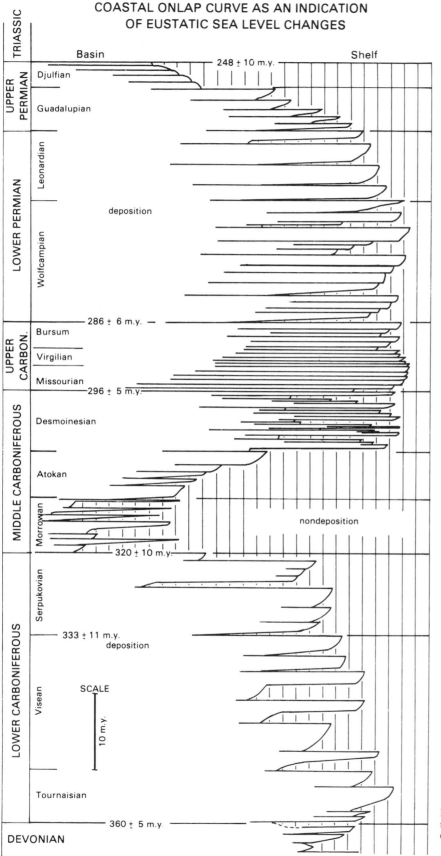

FIG. 9.—Coastal onlap curve for Carboniferous and Permian shelf sediments. Note general similarity between this curve and trends in temperature based on Ca/Mg in Figure 8. (After Ramsbottom 1973, 1977, 1981; Ramsbottom and others, 1978; Ross and Ross, 1985b; Heckel, 1986.)

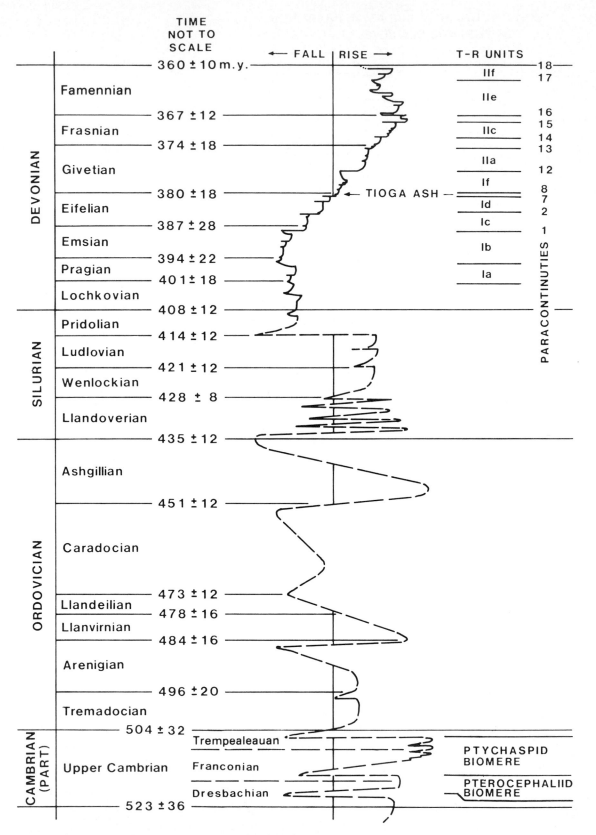

Fɪɢ. 10.—Coastal onlap curve for the Upper Cambrian through Devonian shelf sediments. Of these, the Devonian is the best documented (Dennison, 1985; Johnson, J. G., and others, 1985). The Silurian curve is interpreted from Willman (1973) and McKerrow (1979). The Ordovician curve is based on Ross (1985) and the Upper Cambrian is interpreted from Miller (1984).

million years, or each averages between 6 and 8 million years.

MECHANISM FOR PALEOZOIC EUSTASY

Eustasy is considered the cause for the transgressive-regressive events; however, the reasons behind these changes in sea level are not known. They were, in every likelihood, the result of a combination of several different mechanisms, including changes in seafloor spreading rates, the volume of ocean basins, the rate of ocean trench activities, orogenic activities, and climatic changes (including glaciation). Many of these mechanisms were active and important during the late Paleozoic. Similar changes in sea level during the Pleistocene, as a result of climatic changes causing accumulation of ice sheets, resulted in eustatic changes of at least 100 m. The Middle and Late Carboniferous/Early Permian was a time of both rapid-plate motion and seafloor spreading and, in parts of Gondwana and Angara, widespread glaciation.

The only direct relations between the general patterns of sedimentation and the transgressive-regressive events are that the transgressions and regressions were more frequent during times when carbonate production was low, as in the Namurian and earliest Late Carboniferous, and that at these times, the transgressions did not extend so far onto the cratonic shelves as when carbonate production was higher. These relations imply that clastic-rich intervals were times of generally lower average sea level and cooler temperatures; and therefore Middle Carboniferous through Early Permian glaciations in Gondwana may be one of several causal factors for low-average sea levels at these times.

Two and possibly three mechanisms for these smaller subcycles are generally considered likely: changes in general sea level probably caused by glaciation or climatic changes (Wanless and Shepard, 1936; Heckel, 1986); and changes caused by lateral shifts in distributary and interdistributary depositional facies along a shoreline (Beerbower, 1964). Closely related to this second mechanism are changes in the amount of clastics and in river discharges as a result of climatic changes and topographic changes. Topographic changes resulted from local and regional orogenic or epierogenic movement—in these cases, tectonic activities that affected the whole of the Euramerican craton at the same time.

CONCLUSIONS

About sixty Carboniferous-Permian transgressive-regressive depositional sequences are well developed on stable, warm-water, cratonic shelves where they can be recognized using available biostratigraphic zonations. In cooler water environments, similar transgressive-regressive events are present; however, these are less easy to correlate because of their less diverse temperate and cool-water faunas.

Comparison with Cambrian through Devonian eustatic events suggests that the Carboniferous-Permian events were more rapid, of shorter duration, and approximately of similar vertical change.

The combination of sea-level and climatic fluctuations is considered to be one of the major causes for the maintenance of several distinct, warm, shallow-water marine provinces during the Carboniferous and Early Permian and also for the infrequent and incomplete dispersals between those provinces. These rare dispersals appear to be associated with times of higher sea level and warmer temperatures.

The incompleteness of the transgressive-regressive stratigraphic record and the dispersal of faunas into other provincial areas where they lack immediate ancestors give rise to an incomplete fossil record that could easily be misinterpreted as indicating a punctuated evolution and a high irregular mutation rate.

ACKNOWLEDGMENTS

C. A. Ross thanks Chevron U.S.A., Inc., for permission to publish this article.

REFERENCES

ARMSTRONG, A. K., KOTTLOWSKI, F. E. STEWART, W. J. MAMET, B. L., BALTZ, E. H., JR., SIEMERS, W. T., AND THOMPSON, S., III, 1979, The Mississippian and Pennsylvanian (Carboniferous) Systems in the United States—New Mexico: U.S. Geological Survey Professional Paper 1110-W, p. W1–W27.

————, MAMET, B. L., AND REPETSKI, J. E., 1980, The Mississippian System of New Mexico and southern Arizona, *in* Fouch, T. D., and Magathan, E. R., eds., Paleozoic Paleogeography of the Western-Central United States: Rocky Mountain Section, Society of Economic Paleontologists and Mineralogists, Symposium 1, p. 82–99.

BEERBOWER, J. R., 1964, Cyclothems and cyclic depositional mechanisms in alluvial plain sedimentation, *in* Merriam, D. F., ed., Symposium on Cyclic Sedimentation: State Geological Survey of Kansas, the University of Kansas, Bulletin 169, Lawrence, Kansas, p. 31–42.

BROWN, L. F., JR., 1969, Virgil and Lower Wolfcamp repetitive environments and the depositional model, north-central Texas: University of Texas Bureau of Economic Geology, Geological Circular 69-3, p. 115–134.

————, 1979, Deltaic sandstone facies of the Mid-Continent: Tulsa Geological Society Special Publication No. 1, Tulsa, Oklahoma, p. 35–63.

CHAO, K.-K., 1965, The Permian ammonoid-bearing formations of South China: Scientia Sinica, v. 14, p. 1813–1826.

CHAVE, K. E., 1954, Aspects of the biogeochemistry of magnesium: 1. Calcareous marine organisms: Journal of Geology, v. 62, p. 266–283.

CONIL, R. AND LYS, M., 1977, Les transgressions Dinantiennes et leur influence sur la dispersion et l'evolution des Foraminifers: Université Louvain, Institut Géologie Memoire 29, p. 9–55.

CONKIN, J. E., AND CONKIN, B. M., 1984a, Paleozoic metabentonites of North America, part 1: Devonian metabentonites in the eastern United States and southern Ontario: Their identities, stratigraphic positions, and correlation: University of Louisville Studies in Paleontology and Stratigraphy No. 16, Louisville, Kentucky, 136 p.

————, AND ————, 1984b, Devonian and Mississippian bone beds, paracontinuities, and pyroclastics, and the Silurian-Devonian paraconformity in southern Indiana and northwestern Kentucky: Geological Society of America Annual Meeting of Southeastern and North-Central Sections, Field Trip Guides, p. 25–42.

DENNISON, J. M., 1985, Devonian eustatic fluctuations in Euramerica: Discussions: Geological Society of America Bulletin, v. 96, p. 1595–1597.

DODD, J. R., 1967, Magnesium and strontium in calcareous skeletons: A review: Journal of Paleontology, v. 41, p. 1313–1330.

DUNBAR, C. O., AND SKINNER, J. W., 1937, Geology of Texas: Permian Fusulinidae of Texas: University of Texas Bulletin 3701, v. 3, part. 2, p. 523–825.

GEORGE, T. N., JOHNSON, G. A. L., MITCHELL, M., PRENTICE, J. E., RAMSBOTTOM, W. H. C., SEVASTOPULO, G. D., AND WILSON, R. B., 1976, A correlation of Dinantian rocks in the British Isles: Geological Society of London Special Report 7, 87 p.

HECKEL, P. H., 1986, Sea-level curve for Pennsylvanian eustatic marine transgressive-regressive depositional cycles along midcontinent outcrop belt, North America: Geology, v. 14, p. 330–334.

JOHNSON, J. G., KLAPPER, G., AND SANDBERG, C. A., 1985, Devonian eustatic fluctuations in Euramerica: Geological Society of America Bulletin, v. 96, p. 567–587.

———, AND MURPHY, M. A., 1984, Time-rock model for Siluro-Devonian continental shelf, western United States: Geological Society of America Bulletin, v. 95, p. 1349–1359.

JOHNSON, M. E., JIA-YU, R., AND YANG, X.-C., 1985, Intercontinental correlation by sea-level events in the Early Silurian of North America and China (Yangtze Platform): Geological Society of America Bulletin, v. 96, p. 1384–1397.

KIER, R. S., BROWN, L. F., JR., AND MCBRIDE, E. F., 1979, The Mississippian and Pennsylvanian (Carboniferous) Systems in the United States—Texas: U.S. Geological Survey Professional Paper 1110-S, p. S1–S45.

KING, P. B., 1930, The Geology of the Glass Mountains, Texas, part 1: Descriptive geology: University of Texas Bulletin 3038, 167 p.

———, 1948, Geology of the southern Guadalupe Mountains, Texas: U.S. Geological Survey Professional Paper 215, 183 p.

KUMMEL, B., AND TEICHERT, C., 1970, Stratigraphic boundary problems: Permian and Triassic of West Pakistan: University of Kansas, Department of Geology Special Publication 4, Lawrence, Kansas, 453 p.

LANE, H. R., AND DE KEYSER, T. L., 1980, Paleogeography of the late Early Mississippian (Tournaisian) in the central and southwestern United States, in Fouch, T. D., and Magathan, E. R., eds., Paleozoic Paleogeography of the Western-Central United States: Rocky Mountain Section, Society of Economic Paleontologists and Mineralogists, Symposium 1, p. 149–162.

LIKHAREV, B. K., 1966, Stratigrafiya SSSR: Permskaya Sistema: Moscow, "Nedra," 536 p.

LOCKMAN-BALK, C., 1970, Upper Cambrian faunal patterns on the craton: Geological Society of American Bulletin, v. 81, p. 3197–3224.

LUDVIGSEN, R., AND WESTROP, S. R., 1985, Three new Upper Cambrian stages for North America: Geology, v. 13, p. 139–143.

MAMET, B., 1974, Une zonation Foraminifères du Carbonifère Inferieur de la Tethy Occidentale: Septième Congrès International de Stratigraphie et de Géologie due Carbonifère, Compte Rendu, Geologisches Landesamt Nordrhein-Westfalen, Krefeld, BDR, v. 3, p. 381–408.

MCGUGAN, A., 1984, Carboniferous and Permian Ishbel Group stratigraphy, northern Saskatchewan Valley, Canadian Rocky Mountains, western Alberta: Canadian Petroleum Geology Bulletin, v. 32, p. 372–381.

MCKERROW, W. S., 1979, Ordovician and Silurian changes in sea level: Geological Society of London Journal, v. 136, p. 137–145.

MILLER, J. F., 1984, Cambrian and earliest Ordovician conodont evolution, biofacies, and provincialism: Geological Society of America Special Paper 196, p. 43–65.

MOORE, R. C., AND 27 OTHERS, 1944, Correlation of Pennsylvanian formations of North America: Geological Society of America Bulletin, v. 55, p. 657–706.

———, FRYE, J. C., JEWETT, J. M., LEE, W., AND O'CONNOR, H. G., 1951, The Kansas rock column: Kansas State Geological Survey Bulletin, v. 89, 132 p.

O'CONNOR, H. G., 1963, Changes in Kansas stratigraphic nomenclature: American Association of Petroleum Geologists Bulletin, v. 47, p. 1873–1877.

PALMER, A. R., 1984, The biomere problem: Evolution of an idea: Journal of Paleontology, v. 58, p. 599–611.

POPP, B. N., ANDERSON, T. F., AND SANDBERG, P. A., 1986, Textural, elemental, and isotopic variations among constituents in Middle Devonian limestones, North America: Journal of Sedimentary Petrology, v. 56, p. 715–727.

RAMSBOTTOM, W. H. C., 1973, Transgression and regression in the Dinantian: A new synthesis of British stratigraphy: Yorkshire Geological Society, Proceedings, v. 39, p. 567–607.

———, 1977, Major cycles of transgression and regression (mesothems) in the Namurian: Yorkshire Geological Society, Proceedings, v. 41, p. 261–291.

———, 1979, Rates of transgression and regression in the Carboniferous of N. W. Europe: Journal of the Geological Society of London, v. 136, p. 147–154.

———, 1981, Eustasy, sea level and local tectonism, with examples from the British Carboniferous: Yorkshire Geological Society, Proceedings, v. 43, p. 473–482.

———, CALVER, M. A., EDGAR, R. M. C., HODSON, F., HOLLIDAY, D. W., STUBBLEFIELD, C. J., AND WILSON, R. B., 1978, A correlation of Silesian rocks in the British Isles: Geological Society of London Special Report 10, 81 p.

ROSS, C. A., 1973, Pennsylvanian and Early Permian depositional history, southeastern Arizona: American Association of Petroleum Geologists Bulletin, v. 57, p. 887–912.

———, 1979, Carboniferous, in Robison, R. A., and Teichert, C., eds., part A; Introduction, Fossilization (Taphonomy), Biogeography and Biostratigraphy: Treastise on Invertebrate Paleontology: Geological Society of America and Kanasa University Press, Boulder, Colorado, and Lawrence, Kansas, p. A254–A290.

———, AND ROSS, J. R. P., 1979, Permian in Robison, R. A. and Teichert, C., eds., part A: Introduction, Fossilizatioin (Taphonomy), Biogeography and Biostratigraphy: Treatise on Invertebrate Paleontology: Geological Society of America and Kansas University Press, Boulder, Colorado, and Lawrence, Kansas, p. A291–A350.

———, AND ———, 1981, Biogeographic influences on Late Paleozoic faunal distributions, in Larwood, G. P., and Nielsen, C., eds., Recent and Fossil Bryozoa: Olsen & Olsen, Fredensborg, Denmark, p. 199–212.

———, AND ———, 1985a, Carboniferous and Early Permian biogeography: Geology, v. 13, p. 27–30.

———, AND ———, 1985b, Late Paleozoic depositional sequences are synchronous and worldwide: Geology, v. 13, p. 194–197.

———, AND ———, 1987a, Late Paleozoic sea levels and depositional sequences, in Ross, C. A., and Haman, D., eds., Timing and Depositional History of Eustatic Sequences: Constraints on Seismic Stratigraphy: Cushman Foundation for Foraminiferal Research, Special Publication 24, p. 137–149.

———, AND ———, 1987b, Biostratigraphic zonation of late Paleozoic depositional sequences, in Ross, C. A., and Haman, D., eds., Timing and Depositional History of Eustatic Sequences: Constraints on Seismic Stratigraphy: Cushman Foundation for Foraminiferal Research, Special Publication 24, p. 151–168.

ROSS, J. R. P., 1978, Biogeography of Permian ectoproct Bryozoa: Paleontology, v. 21, p. 341–356.

———, 1981a, Biogeography of Carboniferous ectoproct Bryozoa: Palaeontology, v. 24, p. 313–341.

———, 1981b, Late Palaeozoic ectoproct biogeography, in Larwood, G. P., and Nielsen, C., eds., Recent and Fossil Bryozoa: Olsen & Olsen, Fredensborg, Denmark, p. 213–219.

———, 1984, Biostratigraphic distribution of Carboniferous Bryozoa, in Sutherland, P. K., and Manger, W. L., eds., Neuvième Congrès International de stratigraphie et de géologie du Carbonifère, Compte Rendu, v. 2, Biostratigraphy: Southern Illinois University Press, Carbondale and Edwardsville, Illinois, p. 19–32.

———, 1985, Biogeography of Ordovician ectoproct (bryozoan) faunas, in Nielsen, C., and Larwood, G. P., eds., Bryozoa: Ordovician to Recent: Olsen and Olsen, Fredensborg, Denmark, p. 265–271.

SAUNDERS, W. B., RAMSBOTTOM, W. H. C., AND MANGER, W. L., 1979, Mesothemic cyclicity in the mid-Carboniferous of the Ozark shelf region?: Geology, v. 7, p. 293–296.

SELLARDS, E. H., ADKINS, W. S., AND PLUMMER, F. B., 1932, The geology of Texas, v. 1, stratigraphy: University of Texas Bulletin 3232, 1007 p.

SESLAVINSKII, K. B., 1979, Ordovician and Silurian climates and global climatic belts: International Geology Review, v. 21, p. 140–152.

SHENG, J. C., 1963, Permian Fusulinids of Kwangsi, Kueichow and Szechuan: Palaeontologica Sinica, whole number 149, new series B, no. 10., 247 p.

SINITSYN, I. I., ed., 1975, Field Excursion Guidebook for the Carboniferous Sections of South Urals (Bashkiria): Ministry of Geology of the RSFSR, Bashkirian Territorial Geological Survey, Eighth International Congress on Carboniferous Stratigraphy and Geology, 183 p.

SLOSS, L. L., 1963, Sequences in the cratonic interior of North America: Geological Society of America Bulletin, v. 74, p. 93–114.

SPJELDNAES, N., 1961, Ordovician climatic zones: Norsk Geologisk TidssKrift, v. 41, p. 45–77.

TARAZ, H., AND 10 OTHERS (Iranian-Japanese Research Group), 1981, The

Permian and the Lower Triassic Systems in Abadeh region, central Iran: Kyoto University, Memoirs of the Faculty of Science, Series of Geology and Mineralogy, v. 47, p. 61–133.

TEMPLETON, J. S., AND WILLMAN, H. B., 1963, Champlainian Series (Middle Ordovician) in Illinois: Illinois State Geological Survey Bulletin, v. 89, 260 p.

VAIL, P. R., AND MITCHUM, R. M., JR., 1977, Seismic stratigraphy and global changes of sea level, part 1: Overview, *in* Payton, C. E., ed., Seismic Stratigraphy—Applications to Hydrocarbon Exploration: American Association of Petroleum Geologists Memoir 26, p. 51–52.

WANLESS, H. R., AND SHEPARD, F. P., 1936, Sea level and climatic changes related to late Paleozoic cycles: Geological Society of America Bulletin, v. 47, p. 1177–1206.

WILLMAN, H. B., 1973, Rock stratigraphy of the Silurian System in north- eastern and northwestern Illinois: Illinois State Geological Survey Circular 479, 55 p.

———, ATHERTON, E., BUSCHBACH, T. C., COLLINSON, C., FRYE, J. C., HOPKINS, M. E., LINEBACK, J. A., AND SIMON, J. A., 1975, Handbook of Illinois stratigraphy: Illinois State Geological Survey Bulletin, v. 95, 261 p.

YABLOKOV, V. S., ed., 1975, Field Excursion Guidebook for the Carboniferous Sequences of the Moscow Basin: Ministry of Geology of the RSFR, Territorial Geological Surgey of Central Regions, Eighth International Congress on Carboniferous Stratigraphy and Geology, 176 p.

YASAMANOV, N. A., 1981, Temperatures of Devonian, Carboniferous, and Permian seas in Transcaucasia and the Ural region: International Geology Review, v. 23, p. 1089–1104.

TRIASSIC SEA-LEVEL CHANGES: EVIDENCE FROM THE CANADIAN ARCTIC ARCHIPELAGO

ASHTON F. EMBRY

Geological Survey of Canada, 3303-33rd Street NW, Calgary, Alberta T2L 2A7

ABSTRACT: Triassic sea-level changes are not well documented because of a scarcity of Triassic marine strata over many of the continental interiors and on passive continental margins. An excellent laboratory for studying Triassic sea-level changes is the Sverdrup Basin, which was a major depocenter in the Canadian Arctic Archipelago from the Carboniferous to early Tertiary. Marine Triassic strata are widespread across the basin and are as thick as 4,000 m.

The established stratigraphic pattern for the Triassic succession consists of thick progradational wedges of deltaic and marine strata, alternating with thin, transgressive, clastic units (T-R cycles). On the basin margins, subaerial unconformities cap the progradational wedges, and over much of the basinal area, submarine unconformities form the cycle boundaries. Nine T-R cycles occur in the basin and are interpreted as having been generated by an interplay of eustatic sea-level change, gradually decaying thermal subsidence, and variable rates of sediment supply and load subsidence. In this model, rapid eustatic sea-level rises coincide with major transgressions that occurred in earliest Griesbachian, earliest Smithian, late Smithian, earliest Anisian, early Ladinian, earliest Carnian, mid-Carnian, earliest Norian, earliest Rhaetian, and earliest Jurassic. Progradation occurred in the intervening time intervals under conditions of slow eustatic sea-level rise, stillstand, and fall.

The long duration of each of the sea-level cycles (about 5 million years) and the apparent lack of Triassic glacial deposits indicate the cycles had a tectono-eustatic origin that relates sea-level changes to changes in the volume of the ocean basins. Sea-level rises are related to episodes of increased rates of seafloor spreading and oceanic volcanism that resulted in reduced oceanic-basin volume. The intervals of sea-level fall occurred when seafloor spreading and associated volcanism were subdued and the ocean basins gradually enlarged due to thermal subsidence.

INTRODUCTION

Vail and others (1977) proposed a curve of third-order, relative sea-level changes for the Jurassic, Cretaceous, and Tertiary and suggested that the changes were due mainly to eustatic sea-level variations. Vail and Todd (1981) went a step farther and presented a eustatic sea-level curve for the Jurassic. These sea-level curves were based on seismic and well data from continental shelves along present passive margins, and the curves naturally begin in latest Triassic when these passive margins were in the continental-rifting phase of formation. Following publication of these curves, no attempt has been made to construct a sea-level curve for the Triassic. In this paper, a eustatic sea-level curve for the Triassic is proposed mainly on the basis of data and interpretations of Triassic strata in the Sverdrup Basin.

The Sverdrup Basin of the Canadian Arctic Islands (Fig. 1) is an excellent laboratory for studying Triassic sea-level changes. The basin is 300 × 1,000 km in size, and it formed by rifting in the Late Paleozoic (Balkwill, 1978). Basin fill consists of Carboniferous to Tertiary strata as thick as 13,000 m. Deformation and uplift occurred in the early Tertiary (Eurekan Orogeny), and the intensity of deformation decreases westward. The eastern portion of the basin is folded and cut by thrust faults, and strata are well exposed in the mountainous terrain. Topographic and structural relief are much lower in the west, where broad anticlines occur on low-lying islands.

Triassic strata as thick as 4,000 m occur over most of the Sverdrup basin (Fig. 1) and are mainly of marine origin. Excellent surface exposures occur in the east, and numerous wells have penetrated Triassic strata in the west. Nonmarine to shallow-marine sandstones dominate the basin margins with marine shale and siltstone predominant in the basin center. The author has recently completed a regional stratigraphic and sedimentologic study of the Triassic strata of the Sverdrup Basin, utilizing both surface and subsurface

data (Fig. 1), and the data and interpretations presented in this paper are based on that work.

T-R cycles.—

The basic building block of the Triassic stratigraphic succession is a transgressive-regressive cycle (here referred to as a T-R cycle) following Johnson and others (1985, p. 568), who defined the term as "sedimentary rocks deposited during the time between one deepening event and the beginning of the next, following one of the same scale." The T-R cycles discussed herein are usually hundreds of meters thick. The stratigraphic makeup of a T-R cycle is illustrated in Figure 2. Along the basin margin, a T-R cycle begins with a thin transgressive unit, commonly a calcareous sandstone or arenaceous limestone, which unconformably overlies strata of the preceding cycle. This unit is overlain by a thick, regressive, progradational succession consisting of shale and siltstone in the lower portion and sandstone in the upper portion. A subaerial unconformity usually caps the sandstone unit. A submarine unconformity or hiatal surface (Frazier, 1974) usually occurs on top of the basal transgressive unit. Basinward, the transgressive unit thins and eventually disappears, and the submarine-unconformity surface forms the base of the cycle. The regressive sandstone unit and its capping subaerial unconformity also disappear basinward, and siltstones commonly form the upper portion of the cycle over much of the basin.

Nine regional T-R cycles are recognized in the Triassic succession of the Sverdrup Basin. All nine cycles are clearly identifiable in numerous sections over the basin, and their regional nature is well established. Figure 3 illustrates the Triassic stratigraphy of the Sverdrup Basin (Embry, 1983a, b; 1984a,b) and the recognized T-R cycles. The biostratigraphic control for the age assignments is derived mainly from ammonites and pelecypods that have been identified by E. T. Tozer of the Geological Survey of Canada. These data are supplemented by palynological data in areas where macrofossils are rare or absent. References to the biostra-

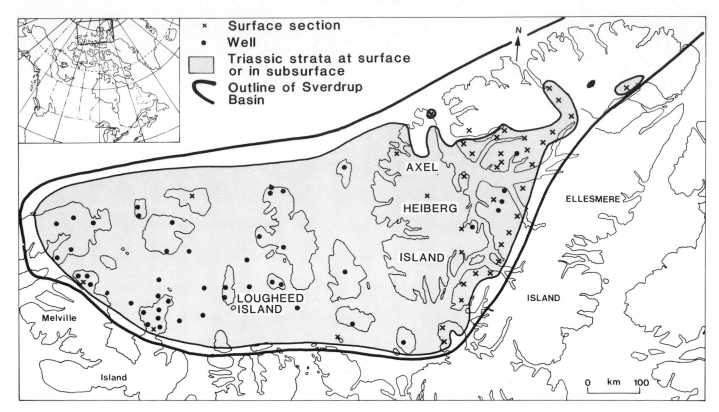

FIG. 1.—Distribution of Triassic strata, Sverdrup Basin and available control points.

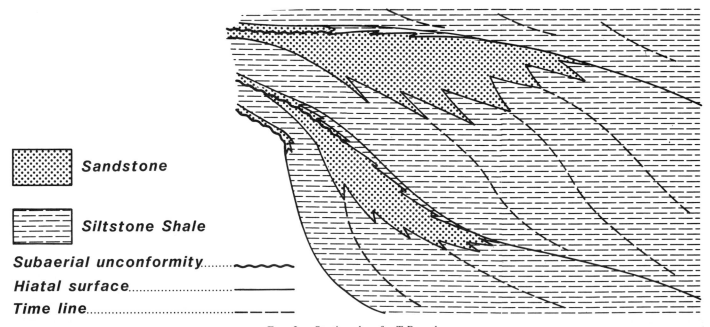

FIG. 2.—Stratigraphy of a T-R cycle.

tigraphic data are found in the stratigraphic papers quoted above.

TRIASSIC T-R CYCLES, SVERDRUP BASIN

Early Triassic.—

Three T-R cycles occur in the Lower Triassic strata of the Sverdrup Basin and their ages are Griesbachian-Di-enerian, Smithian, and late Smithian-Spathian. Lower Triassic strata consist mainly of non-marine to shallow-marine sandstone (Bjorne Formation) on the basin margins and shale and siltstone of marine-shelf to slope origin (Blind Fiord Formation) within the basin (Fig. 4). A unit of marine shale (Blind Fiord Formation) containing earliest Triassic ammonites (*Otoceras*) occurs at the base of the succession both on the margins and within the basin (Fig. 4). These

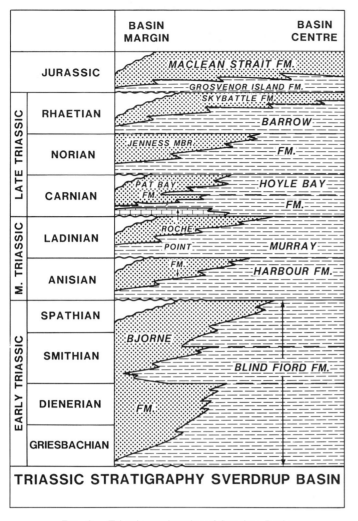

	BASIN MARGIN	BASIN CENTRE
JURASSIC	*MACLEAN STRAIT FM.*	
RHAETIAN	*GROSVENOR ISLAND FM.* *SKYBATTLE FM.*	BARROW
NORIAN	*JENNESS MBR.*	FM.
CARNIAN	*PAT BAY FM.*	HOYLE BAY FM.
LADINIAN	*ROCHE*	
ANISIAN	*POINT FM.*	MURRAY HARBOUR FM.
SPATHIAN		
SMITHIAN	BJORNE	BLIND FIORD FM.
DIENERIAN	FM.	
GRIESBACHIAN		

TRIASSIC STRATIGRAPHY SVERDRUP BASIN

Fig. 3.—Triassic stratigraphy of Sverdrup Basin.

strata unconformably overlie Upper Permian carbonate and sandstone on the basin margin (Fig. 5). The basal shales coarsen upward to sandstones on the margin and to coarse siltstones farther basinward. Griesbachian and Dienerian ammonites occur in these strata.

A second transgression occurred in late Dienerian or earliest Smithian, and Smithian shales overlie Dienerian sandstone or siltstone over much of the basin (Fig. 6). The shales grade upward into coarser clastics at all localities with the height of regression occurring in late Smithian. The third and final regional transgression of the Early Triassic is usually recognizable on the basin margins and is clearly evident in more basinward localities (Fig. 6). Late Smithian ammonites occur in a basal transgressive limestone at one locality, and late Smithian-Spathian fossils occur in the overlying regressive shale and siltstone at numerous localities. The maximum regression appears to have been in latest Spathian.

Middle Triassic.—

Middle Triassic strata contain two regional T-R cycles which are Anisian and Ladinian, respectively. Figure 7 illustrates a subsurface cross section for the Middle Triassic

strata of the western Sverdrup, and the two T-R cycles are apparent. The transgression at the base of the first cycle is well dated by ammonites as early Anisian (Fig. 8). The overlying regressive shales of the Murray Harbour Formation are commonly bituminous and phosphatic and grade upward into shallow-marine sandstone (Eldridge Bay Member, Roche Point Formation) on the basin margins.

Another transgression occurred in late Anisian or early Ladinian, and marine shale and siltstone of the Cape Caledonia Member (Roche Point Formation) unconformably overlie Eldridge Bay sandstones in basin margin sections (Fig. 7). The transgression at the base of the Cape Caledonia Member is not well dated. Overlying siltstone and calcareous sandstones (Chads Point Member) contain Ladinian ammonites, and the transgression is assumed to have occurred near the Anisian-Ladinian boundary. A marked subaerial unconformity commonly caps the Ladinian T-R cycle on the basin margin (Figs. 7, 9), and Middle Triassic strata are overlapped by Carnian strata in some areas.

Late Triassic.—

Four regional T-R cycles have been identified in Upper Triassic strata of the Sverdrup Basin. They are dated as early Carnian, late Carnian, Norian, and Rhaetian. The two Carnian cycles are illustrated in Figure 10, which is a subsurface cross section of the western Sverdrup Basin. The lower Carnian cycle was initiated in earliest Carnian or possibly latest Ladinian by a regional transgression, and a basal limestone (Gore Point Member, Roche Point Formation; Figs. 7, 9) is a distinctive unit on the southern and eastern basin margins. The overlying regressive shales (Eden Bay Member, Hoyle Bay Formation; Figs. 9, 10) coarsen upward into calcareous sandstone of marine-shelf origin (lower Pat Bay Formation; Figs. 9, 11). Early Carnian pelecypods and ammonites occur throughout these strata.

A regional transgression occurred in mid-Carnian, and this was followed by another regressive succession consisting of a lower shale-siltstone unit (Cape Richards Member, Hoyle Bay Formation) and an upper sandstone (Pat Bay Formation; Figs. 9, 11). Late Carnian pelecypods and one ammonite have been recovered from these strata.

The third Upper Triassic T-R cycle began with a major transgression that coincides with the Carnian-Norian boundary. A transgressive oolitic ironstone bed commonly forms the base of the Norian T-R cycle, and Norian strata overlap Carnian strata on the basin margin (Fig. 12). Regressive Norian marine shales (Barrow Formation) coarsen upward into deltaic sandstones (Heiberg Formation) in the eastern Sverdrup (Figs. 12, 13). In the west (Fig. 14), the Barrow shales coarsen upward into shelf sandstones (Jenness Member, Barrow Formation). The strata contain pelecypods and palynomorphs of early to late Norian age (early to latest middle Norian, *sensu* Tozer, 1979).

The last regional Triassic transgression occurred near the Norian-Rhaetian boundary (latest middle Norian, *sensu* Tozer, 1979). In the western Sverdrup, Rhaetian shales overlie Norian sandstones on the northwestern basin margin (Fig. 14), whereas in the east, a Rhaetian shale unit occurs within the sandstone-dominant deltaic deposits (Fig. 15). The Rhaetian shale coarsens upward into deltaic or shelf sandstones (Heiberg Formation, Skybattle Formation; Figs.

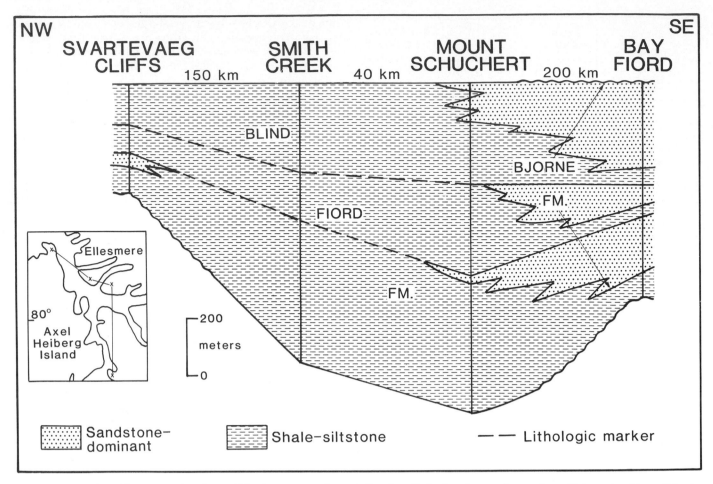

FIG. 4.—Stratigraphic cross section, Lower Triassic strata, northeastern Sverdrup Basin, based on surface sections. Datum—top Lower Triassic. Lithologic markers refer to contacts between thick, resistant siltstone units and overlying thick, recessive shales.

FIG. 5.—Recessive basal Triassic shales containing *Otoceras* unconformably overlying resistant Permian carbonates, northern Ellesmere Island.

FIG. 6.—Three T-R cycles in Blind Fiord Formation: (1) Griesbachian-Dienerian; (2) Smithian; (3) Late Smithian-Spathian. Anisian T-R cycle (4) caps mountain. Arrows indicate cycle boundaries. Otto Fiord area, northern Ellesmere Island.

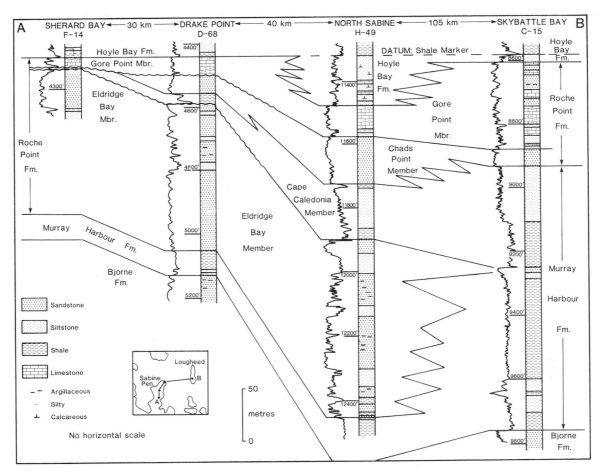

FIG. 7.—Stratigraphic cross section, Middle Triassic and basal Upper Triassic strata, western Sverdrup Basin, based on subsurface data. Two T-R cycles are evident in the Middle Triassic with the contact between them placed at the base of the Cape Caledonia Member. Base Gore Point Member coincides with base Carnian. Gamma ray log shown to left of lithologic column.

14, 15). In the western Sverdrup, the Triassic-Jurassic boundary is marked by a regional transgression and an earliest Jurassic, oolitic ironstone overlies Rhaetian strata (Fig. 14). The contact is a marked unconformity on the basin margin.

Summary.—

Nine regional T-R cycles are recognized in the Triassic strata of the Sverdrup Basin. Each cycle was initiated by a regional transgression; each culminated in marked regression, shoreline progradation, and subaerial exposure of the basin margins. These cycles are directly comparable with the third-order cycles of Vail and others (1977) that characterize the Jurassic-Tertiary stratigraphy of continental shelves along passive margins.

TRIASSIC T-R CYCLES AND EUSTASY

Origin of T-R cycles.—

The origin of the T-R cycles that characterize the Triassic strata of the Sverdrup Basin and the importance of eustatic sea-level changes in their origin are debatable. The simplest interpretation is that the cycles are the product of fluctuating sedimentation rates on a background of regional sub-

sidence and constant sea level. The transgressive portion of the cycle would develop during a time of low rates of sediment supply, when subsidence rate exceeds that of sedimentation. The following regressive portion would occur when the sedimentation rate increased so that it exceeded the subsidence rate. The flaw in this interpretation is that it does not account for subaerial exposure and erosion on the basin margins during regressive phases.

Another interpretation is that the T-R cycles are the result of alternating tectonic uplift and subsidence against a background of relatively constant sediment supply and a stationary sea level. Transgression would occur during times of high subsidence, with regression occurring at times of marginal uplift and lower subsidence rates. This hypothesis explains the observed stratigraphy of the cycles, but it is not favored, for two reasons: (1) The transgressive events recognized in the Sverdrup Basin occur in other basins that developed under different tectonic conditions (Fig. 16). It is highly unlikely that the short-term tectonic events needed for this interpretation would be synchronous over such a wide area. (2) Backstripping analysis of the Sverdrup Basin indicates that the basin was undergoing thermal subsidence during the Triassic (R. Stephenson, pers. commun., 1984). It seems unlikely that numerous short-term tectonic movements occurred in such a tectonic environment.

FIG. 8.—Middle Triassic Murray Harbour Formation overlying Lower Triassic Bjorne Formation (arrow points to contact). Thin transgressive sandstone (1) occurs at the base of the Murray Harbour Formation, which consists mainly of dark grey shale. Northern Ellesmere Island.

FIG. 9.—Thin, resistant Gore Point limestone (Carnian) unconformably overlying siltstone of Murray Harbour Formation (Anisian; lower arrow). A hiatal surface occurs at the top of the limestone and is followed by regressive shale of the Hoyle Bay Formation (early Carnian; upper arrow). Northern Ellesmere Island.

The favored interpretation for the origin of the T-R cycles is that they originated from eustatic sea-level rises and falls on a background of thermal subsidence and fluctuating sediment supply. In this scenario, transgression would occur when combined rates of sea-level rise and subsidence exceeded the rate of sedimentation. Regressive conditions would develop when the rate of sedimentation plus the rate of sea-level fall (or minus the rate of sea-level rise) exceeded the rate of subsidence. This interpretation explains the observed stratigraphic relations, is compatible with the established tectonic environment, and predicts the occurrence of near-synchronous transgressive events in other basins.

Eustatic sea-level changes.—

If geologists accept the major role that eustatic sea-level changes play in the origin of T-R cycles, the next problem is to determine a reasonable eustatic sea-level curve from the succession of T-R cycles established in the Sverdrup Basin. The rate, timing, and magnitude of sea-level changes can all be estimated from the succession of T-R cycles established in the Sverdrup Basin as described here.

The shape of the eustatic curve in part reflects the rate of sea-level change, and five possible shapes have been discussed by Hallam (1978). These shapes plus a sixth alternative are illustrated in Figure 17. The stratigraphic pattern of a typical T-R cycle in the Sverdrup Basin allows a reasonable choice to be made from these six options.

The first shape, which does not include a sea-level fall, can be readily eliminated because the T-R cycles are capped by subaerial unconformities on the basin margins. The second curve is sinusoidal and has been adopted by Exxon in their recent interpretations of eustatic sea-level variations (Vail, pers. commun., 1985). The stratigraphic relationships of the Triassic T-R cycles are not, however, compatible with this shape. In a sinusoidal model, the time of maximum seaward extent of the subaerial unconformity occurs halfway through the interval of eustatic sea-level fall (the inflection point at which the rate of sea-level fall is at a maximum). During the subsequent time of slow fall and slow rise, when regression is still continuing, non-marine strata gradually onlap the subaerial unconformity. This results in wedge of non-marine strata between the subaerial unconformity and the overlying transgressive marine strata. Such a wedge of onlapping non-marine strata does not occur in any of the T-R cycles, so a sinusoidal shape is rejected. The third shape, which is characterized by a slow rise and a fast fall, is also eliminated on the basis of the absence of the onlapping non-marine wedge.

The last three shapes include a rapid sea-level rise, an interpretation favored by the thin transgressive unit that occurs at the base of each T-R cycle. These transgressive strata are interpreted as representing a relatively short interval of time compared with the overlying regressive portion of the cycle, because most of the ammonite zones of the cycle occur in the regressive strata.

The fourth shape, which has a long stillstand between a rapid rise and rapid fall, is rejected because a rapid fall would result in a sudden, short-term regression and in deeply incised channels on the exposed basin margins. Neither of

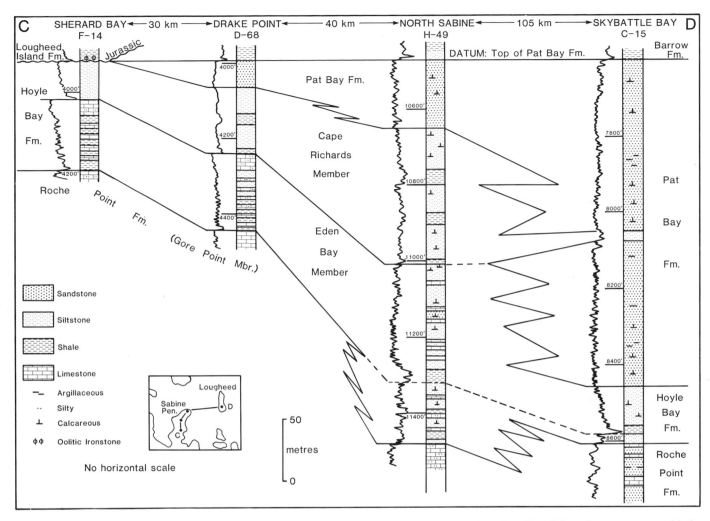

FIG. 10.—Stratigraphic cross section of Carnian strata, western Sverdrup Basin, based on subsurface data. Two T-R cycles are apparent with the contact between them placed at the base of the Cape Richards Member. For basal contact of lower cycle, see Figure 7. Gamma ray log shown to left of lithologic column.

these features occurs in the observed T-R cycles. The fifth shape has a rapid rise followed by a slow fall, whereas the sixth shape is more symmetrical and is characterized by a rapid rise followed by slow rise, stillstand, slow fall, and fast fall. Both of these shapes are compatible with the observed T-R cycles, with transgression occurring only during the interval of rapid rise and regression occurring during the remainder of the cycle. The sixth shape is favored over the fifth because it allows for the occurrence of onlapping fluvial strata on the basin margins during most of the cycle (the exception being the time of fast fall). Thick intervals of onlapping fluvial strata do not occur in the Sverdrup Basin but are common in other basins (Vail and others, 1977).

The timing of the reversals of sea-level movements can be estimated from paleontologic data. The age of the initiation of a sea-level rise for a given cycle is bracketed by the youngest fossil found in the regressive strata of the underlying cycle and the oldest fossil in the basal transgressive strata. The timing of the initiation of sea-level fall cannot be estimated with any confidence, because it occurs within the regressive succession and is not marked by any

notable lithologic change. Its placement is based on the symmetry of the curve shape. As was discussed earlier, the sea-level rises throughout the Triassic are reasonably well dated by macrofossils.

The absolute magnitudes of the recognized sea-level changes are impossible to determine, because the extents of regression and transgression are greatly influenced by both tectonics and sedimentation. The relative magnitudes of the sea-level falls have been estimated by comparing the seaward extent of the subaerial unconformities that cap the T-R cycles. The relative magnitudes of the sea-level rises have been estimated by comparing the water depth interpretations for shales directly overlying the basal transgressive units on the basin margins.

Using the above interpreted relations, a sea-level curve has been drawn for the Triassic (Fig. 18). Nine cycles of rapid rise followed by slow rise, stillstand, slow fall, and rapid fall are postulated for the Triassic. Rapid rises occurred in earliest Griesbachian, earliest Smithian, late Smithian, early Anisian, early Ladinian, earliest Carnian, mid-Carnian, earliest Norian, earliest Rhaetian, and earliest Ju-

FIG. 11.—Light weathering Pat Bay sandstones overlying dark Hoyle Bay shales and overlain by dark Barrow shales (Norian). Mid-Carnian transgressive beds at arrow. Northern Ellesmere Island.

FIG. 13.—Dark, recessive shale of Barrow Formation (Norian) overlying Pat Bay sandstone (Carnian) and coarsening upward into siltstone. Northern Ellesmere Island.

FIG. 12.—Barrow Formation unconformably overlying Carboniferous carbonates (arrow at contact). Light weathering Heiberg sandstones transitionally overlie Barrow Formation. Northern Ellesmere Island.

rassic. The duration of each cycle cannot be closely determined because the absolute ages of the Triassic stage boundaries are poorly known. The cycles are estimated to have been between 4 and 6 million years in duration on the basis of nine cycles for the Triassic, which is estimated to be between 40 and 45 million years long. Thus, these cycles are comparable in length to the third-order cycles recog-

nized by Vail and others (1977) in the Jurassic, Cretaceous, and Tertiary.

Four longer term cycles (second-order?) characterized by major sea-level rises and a significant change in sedimentation pattern are also evident. These cycles are Early Triassic, Middle Triassic, early Late Triassic (Carnian), and late Late Triassic (Norian-Rhaetian). Accepting the 4- to 6-million-year duration for the third-order cycles, these cycles would have been 8 to 12 million years in length.

Origin of the eustatic sea-level-changes.—

The long duration of the sea-level cycles and the apparent absence of Triassic glacial deposits suggest that the cycles are tectono-eustatic in origin and are related to the changing volume of the ocean basins. Following Hallam (1978), the rapid sea-level rises are interpreted as reflecting episodes of high rates of seafloor spreading and oceanic volcanism, which would result in a smaller ocean basin volume. The long intervals of slow rise, stillstand, and fall reflect the times when seafloor spreading was relatively slow and the ocean basin volume decreased slowly or gradually increased due to thermal subsidence.

CONCLUSIONS

Nine third-order eustatic sea-level cycles, each characterized by a rapid sea-level rise followed by slow rise, stillstand, slow fall, and rapid fall, and with a duration of 4 to 6 million years, occurred during the Triassic. Rapid sea-level rises occurred in earliest Griesbachian, earliest Smithian, late Smithian, early Anisian, early Ladinian, earliest Carnian, middle Carnian, earliest Norian, earliest Rhaetian, and earliest Jurassic. The sea-level rises with the greatest

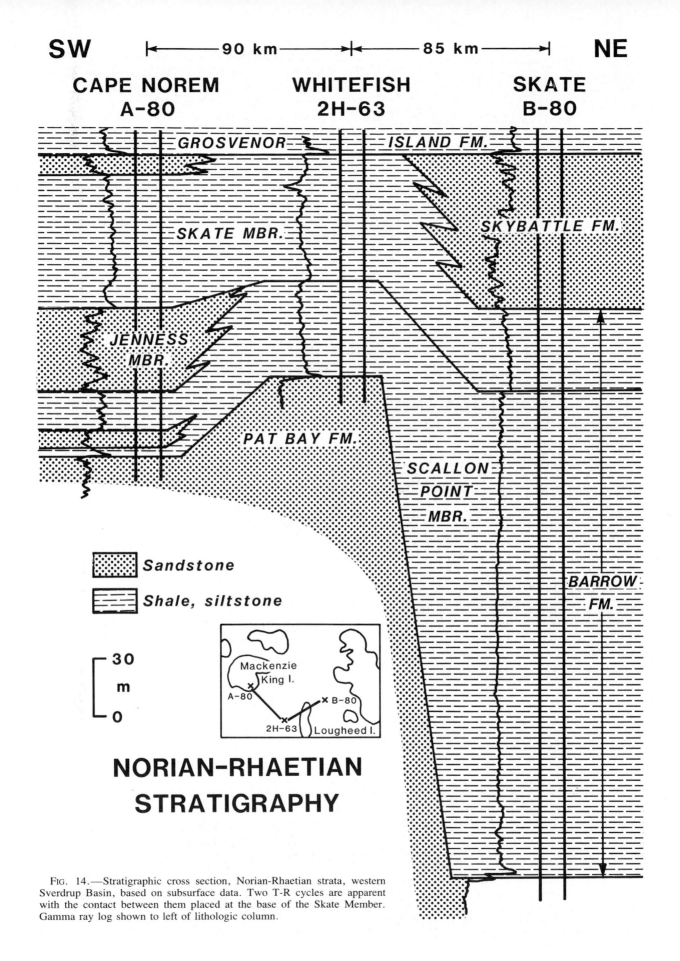

FIG. 14.—Stratigraphic cross section, Norian-Rhaetian strata, western Sverdrup Basin, based on subsurface data. Two T-R cycles are apparent with the contact between them placed at the base of the Skate Member. Gamma ray log shown to left of lithologic column.

magnitude occurred in earliest Griesbachian, early Anisian, earliest Carnian, earliest Norian, and earliest Jurassic. These episodes mark times of major changes in sedimentation patterns and define four second-order cycles for the Triassic. The sea-level cycles are interpreted as tectono-eustatic in origin, reflecting changes in the volume of the ocean basins. Worldwide stratigraphic, sedimentologic, and paleontologic data for the Triassic are still relatively sparse in comparison to those for the Jurassic, Cretaceous, and Tertiary. Significantly more data are needed to refine, or perhaps to reject, this initial attempt at a Triassic eustatic sea-level curve.

ACKNOWLEDGMENTS

I am grateful to the Geological Survey of Canada for encouraging this study and allowing publication of the results. The Polar Continental Shelf Project provided much logistical support during the field seasons. Discussions with Jim Dixon and Peter Vail on the origin of depositional sequences have been very beneficial. Jim Dixon, Kevin Biddle, and Christopher Kendall critically read the manuscript, and their suggestions have improved this paper. Claudia Thompson typed the manuscript with speed and accuracy, and Elspeth Snow kindly drafted the figures.

FIG. 15.—Rhaetian shale unit (about 120 m thick) overlying Norian sandstones and overlain by Rhaetian sandstones, Heiberg Formation, southern Ellesmere Island.

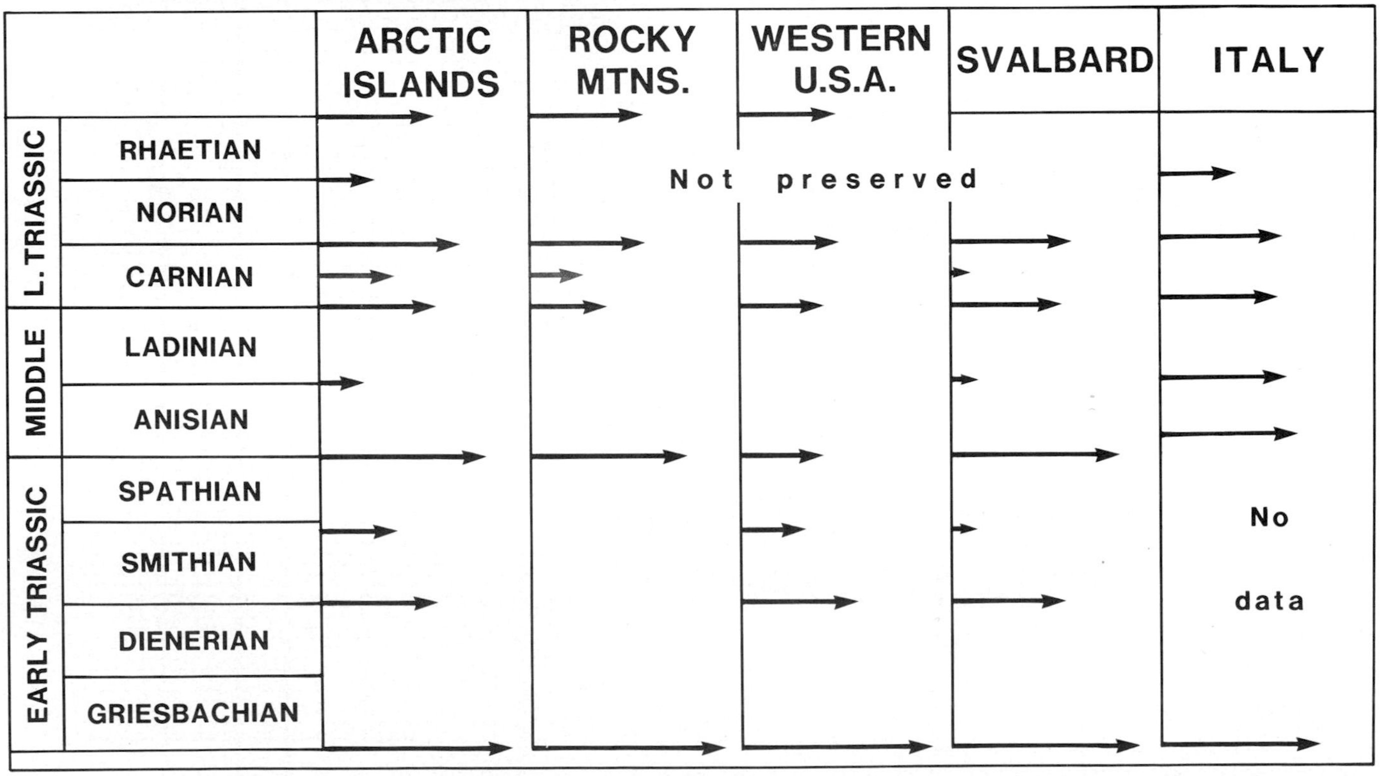

		ARCTIC ISLANDS	ROCKY MTNS.	WESTERN U.S.A.	SVALBARD	ITALY
L. TRIASSIC	RHAETIAN			Not preserved		
	NORIAN					
	CARNIAN					
MIDDLE	LADINIAN					
	ANISIAN					
EARLY TRIASSIC	SPATHIAN					No data
	SMITHIAN					
	DIENERIAN					
	GRIESBACHIAN					

FIG. 16.—Comparison of Triassic transgressions identified in the Canadian Arctic with Triassic transgressions in the Canadian Rocky Mountains (Gibson, 1975), western United States (Silberling and Wallace, 1969; Carr and Paull, 1983), Svalbard (Pchelina, 1977; Mørk and others, 1982), and Italy (Bosellini and Rossi, 1974). The length of the arrows reflect the interpreted relative magnitudes of the transgressions for each locality.

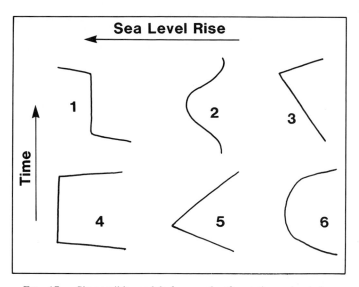

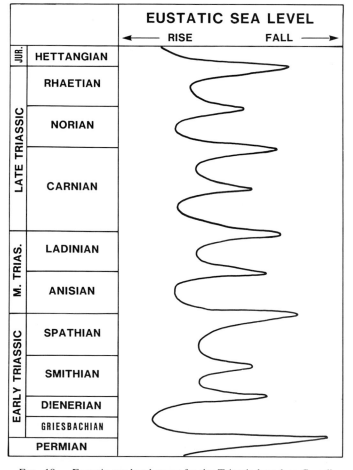

FIG. 17.—Six possible models for a cycle of eustatic sea-level change (modified from Hallam, 1978): (1) rapid rise, slow rise, long stillstand, rapid rise; (2) sinusoidal with maximum rate of rise and fall at the inflection points; (3) slow rise, rapid fall; (4) rapid rise, long stillstand, rapid fall; (5) rapid rise, slow fall; (6) rapid rise, slow rise, stillstand, slow fall, rapid fall.

FIG. 18.—Eustatic sea-level curve for the Triassic based on Canadian Arctic T-R cycles and the proposed relationship between T-R cycles and eustatic sea-level changes. Magnitude of changes are relative.

REFERENCES

BALKWILL, H. R., 1978, Evolution of Sverdrup Basin, Arctic Canada: American Association of Petroleum Geologists Bulletin, v. 62, p. 1004–1028.

BOSSELLINI, A., AND ROSSI, D., 1974, Triassic carbonate buildups of the Dolomites, northern Italy, *in* Laporte, L. F., ed., Reefs in Time and Space: Society of Economic Paleontologists and Mineralogists Special Publication 18, p. 209–233.

CARR, T. R., AND PAULL, R. K., 1983, Early Triassic stratigraphy and paleogeography of the Cordilleran miogeocline, *in* Reynolds, M. W., and Dolly, E. D., eds., Mesozoic Paleogeography of the West-Central United States: Rocky Mountain Section, Society of Economic Paleontologists and Mineralogists, p. 39–55.

EMBRY, A. F., 1983a, The Heiberg Group, western Sverdrup Basin, Arctic Islands: Geological Survey of Canada Paper 83-1B, p. 381–389.

———, 1983b, Stratigraphic subdivision of the Heiberg Formation, eastern and central Sverdrup Basin, Arctic Islands: Geological Survey of Canada Paper 83-1B, p. 205–213.

———, 1984a, The Schei Point and Blaa Mountain Groups (Middle-Upper Triassic), Sverdrup Basin, Canadian Arctic Archipelago: Geological Survey of Canada Paper 84-1B, p. 327–336.

———, 1984b, Stratigraphic subdivision of the Roche Point, Hoyle Bay and Barrow Formations (Schei Point Group), western Sverdrup Basin, Arctic Islands: Geological Survey of Canada Paper 84-1B, p. 275–283.

FRAZIER, D. E., 1974, Depositional Episodes: Their Relationship to the Quaternary Stratigraphic Framework in the Northwest Portion of the Gulf Basin: University of Texas Bureau of Economic Geology, Geological Circular 74-1, 28 p.

GIBSON, D. W., 1975, Triassic Rocks of the Rocky Mountain Foothills and Front Ranges of Northeastern British Columbia and West-Central Alberta: Geological Survey of Canada, Bulletin 247, 61 p.

HALLAM, A., 1978, Eustatic cycles in the Jurassic: Palaeogeography, Palaeoclimatology, Palaeoecology, v. 27, p. 1–32.

JOHNSON, J. G., KLAPPER, G., AND SANDBERG, C. A., 1985, Devonian eustatic fluctuations in Euramerica: Geological Society of America Bulletin, v. 99, p. 567-587.

MØRK, A., KNARUD, R., AND WORSLEY, D., 1982, Depositional and diagenetic environments of the Triassic and Lower Jurassic of Svalbard, *in* Embry, A. F., and Balkwill, H. R., eds., Arctic Geology and Geophysics: Canadian Society of Petroleum Geologists Memoir 8, p. 371–398.

PCHELINA, T. M., 1977, Stratigraphy and some characteristics of the composition of the Mesozoic sediments in the southern and eastern regions of West Spitsbergen, *in* Sokolov, V. N., ed., Stratigraphy of Spitsbergen: The British Library, Lending Division, Wetherby, Yorkshire, England, p. 164–205.

SILBERLING N. J., AND WALLACE, R. E., 1969, Stratigraphy of the Star Peak Group (Triassic) and Overlying Lower Mesozoic Rocks, Humboldt Range, Nevada: U.S. Geological Survey Professional Paper 592, 50 p.

TOZER, E. T., 1979, Latest Triassic ammonoid fauna and biochronology, western Canada: Geological Survey of Canada Paper 79-1B, p. 127–135.

VAIL P. R., MITCHUM, R. M., JR., AND THOMPSON, S., III, 1977, Seismic stratigraphy and global changes of sea-level, part 4: Global cycles of relative changes of sea-level, *in* Payton, C. E., ed., Seismic Stratigraphy, Applications to Hydrocarbon Exploration: American Association of Petroleum Geologists Memoir 26, p. 83–97.

———, AND TODD, R. G., 1981, Northern North Sea Jurassic unconformities, chronostratigraphy and sea-level changes from seismic stratigraphy, *in* Illing, L. B., and Hobson, G. D., eds., Petroleum Geology of the Continental Shelf of Northwest Europe: Heyden, London, p. 216–235.

A REEVALUATION OF JURASSIC EUSTASY IN THE LIGHT OF NEW DATA AND THE REVISED EXXON CURVE

A. HALLAM

School of Earth Sciences, University of Birmingham, P.O. Box 363, Birmingham B15 2TT

ABSTRACT: A comparison is made between the revised Exxon eustatic curve for the Jurassic, based essentially on seismic stratigraphic analysis of North Sea data, and a new curve derived from more conventional stratigraphic analysis. The two curves are broadly similar in that a secular rise of sea level through most of the period is indicated on which about 17 shorter term cycles are superimposed. Both record notable rises in the Sinemurian, Toarcian, Bajocian, Callovian, Oxfordian, and Kimmeridgian. The Exxon curve, however, misses the important event across the Triassic-Jurassic boundary and underestimates the rate of rise in sea level for a number of cycles. In addition, some supposed eustatic events can be discounted as the consequence of regional tectonics. Tectonic activity involving subsidence and uplift, rather than geoid changes, is thought to be the principal cause of regional distortions of the global picture. There is a need for better quantitative data on the amplitude and rate of changes in sea level.

INTRODUCTION

The application of seismic stratigraphy to the study of eustasy by the Exxon group has stimulated an enormous amount of interest, as reflected, for instance, by the intense discussion provoked by their earliest published work (Vail and others, 1977). Much less widely appreciated is the evolution that has taken place in the group's ideas since publication of the 1977 memoir. Thus, the familiar and much cited "sawtooth" eustatic curve, with gradual rises of sea level alternating with sharp, geologically "instantaneous" falls, is now interpreted as reflecting coastal onlap of sediments rather than changes in sea level, although, of course, a correlation exists between the two phenomena. In addition, more attention is now paid to different types of sequence boundary and to the concept of sedimentary "accommodation" that changes through a eustatic cycle. These developments in interpretation are well illustrated by the Jurassic cycles, and have resulted in the production of a new eustatic curve (Vail and others, 1984). The more recent curve published by Haq and others (1987) differs only marginally, and subsequent reference to the Exxon group's work will be confined to discussion of the 1984 paper.

The seismic stratigraphic approach revived my own interest in Jurassic eustasy and led me to adopt an alternative approach, on the basis of more conventional stratigraphic methods, to produce tentative eustatic curves (Hallam, 1978, 1981). These showed a number of similarities to the original Exxon curve of Vail and others (1977) but also some important differences. The present study is a response, on the basis of new data, to the revised Exxon curve. Because of the difficulty, if not impossibility, of directly checking the raw data utilized in producing the Exxon curve, it is imperative to test the curve by reference to independent sources of information and techniques. Agreement between the curves may serve to strengthen confidence in the different methods, whereas disagreement should focus attention on points of difference and stimulate attempts to resolve them. Only by following such a dialectic process will geologists reach a satisfactory consensus.

THE REVISED EXXON CURVE

As a result of a new seismic stratigraphic analysis of North Sea data, Vail and Todd (1981) produced a revised Jurassic eustatic curve that clearly implies an assumption that what is valid for one region is valid for the whole world. Although it was acknowledged in the text that patterns of coastal onlap did not accurately reflect the pattern of changes in sea-level, the eustatic curve illustrated retained the sawtooth shape of earlier sea-level curves. A more recent paper (Vail and others, 1984) however, shows a sinusoidal eustatic curve together with the coastal onlap curve on which it is based (Fig. 1). Despite the claim by Vail and others (1984) that data from different parts of the world have been used, the coastal-onlap curve is virtually identical with that of Vail and Todd (1981) based exclusively on North Sea data.

Vail and others (1984) have refined their analysis of sequence boundaries by distinguishing two types of unconformity. Type 1 unconformities are characterized by downward shift of coastal onlap commonly below the shelf edge, with associated valley entrenchment, canyon cutting, submarine fans, and lowstand deltas. They are held to signify a rate of sea-level fall greater than the rate of tectonic subsidence at the shelf edge. Type 2 unconformities are characterized by downward shift of coastal onlap to positions landward of the shelf edge, associated with subaerial exposure only in landward parts of the shelf, and no canyon cutting. Submarine condensed sections are seen as signifying, commonly at maximum water depth, a rate of sea-level rise significantly greater than the rate of sediment accumulation. Downlap surfaces associated with condensed sections mark a change from the end of transgression to the start of regression, as the rate of sea-level rise decreases and sediments start to prograde. The almost continuous presence of coastal onlap means that most sediments are deposited during relative rise of sea level. Either a rise in sea level or subsidence of the depositional surface, or both together, creates vertical space for the *accommodation* of sediments. In general, the maximum point of regression occurs after the downward shift of coastal onlap.

These criteria may be applied to the coastal onlap curve obtained from seismic stratigraphic analysis to generate a sea-level curve (Fig. 1). Vail and others (1984) recognize 17 global unconformities and correlative conformities (= sequence boundaries) for the Jurassic and earliest Cretaceous, equivalent to 16 eustatic cycles. Eight type 1 and nine type 2 unconformities are distinguished, together with

Sea-Level Changes—An Integrated Approach, SEPM Special Publication No. 42

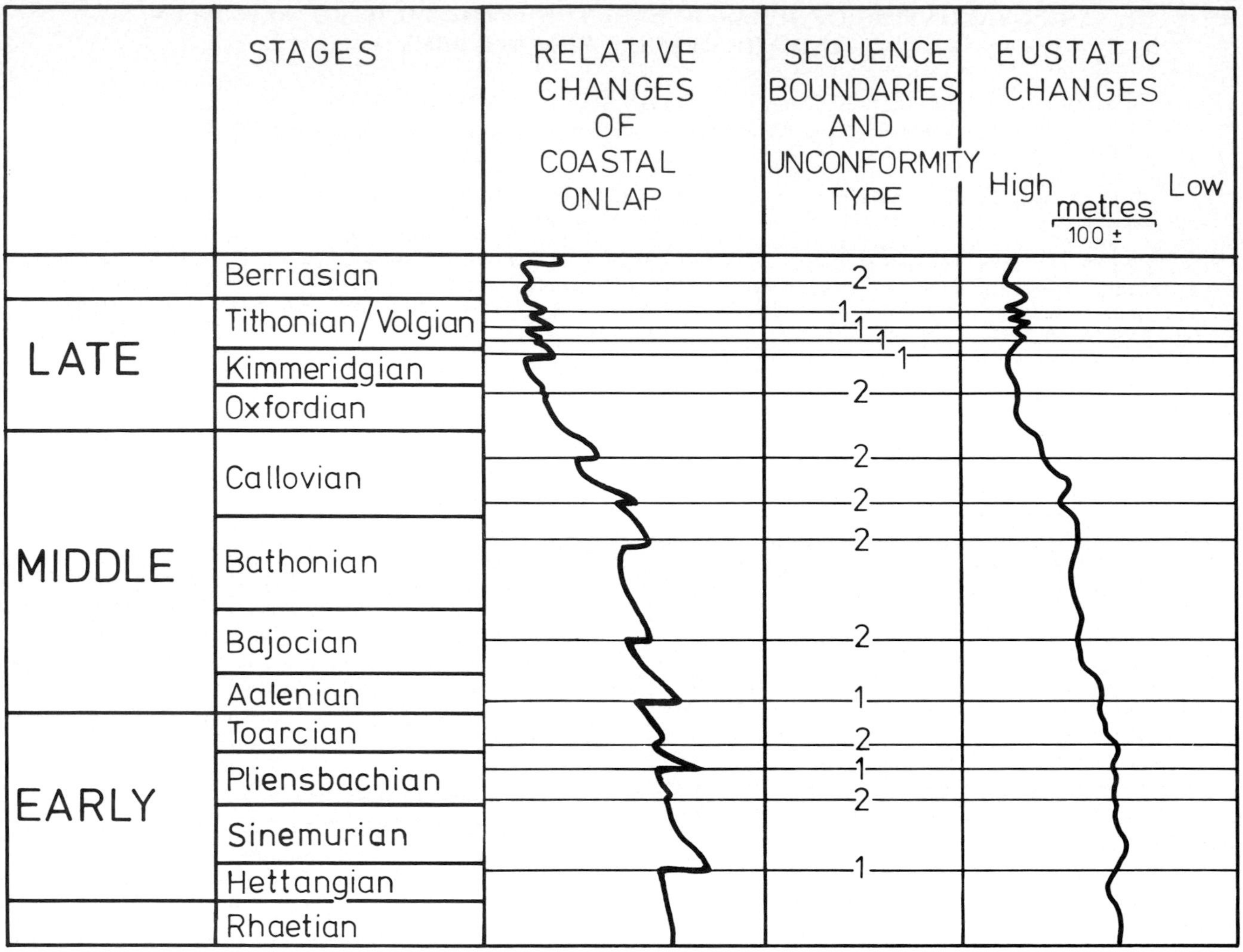

	STAGES	RELATIVE CHANGES OF COASTAL ONLAP	SEQUENCE BOUNDARIES AND UNCONFORMITY TYPE	EUSTATIC CHANGES High Low metres 100 ±
LATE	Berriasian		2	
	Tithonian/Volgian		1 1 1 1	
	Kimmeridgian			
	Oxfordian		2	
MIDDLE	Callovian		2 2	
	Bathonian		2	
	Bajocian		2	
	Aalenian		1	
EARLY	Toarcian		2	
	Pliensbachian		1 2	
	Sinemurian			
	Hettangian		1	
	Rhaetian			

FIG. 1.—The coastal-onlap and sea-level curves for the Jurassic, simplified from Vail and others, 1984.

16 marine condensed sections. Potentially complicating factors of tectonic subsidence and sediment load and supply may be handled by geohistory analysis (Van Hinte, 1978). Tectonic subsidence along passive margins is believed to be long term, decreasing gradually with time, and does not change rapidly enough to cause global unconformities.

There can be no doubt of the importance of this new contribution of the Exxon group, but it is nevertheless open to a number of criticisms.

(1) Perhaps the most important criticism is the assumption that analysis of any one region can give the definitive picture of sequence boundaries for the whole world, and that regional tectonics can be discounted. Not all such tectonics in areas such as the North Sea fall into the category of steady long-term subsidence. Tilted fault-block taphrogenic tectonics and associated sedimentation, involving oscillations of the sort well described for East Greenland by Surlyk (1978), were widespread throughout northwest Europe during the early Mesozoic. This is especially true of the North Sea region, and the exceptionally "noisy" part of the Exxon eustatic curve for the Kimmeridgian-Tithonian interval (Fig. 1) could be the result of such tectonics. The

stratal sequence and geological history of the North Sea have many close resemblances to those of contemporary rocks and events in East Greenland, which was situated much closer to Europe before the opening of the North Atlantic. In both areas, there is evidence of a pronounced increase in fault-controlled tectonic activity toward the end of the Jurassic (Hallam and Sellwood, 1976, Surlyk and others, 1981).

(2) The biostratigraphic data on which the Exxon curve is based are presumably palynological. These data have not been published and presumably they never will be: hence, they cannot be evaluated by independent workers. Whereas there is no reason to doubt the general reliability of the palynological ages, the fine precision of the dating is questionable: dinoflagellates are rarely sufficient to establish resolution at substage level, let alone ammonite zones. For this reason, Vail and his co-workers have recently attempted to recognize their events in classic onshore sections e.g., in Dorset, where the ammonite zonation is well established) in order to achieve a better stratigraphic calibration. Dorset, however, cannot be established as the standard section for eustatic analysis, because tectonic effects

involving fault-controlled, nonperiodic subsidence have been superimposed (Sellwood and Jenkyns, 1975; Hallam and Sellwood, 1976). Some apparently eustatic signals come through clearly, while others are obscured because of either tectonic overprinting or lack of diagnostic facies.

(3) The sedimentation model adopted by the Exxon group is heavily influenced by Gulf Coast geology, with siliciclastics in plentiful supply. In much of the classic Jurassic of western Europe, which must be adequately accommodated in any eustatic scheme, such a model is inapposite. In many cases, sedimentation rates were extremely slow, with little influx of sand; regressions may be marked by sediment starvation, as is the case with sedimentary cycles capped by bored and encrusted hardgrounds (Hallam, 1978; Bayer and others, 1985; Einsele 1985; Gabilly and others, 1985).

(4) It is inappropriate to use the term "shelf" for Jurassic intracratonic regions within western Europe, which were distant from any contemporary ocean. These regions are divided into swells and basins, so the unconformity definitions must be amended correspondingly if they are to remain meaningful. Type 1 unconformities, for instance, would be marked by extending a significant and extensive erosional hiatus basinward. Type 2 unconformities would be confined to basin margins. Despite the claim by Vail and others (1984) that no fewer than eight type 1 unconformities can be recognized, I know of only one example where a convincing case can be made for such an unconformity across western Europe (in the mid-Sinemurian, and it does not correspond with any of the eight that Vail and others, 1984, list. Furthermore, marine condensed intervals do not occur where they should in such classic Jurassic basinal developments as Lower Saxony, West Germany, or Mochras, Wales.

Such problems do not, of course, invalidate the Exxon curve, but they do emphasize the need to test the curve by independent analysis using other data, and to disentangle eustatic "signals" from tectonic "noise."

TOWARD A NEW JURASSIC EUSTATIC CURVE

The methods used to produce my alternative Jurassic curve are fully explained elsewhere (Hallam, 1978) and need not be repeated at length here. The gross secular trends were established by a stage-by-stage analysis of the areal spread of seas over the continents as inferred from paleogeographic studies. The shorter term cyclic changes were inferred from the recognition of extensive shallowing and deepening events in epicontinental marine sequences, together with correlative transgressions and regressions, as established by facies analysis. Because of the much better data base, I concentrated my attention on Europe rather than other continents; however, major sea-level changes were not considered to be established unless events could be traced widely across the world, independent of what appear to be local or regional tectonic events. Ammonite data were used because, in general, ammonites allow intracontinental correlation at the zonal or subzonal level and intercontinental correlation at the substage level (Hallam, 1978).

Figure 2 illustrates the essence of the method by which changes in sea level are inferred from facies changes in shallow-water Jurassic calcareous and siliciclastic epicontinental sequences exposed onshore. A shallowing-upward siliciclastic sequence could be due to increase in sediment supply as a result of delta progradation rather than a fall in sea level or regional uplift. If, however, the shallowing event correlates with a corresponding event in a calcareous sequence, as shown in Figure 2, then the sediment-supply alternative can be eliminated. The model attempts to incorporate the concept of changing accommodation as developed by the Exxon group. As a result of strongly varying sedimentation rates, with the early stages of a transgressive or sea-deepening event being marked by condensed deposits signifying low-sedimentation rates, the sea-level curve inferred from the sedimentary sequences will be cuspate, as shown in Figure 2. I adopted such a cuspate form in my 1978 curve, but following the changes adopted by Vail and his colleagues, the eustatic curve is portrayed more accu-

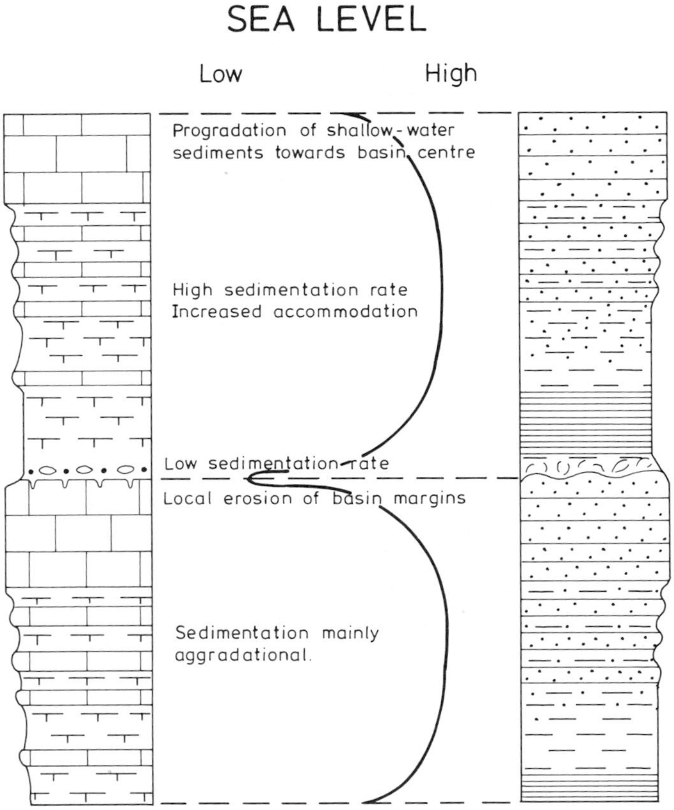

FIG. 2.—Derivation of sea-level curve from facies sequence data. The left-hand column signifies a shallowing-upward marly and calcareous sequence, the right-hand column a shallowing-upward siliciclastic sequence, with conventional lithological symbols. The bases of the sedimentary cycles are characterized by condensed beds, shown as a concentration of eroded pebbles and phosphatic nodules on the left, and a shell concentration on the right; these are interchangeable or may be mixed. Similarly, the laminated (and organic-rich) shale near the base of the right-hand cyclic sequence could also be portrayed in an equivalent position in the left-hand sequence. As the amount of accommodation decreases with falling sea level, coarse siliciclastics tend to prograde toward the basin center, as indicated in the diagram. Within a regime of calcareous sedimentation, the response may be one of sediment starvation, with the development of one or more hardgrounds by submarine and/or subaerial cementation.

rately as sinusoidal, with less contrast between high and low stands of sea level.

It has been argued that shallowing events need not necessarily involve a fall in sea level but could instead signal a condition of stillstand or reduction of rate of rise in sea level (Talbot, 1973; Vail and others, 1977; Pitman, 1978). Whereas this could be true of some "Gulf Coast" sedimentation situations with high-siliciclastic influx into marine basins, it is not the case for the Jurassic of much of Europe, as noted in the previous section. Consider, for instance, the classic Dorset section. Figure 3 portrays three examples where erosional events can be inferred unequivocally and which are most reasonably explained by a fall of sea level or regional uplift (see additional discussion by Einsele, 1985). Although he inferred eustatic control of sedimentation in the Oxfordian of Dorset, Talbot (1973) failed to recognize the erosional truncation of calcareous concretions shown in Figure 3C, which argues against his conclusion of no fall in sea level.

In order to recognize genuine eustatic signals, Hallam (1984) proposed criteria for distinguishing tectonically induced *local* or *regional* subsidence or uplift. Local events, those affecting thousands to tens of thousands of square kilometers, are relatively easy to distinguish by, for example, angular discordances and abrupt facies changes. The identification of regional events, those affecting hundreds of thousands or more square kilometers, is usually a more subtle matter, particularly if biostratigraphic correlation is imprecise. A particular problem arises because of the hierarchical development of sedimentary cycles of the kind portrayed in Figure 2 (cf. Bayer and others, 1985). The thicker and more fully developed a given stratal succession, the more it tends to contain such cycles, corresponding to ammonite "chrons" (either zones or subzones). Generally speaking, the finer the stratigraphic subdivision, the less pronounced the facies changes and less extensive the inferred marine shallowing and deepening events, which are consequently more difficult to attribute to eustasy as opposed to regional tectonics.

In the following discussion, events are considered to be eustatically controlled only if they can be correlated globally as well as extensively within Europe, the best-studied continent. Inevitably, some events are better defined than others, and the stratigraphic precision on a global scale is normally no better than substage.

Most of the relevant information is already documented in extensive reference lists (Hallam, 1978, 1981). More information on Jurassic eustasy can be gleaned from more recently published papers:

Europe–Callomon (1979); Corbin (1980); Enay (1980); Winterer and Bosellini (1981); Loughman (1982); Bayer and others (1985); Brandt (1985); Einsele (1985); Gabilly and others (1985); McGhee and Bayer (1985); Pedersen (1985); Phelps (1985); Jenkyns and Clayton (1986); Mertmann (1986); Morton (1987).

Africa and Asia–Adams (1979); Walley (1983, 1985); Wang and Sun (1985); Livnat and others (1986); Hallam and Maynard (1987).

North America–Surlyk and others (1981); Surlyk and Clemmensen (1983); Embry (1982); Balkwill and others (1983); Taylor and others (1983); Cameron and Tipper (1985).

South America–Riccardi (1983); Gröschke and von Hillebrandt (1985); Hallam and others (1986).

The most significant advance in regional geology is in the Sverdrup Basin of Arctic Canada, where a series of major sedimentary cycles in siliciclastic facies of "Gulf Coast" type, and correlative transgressions and regressions, are recognized in the Mesozoic succession (Embry, 1982; Balkwill and others, 1983). Another notable advance is the recognition by Callomon (1984) that many North American deposits previously assigned to the Callovian belong, in fact, to the Bathonian; the precise location of the Bathonian-Callovian boundary in South America is not yet settled.

Sea-level rises are recorded by deepening events in marine sequences and correlative marine transgressions; sea-level falls by shallowing events in marine sequences and marine regressions. Sea-level rises are easier to recognize (Hallam, 1978; Vail and others, 1984). A list of notable transgressive/deepening events recognizable in more than one continent is presented in Table 1 (see also Figs. 4–9). Because more stratigraphic data are available for Europe, information for this continent is given in more detail than for other continents. Because of the global character of the survey, it is inappropriate to present stratigraphic data to the degree of refinement of ammonite zones.

The events with underlined numbers in Table 1 are considered to be more significant than those that are not underlined. The 18 events cannot be regarded as definitive. Further work may require deletion of certain events that are regional rather than eustatic, whereas other composite events may need splitting into two or more events. The list of 18 transgressive/deepening events is, however, a reasonable approximation to the global picture.

If there were no regressive/shallowing events intervening between the transgressive/deepening events, Jurassic seas would have spread much more widely through the period than they evidently did. As noted earlier, it is much harder to trace regressive/shallowing events globally; however, events that appear to be significant in Europe, of at least regional importance and traceable over wide areas as unconformities or hiatuses, are given in Table 2. Event 9,

---

Fig. 3.—Evidence from the Dorset coast section of unconformities that were produced as a result of erosion rather than nondeposition, with the amount of inferred sediment removal indicated. (A) The White Lias (= Langport Member) and Blue Lias Formation at the Triassic-Jurassic boundary (Hallam, 1960). U-shaped burrows signify the trace fossil *Diplocraterion*; brick ornament = limestone; stippled ornament = shale. (B) Unconformity within the Black Ven Marls Formation, at the level of the Coinstone concretionary limestone horizon (Hallam, 1969). This is the local expression of the widespread intra-Sinemurian regressive event mentioned in the text. Stippling signifies shale. (C) Unconformity within the Middle Oxfordian, marking the base of the Osmington Oolite Formation cycle of Talbot (1973). Stippling indicates sand, either cross-bedded or bioturbated with shells; solid circles = reworked pebbles; elliptical outline indicates calcareous concretion (or 'dogger') in friable sand of Bencliff Grit.

Erosional Event

BEFORE   AFTER

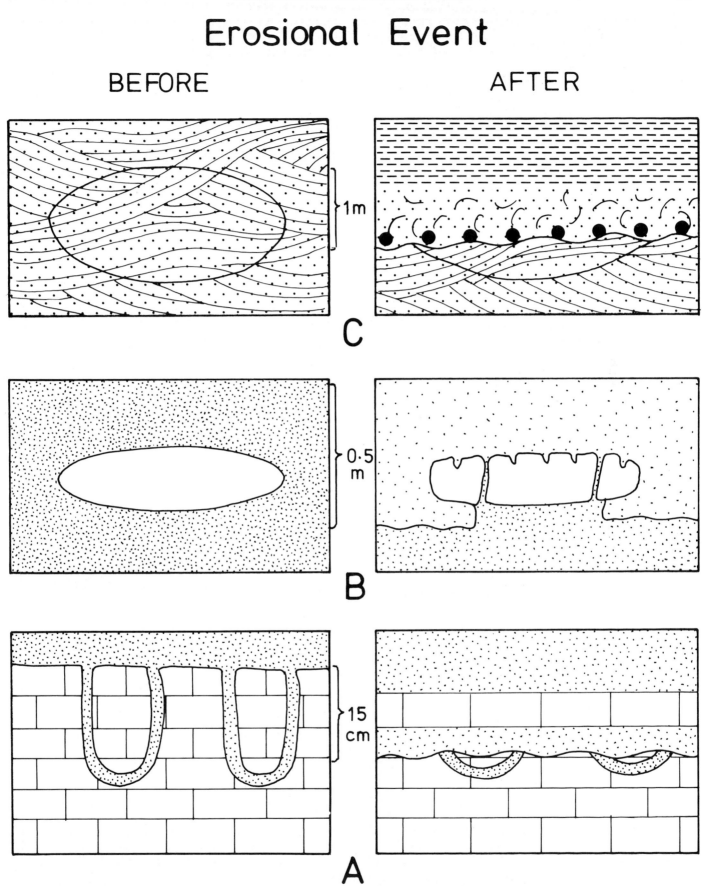

C

B

A

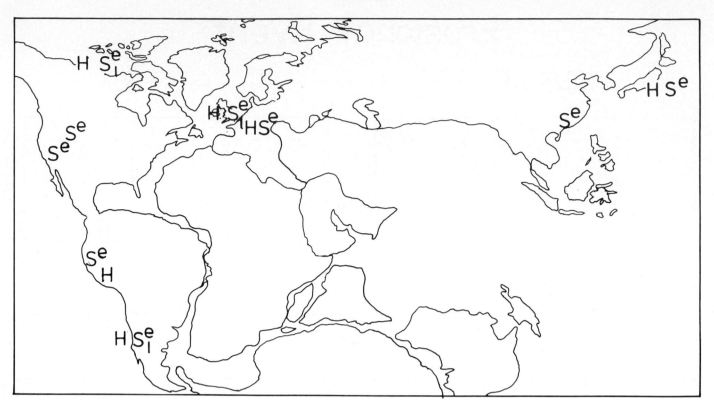

Fig. 4.—Global location of transgressive/deepening events in the Hettangian (H) and Sinemurian (S); e = early; l = late.

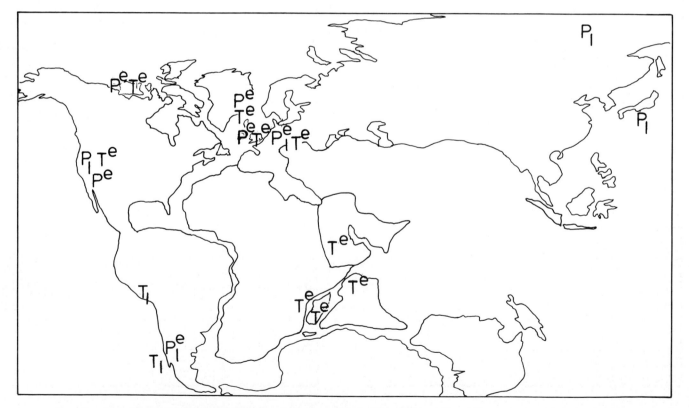

Fig. 5.—Global location of transgressive/deepening events in the Pliensbachian (P) and Toarcian (T); e = early; l = late.

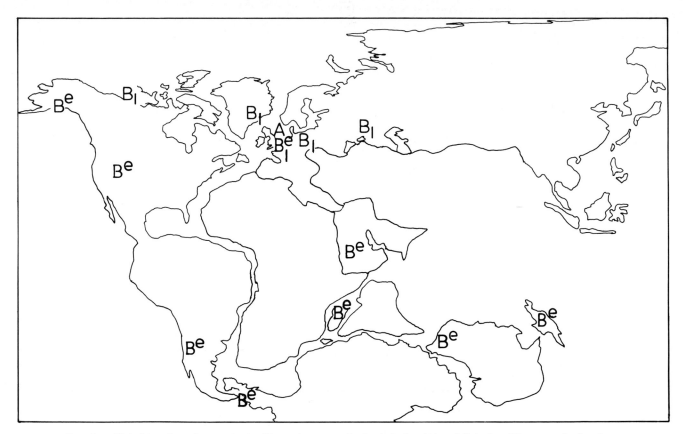

FIG. 6.—Global location of transgressive/deepening events in the Aalenian (A) and Bajocian (B); e = early; l = late.

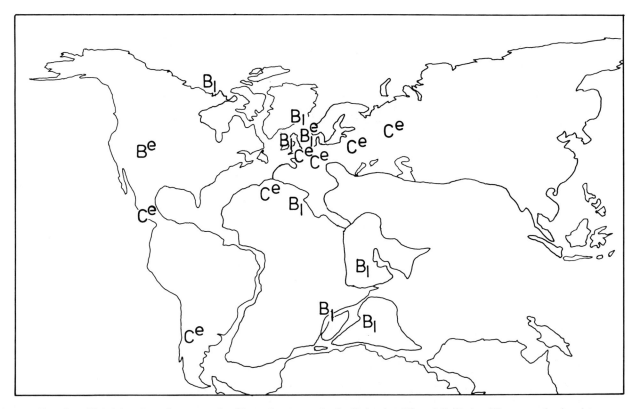

FIG. 7.—Global location of transgressive/deepening events in the Bathonian (B) and Callovian (C); e = early; l = late.

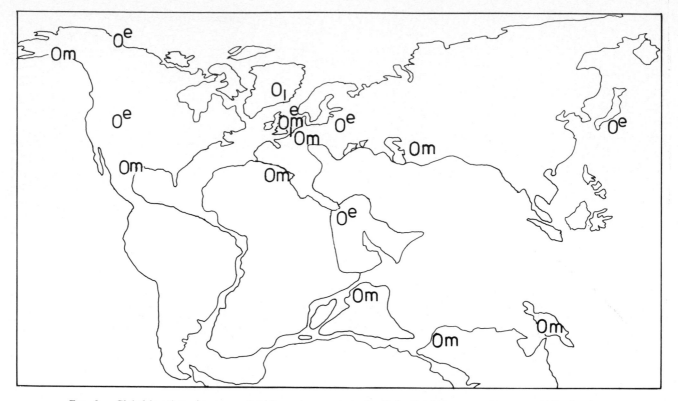

FIG. 8.—Global location of transgressive/deepening events in the Oxfordian (O); e = early; m = middle; l = late.

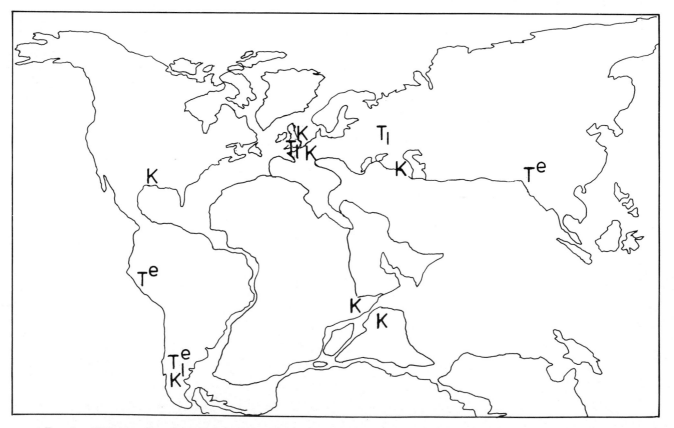

FIG. 9.—Global location of transgressive/deepening events in the Kimmeridgian (K) and Tithonian (T); e = early; l = late.

TABLE 1.—TRANSGRESSIVE/DEEPENING EVENTS RECOGNIZABLE IN MORE THAN ONE CONTINENT

EVENT	EUROPE	OTHER CONTINENTS
1. Early to mid-Hettangian	England, Wales, France, Germany, Austria	Nevada, Peru, Japan
2. Early Sinemurian	Northwest Scotland, south Sweden, south England, northeast France, south Germany	Alberta, Nevada, Arctic Canada, Colombia, Peru, Chile, Mexico, Japan, southeast China
3. Late Sinemurian	England, northeast Scotland, central France, south Germany, Spain, Portugal	Chile
4. Early Pliensbachian	England, Denmark, south Sweden, France, Germany, Poland	East Greenland, Nevada, Arctic Canada, Morocco
5. Late Pliensbachian	England, France, Germany	Arctic North America, Oregon, east Siberia, Japan
6. Early Toarcian	England, northwest Scotland, France, Germany	Arctic North America, Alberta, east Greenland, Saudi Arabia, Madagascar, Pakistan
7. Late Toarcian	France, Germany	Chile, Argentina, Peru
8. Early Bajocian	England, France, Germany, Swiss Jura	Western Interior of North America, Chile, Saudi Arabia, north Madagascar, Australia, New Guinea, Antarctica (Ellsworth Land)
9. Late Bajocian	South England, northwest Scotland, France, south Germany, Poland Crimea (USSR)	Arctic Canada, east Greenland, Chile, Kachchh (India)
10. Early Bathonian	South England, North Sea, northwest France	Western Interior of North America
11. Late Bathonian	Northwest Scotland, England, France	Arctic Canada, east Greenland, north and east Africa, Saudi Arabia, Pakistan, Kachchh
12. Early to mid-Callovian	England, Scotland, France, Germany, Spain, Russian Platform	West Siberia, Morocco, Pakistan, Chile, Argentina, Mexico, Scotia Shelf
13. Early Oxfordian	North England, northwest Scotland, Spain, south Germany, France, Poland, Russian Platform	Western Interior and Arctic North America, Syria
14. Mid-Oxfordian	South and central England, northeast Scotland, Jura (France and Switzerland)	Morocco, Tunisia, Pakistan, Turkmenia (USSR), Mexico, Gulf Coast of the United States, south Alaska, Australia, New Guinea
15. Late Oxfordian	Northwest and northeast Scotland, North Sea, south England, France, south Germany	East Greenland
16. Early to mid-Kimmeridgian	England, North Sea, north France	Azerbaijan (USSR), Pakistan, Ethiopia, Somalia, Saudi Arabia, Gulf Coast of the United States, Argentina
17. Early Tithonian	South England, France	West Siberia, Tibet, Chile, Argentina
18. Late Tithonian (= mid-Volgian)	East and south England, Poland, Russian platform	Chile, Argentina

at least, seems to be clearly recorded outside Europe, in the North American Arctic, Tunisia, and Pakistan.

A number of significant regressive events can be recognized that appear to owe their origin to regional tectonics rather than to global falls in sea level. They include the following:

(1) Mid-Sinemurian, southern England to Germany (Donovan and others, 1979; Brandt, 1985). This event, marked by a striking erosional unconformity in basinal-shale facies, corresponds to event 2 in Table 2. No evidence outside Western Europe has been found to support a eustatic interpretation, but the nature of the facies succession is consistent with this interpretation, unlike most of the examples listed below.

(2) Late Pliensbachian, northwest Europe. The influx of

TABLE 2.—REGRESSIVE/SHALLOWING EVENTS OF REGIONAL IMPORTANCE IN EUROPE

1. End 'Rhaetian'
2. Mid-Sinemurian (sub-*Raricostatum* Zone)
3. End Sinemurian
4. End Early Pliensbachian (Carixian)
5. End Aalenian
6. End Early Bajocian
7. End Bathonian
8. End Early Callovian (sub-*Coronatum* Zone)
9. End Callovian
10. End Early Oxfordian
11. End Mid-Oxfordian
12. End Oxfordian
13. Early Kimmeridgian (sub-*Mutabilis* Zone)
14. Mid-Tithonian

For events 7, 8, 9, 12, and 13, see Gabilly and others, 1985.

sand from a westerly source into shale sequences may relate, at least in part, to tilted fault-block tectonics prior to the opening of the Atlantic (Hallam, 1984).

(3) End Toarcian to Aalenian, margins of London-Brabant Platform, central and northern England, Denmark. Above an erosional unconformity, early to mid-Toarcian clays are abruptly overlain by non-marine mid-Jurassic sands, with local rapid changes of facies suggestive of contemporary fault activity.

(4) Aalenian to Bajocian, eastern and western margins of southern North Atlantic (northwest Africa and Scotia Shelf), marked by depositional gaps and influx of coarse siliciclastic sediments (Jansa and Wiedmann, 1982). This event should probably be extended into the North Sea and its margins (e.g., Yorkshire), where non-marine Middle Jurassic deltaic deposits abruptly succeed marine Liassic clays. It is also noteworthy that in southern England, shallow-water Aalenian-Bajocian limestones abruptly succeed deeper water Liassic shales.

(5) Bathonian, Chile and Argentina (Riccardi, 1983; Gröschke and Hillebrandt, 1985).

(6) Approximately Oxfordian, northwest Iberian Peninsula and Irish Sea. Marine Middle Jurassic strata are abruptly overlain by non-marine redbeds. In both the British Isles and "Iberia," this sharp change disappears rapidly away from the Atlantic margins and is presumably related to tilted fault-block tectonics associated with the opening of that ocean.

(7) Post-Early Oxfordian regression in the United States Western Interior, related to uplift associated with thrusting and folding to the west.

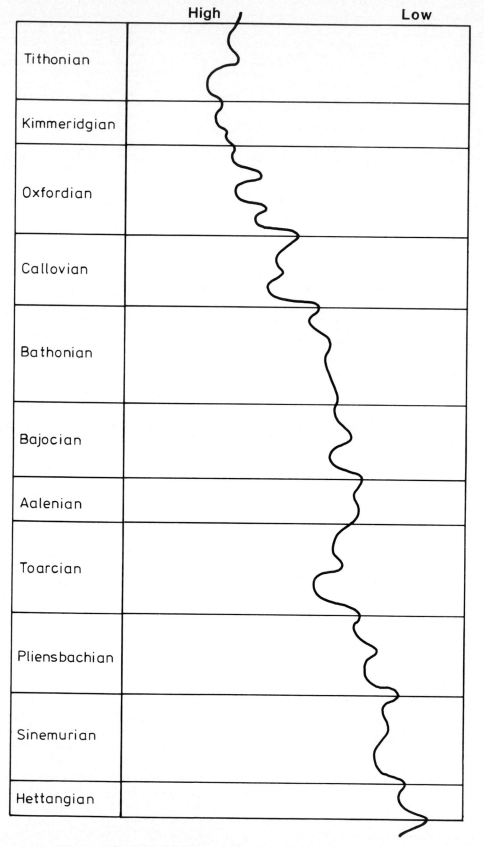

FIG. 10.—Proposed sea-level curve for the Jurassic.

(8) Late Oxfordian to Kimmeridgian, Chile and Argentina, with older marine rocks abruptly overlain by continental gypsiferous and volcaniclastic sediments. This may be related to an eastward migration of the volcanic arc at the continental margin (Hallam and others, 1986).

Figure 10 represents a revised Jurassic eustatic curve. Apart from the replacement of a cuspate by a sinusoidal shape for the successive cycles, the most important differences from the curve published in Hallam (1978) concern the Bathonian and Tithonian, which are no longer seen as relatively low-stand intervals. The secular trend of sea-level rise, starting from its lowest level at the beginning of the period, is now smoother, and reaches its highest level in the Kimmeridgian and early Tithonian rather than the Oxfordian.

COMPARISON WITH THE REVISED EXXON CURVE

If the sea-level curves of Figures 1 and 10 are compared, their overall trends are closely similar, and the cyclicity is portrayed in both curves as more or less sinusoidal. Furthermore, the number of global transgressive/deepening events proposed in this paper (18) is very close to the number of global unconformities (17) recognized by Vail and others (1984), although they do not necessarily correspond. Both curves portray important sea-level rises in the Sinemurian, Toarcian, Bajocian, Callovian, Oxfordian, and Kimmeridgian.

The similarities between the 18 transgressive events and the 17 unconformities, which are derived from different data bases, are encouraging and suggest that genuine global phenomena have been identified, if imperfectly. A number of differences require comment, however. As already noted, the type 1 unconformities recognized by the Exxon group cannot be widely followed across Europe; the important Rhaetian-Hettangian unconformity has been omitted from their list. As previously discussed in Hallam (1981), global stratigraphic evidence indicates strongly that the lowest Jurassic sea-level stand was at the beginning of the Hettangian, not early Sinemurian. The early Sinemurian, early Toarcian, and early Callovian rises in sea level were more rapid and pronounced than indicated in the Exxon curve. The Aalenian regression that gave rise to a type 1 unconformity (event 3 above) is probably the result of regional tectonics. The same is almost certainly the case for the unusually "noisy" Kimmeridgian-Tithonian part of the Exxon curve. Rather than signifying an episode of unusually rapid sea-level oscillations, this part of the curve probably relates to sedimentation controlled by the faulting and tilting of crustal blocks, as noted earlier. My general conclusion is that the Exxon curve presents a distorted, although broadly accurate, view of Jurassic eustasy; the distortions or omissions are due to inadequate biostratigraphic control and underestimation of the importance of regional tectonics.

DISCUSSION AND SUMMARY

The fact that so many Jurassic deepening and shallowing events, with correlative transgressions and regressions, can be recognized across the world, implies events on a global scale and is a strong argument against the view of Mörner (1976, 1981) that geoid changes on the time scale of 10^6 to 10^7 years are the dominant control. Whether geoid changes on this time scale can be dismissed on theoretical rather than empirical grounds is a matter for geophysicists to decide, but it seems more plausible to argue that regional distortions of the global pattern of eustasy are due to tectonic activity. For instance, Cloetingh and others (1985) have proposed a tectonic model in which regional variations in sea level along young continental margins are caused by interactions within the lithosphere of horizontal stresses and vertical deflections as a result of sediment loading.

Comprehensive basin analyses in many parts of the world are necessary to make further advances. Better quantitative data also are needed. Hallam (1975) computed a long-term rise of sea level in Jurassic time of approximately 150 m. This was based on the increase in areal spread of marine sediments on the continents, and the work of Hays and Pitman (1973), who estimated the change in continental freeboard that should result from a given change in mean ocean depth. They arrived at a figure of approximately 500 m for the fall in sea level since the late Cretaceous highstand because of the decline of ocean ridge volume.

Estimates of secular fall in sea level since the Cretaceous highstand vary considerably, from about 100 m to about 650 m. As Pitman's (1978) revised estimate of 350 m, based on changing ocean ridge volume, fell approximately in the middle of this range, I used his estimate, in conjunction with an assumption of no significant change in continental hypsometry through time, as calibration for a Phanerozoic sea-level curve (Hallam, 1984). The latest estimate, taking into account subsiding intra-plate ocean swells as well as ocean ridges, is about 250 m (C. G. A. Harrison, pers. commun., 1986). The figure for the secular Jurassic rise must be correspondingly reduced to one close to the approximate 100-m value derived by Vail and others (1984) from coastal-onlap data.

Superimposed upon the secular trend is a series of cycles of mean duration of about 4 million years, whichever curve is used (Figs. 1, 10). As the evidence for the eustatic origin of at least some of these cycles is strong, it is desirable to have the best possible estimates of amplitude and, hence, rate of sea-level change. These are bound to have an important influence on which controlling mechanism is preferred. At present, I can offer no improvement on my estimate (Hallam, 1978) of amplitudes of a few meters to a few tens of meters and rates of a few centimeters per thousand years. This figure is only slightly higher than the 1 cm/10^3 years computed by Pitman (1978) as the maximum rate of sea-level change producible by the growth and decay of oceanic ridges.

ACKNOWLEDGMENTS

I thank Peter Vail for much stimulating discussion and helpful review of my manuscript.

REFERENCES

ADAMS, A. E., 1979, Sedimentary environments and palaeogeography of the western High Atlas, Morocco, during the Middle and Late Jurassic: Palaeogeography, Palaeoclimatology, Palaeoecology, v. 28, p. 185–196.

BALKWILL, H. R., COOK, D. G., DETTERMAN, R. L., EMBRY, A. F., HÅKANSSON, E., MIALL, A. D., POULTON, T. P., AND YOUNG, F. G., 1983, Arctic North America and Northern Greenland, in Moullade, M., and Nairn, A. E. M., eds., The Phanerozoic Geology of the World, the Mesozoic, B: Elsevier, Amsterdam, p. 1–31.

BAYER, U., ALTHEIMER, E., AND DEUTSCHLE, W., 1985, Environmental evolution in shallow epicontinental seas: Sedimentary cycles and bed formation, in Bayer, U., and Seilacher, A., eds., Sedimentary and Evolutionary Cycles: Springer-Verlag, Berlin, p. 347–381.

BRANDT, K., 1985, Sea-level changes in the Upper Sinemurian and Pliensbachian of southern Germany, in Bayer, U., and Seilacher, A., eds., Sedimentary and Evolutionary Cycles: Springer-Verlag, Berlin, p. 113–126.

CALLOMON, J. H., 1979, Marine boreal Bathonian fossils from the northern North Sea and their palaeogeographic significance: Proceedings of the Geological Association, v. 90, p. 163–169.

———, 1984, A review of the biostratigraphy of the post-Lower Bajocian Jurassic ammonites of western and northern North America: Geological Association of Canada Special Paper 27, p. 143–174.

CAMERON, B. E. B., AND TIPPER, H. W., 1985, Jurassic stratigraphy of the Queen Charlotte Islands, British Columbia: Geological Survey of Canada Bulletin, 365, p. 1–49.

CLOETINGH, S., McQUEEN, H. AND LAMBECK, K., 1985, On a tectonic mechanism for regional sealevel variations: Earth and Planetary Science, Letters, v. 75, p. 157–166.

CORBIN, S. G., 1980, A facies analysis of the Lower-Middle Jurassic boundary beds of north-west Europe: Unpublished Ph.D. Dissertation, University of Birmingham, England, 248 p.

DONOVAN, D. T., HORTON, A., AND IVIMEY-COOK, H. C., 1979, The transgression of the Lower Lias over the northern flank of the London Platform: Journal of the Geological Society of London, v. 136, p. 165–173.

EINSELE, G., 1985, Response of sediments to sea-level changes in differing subsiding storm-dominated marginal and epeiric bands, in Bayer, U., and Seilacher, A., eds., Sedimentary and Evolutionary Cycles: Springer-Verlag, Berlin, p. 68–97.

EMBRY, A. F., 1982, The Upper Triassic-Lower Jurassic Heiberg deltaic complex of the Sverdrup Basin: Canadian Society of Petroleum Geologists Memoir 8, p. 189–217.

ENAY, R., 1980, Indices D'emersion et d'influences continentales dans l'Oxfordian supérieur-Kimmeridgien inférieur en France: Interpretations paléogéographiques et consequences paléobiogéographiques: Bulletin du Societé. Geologique de France, v. 7, p. 581–590.

GABILLY, J., CARIOU, E., AND HANTZPERGUE, P., 1985, Les grandes discontinuites stratigraphiques au Jurassique: Témoins d'évènements eustatiques, biologiques et sedimentaires: Bulletin du Societé Geologique de France, v. 8, p. 391–401.

GRÖSCHKE, M, AND VON HILLEBRANDT, A. 1985, Trias und Jura in der mittleren Cordillera Domeyko von Chile (23°30'–24°30'), Neues Jahrbuch für Geologie und Paläontologie Abhandlungen, v. 170, p. 129–166.

HALLAM, A. 1960, The White Lias of the Devon Coast: Proceedings of the Geological Association, v. 71, p. 47–60.

———, 1969, A pyritised limestone hardground in the Lower Jurassic of Dorset (England): Sedimentology, v. 12, p. 231–240.

———, 1975, Jurassic Environments: Cambridge University Press, Cambridge, England, p. 269.

———, 1978, Eustatic cycles in the Jurassic: Palaeogeography, Palaeoclimatology, Palaeoecology, v. 23, p. 1–32.

———, 1981, A revised sea-level curve for the early Jurassic: Journal of the Geological Society of London, v. 138, p. 735–743.

———, 1984, Pre-Quaternary changes of sea level: Annual Review of Earth and Planetary Science, v. 12, p. 205–243.

———, AND SELLWOOD, B. W., 1976, Middle Mesozoic sedimentation in relation to tectonics in the British area: Journal of Geology, v. 84, p. 302–321.

———, BIRÓ-BAGÓCZKY, L., AND PEREZ, E., 1986, Facies analysis of the Lo Valdés Formation (Tithonian-Hauterivian) of the High Cordillera of Central Chile, and the palaeogeographic evolution of the Andean Basin: Geology Magazine, v. 123, p. 425–435.

———, AND MAYNARD, J. B., 1987, The iron ores and associated sediments of the Chichali Formation (Oxfordian to Valanginian) of the Trans-Indus Salt Range, Pakistan: Journal of the Geological Society of London, v. 144, p. 107–114.

HAQ, B. U., HARDENBOL, J. AND VAIL, P. R., 1987, Chronology of fluctuating sea levels since the Triassic: Science, v. 235, p.1156–1167.

HAYS, J. D., AND PITMAN, W. C., 1983, Lithospheric plate motion, sea level changes and climatic and ecological consequences: Nature, v. 246, p. 16–22.

JANSA, L. F., AND WIEDMANN, J., 1982, Mesozoic-Cenozoic development of the eastern North American and northwest African continental margins: A comparison, in Von Rad, U., Hinz, K., Sarnthein, M., and Seibold, E., eds., Geology of the Northwest African Continental Margin: Springer-Verlag, Berlin, p. 215–269.

JENKYNS, H. C., AND CLAYTON, C. J., 1986, Black shales and carbon isotopes in pelagic sediments from the Tethyan Lower Jurassic: Sedimentology, v. 33, p. 87–106.

LIVNAT, A., FLEXER, A. AND SHAFRAN, N., 1986, Mesozoic unconformities in Israel: Characteristics, mode of origin and implications for the development of the Tethys: Palaeogeography, Palaeoclimatology, Palaeoecology, v. 55, p. 189–212.

LOUGHMAN, D. L., 1982, A facies analysis of the Triassic-Jurassic boundary beds of the world, with special reference to north-west Europe and the Americas: Unpublished Ph.D. Dissertation, University of Birmingham, England, 302 p.

McGHEE, G. R., AND BAYER, U., 1985, The local signature of sea-level changes, in Bayer, U., and Seilacher, A., eds., Sedimentary and Evolutionary Cycles: Springer-Verlag, Berlin, p. 98–112.

MERTMANN, D., 1986, Die regressive Faziesentwicklung im Ober-Toarcium/Aalenium der NW-Iberischen Ketten, Spanien: Neues Jahrbuch für Geologie und Paläontologie Abhandlungen, v. 173, p. 1–46.

MÖRNER, N.-A., 1976, Eustasy and geoid changes: Journal of Geology, v. 84, p. 123–151.

———, 1981, Revolution in Cretaceous sea-level analysis: Geology, v. 9, p. 344–346.

MORTON, N., 1987, Jurassic subsidence history, N.W. Scotland: Marine and Petroleum Geology, v. 4, p. 226–242.

PEDERSEN, G. K., 1985, Thin, fine-grained storm layers in a muddy shelf sequence: An example from the Lower Jurassic in the Stenlille 1 well, Denmark: Journal of the Geological Society of London, v. 142, p. 357–374.

PHELPS, M., 1985, A refined ammonite biostratigraphy for the Middle and Upper Carixian (Ibex and Davoei zones, Lower Jurassic) in northwest Europe and stratigraphic details of the Carixian-Domerian boundary: Geobois, v. 18, p. 321–362.

PITMAN, W. C., 1978, Relationship between eustasy and stratigraphic sequences of passive margins: Geological Society of America Bulletin, v. 89, p. 1389–1403.

RICCARDI, A. C., 1983, The Jurassic of Argentina and Chile, in Moullade, M., and Nairn, A. E. M., eds., The Phanerozoic Geology of the World, the Mesozoic, B: Elsevier, Amsterdam, p. 201–263.

SELLWOOD, B. W., AND JENKYNS, H. C., 1975, Basins and swells and the evolution of an epeiric sea (Pliensbachian-Bajocian of Great Britain): Journal of the Geological Society of London, v. 131, p. 373–388.

SURLYK, E., 1978, Submarine fan sedimentation along fault scarps on tilted fault blocks (Jurassic-Cretaceous boundary, east Greenland): Grønlands Geologiske Undersøgelse Bulletin, v. 128, p. 1–108.

———, AND CLEMMENSEN, L. B., 1983, Rift preparation and eustasy as controlling factors during Jurassic inshore and shelf sedimentation in northeastern East Greenland: Sedimentary Geology, v. 34, p. 119–143.

———, AND LARSEN, H. C., 1981, Post-Paleozoic evolution of the East Greenland continental margin: Canadian Society of Petroleum Geologists memoir 7, p. 421–436.

TALBOT, M. R., 1973, Major sedimentary cycles in the Corallian Beds: Palaeogeography, Palaeoclimatology, Palaeoecology, v. 14, p. 293–317.

TAYLOR, D. G., SMITH, P. L., LAWS, R. A., AND GUEX, J., 1983, The stratigraphy and biofacies trends of the Lower Mesozoic Gabbs and Sunrise formations, west-central Nevada: Canadian Journal of Earth Sciences, v. 20, p. 1598–1608.

VAIL, P. R., HARDENBOL, J., AND TODD, R. G., 1984, Jurassic unconformities, chronostratigraphy and sea-level changes from seismic stratigraphy and biostratigraphy in Schlee, J. S., ed., Interregional Unconformities and Hydrocarbon Accumulation: American Association of Petroleum Geologists Memoir 36, p. 129–144.

———, MITCHUM, R. M., JR., TODD, R. G., WIDMIER, J. M., THOMPSON, S., III, SANGREE, J. B., BUBB, J. N., AND HATLELID, W. G., 1977, Seismic stratigraphy and global changes of sea level, part 2: Application of seismic reflection configuration to stratigraphic interpretations, *in* Payton, C. E., ed., Seismic Stratigraphy—Applications to Hydrocarbon Exploration: American Association of Petroleum Geologists Memoir 26, p. 49–212.

———, AND TODD, R. G., 1981, Northern North Sea Jurassic unconformities, chronostratigraphy and sea-level changes from seismic stratigraphy, *in* Illing, L. V., and Hobson, G., eds., Petroleum Geology of the Continental Shelf of North-West Europe: Heyden, London, p. 216–235.

VAN HINTE, J. E., 1978, Geohistory analysis—Application of micropaleontology in exploration geology: American Association of Petroleum Geologists Bulletin, v. 62, p. 201–222.

WALLEY, C. D., 1983, The palaeoecology of the Callovian and Oxfordian strata of Majdal Shams (Syria) and its implications for Levantine palaeogeography and tectonics: Palaeogeography, Palaeoclimatology, Palaeoecology, v. 42, p. 323–340.

———, 1985, Depositional history of southern Tunisia and northwestern Libya in Mid and Late Jurassic time: Geological Magazine, v. 122, p. 233–247.

WANG, Y.-G., AND SUN, D.-L., 1985, The Triassic and Jurassic paleogeography and evolution of the Qinghai-Xizang (Tibet) Plateau: Canadian Journal of Earth Sciences, v. 22, p. 195–204.

WINTERER, E. L., AND BOSELLINI, A., 1981, Subsidence and sedimentation on Jurassic passive continental margin, Southern Alps, Italy: American Association Petroleum Geologists Bulletin, v. 65, p. 394–421.

EARLY CRETACEOUS SEA–LEVEL CURVES, GULF COAST AND SOUTHEASTERN ARABIA

R. W. SCOTT, S. H. FROST, AND B. L. SHAFFER

Amoco Production Company, P. O. Box 3385, Tulsa, Oklahoma 74102; 12926 Villa Wood Lane, Houston, Texas 77072;
Everest Geotech, Inc., 10101 Southwest Freeway, Houston, Texas 77074

ABSTRACT: Three surface and subsurface sections of Lower Cretaceous strata from the Gulf Coast are correlated with three comparable sections in Oman. Detailed fossil ranges and graphic correlation methods resulted in a biostratigraphic data base that could be related to the geologic time scale.

Two events of relative sea-level rise are synchronous in the Gulf Coast basin and the southeastern Arabian platform and may represent eustatic sea-level rises. The intra-Aptian rise began about 115.8 Ma and in many places is represented by a sharp lithologic change, by submarine hardgrounds, or by onlap. Deep-water deposition resumed from 115.2 to 113.9 Ma. The intraCenomanian rise began approximately 94.6 Ma. In Oman, this rise is locally represented by a submarine hardground that formed after drowning of a carbonate shelf. In the updip Gulf Coast, mid-Cenomanian paralic and deltaic sediments were deposited upon an Albian-early Cenomanian shallow carbonate shelf. In the downdip Gulf Coast, this event either is not recognizable in deep-water muds or is represented by drowning of shallow-water carbonates. A third, intra-Albian event at 104.3 Ma may also be a eustatic sea-level rise; however, it needs to be identified in other tectonic settings.

INTRODUCTION

A number of authors have proposed curves showing Early Cretaceous relative sea level (Fig. 1). One of the earliest curves equated transgressive–regressive cycles with eustatic sea-level changes and periods of intense vulcanism in the United States Western Interior (Kauffman, 1977, 1984). The curves of Vail and others (1977) actually illustrate relative coastal onlap of seismic sequences rather than eustatic curves (Pitman, 1978). The curves of Caldwell (1984) for the Canadian interior and of McFarlan (1977) for the Gulf Coast are clearly indicated as transgressive-regressive events that only indirectly relate to sea level. A sea-level curve for the Cretaceous section on the San Marcos Platform in central Texas may partly reflect eustasy and partly local subsidence along structural zones (Young, 1986). Other curves are based on water depth from which sea-level changes may be inferred (Rey, 1982; Harris and others, 1984; Longoria, 1984; Flexer and others, 1986). Modeling studies suggest that actual eustatic sea-level changes are slightly out of phase with curves of water depth and coastal onlap (T. A. Cross, pers. comm. 1986; Haq and others, 1987; Posamentier and Vail, this volume).

A few consistently coeval trends among the curves from different basins (Fig. 1) suggest that these changes may reflect eustatic sea-level changes, although the precise timing of highstands and lowstands is not yet determined. A transgressive peak in the Valanginian is present in all but two curves. A second major transgression in the Early Aptian is found in all sections, although different facies are present in each section. A third widespread transgression occurred during the Late Albian, followed by a widespread relative drop during the Early Cenomanian. It is likely that these three relative changes in sea level reflect three Early Cretaceous eustatic rises. The timing of the peaks, however, remains to be determined accurately. If the sea-level modeling by other authors in this volume is correct, the highstand follows the deepening event and the coastal-onlap peak. Other perturbations in these curves of Early Cretaceous events most likely reflect local, intrabasinal changes in subsidence rates and/or sedimentation rates. Some of the discrepancies in the timing of various peaks among these curves may re-sult from different criteria used in identifying stadial boundaries. For example, in eastern Mexico planktonic foraminifera in thin sections are not abundant, and the boundaries could be placed within several tens of meters of sections (Longoria, 1984). In Oman, a few benthic foraminifera and nannoplankton occur in specific facies, and total ranges cannot be demonstrated (Harris and others, 1984). Boundaries in the Gulf Coast are based on partly provincial ammonites; and in the Canadian Western Interior, different ammonites are used. A more precise method of correlating sections and consistently placing boundaries is needed.

In this study, paleobathymetric curves were determined for each section by conventional sedimentologic and paleoecologic criteria (Frost and others, 1983; Harris and others, 1984; Connally and Scott, 1985; Simmons and Hart, 1987). These depths are relative, not absolute. Biostratigraphic criteria were used to locate stadial boundaries, although their positions are usually imprecise because of gaps in the fossil recovery. Nannoplankton, nannoconids, calpionellids, benthic and plankontic foraminifera, and calcareous algae were identified in measured sections and drill samples. Their ranges were determined from analyses of numerous samples spaced at least 3 to 10 m apart. The absolute time scales of Palmer (1983) and Kent and Gradstein (1985) were applied to the stadial boundaries in order to provide an estimate of rates.

A useful test of the eustatic nature of a local relative change in water depth is the synchroneity of similar events in other basins and on other continents (Hancock and Kauffman, 1979; Matsumoto, 1980; Mörner, 1980); however, an alternate interpretation of synchroneity between any pair of relative changes is that they are fortuitous.

Sections on two widely separated, relatively stable carbonate platforms, the United States Gulf Coast and the Arabian Peninsula, were selected to test correlation of deepening events (Fig. 2). The North American Gulf Coast basin developed a very large carbonate platform during the Early Cretaceous (Bay, 1977; McFarlan, 1977; Young, 1983; Scott, 1984). The carbonate depositional systems (Fig. 3) first developed during the Oxfordian and recurred during the Barremian-Aptian, progressively expanding in area until the Albian. During the Berriasian-Valanginian, local carbonate

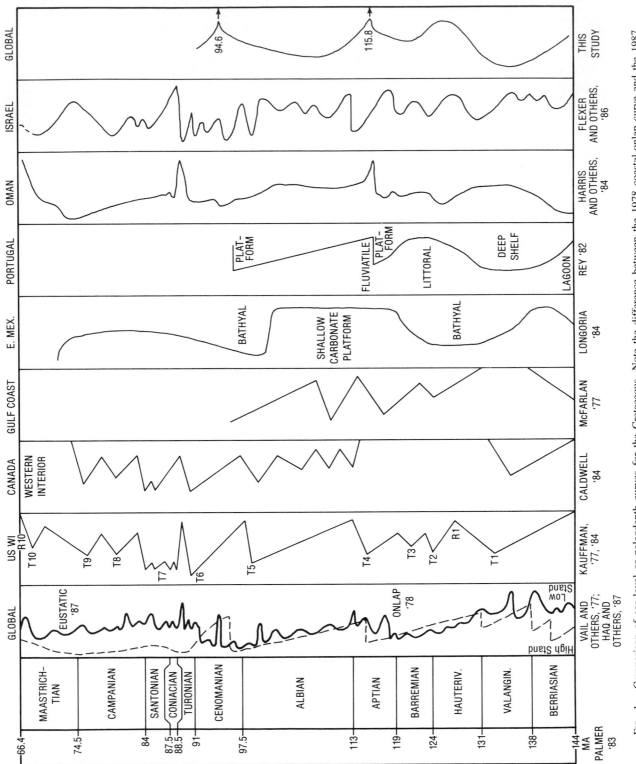
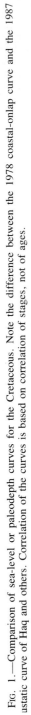

Fig. 1.—Comparison of sea-level or paleodepth curves for the Cretaceous. Note the difference between the 1978 coastal-onlap curve and the 1987 eustatic curve of Haq and others. Correlation of the curves is based on correlation of stages, not of ages.

systems ("Cotton Valley", "Adoue", and "Knowles" lime-stones) were deposited. They were not widespread because of locally high rates of clastic sedimentation. Sections from three areas provide excellent control on the facies successions and biostratigraphic ranges. In eastern Mexico, the Santa Rosa Canyon section south of Monterrey (Fig. 4) spans the Berriasian to Cenomanian with continuously exposed deep-water facies that grade down into nearshore-shelf facies of the La Casita Formation (Blauser and McNulty, 1980; Ice and McNulty, 1980; Ross and McNulty, 1981). The section is punctuated by two diastems represented by sharp contacts between limestone below and dark gray, thin shale beds above. In South Texas, a series of cored wells illustrates the facies succession from forereef shelf lime mudstone/wackestone to coral-rudist reef boundstone/packstone (Bebout and Loucks, 1974; Bebout and others, 1977). In North Texas, a series of outcrops has been stacked to represent the sequences from littoral, mixed carbonate-terrigenous sands to shelf basin shale and limestone during the Early to Late Albian (Scott and others, 1978).

In southeastern Arabia, a number of sections have been measured and numerous wells have penetrated the Lower Cretaceous section (Fig. 5). Three outcrops in the Oman Mountains have been correlated and graphed with the composite standard to show the major stratigraphic events. The stratigraphy has recently been reviewed by Harris and others (1984) and Simmons and Hart (1987). The eastern Arabian Peninsula was a stable carbonate platform from Per-mian to mid-Cretaceous time, about 170 Ma (Saint-Marc, 1978; Murris, 1980). Consequently, eustatic sea-level changes are likely to be recorded in the strata. The Lower Cretaceous Rayda-Salil formations are deep-shelf carbonates deposited during a sea-level rise (Connally and Scott, 1985). The overlying sequence of limestones up to the Shuaiba Formation represents shallow-shelf deposition. The top of the Shuaiba is a widespread disconformity locally marked by a bored, submarine hardground. The overlying Nahr Umr Formation consists of shale and orbitolinid limestones that grade upward into mainly shallow-shelf limestones of the Natih Formation. The top of the section at Wadi Bani Kharus in Jabal Akhdar is faulted and the precise base of the Natih is difficult to determine. The top of the Natih is another regional disconformity, locally marked by a bored submarine hardground that is overlain by Upper Cretaceous terrigenous clastics, Aruma Group, of the foreland basin regime.

CHRONOSTRATIGRAPHY

The timing of the unconformities in the six reference sections must be based on a method that is more precise than the interpolation of dates to zonal boundaries. We use the graphic correlation method of Shaw (1964; Miller, 1977), which can produce a precise data base of fossil ranges in numerous carefully measured and sampled sections. A

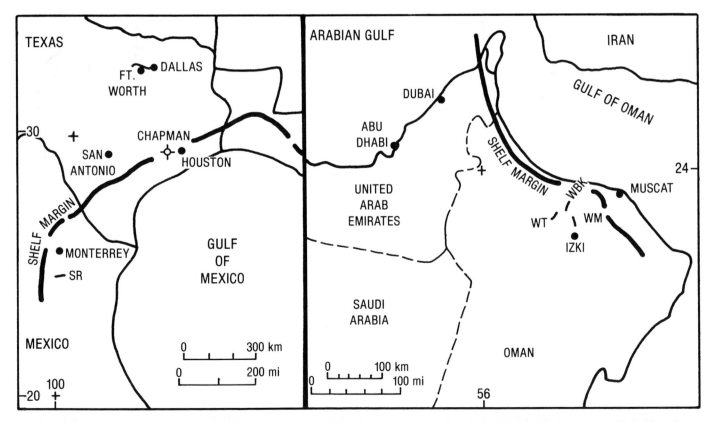

FIG. 2.—Location map of the Texas-Mexico Gulf Coast (left) and southeastern Arabian Platform (right), showing sections studied. SR = Santa Rosa Canyon; Shell No. 1 Chapman well; Trinity River Composite by Fort Worth; WT = Wadi Tanuf; WBK = Wadi Bani Kharus; WM = Wadi Maidin.

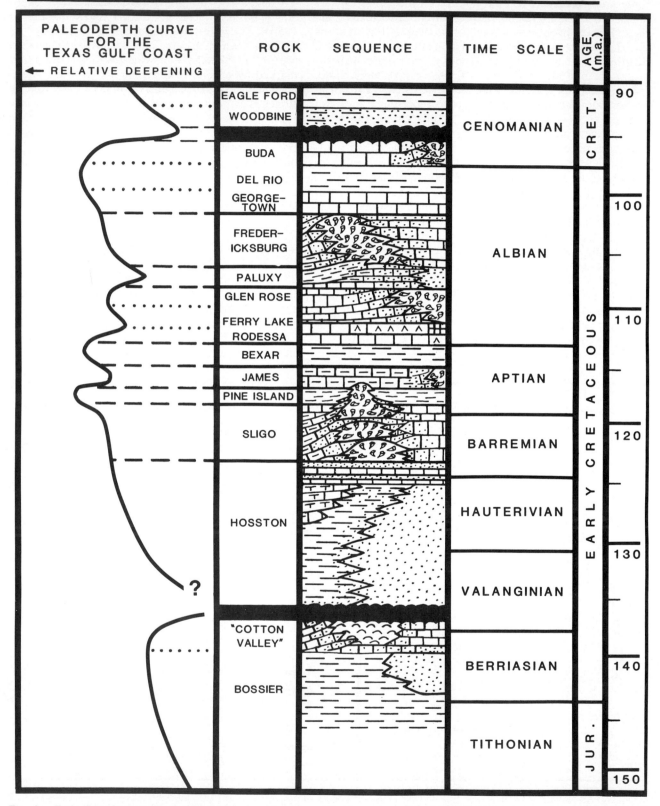

FIG. 3.—Early Cretaceous eustatic sea-level changes and corresponding formations of the Texas Gulf Coast, based on paleodepth curves.

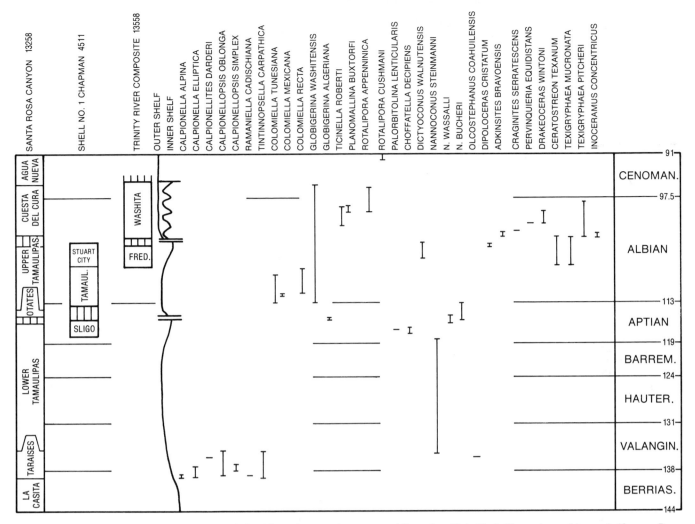

FIG. 4.—Paleodepth curve of the Gulf Coast derived from three sections arranged from the off-shelf, shelf margin, and inner-shelf areas. Ranges of key fossils used in graphic correlation of sections given in ages converted from composite standard units.

composite standard data base is built by first cross-plotting fossil ranges in two stratigraphic sections. The result is a range chart on a scale that is proportional to the thickness of the reference section. Subsequent sections are graphed by cross-plotting to these original range data. Ranges may be extended and new taxa added until new sections do not result in further changes in the ranges. The senior author generated a composite standard data base consisting of 60 outcrops and wells from the Gulf Coast, Europe, North Africa, Lebanon, Oman, the United Arab Emirates, and the Atlantic Ocean basin. More than 1,200 taxa of foraminifera, nannoplankton, calpionellids, ammonites, bivalves, corals, dinoflagellates, and spore-pollen have been graphed to the composite data base in selected reference sections that were either collected and analyzed by the Amoco staff or from the literature.

The six Gulf Coast and Arabian reference sections (Figs. 4, 5) spanning the Berriasian to Cenomanian were graphed to this data base. The relative ages of fossil ranges, formational contacts, and unconformities were determined in terms of composite standard units. These units were then converted to absolute time in million years by cross-plotting the composite standard scale to the absolute time scale (Palmer, 1983). A line of correlation for converting composite standard units to years was based on two points: 91 Ma for top of the Cenomanian and 97.5 Ma for top of the Albian (Palmer, 1983). This line resulted in a remarkably close fit for the ages of other Early Cretaceous stages. The ages of the stage boundaries on Figures 4 and 5 are those inferred by Palmer (1983), assuming 5 to 7 million years for each stage. The identification of the stage boundaries by paleontologic criteria in standard European reference sections also graphed to the composite standard results in slight differences in ages from those of Palmer; but in this study, the precise correlation of the stages is not the objective.

The ranges of the fossils in Figures 4 and 5 are those in the composite standard converted to millions of years. These are the taxa that determined the correlation of the six sections. The ranges are relative to each other in the composite data base and give an indication of the reliability, accuracy, and completeness of this composite standard. The ranges

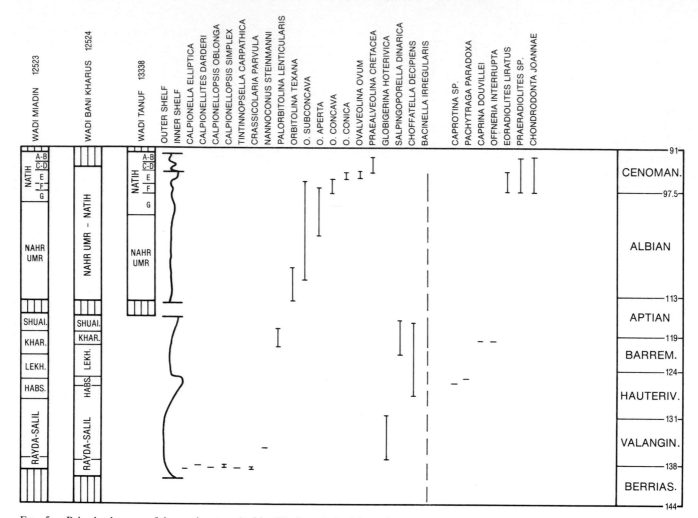

FIG. 5.—Paleodepth curve of the southeastern Arabian Platform derived from three sections in the Oman Mountains just behind the shelf margin. Ranges of key fossils as in Figure 4.

of some taxa may be longer than shown here, but no sections showing that have yet been graphed to the standard.

Santa Rosa Canyon, Mexico.—

The Lower Cretaceous section in Santa Rosa Canyon, some 110 km (65 mi) south of Monterrey, Mexico, consists of 1,310 m (4,300 ft) of inner- to outer-shelf deposits that represent nearly continuous deposition from the Tithonian to the Turonian stage (Diaz and Antonio Rios, 1959; Fig. 4). The lithostratigraphic classification has evolved since 1959, and we accept the modifications used by Ross and McNulty (1981). The ranges of fossils in thin section and a few ammonites were determined by Blauser and McNulty (1980), Ross and McNulty (1981), and Ice and McNulty (1980). This section was graphed to the composite standard, using the key fossils in Figure 4, which are also present in as many as 10 other sections in the Tethys. Two unconformities are evident from the outcrop as well as from the graph: at the base of the Otates Member, Tamaulipas Formation, and at the base of the Cuesta del Cura Formation. Both contacts are sharp; the latter is distinguished

by a 30-cm-thick black shale (C. I. Smith, pers. comm., 1985).

Texas shelf margin.—

The Barremian to Albian section was cored in the Shell No. 1 Chapman Well in Waller County, Texas (Bebout and Loucks, 1974). The 1,070-m-thick (3,500 ft) section consists of shallow-shelf deposits in the Sligo Formation overlain by deep forereef shelf sediments that grade upward into shelf margin facies of the Stuart City Formation. A similar section is reported in Bee County, Texas (Cook, 1979). Key microfossils in the section are published here for the first time and represent part of the succession described by Trejo (1975) and Coogan (1977). Neither the unconformity at the top of the Sligo nor the one at the top of the Stuart City was cored. The Sligo-Pearsall break is very short and shows in the graph. The Stuart City-Austin break is a major lithologic change and is supported by Upper Cretaceous fossils overlying Middle Albian fossils. These breaks also are indicated by onlapping relations of overlying seismic sequences.

North Texas.—

The Trinity River composite section consists of a set of roadcuts and streamcuts exposed from Fort Worth to Grayson Bluff. Most have been described by Perkins (1961) and Scott and others (1978). The distinct unconformity between the Goodland Limestone of the Fredericksburg Group and the Kiamichi Shale of the Washita Group is marked by a bored, iron-stained submarine hardground (Scott and others, 1978). The ranges of the fossils form a distinct terrace, indicating an incomplete lithologic section. At the top of the Washita, the Buda Limestone is overlain by the Woodbine Formation. This contact is an erosional surface with as much as 10 m of relief.

Oman Mountains.—

Three outcrops in Jabal Akhdar, Oman (Fig. 5), illustrate the stratigraphy and the major breaks in the Lower Cretaceous section. This stratigraphic sequence is substantiated by lithologic and paleontologic analyses of 60 additional wells and outcrops from various places on the southeastern Arabian Platform (Harris and Frost, 1984; Simmons and Hart, 1987; Frost and Shaffer, in prep.). The stratigraphic nomenclature of the section at Wadi Bani Kharus was reviewed by Connally and Scott (1985) and is based on the study of Glennie and others (1974). The section at Wadi Tanuf was measured by Connally and Scott, and the biostratigraphy by Scott was based on analyses of thin sections. The Wadi Miadin section is a regional reference section for geologists of Petroleum Development Oman and was measured for Amoco by J. D. Smewing, University of Swansea. These results have been verified by our own data and interpretations. Biostratigraphic documentation by Scott is unpublished. Four major breaks are apparent in the outcrop and on graphs with the composite standard. Significant erosion occurred at the Jurassic-Cretaceous contact and at the Cenomanian-Santonian contact; however, the Shuaiba-Nahr Umr contact and the intra-Natih break between the "E" and "D" members are submarine hardgrounds.

The contact between the Shuaiba and Nahr Umr apparently becomes conformable in western Oman and Abu Dhabi. A suite of Upper Aptian nannoconids is present in the basal Nahr Umr shale (Shaffer, in prep.). Upper Aptian ammonites have been reported from the Bab Member of the Shuaiba in Abu Dhabi (Alsharhan, 1985; Hassan and others, 1975). This suggests that deposition was continuous from the Shuaiba to the Nahr Umr.

CORRELATION OF TRANSGRESSIVE EVENTS

Two unconformities in the Gulf Coast are synchronous with two breaks in Oman: the intra-Aptian and the intra-Cenomanian breaks. The intra-Aptian break in the Gulf Coast is represented by the contact between the Sligo Limestone or the Lower Tamaulipas Member below and the Pearsall, Otates, and La Peña formations above. The lithologic change across this contact represents a change from a shallow, clearwater mass to a deeper, muddier water mass. The hiatus is as much as 500,000 years long and is dated from 115.8 to 115.2 Ma by graphic correlation. No erosion of the off-

shore strata below is evident in these localities. Some cement in the Sligo, however, indicates that it was exposed to the meteoric phreatic environment (Moldovanyi and Lohmann, 1984). The updip equivalent of the Pearsall is the Hammett Shale that onlaps the Sligo and older units (Lozo and Stricklin, 1956; Stricklin and others, 1971). Downdip of the Sligo shelf margin, seismic sections show the Pearsall onlapping the Sligo (Tyrrell and Scott, 1988; Frost, in prep.). This evidence suggests that deposition of the Sligo and its equivalent facies was terminated by a sea-level stillstand during which fresh water altered the Sligo sediments behind the shelf margin. This was followed by a rise in relative sea level, and mud deposition resumed both on the shelf and in the forereef shelf during the later part of the Early Aptian.

During the Late Aptian, a relative drop in sea level in the Gulf Coast is represented by the Upper Aptian, progradational Cow Creek Limestone updip and the James Limestone downdip, which are shallow, clear-water shelf carbonates (Young, 1986). They indicate that one of several changes occurred: a brief eustatic drop or a brief stillstand, a brief intrabasinal upwarp, or a decrease in the rate of subsidence. This limestone was followed by renewed relative deepening in the latest Aptian, represented downdip by the Bexar Shale Member at the top of the Pearsall.

The intra-Aptian hiatus in eastern Oman had a maximum duration of about 2.2 Ma beginning about 116.1 Ma, and deposition resumed about 113.9 Ma. The duration of the hiatus probably diminished westward. At this contact, the Shuaiba Formation below and the Nahr Umr Formation above indicate a dramatic change in depositional regime from clear, shallow-shelf to muddy, somewhat deeper shelf. The Shuaiba represents deposition upon a shallow carbonate shelf (Frost and others, 1983). Following deposition, it was locally exposed to a meteoric phreatic environment (Harris and others, 1984). Erosion was local only and is represented by a few clasts of limestone in the basal bed of the Nahr Umr. Deposition of the overlying orbitolinid clays and lime muds persisted in a repetitive manner throughout much of the Albian. No intra-Albian breaks are yet recognized, although numerous sharp bedding planes may indicate brief diastems. The evidence in southeastern Arabian basins so far indicates a major single rise in sea level during the Late Aptian; at the eastern margin, deposition was renewed about a million years later than in the Gulf Coast; however, Shuaiba deposition ceased at approximately the same time as did Sligo deposition: at 116.1 Ma compared with 115.8 Ma.

The intra-Albian break in the Gulf Coast section is widespread and marks the contact between the Fredericksburg-Washita Groups (Lozo and Stricklin, 1956; Scott and others, 1978; Young, 1986). A synchronous hiatus is not yet recognized in Oman, although a Late Albian hiatus between the Nahr Umr and Mauddud is suggested (Harris and Frost, 1984). On the Comanche shelf, this break is locally a firm ground or a hardground below the Kiamichi Formation that onlaps the Fredericksburg Group. On local structures at the shelf margin, a variable amount of section is absent. For example, at the shelf margin in the Shell No. 1 Chapman Well, this contact is overlain by the upper-most Turonian to Coniacian Austin Chalk. In the forereef deep shelf, the break is sharp and overlain by dark gray Albian-Cenomanian clay. This contact represents a dramatic deepening and change in water mass conditions on the shelf in the same way the Sligo-Pearsall contact does. The hiatus on the shelf was about 1 million years in duration, from about 104.3 to 103.3 Ma, as calculated by averaging the time units in these sections. This disconformity may represent a eustatic sea-level change or a Gulf-wide change of some other origin.

The intra-Cenomanian break may represent a eustatic sea-level change, because it is approximately the same age in the Gulf of Mexico and in south-eastern Arabia. In the Gulf Coast, the hiatus began about 95.2 Ma, but its duration is not yet known precisely. In Oman, the hiatus was very short-lived and occurred about 94.6 Ma. The differences are a result of the variable age of the top of the Buda Limestone and the limits of precision of the graphic correlation method. In the Gulf Coast, the intra-Cenomanian break in the updip shelf is the Washita-Woodbine contact. This is an erosional unconformity between shallow-shelf carbonates and paralic-to-deltaic terrigenous clastic sediments. This contact is normally interpreted to represent a major regressive-transgressive event and is indicated on the charts of Vail and others (1977). In eastern Oman, this hiatus is a hardground between shallow-shelf carbonates of the "E" member of the Natih Formation and deeper shelf marls of the "D" member. It may also correlate with either a suspected hiatus within the Mishrif Formation in northwestern Oman or the Mauddud-Mishrif contact (Harris and Frost, 1984). Thus, this contact indicates a relative rise in sea level without a prior drop. Perhaps locally in the Gulf Coast, a eustatic rise was offset for a short time by a more rapid drop of base level.

In southeast Arabia, carbonate shelf deposition was interrupted following the Cenomanian by a major tectonic uplift (Glennie and others, 1974). The youngest strata in the Oman sections are dated as approximately 91.8 Ma. The top of the Natih Formation was eroded, and in places it is completely absent across post-Cenomanian and pre-Santonian structures. Prior to resumption of deep marine, flysch-type sedimentation of the Aruma Group, local hardgrounds developed on the top surface of the Natih, and local hematitic ooids formed, suggestive of sediment-bypassing or starved deposition. This break is not recognized in the Gulf Coast, where coeval strata are inner- to middle-shelf deposits of the Eagleford Formation and its equivalents.

CONCLUSIONS

Two and possibly three relative sea-level changes during the Aptian-Cenomanian are likely to represent eustatic changes. The intra-Aptian sea-level rise in the Gulf Coast is the same age as a major hiatus in southeastern Arabia, 115.8 to 115.2 Ma. This rise is represented in many sections by an abrupt change in sedimentation from shallow-shelf carbonates to deep-shelf muds, commonly separated by a diastem. A second sea-level rise is between the Early and Middle Cenomanian, 94.6 Ma. Within shelf carbonate sequences, this change is represented by a diastem normally followed by deeper shelf sediments. In undip regions, the hiatus was longer and commonly associated with

erosion prior to resumption of deposition of paralic clastics. A third change, seen in the Gulf Coast but not yet in Arabia, was during the early part of the Late Albian, at 104.3 Ma. In most of these studied sections, the diastem represents sediment bypassing and the deepest conditions followed by gradual or intermittent shoaling. Graphic correlation is a strong tool to test the synchroneity of similar changes in relative sea level. Quantitative stratigraphy enables the discrimination of local events from world-wide events, and it has the potential to provide a precise dating method.

ACKNOWLEDGMENTS

This discussion is the result of independent studies supported by Amoco Production Company, Amoco Oman Oil Company, and Gulf Oil Company. The authors are grateful to all those who made field work in Oman possible, who supported their laboratory analyses, and who aided in the preparation of this report. Discussions with John Smewing, Swansea College, and Michael Simmons, Swansea College and now British Petroleum, were most stimulating and helpful. This report has been released for publication by the Ministry of Petroleum and Minerals of Oman and by Amoco Production Company.

REFERENCES

ALSHARHAN, A. S., 1985, Depositional environment, reservoir units evolution, and hydrocarbon habitat of Shuaiba Formation, Lower Cretaceous, Abu Dhabi, United Arab Emirates: American Association of Petroleum Geologists Bulletin, v. 69, p. 899–912.

BAY, T. A., Jr., 1977, Lower Cretaceous stratigraphic models from Texas and Mexico: University of Texas at Austin, Bureau of Economic Geology, Report of Investigations 89, p. 12–30.

BEBOUT, D. G., AND LOUCKS, R. G., 1974, Stuart City Trend, Lower Cretaceous, South Texas: University of Texas at Austin, Bureau of Economic Geology, Report of Investigations 78, 80 p.

———, SCHATZINGER, R. A., AND LOUCKS, R. G., 1977, Porosity distribution in the Stuart City Trend, Lower Cretaceous, South Texas: University of Texas at Austin, Bureau of Economic Geology, Report of Investigations 89, p. 234–256.

BLAUSER, W. H., AND MCNULTY, C. L., 1980, Calpionellids and nannoconids of the Taraises Formation (Early Cretaceous) in Santa Rosa Canyon, Sierra de Santa Rosa, Nuevo Leon, Mexico: Transactions, Gulf Coast Association of Geological Societies, v. 30, p. 263–272.

CALDWELL, W. G. E., 1984, Early Cretaceous transgressions and regressions in the Southern Interior plains, in Stott, D. F., and Glass, D. J., eds., The Mesozoic of Middle North America: Canadian Society of Petroleum Geologists Memoir 9, p. 173–203.

CONNALLY, T. C., AND SCOTT, R. W., 1985, Carbonate sediment-fill of an oceanic shelf, Lower Cretaceous, Arabian Peninsula: Society of Economic Paleontologists and Mineralogists, Core Workshop No. 6, p. 266–302.

COOGAN, A. H., 1977, Early and Middle Cretaceous Hippuritacea (rudists) of the Gulf Coast: University of Texas at Austin, Bureau of Economic Geology, Report of Investigations 89, p. 32–70.

COOK, T. D., 1979, Exploration history of South Texas Lower Cretaceous carbonate platform: American Association of Petroleum Geologists Bulletin, v. 63, p. 32–49.

DIAZ, T., AND ANTONIO RIOS, Z., 1959, Plate 6 Seccion geologica area Linares-Galeana, in Diaz, T., Mixon, R., Murray, G., Weidie, A., and Wolleben, J., eds., Mesozoic Stratigraphy and Structure, Saltillo-Galeana area, Coahuila and Nuevo Leon, Mexico: South Texas Geological Society, 1959 Field Trip Guidebook, 115 p.

FLEXER, A., ROSENFELD, A., LIPSON-BENITAH, S., AND HONIGSTEIN, A., 1986, Relative sea level changes during the Cretaceous in Israel: American Association of Petroleum Geologists Bulletin, v. 70, p. 1685–1699.

FROST, S. H., BLIEFNICK, D. M., AND HARRIS, P. M., 1983, Deposition and porosity evolution of a Lower Cretaceous rudist buildup, Shuaiba Formation of eastern Arabian Peninsula: Society of Economic Paleontologists and Mineralogists, Core Workshop No. 4, p. 381–410.

GLENNIE, K. W., BOEUF, M. G. A., HUGHES-CLARKE, M. W., MOODY-STUART, M., PILAAR, W. F. H., AND REINHARDT, B. M., 1974, Geology of the Oman Mountains, parts 1, 2, and 3: Verhandeligen van het Koninklijk Nederlands Geologisch, v. 31, Martinus Nijhoff, The Hague, 2123 p.

HANCOCK, J. M., AND KAUFFMAN, E. G., 1979, The great transgressions of the Late Cretaceous: Journal of the Geological Society, v. 136, p. 175–186.

HAQ, B. U., HARDENBOL, J., AND VAIL, P. R., 1987, Chronology of fluctuating sea levels since the Triassic: Science, v. 235, p. 1156–1167.

HARRIS, P. M., AND FROST, S. H., 1984, Middle Cretaceous carbonate reservoirs, Fahud Field and Northwestern Oman: American Association of Petroleum Geologists Bulletin, v. 68, p. 649–658.

———, ———, SEIGLIE, G. A., AND SCHNEIDERMANN, N., 1984, Regional unconformities and depositional cycles, Cretaceous of the Arabian Peninsula, in Schlee, J. S., ed., Interregional Unconformities and Hydrocarbon Accumulation: American Association of Petroleum Geologists Memoir 36, p. 67–80.

HASSAN, T. H., MUDD, G. C., AND TWOMBLEY, B. N., 1975, The stratigraphy and sedimentation of the Thamama Group (Lower Cretaceous) of the Abu Dhabi: Ninth Arab Petroleum Congress, Dubai, United Arab Emirates, Article 107 (B-3), 11 p.

ICE, R. G., AND MCNULTY, C. L., 1980, Foraminifers and calcispheres from the Cuesta del Cura and Lower Agua Nueva(?) Formations (Cretaceous) in east-central Mexico: Transactions, Gulf Coast Association of Geological Societies, v. 30, p. 403–425.

KAUFFMAN, E. G., 1977, Geological and biological overview: Western Interior Cretaceous basin: The Mountain Geologist, v. 14, p. 75–99.

———, 1984, Paleobiography and evolutionary response dynamic in the Cretaceous Western Interior seaway of North America: Geological Association of Canada Special Paper 27, p. 273–306.

KENT, D. V., AND GRADSTEIN, F. M., 1985, A Cretaceous and Jurassic geochronology: Geological Society of America Bulletin, v. 96, p. 1419–1427.

LONGORIA, J. F., 1984, Mesozoic tectonostratigraphic domains in east-central Mexico: Geological Association of Canada Special Paper 27, p. 65–76.

LOZO, F. E., Jr., AND STRICKLIN, F. L., Jr., 1956, Stratigraphic notes on the outcrop basal Cretaceous, Central Texas: Gulf Coast Association of Geological Societies Transactions, v. 6, p. 67–78.

MATSUMOTO, T., 1980, Inter-regional correlation of transgressions and regressions in the Cretaceous Period: Cretaceous Research, v. 1, p. 359–373.

MCFARLAN, EDWARD, JR., 1977, Lower Cretaceous sedimentary facies and sea level changes, U.S. Gulf Coast: University of Texas at Austin, Bureau of Economic Geology, Report of Investigations No. 89, p. 5–11.

MILLER, F. X., 1977, The graphic correlation method in biostratigraphy, in Kauffman, E. G., and Hazel, J. E., eds., Concepts and Methods of Biostratigraphy: Dowden, Hutchinson and Ross, Stroudsburg, Pennsylvania, p. 165–186.

MOLDOVANYI, E. P., AND LOHMANN, K. C., 1984, Isotopic and petrographic record of phreatic diagenesis: Lower Cretaceous Sligo and Cupido Formations: Journal of Sedimentary Petrology, v. 54, p. 972–985.

MÖRNER, N.-A., 1980, Relative sea-level, tectono-eustasy, geoidal-eustasy and geodynamics during the Cretaceous: Cretaceous Research, v. 1, p. 329–340.

MURRIS, R. J., 1980, Middle East: Stratigraphic evolution and oil habitat: American Association of Petroleum Geologists Bulletin, v. 64, p. 593–618.

PALMER, A. R., 1983, The decade of North American geology: 1983 geologic time scale: Geology, v. 11, p. 503–504.

PERKINS, B. F., 1961, Biostratigraphic Studies in the Comanche (Cretaceous) Series of Northern Mexico and Texas: Geological Society of America Memoir 83, 138 p.

PITMAN, W. C., III, 1978, Relationship between eustacy and stratigraphic sequences of passive margins: Geological Society of America Bulletin, v. 89, p. 1389–1403.

REY, J., 1982, Dynamique et paléoenvironnements du Bassin Mésozoique d'Estremadure (Portugal), au Crétacé Inférieur: Cretaceous Research, v. 3, p. 103–111.

ROSS, M. A., AND MCNULTY, C. L., 1981, Some microfossils of the Tamaulipas Limestone (Hauterivian-Lower Albian) in Santa Rosa Canyon, Sierra de Santa Rosa, Nuevo Leon, Mexico: Transactions, Gulf Coast Association of Geological Societies, v. 31, p. 461–469.

SAINT-MARC, P., 1978, Arabian Peninsula, in Moullade, M., and Nairn, A. E. M., eds., The Phanerozoic Geology of the World, II: The Meozoic (A): Elsevier, New York, p. 435–462.

SCOTT, R. W., 1984, Mesozoic biota and depositional systems of the Gulf of Mexico—Caribbean region: Geological Association of Canada Special Paper 27, p. 49–64.

———, FEE, D., MAGEE, R., AND LAALI, H., 1978, Epeiric depositional models for the Lower Cretaceous Washita Group, north-central Texas: University of Texas at Austin, Bureau of Economic Geology, Report Investigations 94, 1–23 p.

SHAW, A. B., 1964, Time in Stratigraphy: McGraw-Hill, New York, 365 p.

SIMMONS, M. D., AND HART, M. B., 1987, The biostratigraphy and biofacies of the Early to Mid-Cretaceous carbonates of Wadi Mi'aiden, central Oman Mountains, in Hart, M. B., ed., Micropalaeontology of Carbonate Environments: The British Micropalaeontological Society, Ellis Horwood, Chichester, England, p. 176–207.

STRICKLIN, F. L., JR., SMITH, C. I., AND LOZO, F. E., 1971, Stratigraphy of Lower Cretaceous Trinity deposits of Central Texas: University of Texas at Austin, Bureau of Economic Geology, Report of Investigations 71, 63 p.

TREJO, M., 1975, Zonificacion del limite Aptiano-Albiano de Mexico: Revista del Instituto Mexicano del Petroleo, v. 7(3), p. 6–29.

TYRRELL, W. W., JR., AND SCOTT, R. W., 1988, Early Cretaceous shelf margins, Vernon Parish, Louisiana, in Bally, A. W., ed., Atlas of Seismic Stratigraphy, v. 3: American Association of Petroleum Geologists, Studies in Geology No. 27.

VAIL, P. R., MITCHUM, R. M., JR., AND THOMPSON, S., III, 1977, Seismic stratigraphy and global changes of sea level, part 4: Global cycles of relative changes of sea level, in Payton, C. E., ed., Seismic Stratigraphy—Applications to Hydrocarbon Exploration: American Association of Petroleum Geologists Memoir 26, p. 83–97.

YOUNG, K., 1983, Mexico. The Phanerozoic Geology of the World, II: The Mesozoic (B). Elsevier, The Netherlands, p. 61–68.

YOUNG, K., 1986, Cretaceous, marine inundations of the San Marcos Platform, Texas: Cretaceous Research, v. 7, p. 117–140.

APPENDIX GLOSSARY

Coastal onlap: the progressive landward onlap of coastal deposits in a depositional sequence.

Composite standard: a data base of fossil ranges that have been compiled by graphic correlation.

Depositional sequence: a stratigraphic unit composed of a succession of genetically related strata bounded at its top and base by unconformities or their equivalent conformities.

Diastem: a depositional break of short duration, usually less than the time of deposition represented by a formation.

Disconformity: a depositional break between parallel strata of relatively long duration.

Eustatic/Eustasy: pertaining to or the condition of worldwide changes in sea level.

Firm ground: substrate composed of stiff, uncemented sediment.

Graphic correlation: a method of correlation in which the stratigraphic section is compared to the composite of fossil ranges in many other sections.

Hardground: substrate composed of indurated, cemented sediment or rock at an omission surface.

Line of correlation: the regression line that compares a

stratigraphic section to the composite of fossil ranges derived from numerous other sections.

Regression: a retreat of the sea resulting in emergence of land and possibly with progradation of sedimentary facies.

Sea level: a datum defined by mean position of the sea surface relative to the land surface.

Seismic sequence: a depositional sequence recognized by seismic reflection terminations.

Transgressions: an advance of the sea resulting in drowning of land.

Unconformity: a discordant stratigraphic contact showing evidence of erosion or nondeposition with stratal termination or with omissions in the biostratigraphic succession.

RECORD OF RELATIVE SEA-LEVEL CHANGES, CRETACEOUS OF WESTERN INTERIOR, USA

ROBERT J. WEIMER

Professor Emeritus, Department of Geology, Colorado School of Mines, Golden, Colorado 80401

A SUMMARY

Two types of criteria are used to recognize relative changes in sea level in the Cretaceous of the Western Interior (Fig. 1). The first type is a highstand condition identified by: (1) highstand regression of the shoreline, depositing widespread shallow-marine sandstone and shale and shoreline sandstone, sometimese overlain by a widespread coal layer (Fig. 2); (2) deposits that fill incised drainage, reflecting rising sea level and landward movement of the shoreline (transgression) associated with coastal onlap; the incised valley fill may be zoned—more freshwater environments in the lower part and brackish to marine environments in the upper part (Figs. 3, 4); (3) recognition of one or more of the following in a marine condensed section: missing faunal zones; concentrations of phosphate nodules and/or glauconite; organic-rich shale with high total organic content; recrystallized shell debris forming thin lenticular limestone layers or shell hash in shale; residual concentrations of coarse-grained sand with chert pebbles and/or bone and teeth fragments on a transgressive surface of erosion. The second type is a lowstand condition recognized by: (1) lowstand surface of erosion with incised drainage; paleosols and root zones (causing zones of early cementation) may be preserved in marine shale or other deposits under an erosional surface (Fig. 3); this type of erosional surface is a major sequence boundary in analyses related to sequence stratigraphy; (2) more or less uniform depth of erosion by streams over large areas because of lowered base level; (3) missing shoreline and shallow marine sandstone facies; freshwater deposits rest on marine shales; (4) correlation with relative lowering of sea level and unconformities on other continents. Intrabasin fault block movement, creating topographic relief, may influence location of incised drainages (Figs. 3, 4).

Recognition of the lowstand surface of erosion is essential to documenting changes in sea level, because most of the deposits result from highstand conditions. Intrabasin tectonics may cause local unconformities that must be separated from breaks caused by global eustasy.

Nine major regional to near-regional unconformities[1] have been identified in the Cretaceous of the Western Interior of the United States. Five of these unconformities have been related to changes in sea level and to well-known regressive-transgressive cycles. Data are incomplete as to whether or not the origin of the other four unconformities is related to tectonic movement, sea-level changes, or reduction in sediment supply to the basin resulting in submarine erosional surfaces and condensed sections.

The unconformities are grouped overall into three types: those completely within non-marine strata, such as at the base and top of the Cretaceous; those involving both marine and non-marine strata; and, those within marine strata as currently mapped.

Uncertainty exists in dating many of the unconformities. By use of the time scale of Obradovich and Cobban (1975), however, with subsequent minor revisions (Fouch, 1983), the approximate dates for unconformities (erosional surfaces) are estimated as follows (formations involved are in parenthesis; numbers are millions of years before the present): (1) 112 ± (base of the lower Mannville, Lakota, Lytle); (2) 100 ± (upper Mannville, Fall River, Plainview); (3) 97 ± (Viking, Muddy, Newcastle, or J Sandstone); (4) 95 ± (lower Frontier-Peay, and D); (5) 90 ± (base of the upper Frontier or upper Carlile); (6) 89 ± (base of the Niobrara or equivalents); (7) 80 ± (Eagle, lower Pierre, and upper Niobrara); (8) 73 ± (mid-Mesaverde, Ericson, base of the Teapot); (9) 66 ± (top of the Lance or equivalents). Variations in the accuracy of the dating are probably within 1 to 2 million years because of problems in defining accurately the biostratigraphic level of the erosional surfaces, in determining the hiatus, and in the precision of radiometric dates. The stratigraphic positions of these major unconformities are shown in an east-west section across the Western Interior Cretaceous basin (Fig. 5). Formation names are from Wyoming and Colorado, except for the Mannville Formation which is from southern Alberta, Canada. A summary of major changes in sea level interpreted from the rock record and the problem of the placement of stage boundaries by different workers are illustrated in Figure 6.

The surfaces of erosion associated with the major unconformities are related to three different processes. The first process, an adjustment of the base level of drainages to a lower sea level (LSE, Fig. 3) produces significant surfaces of erosion associated with subaerial exposure. In the Cretaceous, these surfaces may show as much as 50 m of relief. The second process, water deepening on a rising sea level (or subsiding depositional surface), produces an erosional surface (TSE, Fig. 4) related to transgression of the shoreline-shoreface zone. Processes described on modern shelfs are an analog for shoreface erosion that occurs as wave-generated energy moves over the shallow-shelf, shoreline, and coastal-plain area during the drowning event. Finer materials are removed landward and coarser material is left as a lag on the erosional surface. Sedimentation is slow, or absent; thus, a condensed section, normally with a high total organic content, covers the erosional surface (called a transgressive or ravinement surface). The magnitude of erosion by these processes is as much as 10 m, and the relief on the erosional surface is usually less than 1 m. A third type of surface is related to erosion or nondeposition in marine condensed sections.

---

[1]An unconformity is defined as a sedimentary structure in which two groups of rocks are separated by an erosional surface; the erosion may be by subaerial or submarine processes.

Sea-Level Changes—An Integrated Approach, SEPM Special Publication No. 42

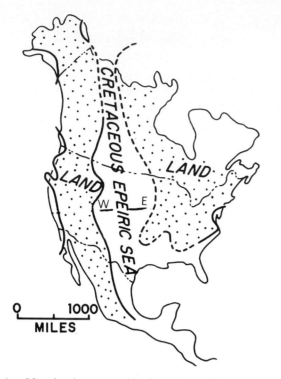

FIG. 1.—Map showing geographic distribution of Cretaceous seaway in interior of continent. Location of cross section W-E (Fig. 5) is indicated (from Weimer, 1984).

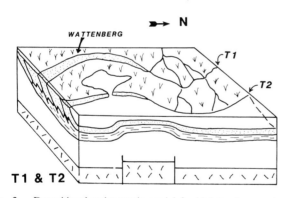

FIG. 2.—Depositional and tectonic model for highstand regression over basement fault blocks with penecontemporaneous fault movement. T_1 = Time 1; T_2 = Time 2. Rate of sediment supply exceeds rate of subsidence or submergence. Not to scale (from Weimer, 1984).

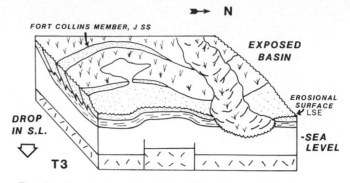

FIG. 3.—Lowstand sea level (T_3 = Time 3) recorded as basin-wide erosional surface (LSE) resulting from subaerial exposure (major sequence boundary). Root zones form on exposed marine shales and sandstones. Not to scale (from Weimer, 1984).

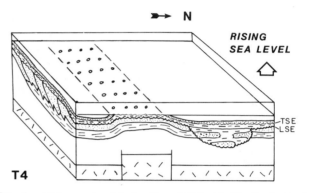

FIG. 4.—Rising sea level during Time 4 (T_4) with fill of incised valley and deposition of marine shale and sandstones. A thin transgressive lag (generally <1 ft thick), sometimes with coarse-grained material, occurs on a surface of erosion (transgressive surface) at top of sequence. Lowstand surface of erosion (LSE) and the transgressive surface of erosion (TSE) are labeled. Not to scale (from Weimer, 1984).

Additional work is needed on the marine unconformities (Fig. 5) in order to determine if erosion or nondeposition was dominant and whether or not an earlier surface of erosion related to a sea-level lowstand may have merged with and been modified by a transgressive event.

For a detailed discussion of the concepts described in this summary, the reader is referred to papers by Weimer and others (1982), Weimer (1984), Weimer and others (1985), Weimer and others (1987), and Hancock and Kauffman (1979).

REFERENCES

FOUCH, T. D., 1983, Patterns of synorogenic sedimentation in Upper Cretaceous rocks of central and northeastern Utah, in Reynolds, M., and Dolly, E., Mesozoic Paleogeography of West-Central United States: Rocky Mountain Section, Society of Economic Paleontologists and Mineralogists, Special Publication, p. 305–336.

HANCOCK, J. M., 1975, The sequence of facies in the Upper Cretaceous of northern Europe compared with that in the Western Interior, in Caldwell, W. G. F., ed., The Cretaceous Systems in the Western Interior of North America: Geological Association of Canada Special Paper 13, p. 82–118.

———, AND KAUFFMAN, E. G., 1979, The great transgressions of the Late Cretaceous: Geological Society of London Journal, v. 136, p. 175–186.

KAUFFMAN, E. G., 1977, Upper Cretaceous cyclothems, biotas, and environments, Rock Canyon anticline, Pueblo, Colorado: The Mountain Geologist, v. 14, p. 129–152.

LANPHERE, M. A., AND JONES, D. L., 1978, Cretaceous time scale for North America, in Cohee, G. V., and Glaessner, M. F., eds., The Geologic Time Scale: American Association of Petroleum Geologists Studies in Geology No. 6, p. 259–268.

McGOOKEY, D. P., 1972, Cretaceous System, in Mallory, W. W., ed., Geologic Atlas, Rocky Mountain Region: Rocky Mountain Association of Geologists Special Publication, p. 190–228.

OBRADOVICH, J. D., AND COBBAN, W. A., 1975, A time scale for the Late

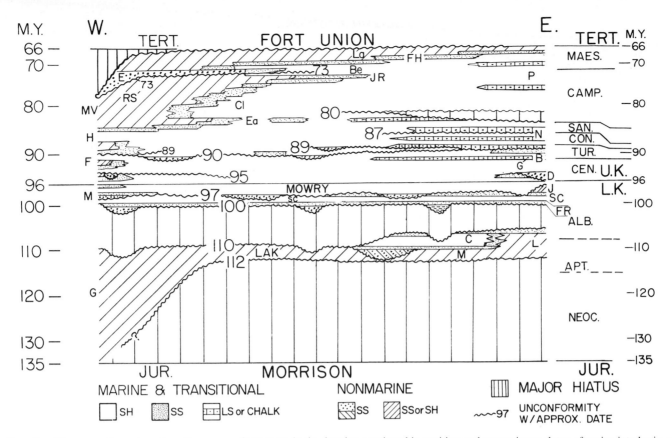

FIG. 5.—Diagrammatic east-west section across Cretaceous basin showing stratigraphic position and approximate dates of major intrabasin unconformities (modified after McGookey, 1972). Formations or groups to the west are: Ga = Gannett; SC = Skull Creek; M = Mowry; F = Frontier; H = Hilliard; MV = Mesaverde; RS = Rock Springs; E = Ericson; Ea = Eagle; Cl = Claggett; JR = Judith River; Be = Bearpaw; FH = Fox Hills; La = Lance. To the east, formations are: L = Lytle; LAK = Lakota; FR = Fall River; SC = Skull Creek; J and D = sandstones of Denver basin; G = Greenhorn; B = Benton; N = Niobrara; P = Pierre; M and C = the McMurray and Clearwater of Canada. The vertical ruled lines represent unconformities where a major hiatus is recognized. When the gap is 1 million years or less, vertical ruling is omitted. Sections removed by erosion under surfaces of erosion have not been restored (from Weimer, 1984).

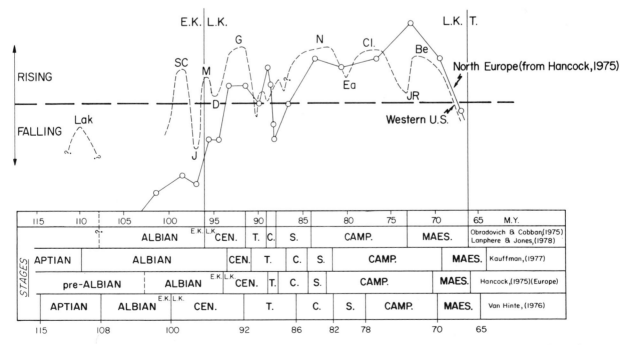

FIG. 6.—Sea-level curves for the United States and Europe (modified from Hancock, 1975). Letters designate the same formations as on Figure 5. Abbreviations for stages: Cen. = Cenomanian; T. = Turonian; C. = Coniacian; S. = Santonian; Camp. = Campanian; Maes. = Maestrichtian; E.K. = Early Cretaceous; L.K. = Late Cretaceous; T. = Tertiary (from Weimer, 1984).

Cretaceous of the Western Interior of North America, *in* Caldwell, W. G. F., ed., The Cretaceous System in the Western Interior of North America: Geological Association of Canada Special Paper 13, p. 31–54.

VAN HINTE, J. E., 1976, A Cretaceous time scale: American Association of Petroleum Geologists Bulletin, v. 60, p. 498–516.

WEIMER, R. J., 1984, Relation of unconformities, tectonics, and sea-level changes, Cretaceous of Western Interior, U.S.A., *in* Schlee, J. S., ed., Interregional Unconformities and Hydrocarbon Accumulation: American Association of Petroleum Geologists Memoir 36, p. 7–35.

————, EMME, J. J., FARMER, C. L., ANNA, L. O., DAVIS, T. L., AND

KIDNEY, R. L., 1982, Tectonic influence on sedimentation, Early Cretaceous, east flank, Powder River basin, Wyoming and South Dakota: Colorado School of Mines Quarterly, v. 77, 61 p.

————, PORTER, K. W., AND LAND, C. B., 1985, Depositional modeling of detrital rocks with emphasis on cored sequences of petroleum reservoirs: Society of Economic Paleontologists and Mineralogists, Core Workshop No. 8, 252 p.

————, SONNENBERG, S. A., AND YOUNG, G. B. C., 1987, Wattenberg field, Denver basin, Colorado, *in* Spencer, C. W., and Mast, R. F., eds., Geology of Tight Reservoirs: American Association of Petroleum Geologists Studies in Geology No. 24, p. 143–164.

FORAMINIFERAL MODELING OF SEA-LEVEL CHANGE IN THE LATE CRETACEOUS OF NEW JERSEY

RICHARD K. OLSSON

Department of Geological Sciences, Rutgers University, New Brunswick, New Jersey 08903

ABSTRACT: Paleoslope models of foraminifera in the Upper Cretaceous of the New Jersey coastal plain are utilized to estimate paleo-bathymetric change during cycles of rising and falling sea level. The paleoslope method estimates change in sea level from the distribution of foraminiferal assemblages and species on a baseline parallel to the regional dip. The paleoslope is the restoration of the original depositional slope of the basin. The paleobathymetry of the foraminifera along the paleoslope is estimated from the gradient of the original depositional slope. Application of the paleoslope model to the Campanian of New Jersey indicates a maximum rise of sea level of 90 m and 80 m, respectively, during two cycles of sea-level change. By extension, a paleodepth curve is derived for the other cycles in the Late Cretaceous. Eight cycles are recognized in the Late Cretaceous section of New Jersey.

INTRODUCTION

The development of sequence stratigraphy (Vail and others, 1977) has focused attention on the cycles of sea-level change within which sequences are deposited. Vail and Hardenbol (1979) used onlap-offlap sequences to derive a sea-level curve showing relative changes in sea level for the Tertiary. These relative changes in sea level are interpreted by Vail and others (1984) and Haq and others (1987) as due to eustatic changes superimposed on a long-term eustatic sea-level curve. If stratigraphic sequences develop during cycles of eustatic rise and fall of sea level, knowledge of the magnitude of eustatic change is important to understanding the mechanism(s) that cause the change. Estimation of the magnitude of eustatic change relative to present sea level has been the most elusive data of all to obtain in analyzing sequence stratigraphy. In this paper, a technique utilizing foraminifera is derived for estimating paleo-sea level relative to the present sea level and to derive a eustatic curve for the Late Cretaceous of the western Atlantic margin of New Jersey.

Foraminifera are the most widely used fossil organisms for estimating paleodepth. Paleobathymetric studies utilizing foraminifera have developed around the concepts of direct comparison with modern distribution when dealing with extant species, homeomorphic comparison of fossil and living species (pioneered by Bandy, 1960), and determination of non-specific characteristics such as benthic/planktonic ratios and diversity trends. All of these approaches have proven useful in making estimates of paleodepth. Even so, the accuracy of such estimates is compromised by the uncertainty of a best match with modern depth distribution and, in the case of increasingly older Cenozoic and Cretaceous assemblages, uncertainty about the ecologic role played by extinct species. The problem is compounded when studies are conducted on isolated sections or a single borehole (as most are).

In such studies it is difficult to establish a frame of reference with which to compare foraminiferal assemblages in a stratigraphic section or well to the lateral basinal distribution of foraminiferal species within a chronostratigraphic interval. Thus, most estimates of the paleobathymetry of a foraminiferal assemblage are abstractions derived from a general comparison with modern distributions. A more meaningful comparison would be one with other assemblages within the same chronostratigraphic interval. The method, described here, develops data on the lateral basinward distribution of foraminiferal species within and between chronostratigraphic units. The paleobathymetric significance of distributional patterns of species is derived from the original slope of the basin floor, called the paleoslope. This method does not rely on comparison with modern foraminiferal assemblages. It measures the bathymetry directly from the paleoslope model.

THE PALEOSLOPE MODEL

A paleoslope model is a graphic reconstruction of the distribution of benthic foraminiferal species and assemblages along a profile parallel to the dip slope of the basin (called the paleoslope). In the paleoslope model, distance downdip is used as a measure of increasing paleodepth. Thus, the paleoslope model relates the abundance and distribution of benthic foraminifera along the profile to paleodepth. This allows a more critical evaluation of the role played by benthic foraminifera in ancient shelf, slope, and deep-sea environments.

The Atlantic passive margin of the United States is particularly well suited for paleoslope studies because it is structurally uncomplicated, and because chronostratigraphic intervals can be traced across an entire basin (Fig. 1). The coastal plain of New Jersey has long been recognized for its well-developed Cretaceous and Tertiary stratigraphy and its record of numerous sea-level changes. Furthermore, the coastal plain, which lies at the edge of the Baltimore Canyon Trough and contains non-marine, nearshore, and shelf lithofacies, offers an opportunity to trace foraminiferal assemblages from shoreline environments to shelf and upper-slope environments. A paleoslope model was first completed in the New Jersey coastal plain for the Campanian and Lower Maestrichtian by utilizing samples from outcrops and wells (Fig. 2) in formations deposited during four events of sea-level change. This paleoslope model (Olsson and Nyong, 1984) shows the abundance distribution of key benthic foraminiferal species for inner-shelf to upper-slope environments (Fig. 3). The paleoslope model was subsequently extended into wells in the Baltimore Canyon Trough and into Deep Sea Drilling Sites (Fig. 4) in the Western Atlantic Basin (Nyong and Olsson, 1984).

The paleoslope model makes the assumption that the bathymetric profile has remained constant throughout the history of subsidence of the basin, at least for shelf envi-

Sea-Level Changes—An Integrated Approach, SEPM Special Publication No. 42

idium peaks are present in the section. The lowest anomaly occurs in the Maastrichtian, the middle anomaly near the K-T boundary, and the upper anomaly at the burrowed/bored (limestone glauconite mudstone) surface at 5.5 m (18 ft). The uppermost anomaly was first reported by Baum and others (1984). At most K-T boundary localities, one iridium peak, located just below the K-T transition, is reported (Smit and Romein, 1985).

SEQUENCE STRATIGRAPHY OF THE BRAGGS LOCALITY

A sequence analysis of the Braggs K-T boundary section is presented in Figure 8. The sequence boundary at the top of the Prairie Bluff Formation is defined by regional truncation (Fig. 2), localized incised-valley fills, and local paleosol development (Donovan, 1986). This sequence boundary is related to a major eustatic fall during the latest Maastrichtian (67 Ma), which coincides with the start of the Tejas supercycle (Fig. 9). This boundary is also dis-

cussed by Greenlee and Moore (this volume). With the subsequent initial rise in sea level, local incised topography filled with sediment. The basal transgressive surface of the Clayton Formation formed when sea level rose enough to flood back across the exposed paleoshelf.

The interbedded muddy sandstones and sandy packstones in the basal portions of the Clayton Formation at Braggs are interpreted to represent two parasequences within the transgressive systems tract. The faunal K-T boundary occurs within this transgressive systems tract, which is dated from the latest Maastrichtian through the earliest Danian. The glauconitic mudstones, which overlie the basal sandstone and limestone strata of the Clayton Formation, mark a condensed section formed during the maximum flooding onto the paleoshelf. The time of maximum flooding of this condensed section occurs within faunal zone NP1. This is in agreement with an age of 66 million years for the downlap surface of the interpreted cycle (Figure 9). The increase in terrigenous sediments above 7 m (23 ft) on the measured

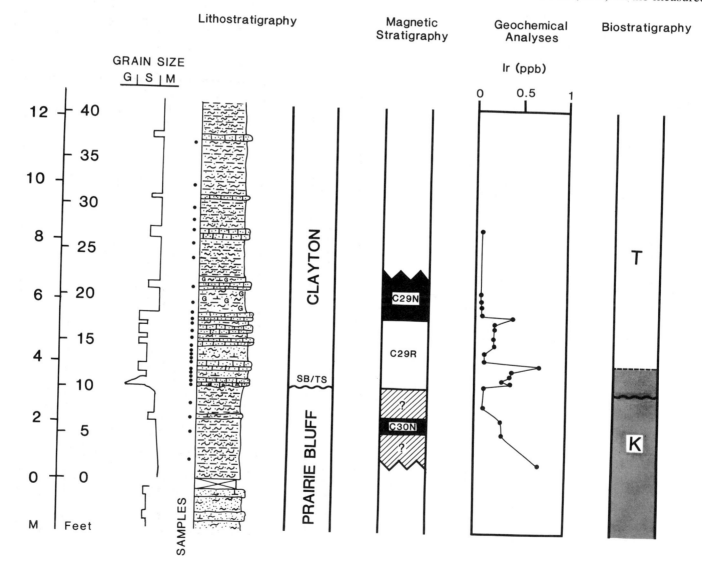

FIG. 7.—Iridium distribution and magnetic stratigraphy of the Braggs locality. Iridium analysis by fire assay on carbonate-free basis. Magnetic stratigraphy from Jones and others (1987).

SUMMARY

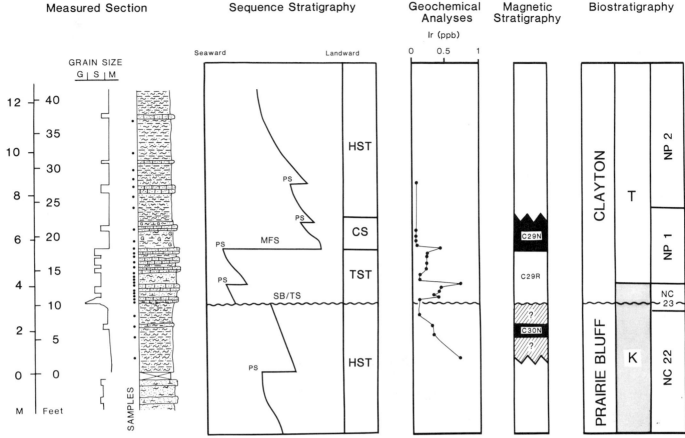

FIG. 8.—Sequence stratigraphic interpretation of the Braggs K-T boundary locality. Higher iridium concentrations occur at flooding surfaces of interpreted parasequence boundaries from the late Maastrichtian through earliest Danian. TST=transgressive systems tract, HST=highstand systems tract, CS=condensed section, SB=sequence boundary, TS=transgressive surface, MFS=maximum flooding surface, PS=parasequence boundary.

section marks the beginning of highstand deposition. Figure 10 is a sequence-keyed interpretation of the geophysical logs for the U.S. Geological Survey Braggs #1 test well drilled at this locality.

GEOCHEMICAL ANOMALIES AND THE K-T BOUNDARY AT BRAGGS: A SUMMARY

Of the three distinct iridium anomalies at Braggs (Fig. 8), the lowest occurs in the late Maastrichtian, the middle near the K-T boundary, and the upper within faunal zone NP1. These anomalies coincide with marine-flooding surfaces interpreted as parasequence boundaries. The uppermost parasequence boundary is also the base of the well-developed condensed section. The presence of iridium at these flooding surfaces suggests that iridium was present in the open ocean from the latest Maastrichtian through earliest Danian and was concentrated on the Alabama paleoshelf during periods of terrigenous-sediment starvation caused by rapid sea-level rises. Thus, it appears that iridium was not introduced into the atmosphere during a unique event

occurring at the K-T boundary, but was present in the atmosphere for a much longer period of time.

Although Alvarez and others (1982) have generally discounted the effects of sea-level changes on the deeper marine record across the K-T boundary, we believe this assumption is incorrect. Eustatic falls and rises produce major changes in basin configuration, stratal patterns, and sedimentation rates, as well as in oceanic geochemistry. The major eustatic fall in the latest Maastrichtian, marked by the withdrawal of epicontinental seas throughout the globe (Graban, 1940; Dott and Batten, 1981), and the subsequent eustatic rise had a tremendous effect on global depositional patterns, as well as on the biotic and geochemical conditions in the world's oceans. Globally, the K-T boundary is marked by a period of marine terrigenous-sediment starvation. Whatever the cause for the increase in iridium concentrations in the water column during this period, the decrease in terrigenous sedimentation associated with a global rise in sea level was a fundamental part of the process that concentrated the iridium and other possible cosmogenic debris in the sediments. Iridium enrichments due to dissolution, such as those observed at Stevns Klint in Denmark,

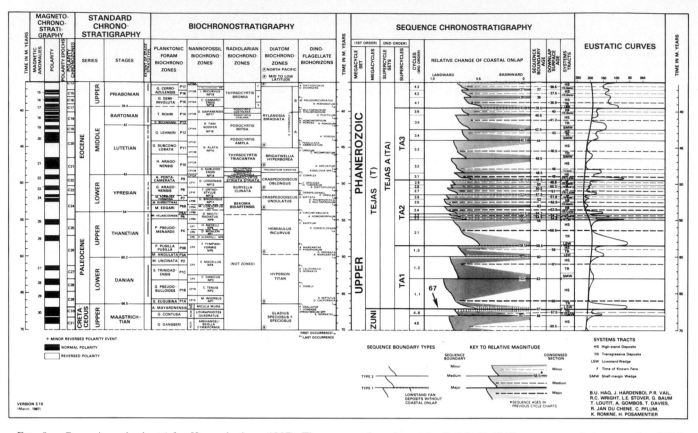

FIG. 9.—Cenozoic cycle chart (after Haq and others, 1987). The sequence boundary at the Prairie Bluff-Clayton contact is interpreted at 67 Ma, whereas the age of the maximum flooding surface of the condensed section (downlap surface) in the basal Clayton Formation is placed at 66 Ma.

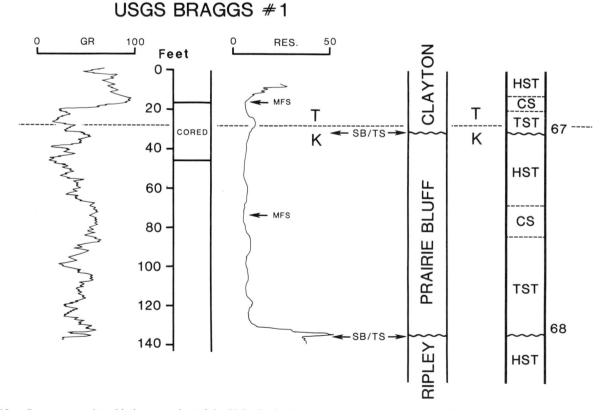

FIG. 10.—Sequence stratigraphic interpretation of the U.S. Geological Survey Braggs #1 Test Well. The log responses have been calibrated to both the U.S. Geological Survey core and the adjacent Braggs K-T boundary locality. The base of the Braggs measured section corresponds to a depth of 38 ft (12 m) on the log.

need also to be considered (Ekdale and Bromley, 1984). Basin margin localities, such as Braggs, offer a clearer, less compressed record of the K-T transition and suggest that sediment supply is an important factor controlling iridium enrichment.

ACKNOWLEDGMENTS

We especially thank John Van Wagoner for the many editorial comments and insightful discussions of sequence stratigraphy, which have greatly improved this manuscript. The aid and cooperation of Doug Jones and Jon Dobson of the University of Florida, Juergen Reinhardt of the U.S. Geological Survey, and Ernie Mancini of the Alabama Geological Survey are also greatly appreciated. Iridium analysis was conducted by Nuclear Activation Services Ltd., Hamilton, Ontario.

REFERENCES

ALVAREZ, L. W., ALVAREZ, W., ASARO, F., AND MICHEL, H. V., 1980, Extraterrestrial cause for the Cretaceous-Tertiary extinction: Science, v. 208, p. 1095–1108.

ALVAREZ, W., ALVAREZ, L. W., ASARO, F., AND MICHEL, H. V., 1979, Experimental evidence in support of an extraterrestrial trigger for the Cretaceous-Tertiary extinctions: EOS, v. 60, p. 734.

———, ———, ———, AND ———, 1982, Current status of the impact theory for the terminal Cretaceous extinction, in Silber, L.T., and Schultz, P. H., eds., Geological Implications of Impacts of Large Asteroids and Comets on the Earth: Geological Society of America Special Paper 190, p. 305–316.

BAUM, G. R., BLECHSCHMIDT, G. L., HARDENBOL, J., LOUTIT, T. S., VAIL, P. R., AND WRIGHT, R. C. 1984, The Maastrichtian/Danian boundary in Alabama: A stratigraphically condensed section: Geological Society of America, Abstracts with Programs, v. 16, p. 440.

BOHOR, B. F., MODRESKI, P. J., AND FOORD, E. E., 1987, Shocked quartz in the Cretaceous-Tertiary boundary clays: Evidence for a global distribution: Science, v. 225, p. 705–709.

COPELAND, C. W., AND MANCINI, E. A., 1986, Cretaceous-Tertiary boundary southeast of Braggs, Lowndes County, Alabama, in Neathery, T. L., ed., Centennial Field Guide Volume 6, Southeastern Section of the Geological Society of America, p. 369–372.

DONOVAN, A. D., 1985, Stratigraphy and sedimentology of the Upper Cretaceous Providence Formation (western Georgia and eastern Alabama): Unpublished Ph.D. Dissertation, Colorado School of Mines, Golden, Colorado, 236 p.

———, 1986, Sedimentology of the Providence Formation, in Reinhardt, J., ed., Stratigraphy and Sedimentology of Continental, Nearshore, and Marine Cretaceous Sediments of the Eastern Gulf Coastal Plain: American Association of Petroleum Geologists Annual Meeting, SEPM Field-trip Guidebook No. 3, p. 29–44.

DOTT, R. H., AND BATTEN, R. L., 1981, Evolution of the Earth: McGraw-Hill, New York, 573 p.

EKDALE, A. A., AND BROMLEY, R. G., 1984, Sedimentology and ichnology of the Cretaceous-Tertiary boundary in Denmark: Implications for the causes of the terminal Cretaceous extinction: Journal of Sedimentary Petrology, v. 54, p. 681–703.

GRABAU, A. W., 1940, The Rhythm of the Earth: Henri Vetch, Peking, 561 p.

HAQ, B. U., HARDENBOL, J., AND VAIL, P. R., 1987, Chronology of fluctuating sea levels since the Triassic: Science, v. 235, p. 1156–1167.

JARZEN, D. M., 1978, The terrestrial palynoflora from the Cretaceous-Tertiary transition, Alabama, USA: Pollen et Spores, v. 20, p. 535–553.

JONES, D. S., MULLER, P. A., BRYAN, J. R., DOBSON, J. P., CHANNELL, J. E., ZACHOS, J. C., AND ARTHUR, M. A., 1987, Biotic, geochemical, and paleomagnetic changes across the Cretaceous-Tertiary boundary at Braggs, Alabama: Geology, v. 15, p. 311–315.

LAMOREAUX, P. E., AND TOULMIN, L. D., 1959, Geology and groundwater resources of Wilcox County, Alabama: Alabama Geological Survey County Report 4, 280 p.

MITCHUM, R. M., 1977, Seismic stratigraphy and global changes of sea level, part 1: Glossary of terms used in seismic stratigraphy, in Payton, C.E., ed., Seismic Stratigraphy-Applications to Hydrocarbon Exploration: American Association of Petroleum Geologists Memoir 26, p. 205–212.

SMIT J., AND ROMEIN, A. J., 1985, A sequence of events across the Cretaceous-Tertiary boundary: Earth and Planetary Science Letters, v. 74, p. 155–170.

SMITH C. C., MANCINI E. A., AND RUSSELL, E. E., 1984, The Cretaceous-Tertiary boundary in eastern Mississippi and western Alabama: Lithostratigraphy and biostratigraphy (Abst.): Geological Society of America, South-Central Section, Abstracts with Programs v. 16, p. 113.

SMITH, J. K., 1978, Ostracoda of the Prairie Bluff Chalk, Upper Cretaceous (Maastrichtian), and the Pine Barren Member of the Clayton Formation, Lower Paleocene (Danian), from exposures along Alabama State Highway 263 in Lowndes County, Alabama: Gulf Coast Assoications of Geological Societies, Transactions, v. 28, p. 539–579.

VAN WAGONER, J. C., 1985, Reservoir facies distribution controlled by sea level change (Abst.): Society of Economic Paleontologists and Mineralogists Annual Midyear Meeting, Abstracts with Programs, p. 91–92.

WORSLEY, T., 1974, The Cretaceous-Tertiary boundary event in the ocean, in Hay, W. W., ed., Studies in Paleo-oceanography: Society of Economic Paleontologists and Mineralogists Special Publication 20, p. 94–125.

SEQUENCE STRATIGRAPHIC CONCEPTS APPLIED TO PALEOGENE OUTCROPS, GULF AND ATLANTIC BASINS

GERALD R. BAUM AND PETER R. VAIL[1]

ARCO Oil and Gas Company, 2300 West Plano Parkway, Plano, Texas 75075;
Exxon Production Research Company, P.O. Box 2189, Houston, Texas 77252-2189

ABSTRACT: Type 1 and type 2 sequence boundaries can be used for regional correlation in seismic, wireline log, and outcrop data. Marine condensed sections (zones of markedly reduced sedimentation) divide these sequences and are recognized seismically as downlap surfaces. Sequence boundaries can be dated at their basinward correlative conformities. Depositional sequences are not synthems or allostratigraphic units. Synthems or allostratigraphic units are extended only as far as both of the bounding unconformities or discontinuities are identifiable. Sequences are bounded by unconformities and their correlative conformities and so are identifiable beyond the extent of their bounding discontinuities. Because most of the exposed Paleogene units in the Gulf and Atlantic basins were deposited landward of their respective shelf slope breaks, evidence of deposition of deep-sea fans common to type 1 unconformities is precluded. Regional mapping, however, generally reveals discontinuous incised valleys that are indicative of type 1 unconformities. Typically, the incised valleys are onlap-filled with reservoir-prone fluvial-to-estuarine sediments. In additonal, sequence boundaries are characterized by abrupt downward shifts in facies with relatively shallower water facies resting sharply on relatively deeper water facies. In carbonates, subaerial unconformities are typically characterized by mesokarst, phosphate pebble conglomerates, and sediment fill of early moldic porosity.

Condensed sections are characterized by anomalous concentrations of mammillated-to-lobate glauconite, planktonic organisms, phosphate, and exotic minerals, and by glauconitized/phosphatized surfaces commonly associated with hardgrounds or burrowed omission surfaces. Hardgrounds are characterized by intercrystalline sediment fill after subaqueous, acicular, bladed, and/or pelloidal marine cements, and by abrupt shifts to more negative $\delta^{13}C$ values of calcite above the hardgrounds associated with condensed sections.

Application of these concepts to outcrop studies reveals that many stage boundaries are typically not placed at sequence boundaries. Rather, they are defined either by micropaleontologic hiatuses and/or planktonic zonal boundaries associated with condensed sections, or by transgressive (flooding) surfaces overlying incised-valley-fill sediments. Also, the currently recognized European and Gulf Coast stages do not adequately reflect the higher frequency coastal-onlap cycles recognized in outcrop.

Because most micropaleontologic zones appear to span sequence boundaries, the current micropaleontologic zonations cannot, at present, precisely define a sequence boundary in time. They can approximate sequence position, however. By intergrating physical stratigraphy, seismic stratigraphy, and paleontology, these higher frequency eustatic events can be resolved and fixed in a relative time framework.

INTRODUCTION

The development and refinement of seismic stratigraphic techniques over the past decade have added the dimension of large-scale stratal geometries to stratigraphic correlations and have given renewed impetus to regional correlations based upon the recognition of depositional sequences (Fig. 1; Vail, 1976; Todd and Mitchum, 1977; Vail and others, 1977; Vail and others, 1980; Vail and others, 1982; Vail and Todd, 1981). Seismic data compensate for the generally incomplete rock record in outcrop; however, due to the limits of seismic resolution, outcrop-based studies have provided a direct method to document the age and physical character of sequence boundaries. The integration of outcrop and seismic observations has provided a framework to divide sequences into component parts, as well as to refine the coastal-onlap chart (Figs. 1, 2, 3). An integrated stratigraphic study is the only method to relate land-based stratotypes to global coastal-onlap cycles.

Sequence boundaries are not synthems (Salvador, 1987) or allostratigraphic units (North American Commission on Stratigraphic Nomenclature, 1983), because sequences can be identified beyond the extent of their bounding unconformities by correlating their correlative conformities. Synthems or allostratigraphic units are recognized only where their bounding unconformities or discontinuities can be identified.

Within the confidence level of paleontologic zonations,

the synchronism of depositional sequences has been demonstrated for the Eocene carbonates of the Carolinas (Baum and others, 1979a; Powell and Baum, 1982, 1984). Work was extended into the Paleogene of Alabama (Baum and others, 1982) and Europe to test the synchronism in clastic depositional settings within basins having different structural histories. The Gulf and Atlantic coastal plains have provided by far the most complete and most fossiliferous outcrops for study. Within the resolution limits of techniques available to define and to fix sequence boundaries in time, the sequence boundaries in the Gulf and Atlantic Tertiary basins are synchronous, thus supporting the idea that they are caused by eustasy (Vail, 1976; Vail and others, 1977; Vail and Hardenbol, 1979). Although structural histories leave an imprint on facies development, sequence distribution, and geometry (e.g., Eocene carbonates of the Carolinas, Baum and others, 1979a; Baum and others, 1979b; Baum, 1981; Powell, 1985), the overriding control of sequence boundaries appears to be eustasy. Likewise, sedimentation rates do not seem to affect the timing of sequences; however, carbonates tend to react more immediately ("give up"; Kendall and Schlager, 1981) to rapid sea-level changes.

SEQUENCE STRATIGRAPHIC CONCEPTS

Sediments record the response to changes in rate of accommodation (space available for sedimentation; Posamentier and others, this volume). Depositional models have suggested that accommodation is primarily a function of

---

[1]Present address: Department of Geology, Rice University, Houston, Texas 77251.

Sea-Level Changes—An Integrated Approach, SEPM Special Publication No. 42

SEQUENCE STRATIGRAPHY DEPOSITIONAL MODEL
SHOWING SURFACES AND SYSTEMS TRACTS

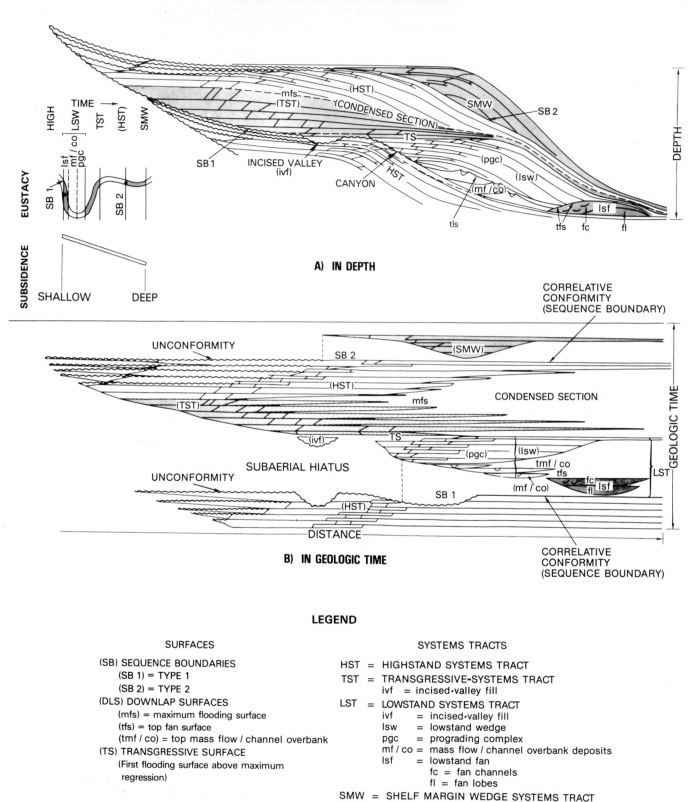

LEGEND

SURFACES

(SB) SEQUENCE BOUNDARIES
  (SB 1) = TYPE 1
  (SB 2) = TYPE 2
(DLS) DOWNLAP SURFACES
  (mfs) = maximum flooding surface
  (tfs) = top fan surface
  (tmf / co) = top mass flow / channel overbank
(TS) TRANSGRESSIVE SURFACE
  (First flooding surface above maximum regression)

SYSTEMS TRACTS

HST = HIGHSTAND SYSTEMS TRACT
TST = TRANSGRESSIVE-SYSTEMS TRACT
  ivf = incised-valley fill
LST = LOWSTAND SYSTEMS TRACT
  ivf = incised-valley fill
  lsw = lowstand wedge
  pgc = prograding complex
  mf / co = mass flow / channel overbank deposits
  lsf = lowstand fan
    fc = fan channels
    fl = fan lobes
SMW = SHELF MARGIN WEDGE SYSTEMS TRACT

FIG. 1.—Linear depth and linear geologic time displays for type 1 and type 2 depositional sequences. Note that the subaerial unconformity is not a time line; however, at its correlative conformity basinward (sequence boundary), it is a time boundary. This is the essence and distinction between the disciplines of allostratigraphy and sequence stratigraphy.

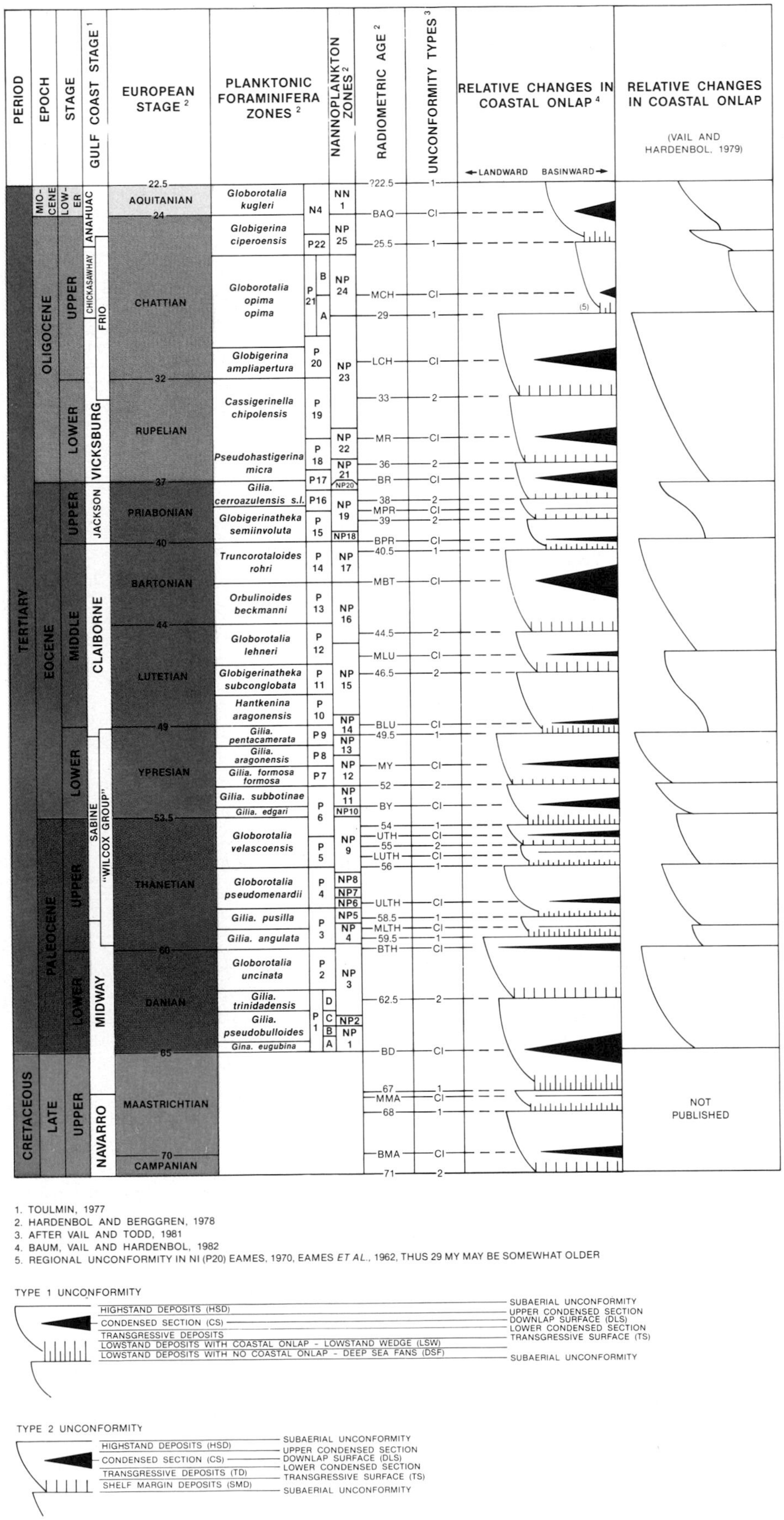

PERIOD	EPOCH	STAGE	GULF COAST STAGE [1]	EUROPEAN STAGE [2]	PLANKTONIC FORAMINIFERA ZONES [2]		NANNOPLANKTON ZONES [2]	RADIOMETRIC AGE [2]	UNCONFORMITY TYPES [3]	RELATIVE CHANGES IN COASTAL ONLAP [4]	RELATIVE CHANGES IN COASTAL ONLAP (VAIL AND HARDENBOL, 1979)
								22.5		←LANDWARD BASINWARD→	
								?22.5	1		
TERTIARY	MIOCENE	LOWER	ANAHUAC	AQUITANIAN	Globorotalia kugleri	N4	NN 1	24 / BAQ	CI		
	OLIGOCENE	UPPER	CHICKASAWHAY / FRIO	CHATTIAN	Globigerina ciperoensis	P22	NP 25	25.5	1		
					Globorotalia opima opima	P21 B / A	NP 24	MCH / 29	CI / 1	(5)	
					Globigerina ampliapertura	P 20	NP 23	LCH / 32	CI		
		LOWER	VICKSBURG	RUPELIAN	Cassigerinella chipolensis	P 19		33	2		
					Pseudohastigerina micra	P 18	NP 22	MR / 36	CI / 2		
						P17	NP 21 / NP20	BR / 37	CI		
	EOCENE	UPPER	JACKSON	PRIABONIAN	Gilia. cerroazulensis s.l.	P16	NP 19	38 / MPR / 39	CI / 2		
					Globigerinatheka semiinvoluta	P 15	NP18	BPR / 40	1 / 2		
		MIDDLE	CLAIBORNE	BARTONIAN	Truncorotaloides rohri	P 14	NP 17	40.5 / MBT	CI		
					Orbulinoides beckmanni	P 13	NP 16	44	2		
				LUTETIAN	Globorotalia lehneri	P 12		44.5 / MLU	CI		
					Globigerinatheka subconglobata	P 11	NP 15	46.5	2		
					Hantkenina aragonensis	P 10					
		LOWER	SABINE "WILCOX GROUP"	YPRESIAN	Gilia. pentacamerata	P9	NP 14	BLU / 49.5	CI / 1		
					Gilia. aragonensis	P8	NP 13	MY	CI		
					Gilia. formosa formosa	P7	NP 12	52			
					Gilia. subbotinae	P 6	NP 11	BY	1		
					Gilia. edgari		NP10				
	PALEOCENE	UPPER		THANETIAN	Globorotalia velascoensis	P 5	NP 9	54 / UTH / 55 / LUTH / 56	1 / 2 / 1		
					Globorotalia pseudomenardii	P 4	NP8 / NP7 / NP6	ULTH	CI		
					Gilia. pusilla	P 3	NP5 / NP 4	58.5 / MLTH / 59.5	1 / 1		
					Gilia. angulata			BTH	CI		
		LOWER	MIDWAY	DANIAN	Globorotalia uncinata	P 2	NP 3	60			
					Gilia. trinidadensis	P1 D / C	NP2	62.5	2		
					Gilia. pseudobulloides	B / A	NP 1	BD	CI		
					Gina. eugubina			65			
CRETACEOUS	LATE	UPPER	NAVARRO	MAASTRICHTIAN				67 / MMA / 68	1 / CI / 1		NOT PUBLISHED
				CAMPANIAN				70 / BMA / 71	CI / 2		

1. TOULMIN, 1977
2. HARDENBOL AND BERGGREN, 1978
3. AFTER VAIL AND TODD, 1981
4. BAUM, VAIL AND HARDENBOL, 1982
5. REGIONAL UNCONFORMITY IN NI (P20) EAMES, 1970, EAMES *ET AL.*, 1962, THUS 29 MY MAY BE SOMEWHAT OLDER

TYPE 1 UNCONFORMITY

HIGHSTAND DEPOSITS (HSD) — SUBAERIAL UNCONFORMITY
CONDENSED SECTION (CS) — UPPER CONDENSED SECTION
— DOWNLAP SURFACE (DLS)
— LOWER CONDENSED SECTION
TRANSGRESSIVE DEPOSITS — TRANSGRESSIVE SURFACE (TS)
LOWSTAND DEPOSITS WITH COASTAL ONLAP - LOWSTAND WEDGE (LSW)
LOWSTAND DEPOSITS WITH NO COASTAL ONLAP - DEEP SEA FANS (DSF) — SUBAERIAL UNCONFORMITY

TYPE 2 UNCONFORMITY

HIGHSTAND DEPOSITS (HSD) — SUBAERIAL UNCONFORMITY
CONDENSED SECTION (CS) — UPPER CONDENSED SECTION
— DOWNLAP SURFACE (DLS)
— LOWER CONDENSED SECTION
TRANSGRESSIVE DEPOSITS (TD) — TRANSGRESSIVE SURFACE (TS)
SHELF MARGIN DEPOSITS (SMD) — SUBAERIAL UNCONFORMITY

FIG. 2.—Coastal-onlap chart with type 1 and type 2 sequences differentiated (from Baum and others, 1982; Baum, 1986b).

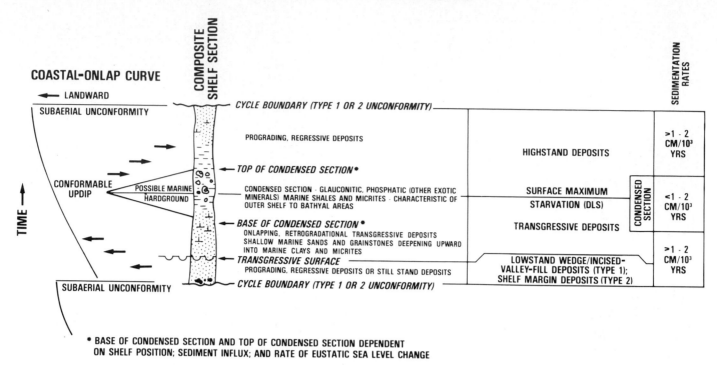

* BASE OF CONDENSED SECTION AND TOP OF CONDENSED SECTION DEPENDENT
ON SHELF POSITION; SEDIMENT INFLUX; AND RATE OF EUSTATIC SEA LEVEL CHANGE

FIG. 3.—Composite shelf section in relationship to coastal-onlap cycle (from Baum and others, 1984). The transgressive surface and surface of maximum starvation (seismic downlap surface, DLS, of Vail and others, 1984) tend to merge with the lower-cycle boundary in landward areas.

two factors: subsidence and change in sea level (Vail and Todd, 1981; Vail and others, 1982, 1984; Posamentier and others, this volume). These authors have suggested that rapid falls and rapid rises in sea level control stratal geometries on continental margins. The basic stratal unit is the depositional sequence defined by Mitchum and others (1977, p. 53) as "a relatively conformable succession of genetically related strata and bounded at its top and base by unconformities or their correlative conformities."

The bounding unconformities of the depositional sequences are interpreted to be produced by rapid falls in sea level. Vail and Todd (1981) classified the bounding unconformities into type 1 and type 2 unconformities. At a type 1 unconformity, the rate of sea-level fall is greater than the rate of subsidence at the depositional shelf edge. As a consequence, there is negative accommodation on the shelf, and shelf dissection commences. Deposition of point-sourced, deep-sea fans in basinal settings is possible at this time. As the rate of sea-level fall decreases and accommodation increases, sedimentation moves landward. At the maximum rate of sea-level rise, terrigenous sediments are trapped landward with concomitant sediment starvation in outer-shelf to bathyal regions. These sediment-starved intervals are termed condensed sections (Loutit and others, this volume). Highstand deposits prograde over the condensed section as the rate of rise of sea level and accommodation decreases.

At a type 2 unconformity, the rate of sea-level fall is less than the rate of subsidence at the depositional shelf edge. As a consequence, there is an abrupt downward shift in coastal onlap landward of the depositional shelf edge. Subsequent condensed-section and highstand deposits are pro-

duced in response to changes in sea level as described for a type 1 sequence.

The integration of seismic and outcrop observations made it possible for Baum and others (1982) to recognize these two types of depositional sequences and to divide a single depositional sequence into component systems tracts (Figs. 1, 3).

Type 1 depositional sequence.—

A type 1 depositional sequence is defined as a relatively conformable succession of genetically related strata and bounded at its base by a type 1 sequence boundary and at its top by either a type 1 or a type 2 sequence boundary (modified from Mitchum and others, 1977).

A type 1 depositional sequence can consist of: basal type 1 unconformity dated at its basinward correlative conformity; lowstand deposits without coastal onlap (deep-sea fans); lowstand deposits with coastal onlap (lowstand wedges; lowstand deltas; slope front fill; incised-valley-fill); transgressive surface; transgressive deposits; condensed section with a surface of maximum starvation; highstand regressive deposits; type 1 or type 2 unconformity dated at its basinward correlative conformity (Fig. 1; Baum and others, 1982). In shelf areas, within a type 1 depositional sequence, two additional boundaries are possible (Figs. 1, 3): transgressive surface and a surface of maximum starvation (characterized by downlapping highstand deposits). These two additional boundaries tend to onlap in a landward direction. Thus, onlapping and downlapping highstand deposits may rest directly on the unconformity if the boundary is traced far enough landward.

Type 2 depositional sequence.—

A type 2 depositional sequence is defined as a relatively conformable succession of genetically related strata and bounded at its base by a type 2 sequence boundary and at its top by a type 1 or type 2 sequence boundary (modified from Mitchum and others, 1977).

A type 2 depositional sequence can consist of: basal type 2 unconformity dated at its basinward correlative conformity; regressive or aggradational shelf margin deposits; transgressive surface; transgressive deposits; condensed section with a surface of maximum starvation; highstand, regressive deposits; type 1 or type 2 unconformity dated at its basinward correlative conformity (Fig. 1; Baum and others, 1982). Within a type 2 depositional sequence, two additional boundaries are possible (Figs. 1, 3): a transgressive surface and a surface of maximum starvation, both of which tend to onlap in a landward direction. Thus, as in a type 1 sequence, onlapping and downlapping highstand deposits may rest directly on the unconformity if the boundary is traced far enough landward.

RELATIONSHIP OF SEQUENCE STRATIGRAPHIC CONCEPTS TO COASTAL-PLAIN OUTCROP STRATIGRAPHY

Stratigraphic surfaces.—

Three significant mapping surfaces are recognized and carried with a degree of confidence on a regional scale in outcrop sections in the Atlantic and Gulf Coastal Plains (Figs. 1, 3). These surfaces are: (1) the depositional sequence boundary, interpreted to be related to a rapid fall in sea level; (2) the transgressive surface at the top of incised-valley-fill sediments, related to an abrupt increase in accommodation and/or a ravinement process as a relative rise in sea level floods the interfluvial areas; and (3) a surface of maximum starvation (condensed section) separating the transgressive deposits from the highstand deposits, also related to a relative rise in sea level.

Depositional sequence boundary.—

The unconformable portions of sequence boundaries, expressed seismically as downward shifts in coastal onlap (Vail and others, 1977), may be characterized by angular discordance, incised valleys (Fig. 4), or minor erosion. In carbonates, the unconformity may be characterized by karst (Fig. 4) and/or sediment from the overlying sequence that infills molds of originally aragonitic molluscs that have been selectively dissolved by the introduction of fresh water during lowstands of sea level (Figs. 5, 6). In all cases, the unconformity is sharp; however, if marine sediments in clastic sedimentary regimes directly overlie the unconformity, the surface is typically burrowed.

The basal lithologies directly overlying the unconformable portion of the sequence boundary may be characterized by lag gravels, rip-up clasts of the underlying lithologies, phosphate pebbles (Fig. 6); and concentrations of bone and sharks' teeth, glauconite, and phosphate. The phosphate and glauconite grains are rounded and/or fractured (Fig. 6). The high terrigenous content and detrital nature of the phosphate

FIG. 4.—(A) Incised-valley-fill deposits confined by 56 Ma unconformity and transgressive surface. The majority of the reservoir-prone sands will be confined to the incised valley. Henry County, Alabama. (B) 59.5- and 58.5-Ma unconformities and incised-valley-fill deposits. Henry County, Alabama.

and glauconite easily differentiate the basal deposits from the overlying condensed section.

The occurrence of rounded, phosphatized rip-up clasts at type 1 unconformities is fairly common within Cretaceous and Tertiary carbonates (Stephenson, 1929; Baum and others, 1979b; Baum, 1980) in the Gulf and Atlantic Coastal Plains. Phosphate and glauconite selectively replace the carbonate matrix and allochems within the clasts. The phosphate pebbles tend to be concentrated in depressions on the karst surface and are sharply overlain by higher energy, shallow-water lithologies. Thus, the phosphatization and rounding processes appear to take place in the intertidal zone during a relative rise in sea level.

Uncomformities are the most continuous and reliable surfaces for mapping on a regional scale. Although unconformities do not represent absolute time lines, they can be paleontologically dated at the correlative conformity. Thus, they are time boundaries which are useful in bracketing systems tracts and separating older rocks from younger rocks (Fig. 1).

Type 1 or type 2 sequence boundaries may not be readily

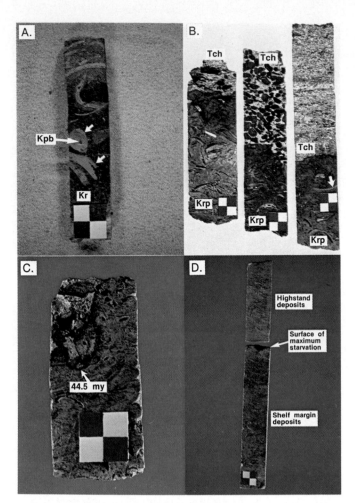

FIG. 5.—Split cores. (A) Ripley Formation (Kr) just below 68-Ma unconformity. Note Prairie Bluff Chalk (Kpb) infilling originally aragonitic
mollusc molds (arrows). Lowndes County, Alabama. (B) Phosphate-pebble conglomerate (center) and bryozoan biosparrudite facies of the middle
Eocene Castle Hayne Limestone (Tch) resting unconformably on Late
Cretaceous Rocky Point Member (Krp) of the Peedee Formation. Note
Castle Hayne Limestone infilling originally aragonitic mollusc molds (arrows). New Hanover County, North Carolina. (C) 44.5-Ma unconformity
within the Castle Hayne Limestone. Pender County, North Carolina. (D)
Surface of maximum starvation, TE 2.3 cycle (44.5 to 40.5 Ma) within
the Castle Hayne Limestone. Pender County, North Carolina.

differentiated in shelf areas; however, regional mapping
generally reveals erosional truncation due to shelf dissection with subsequent incised-valley-fill characteristic of a
Type 1 unconformity (Fig. 1).

Although sequence boundaries are named by a radiometric age (corresponding to the age of the sequence
boundary where it becomes conformable), the sequence
boundaries are dated paleontologically, generally with nannoplankton and planktonic foraminifera. The ranges of the
paleontologic zones are tied to global, composite radiometric-time scales. Currently, there are numerous composite radiometric-time scales, all varying slightly from one
another. Thus, some unnecessary confusion exists because
different authors prefer different time scales. The radiometric age of a sequence boundary may vary from author
to author, but the paleontologic age is the same.

The composite radiometric-time scale of Hardenbol and
Berggren (1978) is used here. Evaluation in a sequence
framework of published radiometric ages from the Gulf and
Atlantic Tertiary basins tends to substantiate this time scale.
There are a few contradictory published and proprietary radiometric ages from middle and late Eocene units that suggest some revision may be necessary. Until more data are
available and evaluated, however, Hardenbol and Berggren's time scale is still useful for naming and placing the
sequence boundaries in a relative time framework.

Incised-valley-fill and transgressive surface.—

Sediments that onlap and fill incised valleys (Fig. 4) consist of a complex system of fluvial, estuarine, and/or marine sands analogous to the J Sandstone (Horsetooth Member) in the Denver basin (Fig. 7; Land and Weimer, 1978).
The more proximal incised-valley-fill sediments are characterized by a complex of anastomosing fluvial to estuarine
channel-fill deposits. The more distal reaches of the incised
valleys are characterized by upward-shoaling, stacked sets
of marine to marginal marine sands. The incised-valley-fill
sediments tend to have a strong aggradational component
with no indication of landward movement of the shoreline
(Clifton, 1982; Mossop and Flach, 1983).

Incised-valley-fill deposits are bounded at their top by a
transgressive surface (Figs. 4, 7, 8) that separates the incised-valley-fill deposits from the overlying deposits of the
same sequence. The transgressive surface appears to represent a highly erosional event and is typically burrowed
(Fig. 8). The overlying lithologies generally contain abundant glauconite and have a deepening-upward component
(referred to here as transgressive deposits), thus indicating
a landward migration of the shoreline. In all cases observed, the reservoir-prone sands are confined to the incised valley below the transgressive surface. In the absence
of an incised valley, the transgressive surface coincides with
the underlying unconformity.

The transgressive surface appears to represent a complex
relationship where the abrupt increase in accommodation

FIG. 6.—Photomicrographs (except B). (A) Prairie Bluff Chalk (Kpb) infilling originally aragonitic mollusc (arrows) in Ripley Formation (Kr).
Thin section from core in Figure 5A. Lowndes County, Alabama. (B) 68-Ma unconformity surface at top of Ripley Formation. Note phosphate-
pebble conglomerates. Lowndes County, Alabama. (C) Phosphate (p) replacing matrix in a phosphate pebble from (B). (D) Rounded glauconite (g)
grains in quartz arenite facies of the Gosport Sand. Clarke County, Alabama. (E) Pyrite (p) replacing and infilling intraparticle porosity of foraminifera
test within the *Nummulites preswichianus* Zone of the Barton Formation. Alum Bay, England. (F) Glauconite (g) replacing and infilling intraparticle
porosity of a foraminifera test within the Matthews Landing Marl Member of the Porter's Creek Formation. Wilcox County, Alabama. (G) Glauconite
directly overlying surface of maximum starvation, TP 1.1 cycle (67 to 62.5 Ma). Lowndes County, Alabama. (H) Same as (G). Note neomorphism
of foraminifera. Outlined area in Figure 11 E.

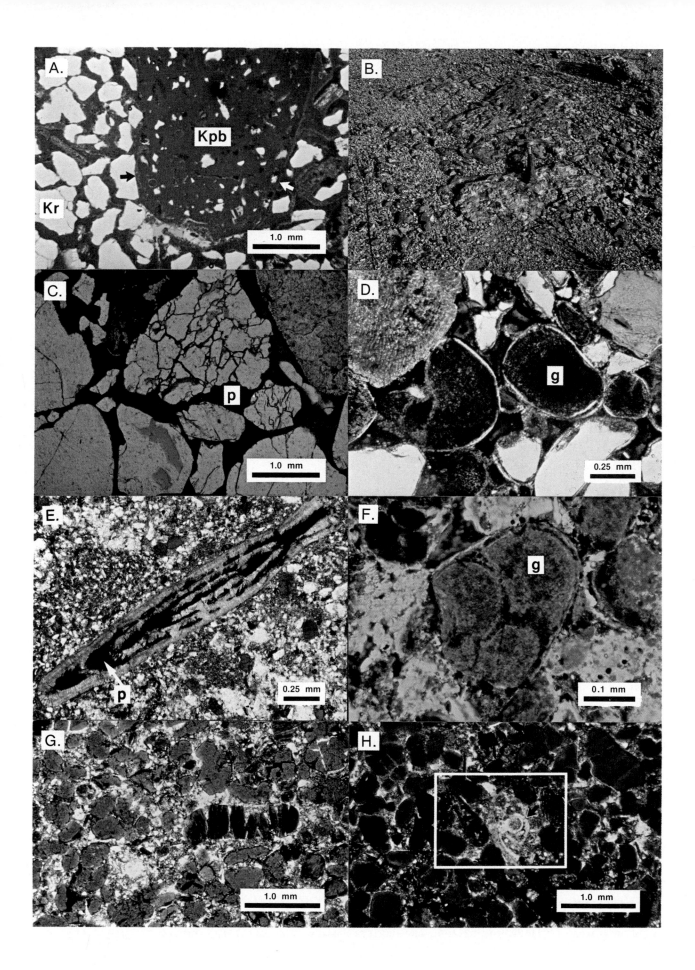

FIG. 7.—(A and B) 102-Ma unconformity between Skull Creek Shale and incised-valley-fill deposits of Horsetooth Member (J-Sand). Transgressive surface separates Horsetooth Member from Mowry Shale. Time scale after Van Hinte (1976). Interpretation after Land and Weimer (1978).

causes the sea to flood the interfluvial areas, resulting in a ravinement surface. The transgressive surface indicates that the shoreline has begun to migrate landward and that sediment influx is insufficient to keep pace with increasing accommodation.

The transgressive surface is generally the most easily recognized surface in outcrop because of the contrast between basal, transgressive, marine sediments resting sharply on non-marine to marginal marine incised-valley-fill sediments and the depositional similarities between incised-valley-fill sediments and the underlying highstand deposits of the previous sequence. For this reason, stage boundaries are typically placed at this surface (e.g., the base of the Sabine Stage at the top of the incised Gravel Creek Sand; the base of the Claiborne Stage at the top of the incised Meridian Sand; the base of the Barton Stage at the top of the Bracklesham Group). Stenzel (1952) used the term "transgressive marine regional disconformities" in essentially the same sense as transgressive surface is used here.

Transgressive deposits.—

As the rate of increasing accommodation exceeds the rate of sediment influx (Posamentier and others, this volume), the shoreline moves in a landward direction, and the sed-

imentary patterns shift from aggradation or regression to transgression (Figs. 1, 3, 9, 10). As a consequence, the basin becomes progressively starved of terrigenous components, water depths increase, and there is an increase in cosmogenic, volcanogenic, and authigenic components. The transgressive deposits are bracketed at the base by the transgressive surface or the unconformity and at the top by the surface of maximum starvation (Figs. 1, 3).

In detail, the transgressive deposits (as well as the incised-valley-fill and highstand deposits) consist of parasequences. Parasequences are bracketed by small-scale, downward shifts in coastal onlap that are superimposed on the more general sedimentary and sea-level changes. The parasequences become thinner as the surface of maximum

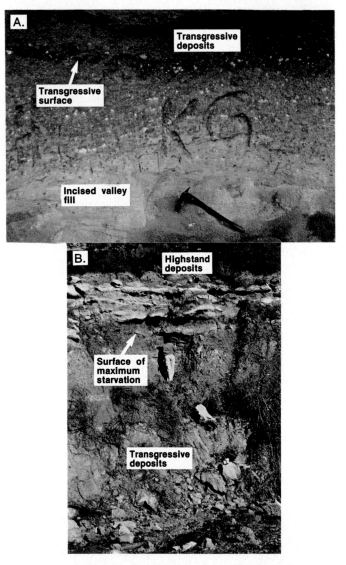

FIG. 8.—(A) Transgressive surface overlying incised-valley-fill deposits of TP 2.2 cycle (56 to 55 Ma; see Fig. 4A). Note bauxite clasts from underlying unit, suggesting exposure on interfluvial divides. Henry County, Alabama. (B) Transgressive deposits, surface of maximum starvation and highstand deposits associated with Eocene/Oligocene (Priabonian/Rupelian) boundary. Note nodular limestones (Bumpnose Limestone) directly overlying surface of maximum starvation. St. Stephen's quarry, Washington County, Alabama.

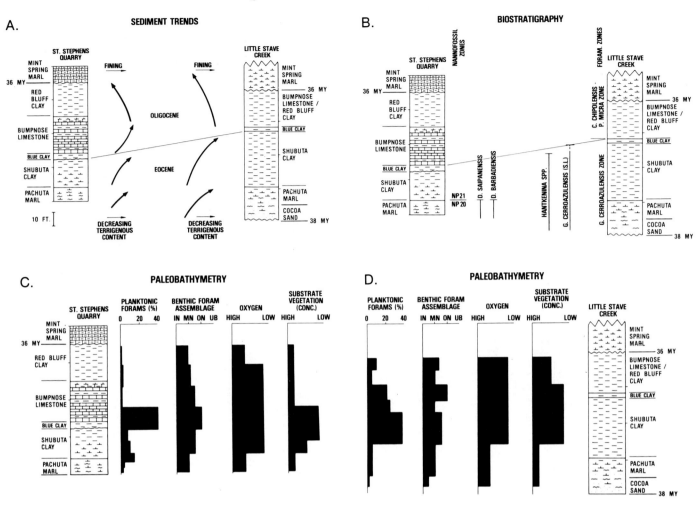

FIG. 9.—(A, B, C, and D) Lithofacies and biofacies analysis across Eocene/Oligocene (Priabonian/Rupelian) boundary. St. Stephen's quarry, Washington County, Alabama (from Loutit and others, 1983).

starvation is approached and gradually expand in thickness above the surface of maximum starvation (Fig. 10).

Condensed section and surface of maximum starvation.—

Condensed sections (Loutit and others, this volume) in the coastal plain are characterized by marine shales or micrites and by anomalously high concentrations of planktonic organisms, glauconite, sulfides, phosphate, and exotic elements such as iridium (Baum and others, 1984; Donovan and others, this volume; Fig. 10). Typically, the glauconite, sulfides, and phosphate occur as replacement products of calcite and aragonite organisms (Fig. 6). They are also found within the intraparticle pore spaces of foraminifera, as partially glauconitized clays, and as mammillate-to-lobate grains (Fig. 6). Because of the unique mineralogy of condensed sections, they generally are easily differentiated on a gamma log as "hot" intervals (Meyer and Nederlof, 1984).

The characteristic lithologies and surfaces associated with condensed sections vary depending on shelf position and sedimentation rates. Except in the most landward positions,

however, there is always a surface of maximum starvation that is typically characterized by a burrowed omission surface or bored marine hardground (Fig. 11; terminology after Kennedy and Garrison, 1975). This surface is expressed seismically as a downlap surface (Vail and others, 1984).

In the outer-shelf area, the surface of maximum starvation is generally clearly expressed. This surface is characterized by burrows (omission surface), borings, and/or marine cementation at the top of the underlying transgressive deposits, and abrupt shifts to more negative $\delta^{13}C$ values of calcite above hardgrounds (e.g., Fig. 12). The surface of maximum starvation marks the last occurrence of the underlying shallower water lithofacies and associated benthic faunas of the transgressive deposits. Typically, the sediments directly below the surface of maximum starvation are lithified and glauconitic, containing a diverse suite of benthic and planktonic organisms (Fig. 11). In all aspects of lithology and faunal abundance and diversity, these sediments indicate well-oxygenated marine waters of normal salinity. The lithologic and faunal changes across the surface of maximum starvation indicate an abrupt deepening event. In general, if the underlying transgressive de-

A.

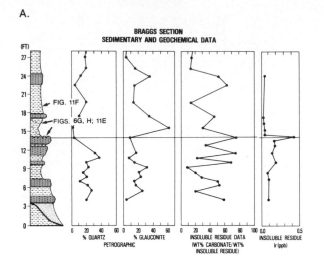

B.

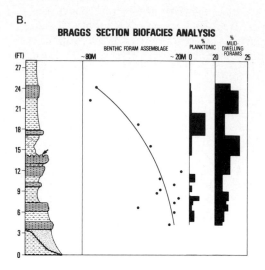

C.

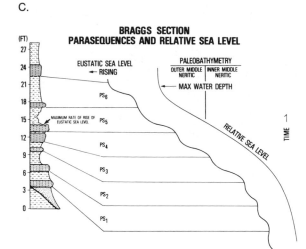

D.

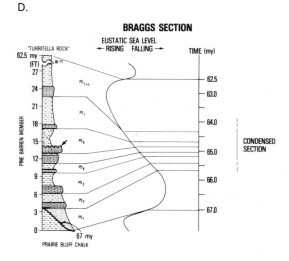

E.

F.

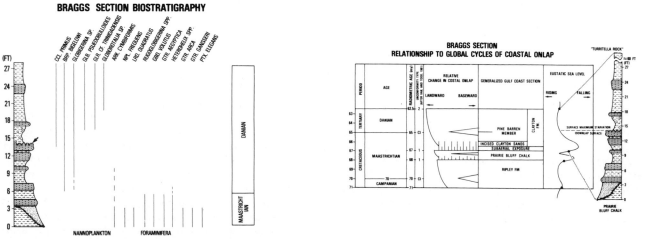

FIG. 10.—(A, B, C, D, E, and F) Lithofacies and biofacies analysis across Cretaceous/Tertiary (Maestrichtian/Danian) boundary. Lowndes County, Alabama (from Baum and others, 1984).

posits are porous at the time of sediment starvation, marine cements precipitate in the interparticle pore spaces and are subsequently filled with intercrystalline sediment from the overlying highstand deposits (Fig. 11). In some cases, the surface of maximum starvation is confused with a subaerial unconformity: this is one of the recurring problems of defining and dating sequence boundaries correctly.

The sediments directly overlying the surface of maximum starvation (Figs. 6, 8) represent the upper condensed section and the initial sporadic resedimentation by the highstand deposits. The upper condensed section, composed of early highstand deposits, is gradational with the later highstand deposits. Typically, the lithologies directly overlying the surface of maximum starvation consist either of framework-supported glauconite; or glauconitic, micritic, planktonic oozes (Fig. 6). In the first case, there is a lack of planktonic and benthic organisms. The few planktonic foraminifera that are present are in various stages of destruction by dissolution, glauconitization, or neomorphism (Figs. 6, 11). Throughout the remainder of the upper condensed section, glauconite and planktonic organisms are progressively diluted by terrigenous or biogenic sediments (Fig. 11). If present, planktonic organisms reveal no dissolution features. Because of its fossiliferous nature and the presence of high concentrations of glauconite, the upper condensed section is lithostratigraphically distinct from the remaining highstand deposits and is sometimes given a formation or member name (e.g., Bumpnose Limestone, Matthews Landing Marl).

In carbonate sedimentary regimes, the condensed section is often represented by only a beveled, glauconitized, phosphatized surface of maximum starvation that may be bored or burrowed by marine organisms (Fig. 11; Schlager, 1981). The absence of distinct transgressive deposits or early highstand deposits is perhaps due to the strong relationship between water depth and intrabasinal carbonate production by marine organisms. Thus, the time represented by the condensed section in carbonates appears to be greater than in equivalent clastic sedimentary regimes. Carbonate transgressive deposits "give up" sooner, and carbonate highstand deposits take longer to prograde basinward.

The surface of maximum starvation represents the interval of time of maximum landward extent of the condensed section and minimum sedimentation rates on the mid- and outer shelf. Seismically, this surface is seen as the seismic discontinuity termed the downlap surface (Vail and others, 1984), and the upper condensed section is represented by the toes of the clinoforms of the prograding highstand deposits.

Along with sediment starvation, this flooding event producing the surface of maximum starvation may cause a concomitant upward movement of the CCD (calcite compensation depth) and oxygen minimum (Worsley, 1974). Paleontologic data from outcrops in the coastal plain suggest that water depths never exceeded those typical of outer-shelf depositional environments. Although it is unlikely that the CCD shallowed to the depths represented by outcrops, the oxygen minimum (or at least the upper-edge effects) could have encroached upon the outer shelf. Abrupt shifts

to more negative $\delta^{13}C$ values of calcite above hardgrounds associated with condensed sections suggest an upward movement of the oxygen minimum during maximum rates of sea-level rise (e.g., Fig. 12). The rate of change in $\delta^{13}C$ of calcite from higher to lower values is apparently directly proportional to sedimentation rates and rates of change of sea level.

The upward movement of the oxygen minimum could explain the high concentration of pelleted glauconite (Fig. 6) associated with the condensed sections. Also, the combination of disaerobic or anaerobic conditions and low-sedimentation rates could produce the acidic conditions (Waples, 1983) observed in some outcrops (Figs. 6, 11) and explain the apparent dissolution of planktonic organisms. The syndepositional removal of organisms (Thierstein, 1981; Moore and others, 1983) coupled with the loss of stratigraphic resolution due to the convergence of paleontologic datums within the condensed section may, in part, account for the paleontologic "hiatus" associated with many hardgrounds.

RELATIONSHIP OF SEQUENCE STRATIGRAPHIC CONCEPTS TO GULF COAST STAGES

In the classical sense of the early Gulf Coast stratigraphers (e.g., summarized in Toulmin, 1977), the Gulf Coast Paleogene stages were generally defined on major drops in sea level (type 1 unconformities) and associated benthic faunal changes, and are essentially the allothemic equivalents to the supercycles of Vail and Hardenbol (1979; Fig. 13). The Midway Stage (67 Ma, Late Cretaceous, to 58.5 Ma, early Thanetian) extends from the type 1 unconformities at the base of the Clayton Formation (including the incised-valley-fill "Clayton sands") to the base of the Gravel Creek Sand Member of the Nanafalia Formation.

The Sabine Stage (early Thanetian, 58.5 Ma, to late Ypresian, 49.5 Ma), initially considered entirely Lower Eocene, extends from the type 1 unconformity at the base of the incised-valley-fill sands of the Gravel Creek Sand Member to the type 1 unconformity at the base of the incised Meridian Sand Member of the Tallahatta Formation.

The Claiborne Stage extends from the type 1 unconformity at the base of the incised-valley-fill sands of the Meridian Sand (49.5 Ma) to the transgressive surface overlying the incised Gosport Sand. The major unconformity, however is at the base of the Gosport Sand (40.5 Ma; Powell and Baum, 1982, 1984). The stratotype of the Claiborne Stage is the Gosport Sand. If the base of the overlying Jackson Stage is shifted to the unconformity (40.5 Ma) to include the Gosport Sand, the Claiborne and lowermost part of the Jackson stages are synonyms.

If an unconformity is recognized at the base of the Gosport Sand, the combined Jackson and overlying Vicksburg stages are bracketed by major falls in sea level in the late Bartonian (40.5 Ma) and middle Chattian (29 Ma, possibly older) and extend from the type 1 unconformity at the base of the incised Gosport Sand to the type 1 unconformity at the base of the incised Waynesboro Sand Member of the Chickasawhay Limestone. Within this supercycle, the

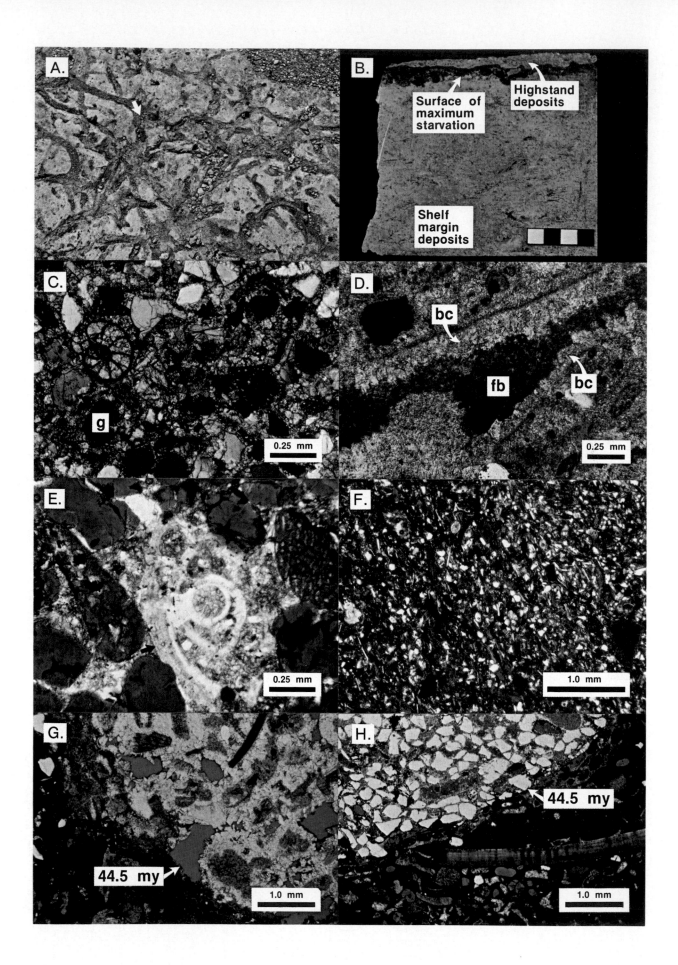

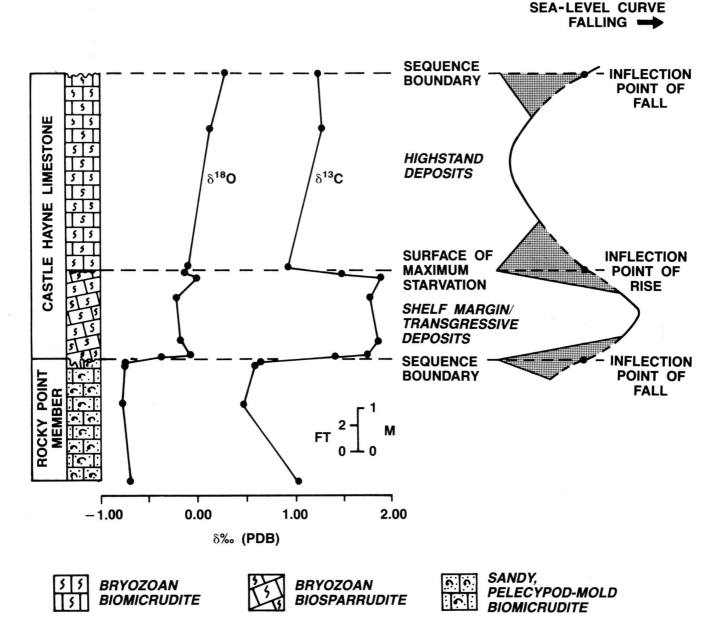

FIG. 12.—Carbonate stable isotope compositions and TOC (Total Organic Carbon) percentages for the Cretaceous Rocky Point Member of the Peedee Formation and middle Eocene Castle Hayne Limestone, Pender County, North Carolina. Samples were selectively drilled to avoid subaerial cements and, where possible, large allochems. Results have been reproduced in quarry site and several adjacent cores. The abrupt change to more negative values for $\delta^{13}C$ stable isotopes occurs at the surface of maximum starvation, which is coincident with the inflection point of sea-level rise and minimum sedimentation rates (from Baum, 1986a).

FIG. 11.—(A) Burrowed omission surface on Glendon Limestone. St. Stephen's quarry, Washington County, Alabama. (B) Split core across phosphatized hardground representing the surface of maximum starvation, TE 2.3 cycle (44.5 to 40.5 Ma; from Powell, 1985). Santee Limestone, Georgetown County, South Carolina. (C–H) Photomicrographs: (C) Glauconitic (g), calcareous-quartz arenite facies of Bashi Marl Member of Hatchitigbee Formation just below surface of maximum starvation, TE 1.1 cycle (54 to 52 Ma). Clarke County, Alabama. (D) "Dirty" bladed submarine cements (bc) followed by foraminiferal biomicrite (fb) from the overlying highstand deposits infilling intercrystalline porosity (from Powell, 1985). (E) Neomorphosed foraminifera, just above surface of maximum starvation, TP 1.1 cycle (67 to 62.5 Ma). Lowndes County, Alabama. (F) 1.5 m (5 ft) above (E). (G) Downward shift in coastal onlap at 44.5 Ma characterized by relatively deeper water facies (bryozoan biomicrudite) sharply overlain by relatively shallower water facies (bryozoan biosparrudite). Castle Hayne Limestone, Pender County, North Carolina. (H) Downward shift in coastal onlap at 44.5 Ma characterized by abrupt increase in sand-size quartz. Castle Hayne Limestone, Pender County, North Carolina.

placement of the Jackson/Vicksburg Stage (Eocene/Oli-
gocene) boundary is a matter of discussion (compare Mancini, 1979, and Keller, 1985), principally because there is
no major drop in sea level associated with the micropa-
leontologically defined Priabonian/Rupelian boundary. This
boundary is bracketed by minor drops in sea level (type 2
sequence boundaries?) that occur at the base of the Cocoa
Sand (38 Ma) and at the base of the Mint Springs Marl (36
Ma; Baum and others, 1982). Typically, the Eocene/Oli-
gocene boundary is placed near the Shubuta Clay and Red
Bluff Clay contact, which is, in fact, a condensed section
associated with a rapid rise in sea level (Loutit and others,
1983; Fig. 9).

The Chickasawhay Stage extends from the type 1 unconformity at the base of the Waynesboro Sand Member to
the type 1 unconformity at the top of the Chickasawhay
Limestone (29 to 25.5 Ma). The confusion on the placement of the Oligocene/Miocene contact is compounded by
the fact that more typically "Miocene" benthic faunas appear about the 29 Ma sequence boundary, and the nummulites become extinct at the 29 Ma sequence boundary;
however, these sequences are zoned well within the micro-
paleontologically defined Chattian (Oligocene). Thus, there
may be some merit to the suggestion that definition of the
Oligocene Epoch is somewhat arbitrary (Mayer-Eymar, 1893;
Pomerol, 1982).

There are essentially two other approaches to defining
stages other than as bounded by unconformities. The first
approach is the tendency to place stage boundaries at the
transgressive surface overlying incised-valley-fill sediments
(e.g., Murray, 1961). The transgressive surface does merge
with the unconformity, particularly in the interfluvial areas;
however, this approach excludes a portion of the coastal-
onlap cycle (Fig. 1; Loutit and others, this volume). As
previously discussed, the study of depositional sequences
cannot be addressed by dating transgressions and regressions. The chronostratigraphic significance of the transgressive surface at present has not been evaluated.

The second approach is to extend paleontologically the
physical limits of stratotypes to coincide with planktonic
zonal boundaries and/or paleontological hiatuses (e.g.,
Hardenbol and Berggren, 1978). These boundaries are accepted as chronostratigraphic; however, they are generally
associated with condensed sections caused by rapid relative
rises in sea level rather than sequence boundaries associated
with rapid relative falls in sea level (Baum and others, 1982,
1984; Loutit and others, 1983). An additional factor which
must be addressed when using planktonic zonal boundaries
to define stages is that paleo-latitude may cause paleontologic datums (biozones) to shift in time. Thus, at any given
locality, paleontologic datums may not always represent
biochronozones (see Loutit and others, this volume, for discussion).

In order to resolve these boundary problems and to document eustasy, the inconsistencies in chronostratigraphy must
be recognized and resolved by adopting a consistent method
of fixing outcrop stratotype boundaries in time, using a sequence stratigraphic system or a micropaleontologic definition. If outcrop sections are placed within a sequence
stratigraphic framework, the accepted definition of stage must

be amended, for sequence boundaries are not time lines but
time-bounding surfaces that separate older rocks from
younger rocks. The sequence boundary can be dated by paleontology at its correlative conformity. Thus, sequence
stratigraphy offers a unifying concept to divide the rock
record into chronostratigraphic units, avoids the weaknesses and incorporates the strengths of other methodologies, and provides a global framework for geochemical,
geochronological, paleontological, and facies analyses. It
is imperative, however, to recognize that the transgressive
surface and surface of maximum starvation (condensed section; generally associated with a micropaleontologic hiatus)
are not sequence boundaries.

RELATIONSHIP OF SEDIMENTATION RATES AND TECTONICS TO
SEQUENCE STRATIGRAPHIC CONCEPTS—TE 2 SUPERCYCLE

Two of the principal concerns regarding sea-level control
of stratal patterns and sequence boundaries are sedimentation rates and tectonics. The carbonates of North and South
Carolina and the clastics of Alabama within the TE 2 supercycle can be used to examine these concerns.

The TE 2 supercycle of Vail and Hardenbol (1979; with
modifications by Baum and others, 1982; Powell and Baum,
1982, 1984) is bounded below by a type 1 unconformity in
the late Ypresian (49.5 Ma) and above by a type 1 unconformity in the late Bartonian (40.5 Ma; Figs. 13, 14). This
supercycle can be further divided into three sequences (TE
2.1, TE 2.2, TE 2.3). The TE 2.1 sequence is characterized
by *Cubitostrea perplicata*; the TE 2.2 by *Cubitostrea lis-
bonensis/Santeelampas oviformis*; and the TE 2.3 by *Cubitostrea sellaeformis* (Baum and others, 1980; Powell and
Baum, 1982, 1984). All three sequences have been recognized in Alabama. The TE 2.1 sequence in the Carolinas
has not been studied; however, updip outliers containing
"*Anodontia*?" *augustana* Gardner suggest equivalence to
the Meridian Sand/Tallahatta Formation and the TE 2.1
sequence.

In basinward areas, the Castle Hayne Limestone of North
Carolina and the Santee Limestone of South Carolina consist of the following three basic lithofacies in ascending order (Baum, 1980; Baum and others, 1980; Powell and Baum,
1982; Powell, 1985): a basal phosphate-pebble biomicrudite facies; a bryozoan biosparrudite facies; and a bryozoan
biomicrudite facies. When present, the basal phosphate-
pebble biomicrudite facies marks the unconformity of both
the TE 2.2 and TE 2.3 sequences (Fig. 5). In the absence
of this facies, the sequence boundary is characterized by a
relatively shallower water facies resting on top of a relatively deeper water facies (Figs. 5, 11) and/or an abrupt
increase in sand-size quartz in a very basal portion of the
overlying sequence (Figs. 5, 11). The surface is typically
sharp or burrowed and is commonly phosphatized.

A distinct surface of maximum starvation has not been
observed within the TE 2.2 sequence of the Carolinas. The
surface of maximum starvation of the TE 2.3 sequence is
very distinctive but is sometimes interpreted as a subaerial
unconformity (Figs. 5, 11). The following criteria, however, suggest a surface of maximum starvation caused by
a rapid rise in sea level: burrowing, boring; phosphatized

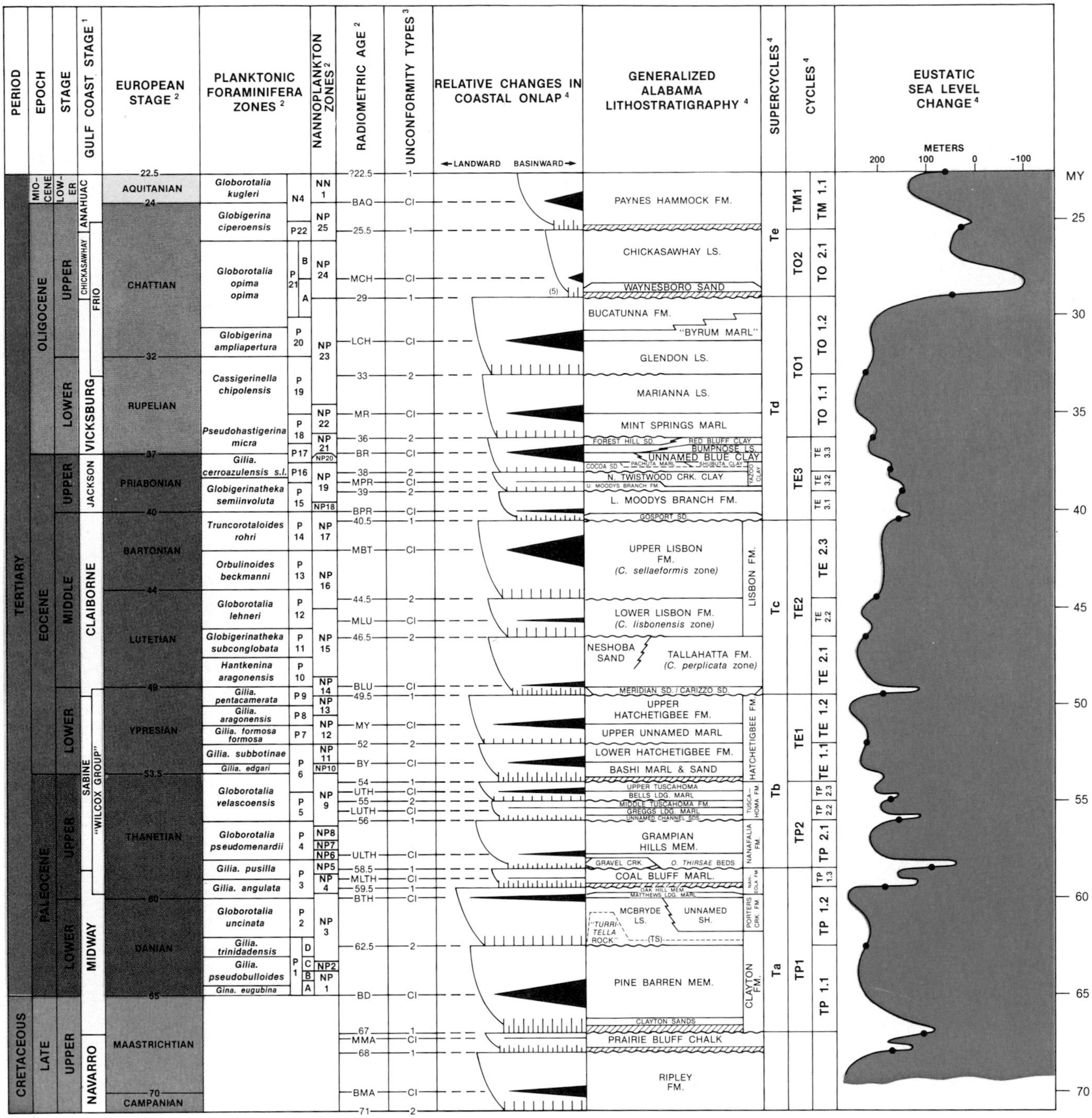

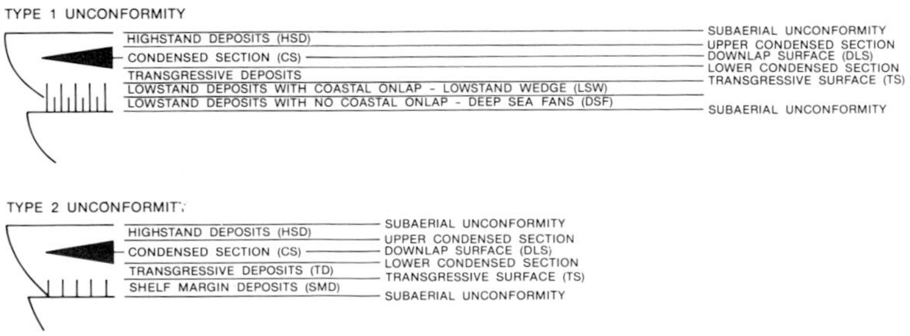

FIG. 13.—Coastal-onlap chart and eustatic sea-level curve (from Baum and others, 1982; Baum, 1986b).

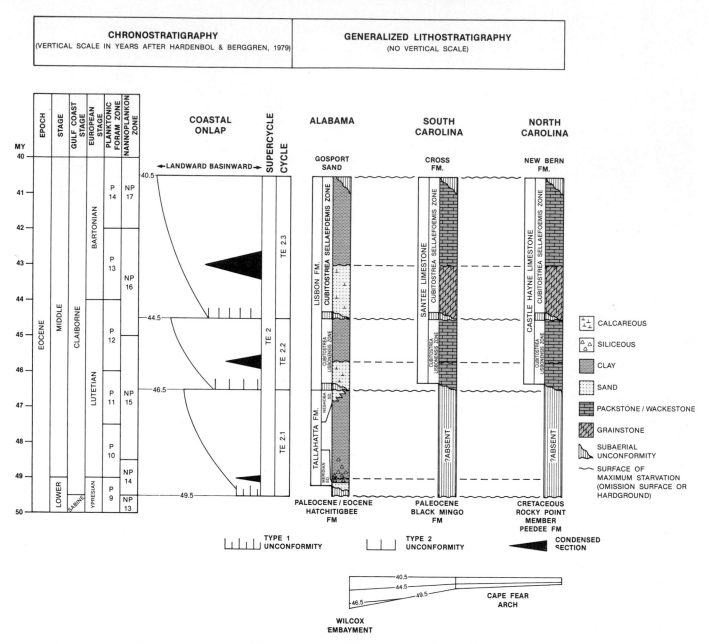

FIG. 14.—Comparison of chronostratigraphy and lithostratigraphy, TE 2 supercycle (49.5 to 40.5 Ma), Alabama, South Carolina, and North Carolina.

surface; and micrite and planktonic organisms from the highstand deposits infilling intercrystalline porosity formed by the precipitation of isopachous, acicular, peloidal, or "dirty" bladed submarine cements (Fig. 15). Within the TE 2.3 sequence of the Carolinas and Alabama and the type Barton, the surface of maximum starvation approximates the NP16/NP17 nannoplankton boundary.

Figure 14 illustrates the time and thickness distribution of the TE 2 supercycle in the Carolinas and Alabama. Although faulting has played a prominent role in the facies development and distribution of the carbonates in the Carolinas (Baum and others, 1979a; Baum and others, 1979b; Baum, 1981; Powell, 1985), and the Cape Fear Arch has

been a prominent tectonic element since the Late Cretaceous, tectonics do not seem to have affected the time distribution of the sequences and sequence boundaries. Likewise, the contrast in sedimentation rates between the carbonates of the Carolinas and the clastics of Alabama does seem to have been a factor. Limited quantitative data, however, suggest that the duration of the condensed section is greater in carbonates than in equivalent clastics.

CONCLUSIONS

—Stage boundaries are inconsistently picked at sequence boundaries, transgressive surfaces, and/or planktonic zonal boundaries.

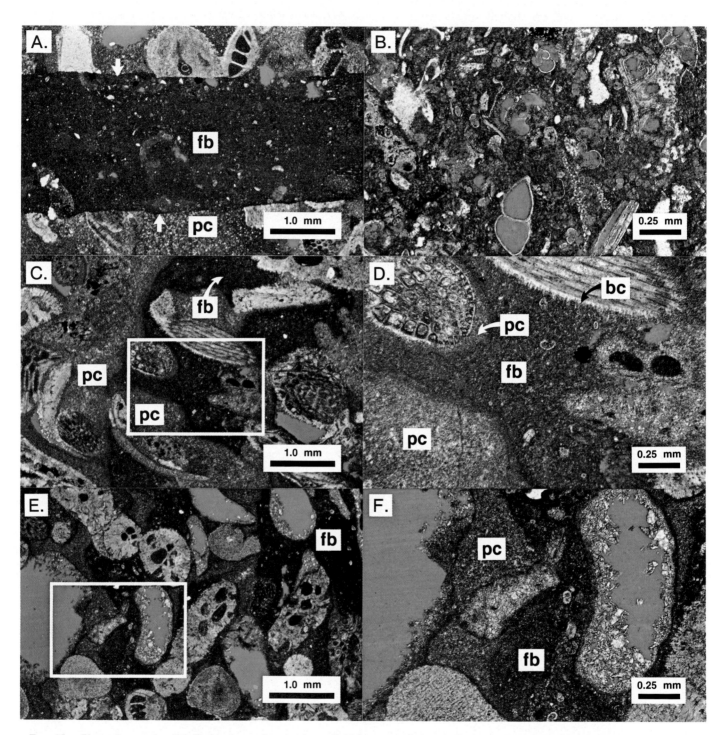

Fig. 15.—Photomicrographs: (A) Boring (arrows) truncating pelloidal cements (pc) and allochems. Foraminiferal biomicrite (fb) facies of the early highstand deposits subsequently infilling boring and intercrystalline porosity. Thin section just below surface of maximum starvation, TE 2.3 cycle (44.5 to 40.5 Ma). Castle Hayne Limestone, Pender County, North Carolina. (B) Basal highstand deposits. Foraminiferal biomicrite. Thin section just above surface of maximum starvation, TE 2.3 cycle (44.5 to 40.5 Ma). Castle Hayne Limestone. Pender County, North Carolina. (C and D) Bryozoan biosparrudite facies cemented by pelloidal (pc) and bladed marine cements (bc). Intercrystalline porosity subsequently infilled with highstand foraminiferal biomicrite (fb). Thin section just below surface of maximum starvation. Castle Hayne Limestone, Pender County, North Carolina. (E and F) Same as (C) and (D). Note mollusc molds contain no infilling of the highstand foraminiferal biomicrite, suggesting subaerial exposure, introduction of fresh water, and dissolution of unstable aragonite after the precipitation of marine cements and subsequent infilling of highstand foraminiferal biomicrite.

—Most major planktonic zonal boundaries occur within condensed sections that are related to rapid rises in sea level.

—Although the classically defined Gulf Coast stages generally reflect the allothemic equivalents of supercycle boundaries, the currently recognized Gulf Coast and European stages do not adequately reflect the higher frequency coastal-onlap cycles.

—Structural histories leave an imprint on facies development, sequence distribution, and geometry. Data from the Gulf and Atlantic basins, however, support the contention of Vail and his colleagues that sequence boundaries are caused by rapid eustatic sea-level falls.

—Sedimentation rates do not seem to affect the time distribution of sequences; however, carbonates tend to react more immediately ("give up") to rapid rises in sea level.

—Sequence boundaries are not synthems or allostratigraphic units, because sequences can be identified beyond the extent of their bounding unconformities.

—It is necessary to integrate all aspects of geology, paleontology, and geophysics in order to resolve these higher frequency sea-level cycles and to fix them in a relative-time framework.

ACKNOWLEDGMENTS

The fundamentals of the stratigraphic approach to recognizing and mapping unconformities in the Gulf and Atlantic Coastal Plains can be found in the classic published works of L. W. Stephenson, H. B. Stenzel, and L. D. Toulmin, and (in later years) in the first author's discussions with F. S. MacNeil. We benefited from discussions with T. S. Loutit and J. Hardenbol.

We thank Martin Marietta Aggregates for allowing access to 2-in. exploration cores and to quarry sites. We also thank W. B. Harris, D. K. Goodman, R. K. Suchecki, and J. S. Baum for providing friendly criticism for this manuscript. E. A. Mancini reviewed the coastal onlap chart.

REFERENCES

BAUM, G. R., 1980, Petrography and depositional environments of the middle Eocene Castle Hayne Limestone, North Carolina: Southeastern Geology, v. 21, p. 175–196.

———, 1981, Lithostratigraphy, depositional environments and tectonic framework of the Eocene New Bern Formation and Oligocene Trent Formation, North Carolina: Southeastern Geology, v. 22, p. 171–191.

———, 1986a, The recognition of allothemic (unconformity bounded) units in Paleogene outcrops, Gulf and Atlantic Tertiary basins: Society of Economic Paleontologists and Mineralogists Third Annual Midyear Meeting, Raleigh, North Carolina, Abstracts, v. 3, p. 6–7.

———, 1986b, Sequence stratigraphy of the Eocene carbonates of the Carolinas, *in* Harris, W. B., Zullo, V. A., and Otte, L. J. eds., Eocene carbonate facies of the North Carolina Coastal Plain, *in* Textoris, D. A., ed., Society of Economic Paleontologists and Mineralogists Field Guidebooks, Southeastern United States: SEPM Third Annual Midyear Meeting, Raleigh, North Carolina, p. 264–269.

———, BLECHSCHMIDT, G. L. HARDENBOL, J., LOUTIT, T. S., VAIL, P. R., AND WRIGHT, R. C., 1984, The Maastrichtian/Danian boundary in Alabama: A stratigraphically condensed section: Geological Society of America, Abstracts with Programs, v. 16, p. 440.

———, COLLINS, J. S., JONES, R. M., MADLINGER, B. A., AND POWELL, R. J., 1979a, Tectonic history and correlation of the Eocene strata of the Carolinas: Preliminary report, *in* Baum, G. R., Harris, W. B., and

Zullo, V. A., eds., Structural and stratigraphic framework of North Carolina: Carolina Geological Society Field Trip Guidebook, p. 87–94.

———, ———, ———, ———, AND ———, 1980, Correlation of the Eocene strata of the Carolinas: South Carolina Geology, v. 24, p. 19–27.

———, HARRIS, W. B., AND ZULLO, V. A., 1979b, Stratigraphic history of the Eocene to Early Miocene strata of North Carolina, *in* Baum, G. R., Harris, W. B., and Zullo, V. A., eds., Structural and stratigraphic framework for the Coastal Plain of North Carolina: Carolina Geological Society Field Trip Guidebook, p. 1–15.

———, VAIL, P. R., AND HARDENBOL, J., 1982, Unconformities and depositional sequences in relationship to eustatic sea level change, Gulf and Atlantic Coastal Plains: Inter-regional Geological Correlation Program, Project 174, Baton Rouge, Louisiana, 3 p.

CLIFTON, H. E., 1982, Estuarine deposits, *in* Scholle, P. A., and Spearing, D., eds., Sandstone depositional environments: American Association of Petroleum Geologists Memoir 31, p. 179–189.

EAMES, F. E., 1970, Some thoughts on the Neogene/Paleogene boundary: Palaeogeography, Palaeoecology, Palaeoclimatology, v. 8, p. 37–48.

———, BANNER, F. T., BLOW, W. H., AND CLARKE, W. J., 1962, Fundamentals of mid-Tertiary stratigraphical correlation: University Press, Cambridge, 163 p.

HARDENBOL, J., AND BERGGREN, W. A., 1978, A new Paleogene numerical time scale, *in* Cohee, G. V., Glaessener, M. F., and Hedbert, H. D., eds., The geologic time scale: American Association of Petroleum Geologists Studies in Geology No. 6, p. 213–234.

KELLER, G., 1985, Eocene and Oligocene stratigraphy and erosional unconformities in the Gulf of Mexico and Gulf Coast: Journal of Paleontology, v. 59, p. 882–903.

KENDALL, C. G. ST. C., AND SCHLAGER, W., 1981, Relative changes in sea level and its effect upon carbonate deposition: Marine Geology, v. 4, 0

KENNEDY, W. J., AND GARRISON, R. E., 1975, Morphology and genesis of nodular chalks and hardgrounds in the Upper Cretaceous of southern England: Sedimentology, v. 22, p. 311–386.

LAND, C. B., AND WEIMER, R. J., 1978, Peoria Field, Denver Basin, Colorado—J Sandstone distributary channel reservoir: Rocky Mountain Association of Geologists, Symposium, p. 81–104.

LOUTIT, T. S., BAUM, G. R., AND WRIGHT, R. C., 1983, Eocene-Oligocene sea level changes as reflected in Alabama outcrop sections: American Association of Petroleum Geologists Bulletin, v. 67, p. 506.

MANCINI, E. A., 1979, Eocene-Oligocene boundary in southwest Alabama: Gulf Coast Association of Geological Societies, Transactions, v. 29, p. 282–289.

MAYER-EYMAR, K., 1893, Le Ligurien et le Tongrien en Egypte: Bulletin of the Society of the Geology of France, v. 21, p. 7–43.

MEYER, B. L., AND NEDERLOF, M. H., 1984, Identification of source rocks on wireline logs by density/resistivity and sonic transit time/resistivity crossplots: American Association of Petroleum Geologists Bulletin, v. 68, p. 121–129.

MITCHUM, R. M., VAIL, P. R., AND THOMPSON, S., III, 1977, Seismic stratigraphy and global changes of sea level, part 2: The depositional sequence as a basic unit for stratigraphic analysis, *in* Payton, C. E., ed., Seismic stratigraphy—Applications to hydrocarbon exploration: American Association of Petroleum Geologists Memoir 26, p. 53–62.

MOORE, T. C., JR., RABINOWITZ, P. D., BOERSMA, A. BORELLA, P. E., CHAVE, A. D., DUEE, G., FUTTERER, D. K., JIANG, M. J., KLEINERT, K., LEVER, A., MANIUIT, H., O'CONNELL, S., RICHARDSON, S. H., AND SHACKLETON, N. J., 1983, The Walvis Ridge transect, Deep Sea Drilling Project Leg 74: The geologic evolution of an oceanic plateau in the South Atlantic Ocean: Geological Society of America Bulletin, v. 94, p. 907–925.

MOSSOP, G. D., AND FLACH, P. D., 1983, Deep channel sedimentation in the Lower Cretaceous McMurray Formation, Athabasca oil sands, Alberta: Sedimentology, v. 30, p. 493–509.

MURRAY, G. E., 1961, Geology of the Atlantic and Gulf Coastal Province of North America: Harper and Brothers, New York, 692 p.

NORTH AMERICAN Commission on Stratigraphic Nomenclature, 1983, The North American stratigraphic code: American Association of Petroleum Geologists Bulletin, v. 67, p. 841–875.

POMEROL, C., 1982, Cenozoic Era: Ellis Horwood, Chichester, 272 p.

POWELL, R. J., 1985, Lithostratigraphy, depositional environment, and sequence framework of the middle Eocene Santee Limestone, South Carolina Coastal Plain: Southeastern Geology, v. 25, p. 79–100.

————, AND BAUM, G. R., 1982, Eocene biostratigraphy of South Carolina and its relationship to Gulf Coastal Plain zonations and global changes in coastal onlap: Geological Society of America Bulletin, v. 93, p. 1099–1108.

————, AND ————, 1984, Eocene biostratigraphy of South Carolina and its relationship to Gulf Coastal Plain zonations and global changes of coastal onlap: Reply: Geological Society of America Bulletin, v. 95, p. 983–984.

SALVADOR, A., 1987, Unconformity-bounded stratigraphic units, International Subcommission on stratigraphic classification: Geological Society of America Bulletin, v. 98, p. 232–237.

SCHLAGER, W., 1981, The paradox of drowned reefs and carbonate platforms: Geological Society of America Bulletin, v. 92, p. 197–211.

STENZEL, H. B., 1952, Boundary problems: Ninth Field Trip, Mississippi Geological Society Guidebook, p. 11–33.

STEPHENSON, L. W., 1929, Unconformities in Upper Cretaceous Series in Texas: American Association of Petroleum Geologists Bulletin, v. 13, p. 1323–1334.

THIERSTEIN, H. R., 1981, Late Cretaceous nannoplankton and the change at the Cretaceous-Tertiary boundary, *in* Warme, J. E., Douglas, R. G., and Winterer, E. L., eds., The Deep Sea Drilling Project: A decade of progress: Society of Economic Paleontologists and Mineralogists Speical Publication 32, p. 355–394.

TODD, R. G., AND MITCHUM, R. M., Jr., 1977, Seismic stratigraphy and global changes of sea level, part 8: Identification of Upper Triassic, Jurassic, and Lower Cretaceous seismic sequences in Gulf of Mexico and offshore West Africa, *in* Payton, C. E., ed., Seismic stratigraphy-Applications to hydrocarbon exploration: American Association of Petroleum Geologists Memoir 26, p. 145–163.

TOULMIN, L. D., 1977, Stratigraphic distribution of Paleocene and Eocene fossils in the Eastern Gulf Coast region: Alabama Geological Survey Monograph 13, v. 1, 602 p.

VAIL P. R., 1976, Phanerozoic eustatic cycles and global unconformities, *in* Sequence Concepts in Petroleum Exploration: National Conference On Earth Science, p. 93–110.

————, AND HARDENBOL, 1979, Sea-level changes during the Tertiary: Oceanus, v. 22, p. 71–79.

————, ————, AND TODD R. G., 1982, Jurassic unconformities, chronostratigraphy, and sea-level changes from seismic and biostratigraphy: Proceedings of the Joint Meeting of the China Geophysical Society and the Society of Exploration Geophysicists to Discuss Geophysical Exploration for Petroleum, Beijing, China, September 7–11, 1981, 17 p.

————, ————, AND ————, 1984, Jurassic unconformities, chronostratigraphy and sea-level changes from seismic stratigraphy and biostratigraphy, *in* Ventress, W. P. S., Bebout, D. G., Perkins, B. F., and Moore, C. H., eds., The Jurassic of the Gulf Rim: Society of Economic Paleontologists and Mineralogists Proceedings, Baton Rouge, Louisiana, p. 347–364.

————, MITCHUM, R. M., Jr., SHIPLEY, T. H., AND BUFFLER, R. T., 1980, Unconformities of the North Atlantic: Philosophical Transactions of the Royal Society of London, v. 294, p. 137–155.

————, ————, AND THOMPSON, S., III, 1977, Seismic stratigraphy and global changes of sea level, part 4: Global cycles of relative changes of sea level, *in* Payton, C. E., ed., Seismic stratigraphy—Applications to hydrocarbon exploration: American Association of Petroleum Geologists Memoir 26, p. 83–97.

————, AND TODD, R. G., 1981, Northern North Sea Jurassic unconformities, chronostratigraphy and sea-level changes from seismic stratigraphy: Proceedings, Petroleum Geology, Continental Shelf, Northwest Europe, p. 216–235.

VAN HINTE, J. E., 1976, A Cretaceous time scale: American Association of Petroleum Geologists Bulletin, v. 60, p. 498–516.

WAPLES, D. W., 1983, Reappraisal of anoxia and organic richness, with emphasis on Cretaceous of North Atlantic: American Association of Petroleum Geologists Bulletin, v. 67, p. 963–978.

WORSLEY, T., 1974, The Cretaceous-Tertiary boundary event in the ocean, *in* Hay, W. W. ed., Studies in paleo-oceanography: Society of Economic Paleontologists and Mineralogists Special Publication 20, p. 94–122.

RECOGNITION AND INTERPRETATION OF DEPOSITIONAL SEQUENCES AND CALCULATION OF SEA-LEVEL CHANGES FROM STRATIGRAPHIC DATA—OFFSHORE NEW JERSEY AND ALABAMA TERTIARY

STEPHEN M. GREENLEE AND THEODORE C. MOORE

Exxon Production Research Company, P.O. Box 2189, Houston, Texas 77252-2189

ABSTRACT: Tertiary depositional sequences beneath the continental shelf and slope off New Jersey and Alabama have been studied using seismic-reflection data that have been tied to available wells. These data illustrate second- (10–20 Ma) and third-order (1–5 Ma) depositional sequences in areas close to and distal to progradational siliciclastic depocenters. Paleogene deposition is characterized by sediment-starved deep-water conditions. Second-order sequence boundaries divide these sediments and are recognized by local erosion of underlying strata and deep-marine onlap. Closer to the depocenter, third-order sequences are noted by basinward shifts in coastal onlap, local erosional incision of shelf and slope strata, and planar erosion of basinal sediments. The stacking pattern of third-order sequences within the second-order sequences is similar to the stacking of systems tracts within the third-order depositional sequences modeled using a sinusoidally varying sea-level curve. Upper Oligocene to uppermost middle Miocene third-order sequences are interpreted to compose a second-order supersequence.

Neogene strata in these areas are further analyzed to estimate sea levels. Subsidence is isolated using geohistory analysis and by calculating the average angular-tilt rate of the continental margin. During the early Tertiary, the subsidence rate was slow in both areas. An increase in subsidence rate in the offshore Alabama area during the Neogene is attributed to loading of the lithosphere adjacent to the study area. Short-term falls in sea level are estimated by measuring the vertical shift in onlapping paralic strata from highstand to lowstand position. Results show a lower overall Neogene sea-level position than the position represented on the Exxon curve but similar magnitudes of short-term fall.

INTRODUCTION

This paper addresses the recognition, description, and interpretation of Tertiary depositional sequences on high-quality seismic-reflection profiles from the Baltimore Canyon Trough and the offshore Alabama area of the northeastern Gulf of Mexico. These data sets are used to estimate eustasy in the Neogene. The study areas were selected for analysis for two reasons. First, the basins have been studied extensively, and their geologic history and biostratigraphic zonations are well known to academic and industry geologists. Because of this, the stratal geometry, sequence interpretation, and age relationships presented here may be compared with published literature or industry file reports. Second, these areas are characterized by a Tertiary sedimentary section that prograded basinward in a relatively quiescent tectonic setting. Neither area has undergone extensive tectonism due to thermal perturbation or salt withdrawal since the lowermost upper Cretaceous. Thus, subsidence in these areas is primarily due to sediment loading and lithospheric flexure superimposed on slow, background, crustal cooling. In this tectonic setting, the effects of third-order (approximately 1–5 Ma) and second-order (10–20 Ma) cycles of eustatic change are most easily identified.

In order to study the effects of eustasy, well-log, seismic, and biostratigraphic data were analyzed in three steps: (1) interpretation of the sequence stratigraphy of the areas; (2) isolation of the tectonic and isostatic subsidence components; and (3) measurement of the total amount of accommodation, or new space made available through time. We utilized the following procedures: first, depositional-sequence boundaries were recognized and traced throughout the seismic grids. On the basis of stratigraphic information from wells, the sequences were dated and major facies packages were identified and summarized on chronostratigraphic charts. Long-term subsidence trends were first isolated using geohistory analysis. Relative changes in coastal onlap were compared for the two areas. We used a technique described in Moore and others (1987) to isolate tectonic subsidence for critical time slices using the calculated angular-tilt rate of the continental margin. Decompacted and isostatically corrected sediment thicknesses (tied to our stratigraphic framework) were used to calculate Tertiary sea levels.

The scope of this paper is not intended to document the global geologic synchroneity of eustatic fluctuations in sea level; however, stratigraphic analysis of the Tertiary strata in offshore New Jersey and Alabama shows the major similarities in the age and character of the depositional sequences in these areas. Tertiary sea levels calculated from these data suggest an overall falling sea level in the Miocene, an early Pliocene highstand, and falls of approximately 10 to 100 m at third-order sequence boundaries.

TERTIARY STRATIGRAPHIC FRAMEWORK—BALTIMORE CANYON TROUGH

The stratigraphy of the Tertiary sediments beneath the continental shelf off New Jersey was studied through the analysis of a regional grid of seismic data tied to available wells, shown in Figure 1. This area has been a focus of research on the effects of sea-level changes on the sedimentary development of basins as a result of the availability of outcrop, well, and seismic data as well as the lack of pervasive structural overprint (Poag and Schlee, 1984; Poag, 1985a, b; and references therein). Seismic stratigraphic analysis of the Tertiary strata beneath the New Jersey continental shelf is discussed in Greenlee and others (1988). Biostratigraphic analysis of wells is from published (Poag, 1980, 1985a) and Exxon reports (M. Crane, pers. commun., 1985). Figure 2 shows an interpretation of the seismic section in our grid that best illustrates Tertiary depositional-sequence development in the study area. Beneath it is a chronostratigraphic chart that summarizes the geographic extent, ages, position of the depositional-shoreline

STEPHEN M. GREENLEE AND THEODORE C. MOORE

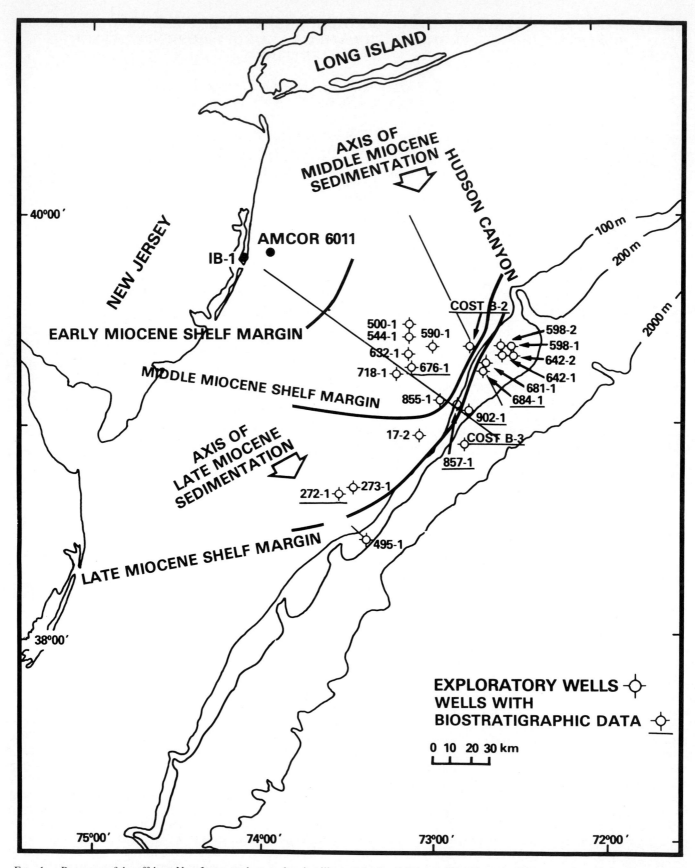

FIG. 1.—Base map of the offshore New Jersey study area showing illustrated seismic lines, well locations, and Neogene depositional-shelf margins.

break,[1] and interpreted facies of the depositional sequences noted on this line. The position of key planktonic and several benthic foraminiferal occurrences and their position relative to sequence boundaries are recognized in four wells (arranged in an approximate dip direction) and are illustrated in Figure 3 (Poag, 1980, 1985a; Exxon Company reports). Figure 1 also shows the location of the major shelf margins and prominent physiographic features within the Baltimore Canyon Trough study area.

Paleocene-Eocene.—

The Cretaceous-Tertiary boundary is characterized by a high-amplitude reflection on which an overlying prograding wedge downlaps. The prograding wedge thins seaward by downlap and erosional truncation and is dated by offshore wells as late Paleocene (Fig. 3). Two thin sequences overlie the Paleocene sequence and show local evidence of erosional truncation (Fig. 2). These two sequences and the upper Paleocene sequence consist of deep-water carbonate-rich sediments. They are dated in COST B-2 and industry wells as early and middle Eocene. During the Paleocene and Eocene, the shoreline was located well landward of the study grid. Because of this, downward shifts in coastal onlap patterns were not observed, making the recognition of minor sequence boundaries difficult. Thus, only major sequence boundaries (Haq and others, 1987) near the top of the Paleocene (55 Ma), near the top of the lower Eocene (49.5 Ma), and near the top of the middle Eocene (39.5 Ma) were recognized. These sequence boundaries are recognized by very local erosional truncation on an otherwise flat surface and by low-angle, deep-marine onlap (Figs. 2, 4).

The seismic data show evidence of onlap of upper Eocene beds at the most landward edge of the seismic grid (Fig. 2). This is consistent with the absence of upper Eocene rocks at the Island Beach well to the west of the seismic line in Figure 2 (Poag, 1985a).

Oligocene.—

One sequence, dated as late Oligocene (Poag, 1980, 1985a; Fig. 2), is identified in the offshore wells. Throughout most of the study area, this sequence overlies upper Eocene sediments. The upper Oligocene onlaps an erosional unconformity that cuts the older section out basinward. This Oligocene unconformity is correlated with the major fall of sea level in the Oligocene (30 Ma). The seismic section in Figure 2 shows a partially eroded wedge that is composed of onlapping upper Eocene rocks and possibly some younger sediment landward of shot point 6. These have been interpreted to be of early Oligocene or earliest late Oligocene age, although the existence of Oligocene deposits at Island Beach is uncertain (see Poag, 1985b, p. 237–238). Bybell and others (1986) have described a complete lower Oli-

gocene section onshore of the seismic grid. The absence or near-absence of lower Oligocene sediments on the offshore portion of the seismic lines is believed to be caused by sediment starvation due to a sea-level rise and increase in shelf accommodation (see discussion in Posamentier and others, this volume) in the early Oligocene combined with erosion associated with the subsequent 30 Ma sea-level fall.

Early Miocene.—

Four early Miocene sequences (Fig. 2) have been recognized on the seismic sections from the study area. Each sequence shows evidence of onlap in a landward direction and downlap in a seaward direction. The lower Miocene sequences show a series of stacked shelf edges just seaward of the upper Oligocene shelf margin. The sequences thin landward by onlap and seaward by downlap toward the offshore well control. The lower Miocene section is largely starved or eroded in seaward portions of the line, as illustrated on the chronostratigraphic chart in Figure 2. Exxon paleontologic reports (M. Crane, pers. commun., 1985) suggest a thin zone of lower Miocene is present over Oligocene deposits at COST B-2. COST B-3 recovered 64 m of lower Miocene bathyal silty clays (Poag, 1985a). Other early Miocene fossil occurrences are shown in Figure 3.

Middle Miocene.—

Four middle Miocene sequences (Fig. 2) are recognized. These sequences are strongly progradational with each successive middle Miocene shelf margin seaward of the previous shelf margin. Figure 1 shows the position of the shelf edge at the end of the middle Miocene and the axis of maximum thickness of middle Miocene progradational lobes. The middle Miocene depocenter is in the vicinity of the COST B-2 well, which sampled a shoaling-upward middle Miocene section spanning nannofossil zones N10–N14 overlying late Oligocene deposits (Poag, 1980, 1985a). Tertiary depositional sequences at COST B-2 are illustrated in Figure 4, which shows a seismic section passing through the COST B-2 well and the Exxon 624-1 well. Miocene chronostratigraphy is shown below the line. Two sequence boundaries are found within sediment dated as zones N10–N12 by Poag (1980) and are correlated with sea-level falls in the mid-Serravalian (13.8 Ma and 12.5 Ma). A sequence boundary near the top of the middle Miocene (10.5 Ma) overlies sediment dated as zone N14 (Poag, 1980) and is characterized by a major downward shift in coastal onlap and submarine canyon erosion (Miller and others, 1987).

Late Miocene-Pliocene.—

Upper Miocene depositional sequences are thickest south of the middle Miocene depocenter (Fig. 1). These deposits thin by onlap against the middle Miocene progradational wedges in a landward direction. These wedges also onlap the middle Miocene deltaic lobes along strike lines to the north. On the seismic line shown in Figure 2, the uppermost portion of the sequence below the 6.3-Ma sequence boundary extended over the latest middle Miocene shelf margin. This sequence continues to thin by onlap to the north and onlaps the middle Miocene shelf edge on the seis-

---

[1]The depositional-shoreline break is a position on the shelf landward of whch the depositional surface is at or near base level, usually sea level, and seaward of which the depositional surface is below base level. This position coincides approximately with the seaward end of the stream mouth bar in a delta or with the seaward end of the upper shoreface in a beach (see Van Wagoner and others, and Posamentier and Vail, this volume).

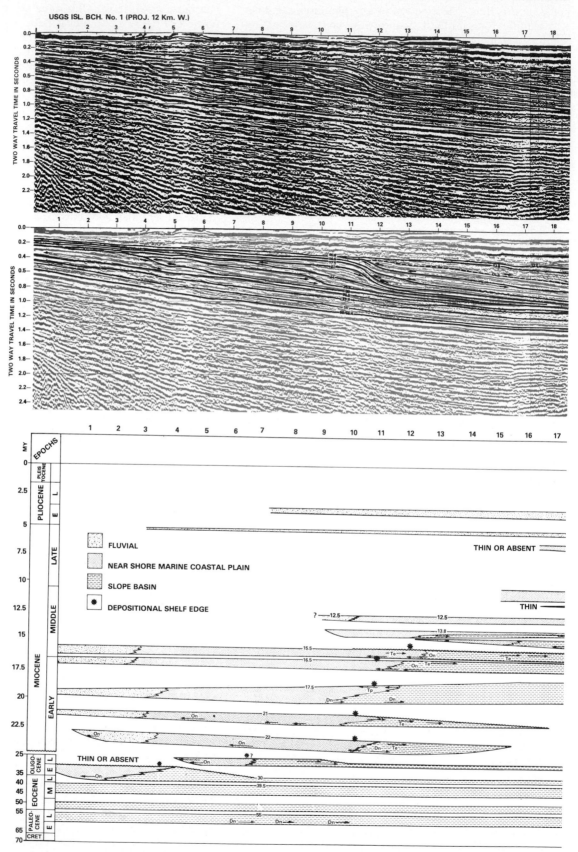

Fig. 2.—Uninterpreted seismic line, interpreted seismic

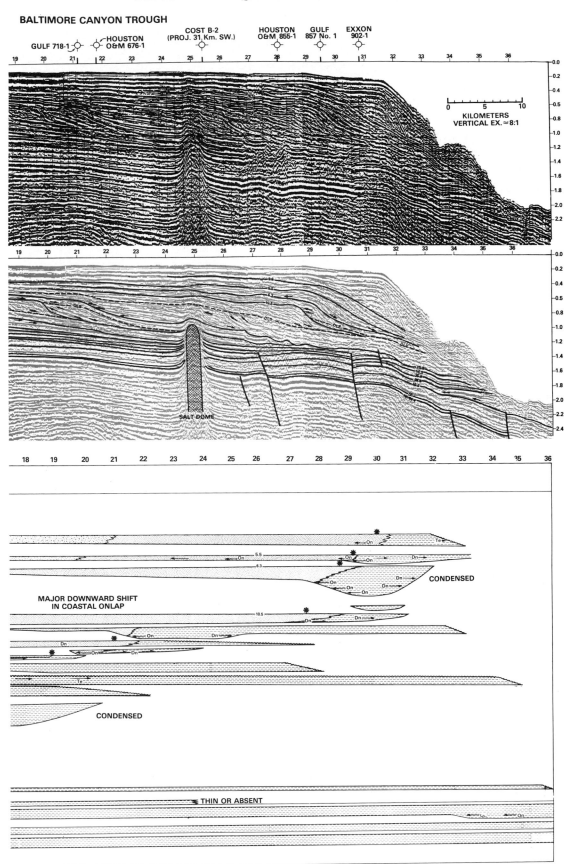

line, and chronostratigraphic chart, offshore New Jersey.

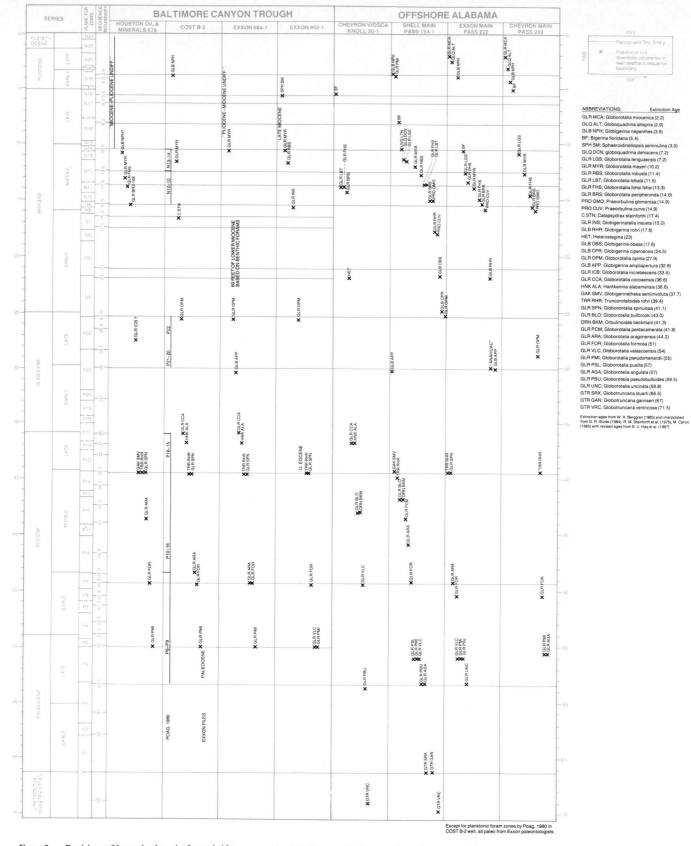

FIG. 3.—Position of key planktonic foraminifera recognized in four wells from each study area relative to sequence boundary position. Occurrences of several benthic foraminifera are also shown, although most have been omitted due to proprietary range information and range uncertainty.

mic line in Figure 4. Uppermost Miocene sediments are widespread in a landward direction, as are the two lower Pliocene depositional sequences.

TERTIARY STRATIGRAPHIC FRAMEWORK—OFFSHORE ALABAMA

Strata from the offshore Alabama area were studied using a grid of regional high-quality seismic lines in conjunction with well control (Fig. 5). Biostratigraphic information is from Exxon company reports (M. Crane, pers. commun., 1985). The positions of microfossils in several key wells relative to the positions of depositional-sequence boundaries recognized on the seismic sections are shown in Figure 3. Seismic sections and chronostratigraphic interpretations of the depositional sequences recognized in the study area are illustrated in Figures 6 and 7. Locations of Tertiary shelf margins and significant physiographic features are shown in Figure 5.

Little information has been published concerning the Tertiary stratigraphy of the offshore Alabama area. These rocks have been studied extensively in coastal-plain outcrops and wells (Murray, 1961; Mancini, 1981; Raymond, 1985; and references therein). Fluker (1983) illustrates a seismic section from the area.

Paleocene-Eocene.—

The base of the Tertiary lies on an erosional surface that removed successively older Cretaceous deposits in a basinward direction. Clinoforms within a thick Paleocene wedge downlap onto this surface. The progradational wedge is composed of lower Paleocene shales. This shale-prone wedge is separated from thinner deposits of late Paleocene and early Eocene age by an erosional surface in the upper Paleocene (Figs. 6, 7) showing erosional truncation and onlap. These depositional sequences are thin but present in wells located in a basinward position. Early and middle Eocene sequences are thin units separated by locally erosive unconformities. Upper Eocene deposits show onlap onto the middle Eocene (Fig. 6). It is unclear whether this is deep-marine or true coastal onlap (for a discussion of coastal and marine onlap, see Vail and others, 1977b, p. 65–72).

Oligocene.—

The seismic profiles show no pronounced unconformity surface separating upper Eocene from lower Oligocene deposits in the offshore Alabama area. Reflections from lower Oligocene rocks either toplap or are erosionally truncated by a major erosional unconformity that has completely cut out the lower Oligocene deposits in a basinward direction (Fig. 6). Overlain by sediments dated as late Oligocene, this unconformity is correlated with the mid-Oligocene (30 Ma) sea-level fall. The late Oligocene section onlaps the mid-Oligocene unconformity and records the first definite incursion of coastal environments into the study area during the Tertiary. The upper Oligocene sequence appears to lap out near the landward limit of the seismic grid. Its upper boundary separates Oligocene from Miocene sediment and is characterized by local erosion.

Miocene.—

Lower Miocene deposits are areally widespread and extend landward beyond the study grid. Shallow-marine environments existed in the landward portion of the study area, as indicated by the presence of a lower Miocene (Aquitanian) coral reef drilled in the Viosca Knoll 30 well (Fig. 7). This reef grew near the most updip onlap position of the underlying upper Oligocene depositional sequence and is overlain by middle Miocene deep-marine shales. Lower Miocene deposits thicken in a seaward direction and show low-relief, seaward-dipping clinoforms. Upper lower Miocene (Burdigalian) deposits onlap against the edge of the reef and are not present in the Viosca Knoll 30 well.

Middle Miocene deposits consist of progradational wedges that moved the depositional shoreline break well into the study area (Fig. 5). Also during this time, thick onlapping wedges of sediment were deposited in the basin. The middle Miocene prograding wedges extend landward beyond the seismic grid.

A major erosional unconformity separates middle and upper Miocene sediments in the study area. Local escarpments are onlapped by packages of mounded and chaotic reflections interpreted as debris shed during lowstands of sea level (Fig. 8). Overlying upper Miocene depositional sequences prograde rapidly seaward and are characterized by thin packages of parallel reflections extending landward, and thicker sigmoidal and oblique-progradational clinoforms seaward of the previous shelf edge. Each of the three late Miocene depositional sequences shows a pronounced downward shift in coastal onlap. A package of oblique progradational clinoforms onlaps the uppermost Miocene sequence boundary (5.5 Ma) and is overlain by a downlap surface at the Miocene-Pliocene boundary.

Pliocene.—

Pliocene biostratigraphic resolution is poor in the Main Pass study area. Four well-defined early Pliocene sequences are recognized within this area. This suggests a higher frequency of sea-level fluctuations during this time than is noted on the global-cycle chart (Fig. 9; Haq and others, 1987). Recognition of these Pliocene depositional sequences is aided by the unusually thick Pliocene section preserved in the area.

GEOHISTORY ANALYSIS

Geohistory analysis (Van Hinte, 1978) was performed at several well sites in both study areas in order to determine the subsidence history of the two basins. This technique calculates the total subsidence of the basement based on the ages, depth, and paleobathymetric values of stratigraphic horizons at a particular site. In this study, age-depth pairs are derived from the stratigraphic surfaces recognized in the well tied to our regional stratigraphic framework and depth-converted seismic picks below the depth of the well. Estimates of paleobathymetry are obtained from paleoenvironmental interpretations of microfossil assemblages, facies interpretation from logs, and seismic and clinoform measurements. When the total subsidence values are cor-

TERTIARY DEPOSITIONAL SEQUENCES AND MIOCENE

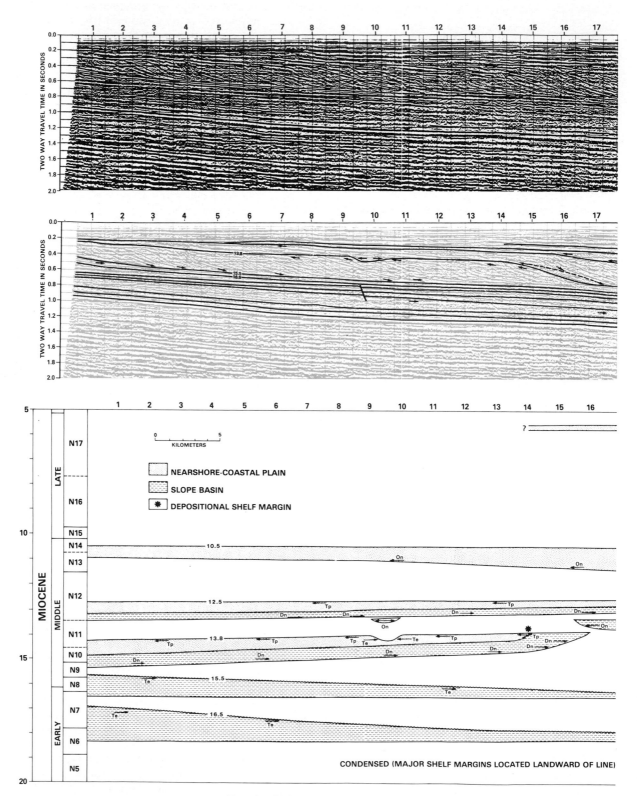

FIG. 4.—Uninterpreted seismic line, interpreted seismic line, and Miocene chrono-

CHRONOSTRATIGRAPHY BALTIMORE CANYON TROUGH

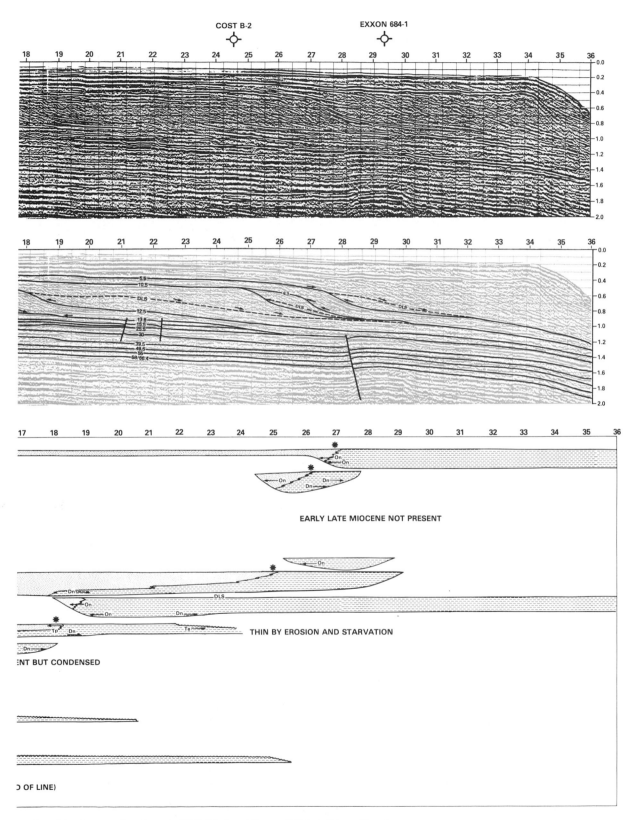

EARLY LATE MIOCENE NOT PRESENT

THIN BY EROSION AND STARVATION

NT BUT CONDENSED

OF LINE)

stratigraphic chart passing through COST B-2 and Exxon 684-1, offshore New Jersey.

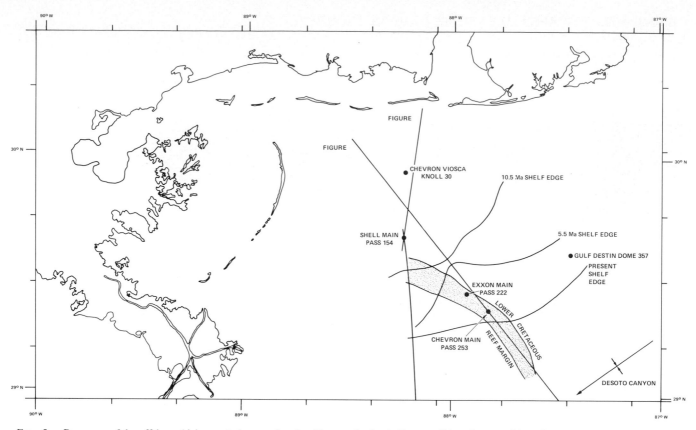

FIG. 5.—Base map of the offshore Alabama study area showing illustrated seismic lines, well locations, position of lower Cretaceous shelf margin, and Neogene depositional-shelf margin.

rected for compaction effects and isostacy due to sediment loading, the resultant subsidence curve represents the contribution of tectonic processes other than Airy isostacy and changes in eustasy. This curve is then corrected for long-term eustatic change using an estimate of this value (Haq and others, this volume). The values obtained illustrate the space made available for sediment accumulation (accommodation) due to thermo-tectonic, salt-related, fault-related, or flexural subsidence. Rapid landward and basinward shifts in coastal sedimentation and erosion are attributed to rapid changes in eustasy.

Baltimore Canyon Trough.—

Geohistory analysis in the Baltimore Canyon Trough was analyzed at COST B-2 and several sites along the seismic section in Figure 2 (Greenlee and others, 1988). Figure 10 shows the curves representing total subsidence, paleobathymetry, and long-term sea level at COST B-2.

Two episodes of uplift followed by rapid subsidence (Fig. 10) are recognized on the thermo-tectonic subsidence curve. The first occurred in the earliest Jurassic and is associated with rifting of North America from Africa and the onset of seafloor spreading and rapid subsidence. A second event occurred in the latest Early Cretaceous and is characterized by an inflection point on the tectonic subsidence curve. This is interpreted as a regional heating event associated with volcanic intrusives recognized in the area (e.g., the Great Stone Dome; Grow, 1980). Thermo-tectonic subsidence

continues into the Tertiary at a constant rate of approximately 6.5 m per million years on the basis of this analysis.

Offshore Alabama.—

Two wells were selected to illustrate the subsidence history of the offshore Alabama study area. Both wells are situated along the seismic line illustrated in Figure 7. The geohistory plots from these wells are shown in Figures 11 and 12. Note that shallow-water conditions dominated by shelf carbonates and siliciclastics existed in both wells until the Late Cretaceous when deep-marine conditions prevailed. The increase in total subsidence in the late Early Cretaceous reflects rapid carbonate deposition associated with the development of rapidly growing Early Cretaceous carbonate platforms rimming the Gulf of Mexico.

Both well locations underwent smooth, continuous subsidence punctuated by a minor event in the Early Cretaceous that may reflect a thermal event concurrent with the event noted in the Baltimore Canyon Trough. From the beginning of the middle Miocene until the present, the thermo-tectonic subsidence curves show a marked increase in slope characterized by a convex-upward shape. This contrasts with the generally concave-upward shape of the exponentially decreasing thermal cooling cycles associated with post-rift subsidence. We interpret this rapid increase in post-early Miocene subsidence as a result of a flexural loading event due to rapid middle Miocene and younger sedimentation patterns immediately seaward of the Lower Cretaceous platform margin (see Fig. 7). The rate of tectonic subsidence increases from approximately 10 to 30 m per million years in association with this event.

COMPARISON OF THE TERTIARY DEPOSITIONAL SEQUENCES RECOGNIZED IN OFFSHORE NEW JERSEY AND OFFSHORE ALABAMA

Continental margins of these two areas are both considered as old, slowly subsiding passive margins. Geohistory analysis suggests that significant uplift or thermal perturbation has not occurred in either area since the mid-Cretaceous. This interpretation is supported by a lack of faulted angular unconformities, or "tectonically enhanced" unconformities recognized in these areas within the Tertiary. The similarity in the overall Tertiary geologic history of both study areas is rooted in the fundamental geologic attributes shared by the two regions. These include: (a) similar crustal types and Mesozoic geologic development; (b) absence of extensive Tertiary uplift; (c) comparable paleogeographic positions; and (d) similar depositional systems.

Although salt underlies both areas, the effects of salt withdrawal on the Tertiary tectonic history of the Baltimore Canyon Trough, as well as offshore Alabama, appear to be small. Tectonically, the major difference between the two areas occurs in mid-Miocene times, when the offshore Alabama study area experienced a marked increase in subsidence rate in the middle Miocene as thick sedimentary wedges caused flexure of the crust. This led to an overall increase in resolution of late Miocene and Pliocene sequence development in the Alabama area.

In terms of depositional history, the present continental-shelf edge in both study areas reflects previous Mesozoic carbonate-platform development. This relict feature was located in bathyal water depths during the Paleogene. In the Baltimore Canyon Trough, the shelf slope break approximately overlies an Upper Jurassic to Lower Cretaceous reef margin situated immediately seaward of the continental-to-oceanic crustal boundary (Schlee and others; 1979). In the northeastern Gulf of Mexico, this major physiographic discontinuity is associated with Lower Cretaceous reef deposits that overlie the thinned continental- to "transitional"-crust boundary (Corso and others, 1987).

Both areas were incised by major submarine canyons when Miocene progradation moved coastal-facies belts across the Mesozoic carbonate platform and the depositional-shoreline break became approximately coincident with the relict edge of the continental shelf. Prior to this time, sequence boundaries are characterized by minor channeling landward of the shelf edge and planar-erosional surfaces that erode successively older strata in a basinward direction (Figs. 2, 4, 6, 7). The flat, eroded surfaces seaward of the depositional-shoreline breaks are interpreted to have been formed by erosion caused by wave and submarine-current processes during periods of lowered sea level. Prior to the late Miocene, the shorelines of lowstand system tracts are located seaward of the associated highstand shelf margin, but well landward of the relict Mesozoic shelf margin.

Tertiary coastal-onlap patterns.—

Exxon seismic stratigraphers have long used coastal-onlap patterns to compare the shifts in areal extent of depositional sequences with time on widely separated continental margins (see Vail and others, 1977b). Posamentier and others (this volume) present a discussion of the mechanics of the relationship between depositional stratal patterns and eustatic change. Examination of offshore New Jersey and Alabama coastal-onlap patterns (Fig. 9), downlap surfaces, and basinal erosional unconformities reveals both similarities and differences. Stratal characteristics common to both areas include: (1) a major downlap surface corresponding to the Cretaceous-Tertiary boundary; (2) low-relief, planar-erosional surfaces at several points in the Paleogene, which are the same age on the basis of available paleontologic data; and (3) generally similar coastal-onlap patterns recognized in upper Oligocene and Neogene deposits. Within the Neogene, however, differences in the areal extent of certain middle Miocene sequences, as well as the number and areal extent of early Pliocene sequences, are recognized.

Both study areas were in a basinal position throughout the Paleogene; thus, coastal-onlap patterns could not be derived. Although a comparison of Paleogene coastal onlap is not possible, certain observations are common to both study areas. The Cretaceous-Tertiary boundary is an erosional unconformity on both the Atlantic and Gulf of Mexico margins with more severe erosion having occurred in the offshore Alabama area. A major downlap surface at this boundary indicates a time of maximum flooding, which we interpret as a result of the rapid rise in sea level in the latest Maastrichtian and earliest Paleocene. In the offshore New Jersey area, the entire lower Paleocene section is absent due to condensed sedimentation followed by mid-Paleocene erosion. The seaward edge of the lower Paleocene prograding Midway shale is evident in offshore Alabama. This unit consists of condensed, basinal shales that are dated by planktonic foraminifera in offshore wells. The condensed section at the Cretaceous-Tertiary boundary is discussed more fully by Loutit and others (this volume). Donovan and others (this volume) discuss the onshore late Maastrichtian and early Paleocene outcrops of Alabama.

Sequence boundaries having planar-erosional surfaces with varying degrees of deep-marine onlap occur near the top of the lower and middle Eocene in both areas. These surfaces are correlated with second-order sequence boundaries at 49.5 Ma and 39.5 Ma. A late Paleocene sequence boundary (58.5 Ma) is recognized in the offshore Alabama grid but has cut down to the Cretaceous-Tertiary boundary in the offshore New Jersey grid. Both areas exhibit onlap of upper Eocene sediments against the top-middle Eocene sequence boundary.

The first clear indication of coastal onlap seen in the two areas is associated with a major downward shift in coastal onlap in mid-Oligocene time. This shift brings nearshore marine and coastal-plain depositional environments into both study areas and is followed by progressive landward onlap of upper Oligocene and lower Miocene (Aquitanian) deposits. Another major downward shift is noted in the mid-early Miocene (21 Ma), when the lower Miocene reef at Viosca Knoll 30 was terminated. Burdigalian sequences (17.5 and 16.5 Ma) are thin in the Alabama grid; however, they are thicker and better developed on the New Jersey grid.

A significant departure between the observed coastal on-

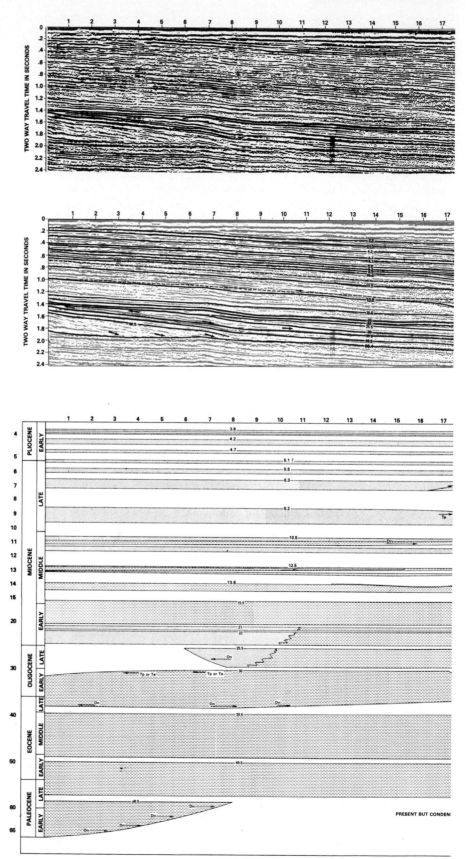

FIG. 6.—Uninterpreted seismic line, interpreted seismic

OFFSHORE ALABAMA

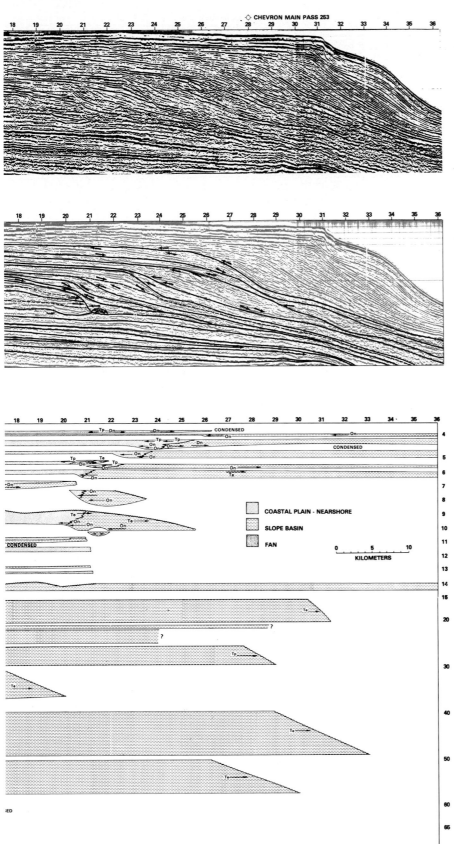

line, and chronostratigraphic chart, offshore Alabama.

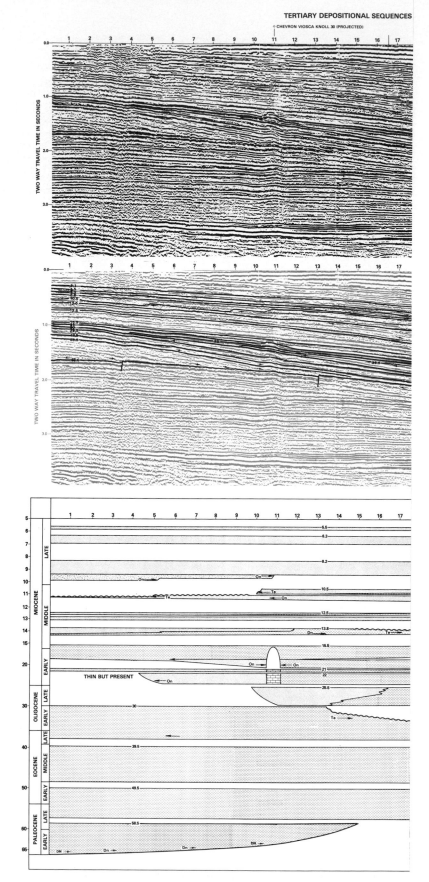

FIG. 7.—Uninterpreted seismic line, interpreted seismic

OFFSHORE ALABAMA

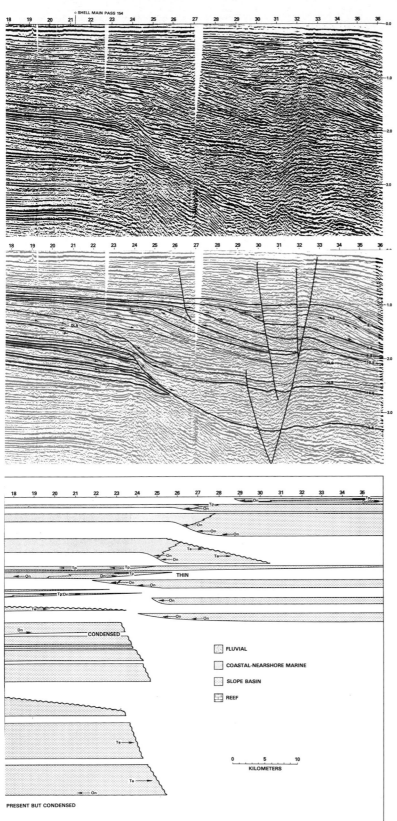

line, and chronostratigraphic chart, offshore Alabama.

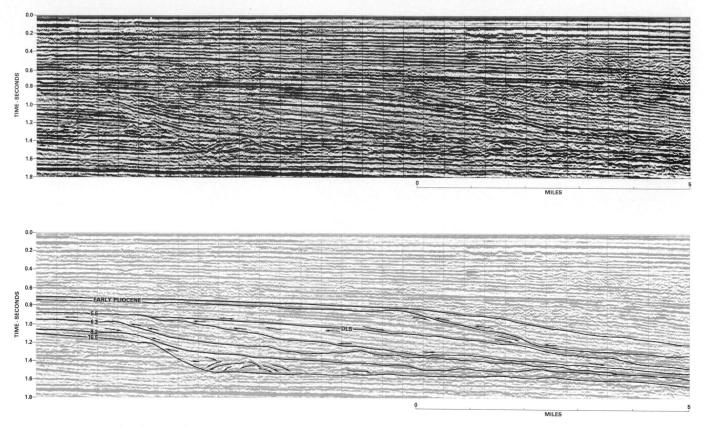

FIG. 8.—Detail of Figure 6 showing extensive slope erosion and onlap of upper Miocene sequences.

lap in the two areas occurs in the early middle Miocene. In the New Jersey area, the unconformity of earliest middle Miocene age (15.5 Ma) is characterized by severe erosion that truncates successively older strata in a basinward direction (Fig. 2). It is followed by a basinally restricted sequence (13.8 Ma to 15.5 Ma; Fig. 2) and a seaward progradation of upper middle Miocene depositional-shelf margins. In the offshore Alabama area, the sequence deposited between 15.5 and 13.8 Ma is areally extensive. This onlap pattern is similar to that depicted on the global coastal-onlap curve (Fig. 9), which shows a pronounced landward shift in onlap during the early Serravalian (13.8 to 15.5 Ma). As in the offshore New Jersey area, Serravalain sequences are highly progradational in the offshore Alabama area.

This difference in regional-onlap patterns between the two areas has most likely been enhanced by the mid-Miocene increase in basin subsidence in the offshore Alabama area, interpreted to be caused by flexural loading. On the other hand, this anomalous coastal-onlap pattern in the Baltimore Canyon Trough could be due to extensive erosion of the 13.8- to 15.5-Ma sequence (Greenlee and others, 1988) or a brief period of regional uplift not recognized in the geohistory analysis.

Both study areas show a pronounced downward shift in onlap at the end of the middle Miocene. Upper Miocene depositional sequences progressively onlap landward. Higher basin subsidence in the offshore Alabama area has resulted in thin portions of the sequences extending farther land-

ward. Major submarine canyon formation during sea-level lowstands in both locations occurred only after the depositional-shoreline break had prograded to the relict, structurally controlled edge of the continental shelf.

The greater rate of subsidence in the Gulf of Mexico area allowed for a finer resolution of uppermost Miocene and Pliocene sequences. In fact, there are more individual cycles identified in this area than are indicated on the global-cycle chart of Haq and others (this volume). To resolve this apparent discrepancy, other areas that have high sedimentation rates (and therefore afford high seismic resolution) need to be investigated in the future.

The high resolution of the Gulf of Mexico stratigraphy might also account for the differences in the estimated age of the Pliocene highstand maximum; however, its timing in the Gulf of Mexico is very similar to that found in the Baltimore Canyon area. It is unclear whether the difference in the age of the early Pliocene maximum highstand seen in Gulf of Mexico and Baltimore Canyon data represents a true disagreement with the study of Haq and others (this volume) or a slight (~0.7 Ma) difference in age estimates of the associated sequence boundaries.

Second-order supersequences.—

The earlier discussion has noted two characteristic patterns of sequence development recognized beneath the New Jersey and Alabama continental shelves. In the Paleocene and Eocene, relatively deep-water deposits are divided by submarine erosional surfaces interpreted to be correlative

TERTIARY COASTAL ONLAP

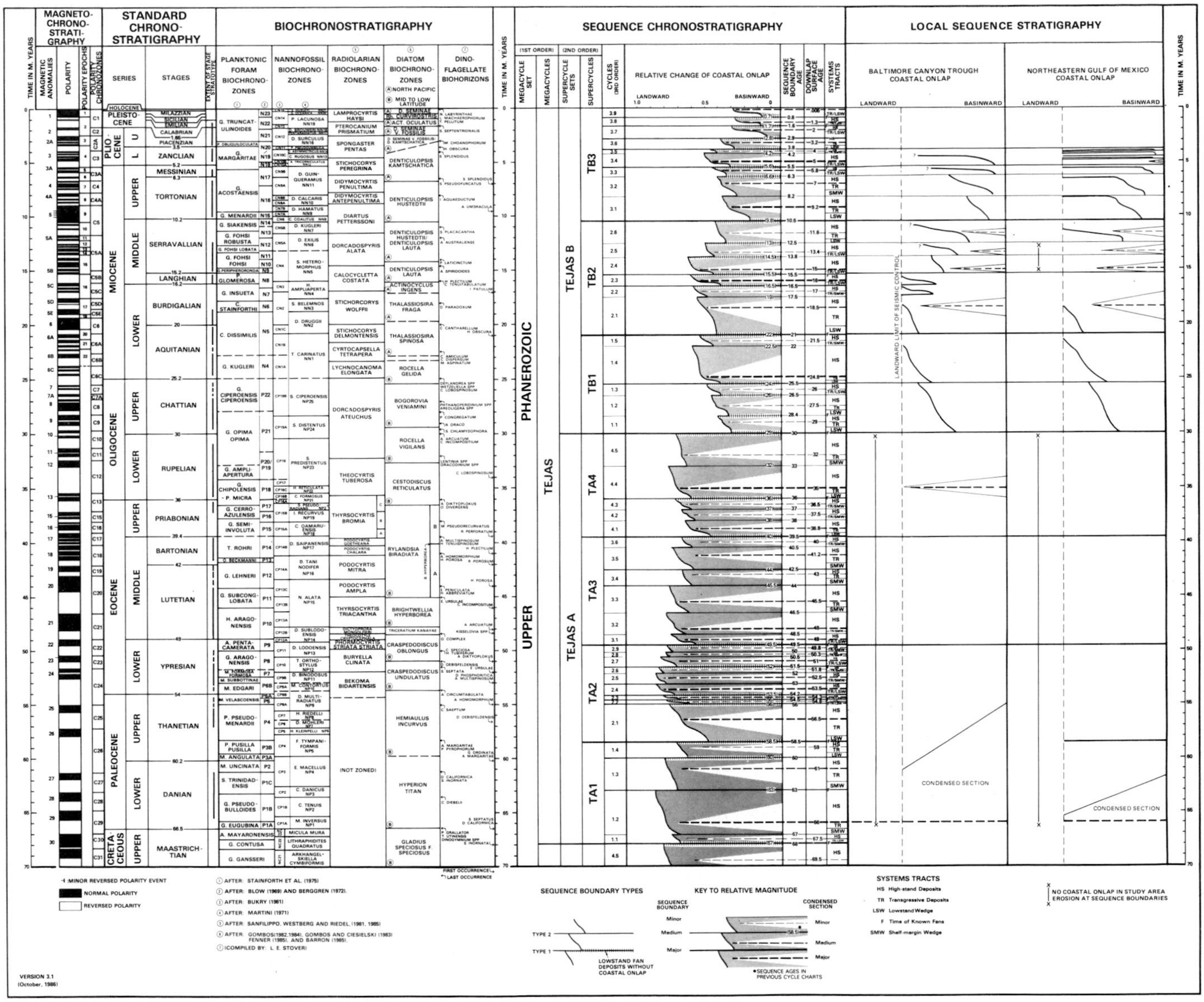

FIG. 9.—Tertiary chronostratigraphy and global coastal-onlap chart from Haq and others (this volume) with added offshore New Jersey and Alabama Coastal onlap. References noted on the eustatic-cycle chart are found in Haq and others (1987).

with second-order or "supercycle" boundaries (Haq and others, 1987) that have a time scale of approximately 10 Ma. A similar correlation based on a regional biostratigraphic analysis of wells and core holes in this area was made by Poag and Ward (1987). During Neogene deltaic progradation, a higher frequency series of sequences was recognized on the basis of basinward shifts in onlap at approximately 1-Ma intervals. These have been correlated to third-order cycle boundaries (Haq and others, 1987).

Our data suggest that the third-order depositional sequences noted in both areas may be grouped into larger packages, or "supersequences," which are bounded by second-order or "supersequence" boundaries (see Vail and others, 1977b, p. 64). The stacking pattern of depositional sequences within the supersequences appears to be similar to the stratal patterns of systems tracts found within third-order depositional sequences (see discussion in Posamentier and others, this volume).

For example, the second-order sequence boundaries at 30 and 10.5 Ma are characterized by a great deal of erosion and an ensuing period of basinally restricted sedimentation (Fig. 13). Overlying these second-order lowstand deposits (upper Oligocene-lower upper Miocene) is a series of backstepping, retrogradational, or aggradational third-order sequences that continue onlapping the margin (lower Miocene). These strata would equate to the transgressive-systems tract within third-order sequences. Finally, a regressive series of third-order sequences (middle Miocene) progrades rapidly seaward over a surface of condensed sedimentation. These are analogous to the strata of a highstand systems

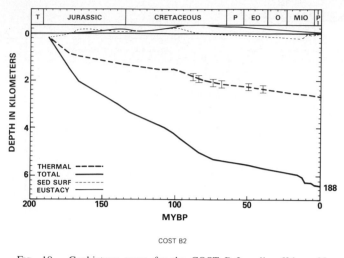

FIG. 10.—Geohistory curve for the COST B-2 well, offshore New Jersey.

tract within a third-order depositional sequence. This pattern of sedimentation reflects the overall (post-30 Ma) landward shift in coastal onlap represented on the global chart (Fig. 9) beginning in the late Oligocene and early Miocene and the progressive basinward shift in the depositional-shoreline break during the middle Miocene. Although the surface interpreted as the 21-Ma sequence boundary is characterized by a large basinward shift in coastal onlap in the offshore Alabama study area, we do not interpret this as a supersequence boundary, as is shown on the global eustatic-cycle chart.

Due to the longer time scale represented by second-order cycles, the physical characteristics of second-order supersequences are more sensitive to basin tectonism. At time scales on the order of 10–20 Ma, relative change in sea level may be affected more by episodes of uplift and sub-

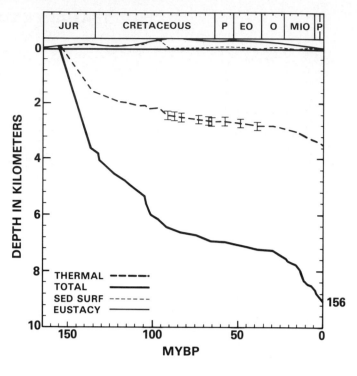

FIG. 12.—Geohistory curve for the Shell Main Pass 154 well, offshore Alabama.

sidence than by eustatic changes in certain continental-margin settings. Recognition of supersequence boundaries and second-order condensed sections (see Loutit and others, this volume), however, has proven useful in the establishment of age models in frontier basins.

INTERPRETATION OF SEA-LEVEL CHANGES FROM THE STRATIGRAPHIC DATA BASE

The effects of relative changes in sea level on the stratal configuration of clastic and carbonate sedimentary sequences are discussed in Posamentier and others (this volume), Van Wagoner and others (this volume), and Sarg (this volume). A great deal of research has gone into the timing and effects of third-order eustatic changes from a stratigraphic data base because of the importance in exploration of prediction of lithofacies and age in the subsurface prior to drilling. Little has been done, however, to estimate the magnitude of these eustatic fluctuations.

The examples presented here illustrate siliciclastic depositional sequences deposited in a relatively quiet, slowly subsiding, tectonic setting. We believe this environment allows the most direct interpretation of the magnitude of second- and third-order eustatic cycles. Because the shoreline is close to the break in slope between relatively flat-lying coastal-plain and shoreline strata and more steeply dipping slope (delta front) strata, eustatic falls of relatively small magnitude may move the shoreline over the break in slope.

We calculated the magnitude of third-order sea-level falls in an identical fashion to the method of Greenlee and others (1988; Fig. 14). The magnitude of a relative fall in sea level

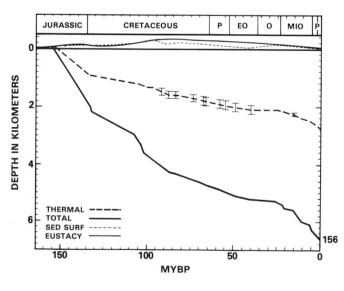

FIG. 11.—Geohistory curve for the Chevron Viosca Knoll 30 well, offshore Alabama.

LATE OLIGOCENE TO LATE MIOCENE 2nd ORDER SEQUENCE

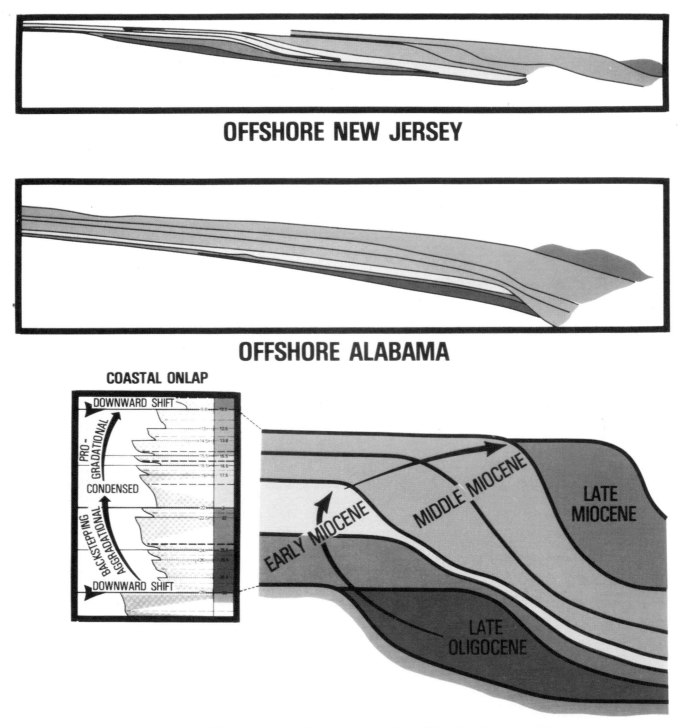

FIG. 13.—Schematic illustration of second-order sequences from offshore New Jersey and Alabama.

is calculated by measuring the amount of vertical shift of the coastal-plain/delta plain surface that occurs between the end of a highstand and the beginning of a subsequent lowstand. It is important to determine whether the observed onlap is true coastal onlap or deep-marine onlap. This is accomplished, where possible, through facies interpretation from well data (Van Wagoner, 1985). The thickness of sediment from the first coastal onlap is then measured from a datum representing the level of the previous highstand shelf (Fig. 14). Factors that complicate the establishment of a shelf datum include post-depositional compaction effects and erosional alteration of the shelf margin area. Because of these complications, we averaged measurements from several locations. The measured distance from the shelf datum to the first coastal onlap must be corrected for sediment-loading effects and compaction. The calculated magnitude of the short-term sea-level fall is then subtracted from the associated highstand value calculated for the long-term highstand sea-level curve to obtain lowstand sea-level values.

Long-term changes in sea level have been studied by numerous authors (summarized in Kominz, 1984) using estimates of mid-ocean spreading rates and corresponding volume of mid-ocean ridge volume. Hardenbol and others (1981) and Greenlee and others (1988) applied a thermal cooling model to isostatically corrected subsidence plots to derive a long-term sea-level curve. All of these studies indicate a long-term sea-level fall during the Cenozoic.

Moore and others (1987) described an alternative method for calculating long-term sea-level change, which we use here. It is similar to that used by Hardenbol and others (1981), in that we calculate the decompacted and isostatically corrected thickess of sediment, add paleo-water depth, and then correct this value for tectonic subsidence. This relationship (Hardenbol and others, 1981) is summarized in the following equation:

$$1.446 \, SL = SEDLOAD + UNCOMP$$
$$+ SUB + PWD - Z \quad (1)$$

where SL = the estimated sea level for horizon "h", and

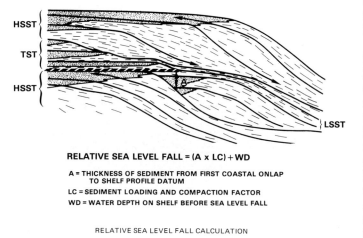

RELATIVE SEA LEVEL FALL = (A x LC) + WD

A = THICKNESS OF SEDIMENT FROM FIRST COASTAL ONLAP TO SHELF PROFILE DATUM
LC = SEDIMENT LOADING AND COMPACTION FACTOR
WD = WATER DEPTH ON SHELF BEFORE SEA LEVEL FALL

RELATIVE SEA LEVEL FALL CALCULATION

FIG. 14.—Method used to calculate short-term sea-level falls from stratigraphic data.

1.446 is the correction factor for the isostatic load effect of any change in sea level; SEDLOAD = the amount of subsidence caused by the isostatic load of sediments overlying a horizon (h); UNCOMP = the decrease in thickness of the section underlying horizon "h" caused by compaction effects; SUB = the amount of subsidence caused by tectonic and thermal effects; PWD = the paleo-water depth of the horizon at the time of deposition; and Z = the depth of the horizon below present sea level.

Unlike Hardenbol and others (1981), we calculate SUB, or the tectonic subsidence, using the method illustrated in Figure 15. Along a passive continental-margin profile, thermal tectonic subsidence increases from zero at some hinge point to progressively higher values in a seaward direction; (Pitman and Golovchenko, 1983; Posamentier and others, this volume). In the absence of local tectonic perturbations such as salt movement, this subsidence rate change should vary smoothly and is recognized by the seaward divergence of strata in the subsurface. By calculating the decompacted and isostatically corrected subsidence at several points along the profile, we use simple trigonometry (Fig. 15) to determine the average angular-tilt rate and the location of the apparent hinge-point locations on a horizon interpreted to represent a flat surface. Because the average angular-tilt rate is unaffected by sea-level change, it represents a better estimate of thermo-tectonic subsidence rates.

In the offshore Alabama area, the 21-Ma sequence boundary is probably the youngest boundary that predated the mid-Miocene increase in subsidence rates and was nearly flat during deposition. At present, the reflector associated with this boundary shows very little local relief and diverges downdip from the overlying reflectors (Figs. 6, 7). We selected three locations on each of the dip lines (Figs. 6, 7) for geohistory analysis of the 21-Ma horizon. These locations were chosen so as to bound the regions of relatively slow subsidence (updip) and relatively rapid subsidence (downdip). At these locations, the effects of sediment loading of the section overlying the 21-Ma horizon and the compaction of the section underlying this horizon were removed (assuming Airy isostacy). A tilt remained in the boundary that indicated a basinward increase in subsidence rate (corrected for sediment loading and compaction). Tilt rates ranged between .010 and .006 degrees per Ma.

We have assumed that this average rate and the apparent hinge-point location have remained constant during the last 21 Ma. The angular-subsidence rate, the age of each highstand, and the location of each depositional-shoreline break relative to the hinge point (Fig. 15) can be used to calculate the amount of subsidence that is in excess of sediment loading and compaction effects (i.e., the apparent tectonic thermal subsidence). By assuming Airy isostacy, we may be overestimating the sediment-loading effect in equation 1; however, such an error would be at least partially offset by underestimating the unloaded tilt rate, which also assumes Airy isostacy. The average rate of tilt may not have remained constant over the past 21 Ma; however, a check on this calculation was made using the 2.4-Ma horizon to estimate average subsidence rates on one of the sections. The results were within about 10 percent of the amount of tec-

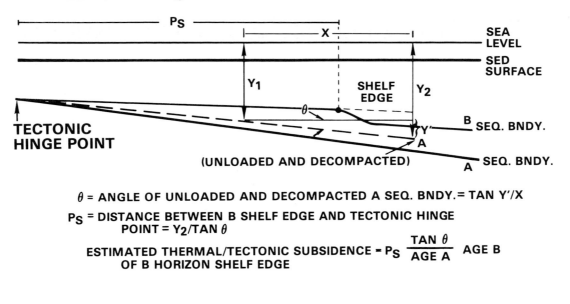

θ = ANGLE OF UNLOADED AND DECOMPACTED A SEQ. BNDY. = TAN Y'/X

P_S = DISTANCE BETWEEN B SHELF EDGE AND TECTONIC HINGE POINT = Y_2/TAN θ

ESTIMATED THERMAL/TECTONIC SUBSIDENCE = $P_S \dfrac{\text{TAN } \theta}{\text{AGE A}}$ AGE B
OF B HORIZON SHELF EDGE

SUBSIDENCE CALCULATION

FIG. 15.—Method for calculating angular-rotation rates and apparent hinge point of continental margin.

tonic thermal subsidence estimated using the 21-Ma horizon.

Each calculation is derived from a measurement at the depositional-shoreline break, because this point is the most easily recognizable on seismic-reflection data. A significant change in the sea-level calculation is affected by paleo-water-depth estimates at the depositional-shoreline break. Although modern shelf breaks lie at depths of less than 20 m to over 200 m (Vanney and Stanley, 1983), many of these are drowned relict features, far from the sites of active sediment deposition. For example, during much of the Tertiary, both study areas had two physiographic discontinuities—a relict linear shelf margin submerged beneath hundreds to thousands of meters of water and an active lobate depositional-shoreline break that defined the zone of active deltaic progradation. Water depth at the depositional-shoreline break was on the order of several meters to tens of meters, according to facies analysis of wells within the central portion of the depocenter. According to Poag (1980, 1985a), the topset beds of the latest middle Miocene wedge sampled at COST B-2 were deposited in "inner to middle sublittoral" environments (20 m or less). For this reason, we try to make our measurements as close to the main axis of the depocenter as possible. To illustrate the effect of the depositional-shoreline break paleo-water depth estimate, we have plotted our sea-level highstand curves using a zero and a 100-m estimate for this value (Fig. 16, Table 1).

RESULTS OF SEA-LEVEL CALCULATIONS

Sea-level highstands.—

Figure 16 summarizes the eustatic highstand values calculated for the offshore Alabama study area and offshore New Jersey (Moore and others, 1987) using both a 100-m estimate of paleo-water depth at the depositional-shoreline break, considered a maximum estimate, and a 0-m esti-

mate. Calculations from both study areas show similar results with sea levels within 20 m. Also plotted is a sea-level curve derived from ocean ridge spreading volumes (Kominz, 1984) and the sea-level estimate presented by Haq and others (1987). The Kominz (1984) and Haq and others (1987) estimates show a similar trend toward sea-level fall in the middle Miocene, which is recorded in our stratigraphic data sets. Our data also indicate an early Pliocene sea-level rise. Our minimum and maximum limits on calculated sea-level highstands bound a 69-m (100 ÷ 1.446) interval which encompasses the Kominz estimate. In order to match the middle Miocene estimate of Haq and others (1987), we would need to add 202 m of paleo-water depth to our calculations.

Sea levels calculated here are low relative to other oceanic-ridge volume estimates (Pitman, 1978) and to geohistory-based studies, as well as estimates based on measurements from Midway Atoll (Moore and others, 1987; Lincoln and Schlanger, 1987). We suggest three possible explanations to account for these low estimates. First, the low sea-level estimates are correct, and previous studies have failed because of the use of incorrect time scales used in the earlier ridge volume calculations (see Kominz, 1984, p. 121), incorrect subsidence calculations at Pacific atolls (Moore and others, 1987), or incorrect paleo-water-depth estimates and theoretical thermo-tectonic subsidence assumptions (Hardenbol and others, 1981; Greenlee and others, 1988). Second, Paleo-water depth at the depositional-shoreline break is higher than that used in this study, and our calculations need to be corrected for these low values. Finally, it is possible that significant shelf edge progradation is occurring during sea-level fall and concurrent updip erosion. The record of the highest sea-level stand would be eroded during the early portion of the third-order sea-level falls (Fig. 17). The effects of this third possibility are most difficult to assess. Without detailed facies analysis, it is impossible to

ESTIMATES OF NEOGENE EUSTATIC HIGHSTANDS

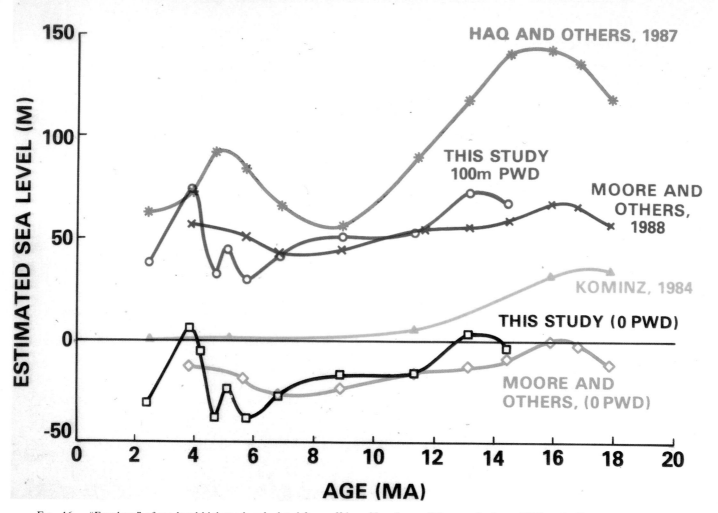

FIG. 16.—"Envelope" of sea-level highstands calculated from offshore New Jersey (Moore and others, 1987) and offshore Alabama (this study), spreading-rate changes (Kominz, 1984), and the curve of Haq and others (this volume). PWD refers to the paleo-water depth value used in the calculation of eustatic sea-level values.

determine whether apparent toplaps beneath the overlying sequence boundaries (see Figs. 3, 4, 6, 7, and 8) represent true depositional toplap, as described by Mitchum and others (1977) and shown in the late highstand systems tract by Posamentier and Vail (this volume), or are erosionally truncated sigmoidal or oblique clinoforms (Fig. 17). The magnitude of this fall would appear to be consistent in both study areas, as the curves calculated from these study areas are within about 20 m throughout the late Neogene.

Sea-level lowstand calculations.—

A complete eustatic curve showing the magnitudes of sea-level falls in relation to each preceding highstand is shown in Figure 18, and values are recorded in Table 1. These sea-level falls show magnitudes of 0 to 116 m, close to the range of possible glacial eustatic changes that might be expected to occur in this time interval. They are slightly smaller in the early Miocene, larger in the middle Miocene, and

are the same or slightly larger in the late Miocene relative to the curve of Haq and others (1987) (with the exception of the fall at 8.2 Ma in the offshore Alabama area, which is smaller). Again, the two areas analyzed show fairly close agreement.

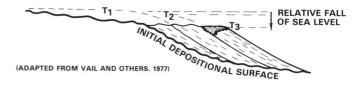

(ADAPTED FROM VAIL AND OTHERS, 1977)

SLOW RELATIVE SEA LEVEL

FIG. 17.—Illustration of the effect of a slowly falling sea level on stratal geometry (from Vail and others, 1977a). Note that measurement of final sea level at shelf edge would not be representative of highest sea-level stand within the eustatic cycle.

TABLE 1.—ESTIMATED SEA LEVEL RELATIVE TO PRESENT FOR HIGHSTANDS AND LOWSTANDS FOR OFFSHORE ALABAMA AND NEW JERSEY USING ESTIMATES OF 100 AND 0 M FOR PALEO-WATER DEPTH (PWD)

Age (Ma)	Moore and others (1987) (100 m PWD)[+]	Moore and others (1987) (0 PWD)	This study (100 m PWD)	This study (0 PWD)	Haq and others (1987)	Kominz (1984)	Lincoln and Schlanger (1987)
2.36			−4	−73			
*2.43			38	−31	62	2.1	
3.6	−9	−88	−1	−70	−29		
*3.85	57	−12	75	6	71		
4.03			−4	−73	67		
*4.2			64	−5			
4.45			−35	−107			
*4.7			31	−38	92		
4.9			−38	−107			
*5.1			45	−24			
5.4	−20	−89	−62	−131	21		−75 to −125
*5.7	51	−18	30	−39	83		
6.1	−18	−87	−13	−82	42		
*6.8	43	−26	41	−28	67		
7.7	19	−50	39	−30	17		
*8.9	45	−24	51	−17	56		
10.1	−32	−101	−64	−132	−28		
*11.3	53	−16	52	−16	88	6	
12.1	−9	−78	2	−67	78		
*13.1	56	−13	72	3	117		
13.6	−4	−73	−34	−103	103		
*14.4	60	−9	66	−3	141		
15.2	14	−55			52		
*15.8	68	−1			142	32	
16.1	−10	−79			40		
*16.7	67	−2			136		
17.4	?	?			112		
*17.8	57	−12			119	35	

Depths are in meters relative to present sea level.

*Indicates highstand value.

†PWD refers to paleo-water depth value used in calculation. Differences between these maximum estimates and estimates presented in Table 2 of Moore and others (1987), result from assuming that accommodation at maximum highstand was not filled in with late highstand sediments.

CONCLUSIONS

This study has shown the characteristics of second- and third-order depositional sequences in areas that are distal with respect to active siliciclastic depocenters as well as areas within the depocenter. Second-order cycles shown on the eustatic cycle chart (Haq and others, 1987) are often the only recognizable cyclicity in distal, starved environments. These supersequence boundaries, noted as "major" on the eustatic cycle chart, are recognized in offshore areas by local erosion and deep-marine onlap. Second-order cycles are also interpreted to be a control on stratal patterns within the depocenter by influencing the stacking patterns of third-order depositional sequences. Upper Oligocene to uppermost middle Miocene third-order depositional sequences are interpreted to compose a second-order supersequence in both study areas.

Within the depocenter, third-order depositional sequences are recognized by basinward shifts in coastal onlap and channelized erosion in exposed shelf areas, and flat, planar-erosional surfaces in slope and basin areas. Recognition of these depositional sequences, correlation to the third-order cycles shown on the eustatic-cycle chart, seismic-facies analysis, and sequence stratigraphic analysis can provide a powerful tool for predrill subsurface prediction of the age and lithologic character of the section encountered.

Coastal-onlap charts prepared from both seismic grids are in generally close agreement to the global coastal-onlap chart of Haq and others (1987). Our analysis suggests, however, that their eustatic sea-level curve is too high in the Neogene, perhaps by over 100 m in the middle Miocene. Estimates of sea-level falls are, in most cases, close to those represented on the curve of Haq and others (1987).

ACKNOWLEDGMENTS

Study of these two areas relied heavily on the cooperation of members of the Offshore Division of Exxon U.S.A. In particular, support from Kempner Scott, Marilyn Crane, Robert Eby, and Gerald Stude is gratefully acknowledged. Discussions with colleagues at Exxon Production Research Company, especially T. R. Nardin, C. R. Tapscott, and P. R. Vail, contributed to this work. Review by H. W. Posamentier and J. C. Van Wagoner significantly improved the manuscript.

REFERENCES

BYBELL, L. M., POORE, R. G., AND AGER, T. A., 1986, Paleogene biostratigraphy of New Jersey core, ACGS #4 (abst.): Society of Economic Paleontologists and Mineralogists Annual Midyear Meeting, Raleigh, North Carolina, p. 17.

CORSO, W., BUFFLER, R. T., AND AUSTIN, J. A., 1987, Controls on early Cretaceous carbonate platform margin development: Northeastern Gulf of Mexico (abst.): American Association of Petroleum Geologists Bulletin, v. 71, p. 543.

FLUKER, J. C., III, 1983, Main Pass and Viosca Knoll, stratigraphic and structural study, *in* Bally, A. W., ed., Seismic Expression of Structural Styles—A Picture and Work Atlas: American Association of Petro-

ESTIMATED SHORT TERM EUSTACY

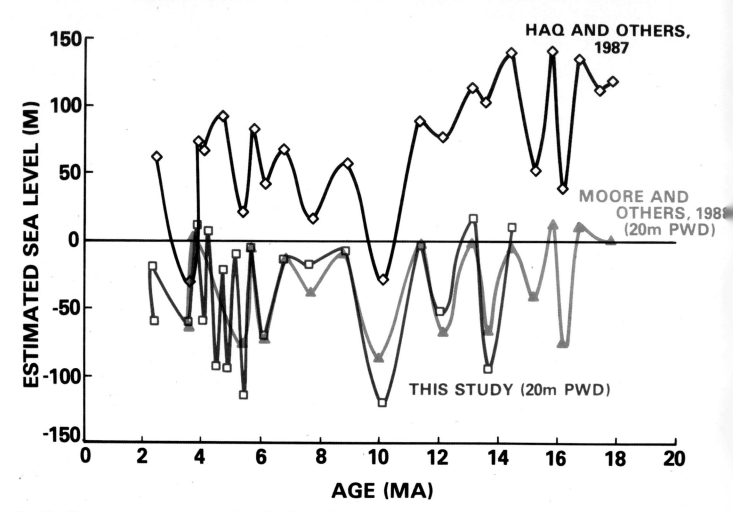

FIG. 18.—Calculated sea-level curves from offshore New Jersey (Moore and others, 1987), offshore Alabama (this study), and the global curve of Haq and others (this volume). PWD refers to the paleo-water depth value used in the calculation of eustatic sea-level values.

leum Geologists Studies in Geology No. 15, v. 1, p. 15–21.

GREENLEE, S. M., SCHROEDER, F. W., AND VAIL, P. R., 1988, Seismic stratigraphic and geohistory analysis of Tertiary strata from the continental shelf off New Jersey—Calculation of eustatic fluctuations, *in* Sheridan, R. E., and Grow, J. A., eds., The Atlantic Margin, U.S.: Geological Society of America, The Geology of North America, v. 1–2, p. 437–444.

GROW, J. A., 1980, Deep structure and evolution of the Baltimore Canyon Trough in the vicinity of the COST No. B-3 well, *in* Scholle, P. A., ed., Geological Studies of the COST No. B-3 Well, United States and Mid-Atlantic Continental Slope Area: U.S. Geological Survey Circular 833, p. 117–132.

HAQ, B. U., HARDENBOL, J., AND VAIL, P. R., 1987, Chronology of fluctuating sea levels since the Triassic: Science, v. 235, p. 1136–1167.

HARDENBOL, J., VAIL, P. R., AND FERRER, J., 1981, Interpreting paleoenvironments, subsidence history and sea-level changes on passive margins from seismic biostratigraphy: 26th International Geological Congress, Geology of Continental Margins: Oceanologica Acta, Supplement, v. 4, p. 33–44.

KOMINZ, M. A., 1984, Oceanic ridge volumes and sea-level change—An error analysis, *in* Schlee, J. B., ed., Interregional Unconformities and Hydrocarbon Accumulation: American Association of Petroleum Geologists Memoir 36, p. 109–127.

LINCOLN, J. M., AND SCHLANGER, S. O., 1987, Miocene sea level falls related to the geologic history of Midway Atoll: Geology, v. 15, p. 454–457.

MANCINI, E. A., 1981, Lithostratigraphy and biostratigraphy of Paleocene subsurface strata in southwest Alabama: Gulf Coast Association of Geological Societies, Transactions, v. 31, p. 359–367.

MILLER, K. G., MELILLO, A. J., MOUNTAIN, G. S., AND FARRE, J. A., 1987, Middle to Late Miocene canyon cutting on the New Jersey continental slope: Biostratigraphic and seismic stratigraphic evidence: Geology, v. 15, p. 509–512.

MITCHUM, R. M., JR., VAIL, P. R., AND SANGREE, J. B., 1977, Seismic stratigraphy and global changes of sea level, part 6: Stratigraphic interpretation of seismic reflection patterns in depositional sequences, *in* Payton, C. E., ed., Seismic Stratigraphy—Applications to Hydrocarbon Exploration: American Association of Petroleum Geologists Memoir 26, p. 117–133.

MOORE, T. C., LOUTIT, T. S., AND GREENLEE, S. M., 1987, Estimating short-term changes in sea level: Paleoceanography, v. 2, p. 625–637.

MURRAY, G. E., 1961, Geology of the Atlantic and Gulf Coast Province of North America: Harper and Brothers, New York, 692 p.

PITMAN, W. C., III, 1978, Relationship between eustacy and stratigraphic sequences of continental margins: Geological Society of America Bulletin, v. 89, p. 1389–1403.

————, AND GOLOVCHENKO, X., 1983, The effect of sea level change on the shelf edge and slope of passive margins, *in* Stanley, D. J., and Moore, G. T., eds., The Shelfbreak: Critical Interface on Continental Margins: Society of Economic Paleontologists and Mineralogists Special Publication 33, p. 41–58.

POAG, C. W., 1980, Foraminiferal stratigraphy, paleoenvironments, and depositional cycles in the outer Baltimore Canyon trough, *in* Scholle, P. A., ed., Geological Studies of the COST No. B-3 Well, United States Mid-Atlantic Continental Slope Area: U.S. Geological Survey Circular 833, p. 44–65.

————, 1985a, Depositional history and stratigraphic reference section for central Baltimore Canyon trough, *in* Poag, C. W., ed., Geologic Evolution of the United States Atlantic Margin: Van Nostrand Reinhold, New York, p. 217–264.

————, 1985b, Cenozoic and Upper Cretaceous sedimentary facies and depositional systems of the New Jersey slope and rise, *in* Poag, C. W., ed., Geologic Evolution of the United States Atlantic Margin: Van Nostrand Reinhold, New York, p. 217–264.

————, AND SCHLEE, J. S., 1984, Depositional sequences and stratigraphic gaps on submerged United States Atlantic margin, *in* Schlee, J. S., ed., Inter-Regional Unconformities and Hydrocarbon Accumulation: American Association of Petroleum Geologists Memoir 36, p. 165–182.

————, AND WARD, L. W., 1987, Cenozoic unconformities and depositional super sequences of North Atlantic continental margins: Testing the Vail model: Geology, v. 15, p. 159–162.

RAYMOND, D. E., 1985, Depositional Sequences in the Pennsacola Clay (Miocene) of Southwest Alabama: Geological Survey of Alabama Bulletin 114, 87 p.

SCHLEE, J. S., DILLON, W. P., AND GROW, J. A., 1979, Structure of the continental slope off the eastern United States, *in* Doyle, L. J., and Pilkey, O. H., eds., Geology of Continental Slopes: Society of Economic Paleontologists and Mineralogists Special Publication 27, p. 95–117.

VAIL, P. R., MITCHUM, R. M., JR., AND THOMPSON, S., III, 1977a, Seismic stratigraphic and global changes of sea level, part 4: Global cycles of relative changes of sea level, *in* Payton, C. E., ed., Seismic Stratigraphy—Applications to Hydrocarbon Exploration: American Association of Petroleum Geologists Memoir 26, p. 83–97.

————, ————, AND ————, 1977b, Seismic stratigraphy and global changes of sea level, part 3: Relative changes of sea level from coastal onlap, *in* Payton, C. E., ed., Seismic Stratigraphy—Applications to Hydrocarbon Exploration: American Association of Petroleum Geologists Memoir 26, p. 63–81.

VAN HINTE, J. E., 1978, Geohistory analysis—Application of micro-paleontology in exploration geology: American Association of Petroleum Geologists Bulletin, v. 62, p. 201–222.

VANNEY, J. R., AND STANLEY, D. J., 1983, Shelfbreak physiography; An overview, *In* Stanley, D. J., and Moore, G. T., eds., The Shelfbreak Critical Interface on Continental Margins: Society of Economic Paleontologists and Mineralogists Special Publication 33, p. 1–24.

VAN WAGONER, J. C., 1985, Reservoir facies distribution controlled by sea level change: Society of Economic Paleontologists and Mineralogists Annual Midyear Meeting, Abstracts with Programs, p. 91–92.

PART IV
APPLICATION CONCEPTS
OF SEA-LEVEL CHANGE

SHARP-BASED SHOREFACE SEQUENCES AND "OFFSHORE BARS" IN THE CARDIUM FORMATION OF ALBERTA: THEIR RELATIONSHIP TO RELATIVE CHANGES IN SEA LEVEL

A. G. PLINT

Department of Geology, University of Western Ontario, London, Ontario N6A 5B7

ABSTRACT: Under conditions of stable sea level, the progradation of a wave-dominated clastic shoreface will give rise to a coarsening-upward sequence, reflecting an increase in the frequency and volume of sand transport with time. Core and well-log data from the Cardium Formation (Turonian) of Alberta reveal two types of shelf-to-shoreface sequences: (1) gradational-based sequences that steadily coarsen upward from thin-bedded, wave-rippled sandstone and mudstone through hummocky cross-stratified (HCS) sandstone and mudstone into mud-free, swaley cross-stratified (SCS) sandstone capped by a root bed, and (2) sharp-based sequences that consist of SCS sandstone, which, near the base, may contain large mudstone intraclasts, sharply overlying thin-bedded sandstone and mudstone. The HCS interval is thin or absent. Log cross sections show that the change from a gradational to a sharp-based sequence is accompanied by a lowering of both the top and bottom of the SCS sandstone, relative to upper and lower markers. Simultaneously, the SCS sandstone may thin from 15 to 18 m to as little as 6 m. The sharp base, the presence of intraclasts, and the relative lowering of the SCS unit suggest deposition during a rapid fall of relative sea level during which the shoreface prograded over an erosion surface cut into the inner shelf by fair-weather wave scour.

Tens of kilometers seaward of Cardium shoreface sandstones lies a series of shore-parallel, lenticular bodies ("offshore bars") of conglomeratic muddy sandstone. The lenticular bodies rest on regional erosion surfaces that can be traced landward into slightly older shoreface deposits. The conglomerates are here interpreted as lowstand shoreface deposits, which lie on erosion surfaces cut into the shelf by wave scour during a relative sea-level fall. The stratigraphic and lithologic relationships demonstrable from the Cardium Formation suggest that a number of sharp-based "offshore bar sandstones" in other parts of the Western Interior Seaway may also be more satisfactorily explained as lowstand shoreface deposits.

INTRODUCTION

It is widely recognized that the progradation of a wave-dominated clastic shoreface onto a relatively muddy shelf will give rise to a coarsening-upward sedimentary sequence in which sandstone beds progressively thicken upward. Such coarsening-upward sequences are easily recognizable, both in outcrop and on well logs, and record an increase with time in the frequency and volume of sand transport across the shoreface and inner shelf by storm-driven currents.

The Cardium Formation (Turonian) of Alberta provides an excellent example of a prograding, storm-dominated shoreface and shallow-shelf deposit in which the resultant coarsening-upward sequence may be studied both in outcrop and in abundant cores and well logs. In this study, 180 cores and about 1,000 well logs were used to map the formation over a study area of about 44,000 km^2 (Fig. 1).

In most areas, the self-to-shoreline sequence in the Cardium shows a gradual coarsening upward from muddy sandstones through hummocky cross-stratified (HCS) sandstones into mud-free, swaley cross-stratified (SCS) sandstones capped by a root bed. A significant minority of the sequences differs markedly from this norm, however, in that mud-free SCS sandstones of the shoreface rest *abruptly* on thin-bedded, relatively muddy offshore sediments. The HCS interval is thin or absent.

Similar sharp-based shoreface sequences are occasionally mentioned in the literature (Roep and others, 1979; Cant, 1984; McCrory and Walker, 1986; Rosenthal and Walker, 1987), but little has been published about their genesis and relationship to adjacent "normal," gradationally based shoreface sequences.

Tens of kilometers seaward of the progradational margin of the Cardium shoreface, coeval shelf sediments contain several bodies of conglomeratic muddy sandstone, which are elongate shore-parallel and lenticular perpendicular to shore. These sediment bodies are of the order of tens of kilometers long, a few kilometers wide and as thick as 11

m. They have traditionally been interpreted as "offshore bars."

This study shows that each "bar" is *underlain* by an *erosion surface* of regional extent. In some cases, the erosion surface beneath each "bar" can be correlated landward with an erosion surface beneath a sharp-based shoreface, suggesting that the genesis of both sharp-based shorefaces and "offshore bars" is related to a relative fall in sea level.

The evidence from the Cardium Formation strongly suggests that these "classic offshore bars" are, in fact, shoreface deposits laid down on wave-scoured erosion surfaces cut into the shelf during periods of relative lowstands of sea level. These conclusions prompt a critical reassessment of some other "offshore bars" in the Western Interior Seaway.

STRATIGRAPHY

Plint and others (1986) developed a comprehensive allostratigraphic nomenclature for distinctive, unconformity-bounded members of the Cardium Formation in the subsurface of west-central Alberta, and their summary stratigraphic diagram is reproduced here (Fig. 2). In the western part of the basin, shoreface sandstones of the Kakwa Member and lagoonal and alluvial sediments of the Musreau Member prograded northeastward over coeval shelf mudstones of the Nosehill, Bickerdike, and Hornbeck members (Fig. 2). The sedimentology of these members has been described elsewhere (Plint and others, 1986; Plint and Walker, 1987a).

Lateral progradation of the shoreface sandstones and simultaneous vertical aggradation of adjacent shelf mudstones were periodically interrupted by relative falls of sea level that generated erosion surfaces extending from the shoreline for more than 130 km across the shelf. Plint and others (1986) recognized that lateral progradation of the shoreface (Kakwa Member) had been interrupted at least three times by relative falls and subsequent rises in sea level,

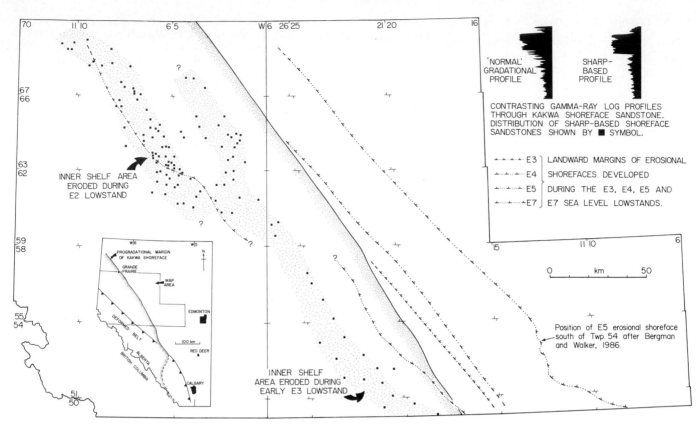

FIG. 1.—Location map showing maximum progradational margin of Cardium shoreface sandstones (Kakwa Member) and simplified distribution of erosional lowstand shoreface bevels cut during the E3, E4, E5, and E7 lowstands. See text for detailed discussion.

resulting in the generation of erosion/transgression surfaces E2/T2, E3/T3, and E4/T4 (Fig. 2). The lowstand erosion surfaces are usually mantled by a layer of chert pebbles or granules, a few centimeters to a few decimeters thick. Locally, however, these pebble beds thicken to as much as 18 m (Plint and others, 1986; Bergman and Walker, 1986, 1987).

Conglomerates are overlain by mudstones deposited during the next transgression. Thus, where conglomerates are thin, each surface may be regarded as the product of both lowstand and transgressive erosion. Where conglomerates are of significant thickness, however, they are bounded below by the erosion (E) surface and above by the transgression (T) surface.

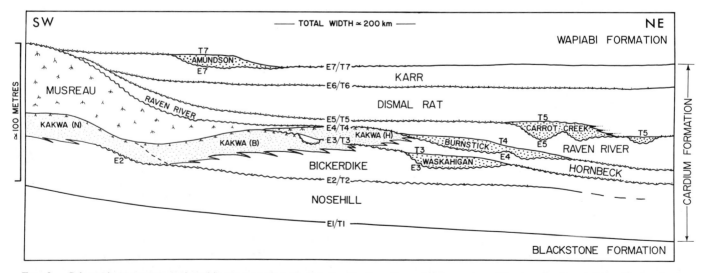

FIG. 2.—Schematic west-east stratigraphic cross section through the Cardium Formation in western Alberta, showing stratigraphic terminology and principal sequence-bounding erosion surfaces, numbered E1 to E7 (after Plint and others, 1986).

SHOREFACE SEQUENCES

Two types of shoreface sequences are recognized—gradationally based and sharp-based. Comparative schematic vertical sections are given in Figure 3.

Gradationally based Shoreface Sequences

Gradationally based shoreface sequences are here considered in terms of four main facies associations (Fig. 3a). Facies association A occupies the lowest part of the sequence and consists of centimeter-scale interbeds of siltstone or very fine sandstone and mudstone (facies 3, 4, and 5 of Walker, 1983). Sandstones are invariably erosively based and often display symmetrical wave ripple-forms. Cross-lamination may be uni- or bidirectional and is usually form-discordant. Pervasive bioturbation has destroyed most primary sedimentary structures in facies association A.

Facies association B gradationally overlies association A and ranges in thickness from 5 to 15 m. It consists of fine-to very fine-grained sandstone beds, a few millimeters to a few decimeters thick, interbedded with centimeter-thick mudstones. Sandstone less than 5 cm thick contain well-developed, symmetrical ripple forms and form-discordant cross-lamination (facies 15, Plint and Walker, 1987a). Thicker sandstone beds display fine, low-angle lamination with dips of less than 15°. This structure represents hummocky cross-stratification (HCS; facies 7, Walker, 1983). Upward, HCS sandstone beds thicken and amalgamate into units as thick as 4 m. The HCS beds are separated by thin mudstone beds.

Facies association C consists of fine- to medium-grained, mud-free sandstone and forms units from 6 to 23 m thick (average 15 m). The base of association C is taken at the point at which mudstone interbeds disappear. The lower part of this association consists of fine-grained sandstone with delicate, low-angle (<15°) lamination, interpreted as swaley cross-stratification (SCS; facies 16 of Plint and Walker, 1987a). The SCS is usually abruptly overlain by cross-bed-

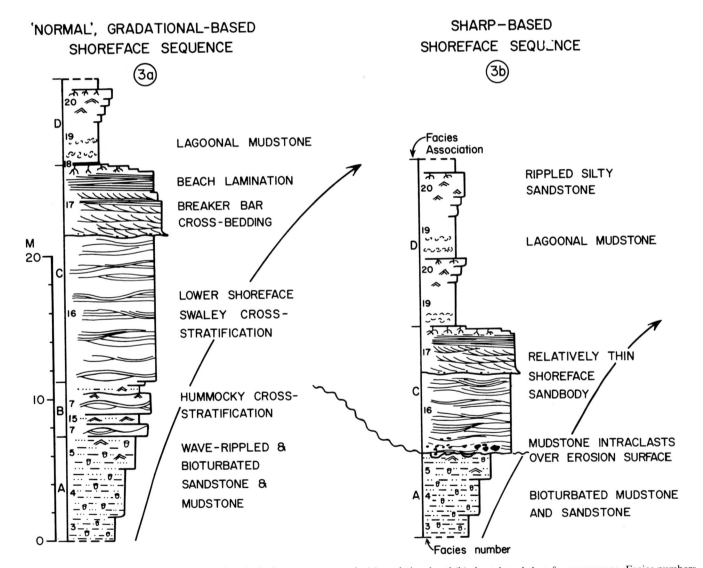

Fig. 3.—Schematic lithologic logs contrasting the facies sequence seen in (a) gradational and (b) sharp-based shoreface sequences. Facies numbers after Plint and Walker (1987a).

FIG. 4.—(a) Core through a sharp-based shoreface showing black lagoonal mudstones with roots at "R," underlain by SCS sandstones of facies association C resting abruptly (arrow) on bioturbated muddy sandstones of association A. Note layer of large mudstone intraclasts at "M." Well 10-11-65-7W6, 1,635.5–1,653 m. (b) Detail of large mudstone intraclast, suggestive of substantial shoreface erosion at base of a sharp-based shoreface. Well 2-16-63-5W6, 1,804.7 m. Note differential erosion of mudstone and sandstone layers.

ded and parallel-laminated medium- and fine-grained sandstone, interpreted as breaker-bar, rip-channel, and beach deposits (facies 17, Plint and Walker, 1987a). The top of the laminated beach sandstone usually contains carbonaceous roots and disseminated organic matter.

Beach and backshore deposits of facies association C grade upward over a few centimeters into non-marine sediments of facies association D, which consists of a variety of black carbonaceous mudstones; laminated silty mudstones; cross-bedded sandstones; rooted, structureless siltstones; and coals (facies 18 to 22, Plint and Walker, 1987a). The fauna in association D indicates brackish to freshwater conditions. Facies association D represents various lagoonal, lacustrine, fluvial-channel and flood-plain environments (Fig. 3a).

The gradational relationships between the four facies associations described above suggests a progressive, gradual shallowing. The preservation potential of fair-weather mud deposits diminishes upward as storm-related erosion becomes progressively more effective. For practical purposes, the disappearance of mud at the junction between facies associations B and C approximates fair-weather wave base at a water depth of about 15 m (Plint & Walker, 1987a).

Sharp-Based Shoreface Sequences

Sharp-based sequences (Fig. 3b) differ from the gradational-based type in that the mud-free sandstones of facies association C rest *abruptly* on the thinly interbedded sandstones and mudstones of association A. These two sequence types are contrasted in Figure 3. The transition between associations A and C, consisting of a thickening-upward sequence of HCS sandstone beds (facies association B), is *thin* or *absent*. In some cores, a bed of mudstone clasts as much as 150 mm in diameter lies at or just above the base of the shoreface sandstones of association C (Fig. 4a, b).

Distribution of sharp-based shoreface sequences.—

Sharp-based shoreface sequences are easily recognizable on gamma ray logs, because the sharp contact between the relatively muddy sediments of association A and the mud-free sandstones of association C produces a very abrupt deflection on the log (typical gamma ray log profiles are shown in Fig. 1). This contrasts markedly with the funnel-shaped log profile that is typical of gradationally based shoreface sequences.

By using the sharp-versus-gradational base to the shoreface as a distinguishing feature, gamma ray log profiles were divided into two categories. All sharp-based shoreface sequences were then plotted on a map (Fig. 1). It is clear from Figure 1 that the distribution of sharp-based shoreface sequences is not random. Sharp-based shoreface sequences

are concentrated in two belts that are approximately parallel to the maximum progradational margin of the shoreface sandstones of the Kakwa Member; the latter was determined from regional subsurface mapping (Fig. 1; Plint and Walker, 1987a).

Cross section through sharp-based shoreface sequence.—

A cross section across the more westerly of the two "sharp-based" belts is shown in Figure 5. The section is hung on a regionally extensive upper datum and utilizes as many cored wells as possible.

In well 7-10-62-7, shoreface sandstones of the Kakwa Member show a regionally typical, well-developed coarsening-upward sequence. Core control shows SCS sandstones grading downward into interbedded HCS sandstones and bioturbated muddy sandstones. Between wells 7-10-62-7 and 7-18-62-6, the Kakwa Member thins from 13 to 6 m at the same time as the top of the member drops 5 m relative to the datum. The top of the Kakwa Member, which contains roots, continues to fall, relative to the datum, through well 7-10-62-6 to rest in well 6-8-62-5 some 14 m lower than it was in well 7-10-62-7. The base of the Kakwa Member also drops about 13 m, relative to the datum, between wells 7-10-62-7 and 6-8-62-5. Note also that between these two wells, the uppermost of three log markers below the Kakwa Member (labelled a, b, and c and shown by dashed lines in Fig. 5) is cut out, apparently by erosion at the base of the shoreface. Cores in wells 7-10-62-6 and 6-8-62-5 show SCS shoreface sandstones resting *abruptly* on bioturbated muddy sandstones, with no evidence of a coarsening-upward HCS sandstone sequence.

It is clear that the change from a gradational to a sharp-based shoreface and the drop of both the top and bottom

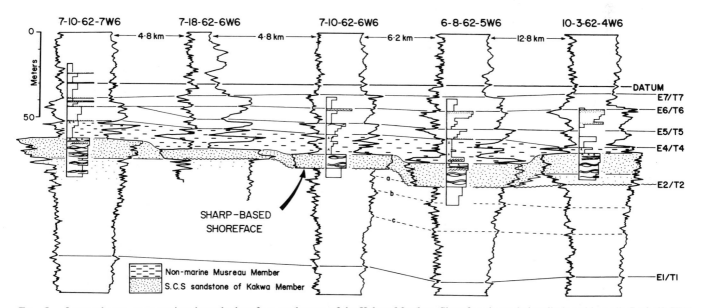

FIG. 5.—Log and core cross section through shoreface sandstones of the Kakwa Member. Shoreface is gradationally based in wells 7-10-62-7W6 and 10-3-62-4W6 and very sharp-based in wells 7-10-62-6W6 and 6-8-62-5W6. Note (1) relative drop in level of shoreface sandstone between wells 7-10-62-7W6 and 6-8-62-5W6; (2) concomitant change from gradational to sharp-based profile; (3) erosion of log markers between wells 7-10-62-6 and 6-8-62-5; and (4) relative rise of shoreface sandstone between wells 6-8-62-5 and 10-3-62-4 in response to the T2 transgression. See Figure 7 for interpretation.

of the Kakwa Member, relative to the datum, take place simultaneously (Fig. 5). Similarly, the thickness of the Kakwa Member tends to vary according to the character of the basal contact. Thus, gradationally based sequences are generally from 14 to 18 m thick, whereas sharp-based sequences are generally thinner, from 14 m to as little as 6 m thick.

Between wells 6-8-62-5 and 10-3-62-4, the top of the Kakwa Member rises by about 7 m and the base rises by 9 m. This is accompanied by a dramatic lateral thinning of the non-marine Musreau Member from 13 to 3 m and a change in the log profile from a sharp to a gradational-based shoreface (Fig. 5).

The downward-cutting E2/T2 erosion surface (Figs. 2, 5) that first appears beneath the sharp-based Kakwa Member in well 7-18-62-6 can be traced seaward (northeastward) from the base of the shoreface in well 6-8-62-5 as an unconformity that marks the top of a coarsening-upward offshore sequence (the Nosehill Member; Fig. 2). The E2/T2 surface is locally mantled by a veneer of chert granules and sideritized- and bored mud intraclasts (Fig. 6). This surface can be traced eastward across the shelf for about 120 km before it becomes unrecognizable (see Fig. 5 of Plint and others, 1986).

Interpretation of Sharp-Based Shoreface

An interpretation of sharp-based shoreface sequences must take into account (1) the sharp base to the shoreface sandstones; (2) the drop of both the top and bottom of the shoreface, relative to adjacent gradational-based sequences; (3) the presence of large mud intraclasts at or just above the base of sharp-based shoreface sequences; and (4) the marked thinning of some sharp-based shoreface sandstones relative to gradational-based sequences.

General relationships.—

It is evident from Figure 5 that the lateral change from a gradational to a sharp-based shoreface sequence and thinning of the SCS sandstones coincide with the drop in the level of both the top and bottom of the shoreface sandstone. The sharp contact at the base of the sandstone, the absence of an intervening, coarsening-upward sequence of HCS sandstones, and the presence of large, obviously locally derived mudstone intraclasts near the base suggest that the relative lowering of the Kakwa Member was due to erosion at the base of the shoreface. Erosion removed all or most of the transitional HCS sandstones.

The presence of sharp-based shoreface sequences in two belts that parallel the regional shoreline trend (Fig. 1) suggests that erosion of the inner shelf and concomitant shoreface progradation occurred on a broad front.

Erosion of the shelf.—

Under conditions of stable sea level and ample sediment supply, essentially mudstone-free, swaley cross-stratified sandstones were deposited as part of the prograding shoreface under conditions of almost continuous sediment movement (Fig. 7a). The depth at which SCS sandstone (facies association C) passes into interbedded HCS sandstone and mudstone (facies association B) is interpreted to approximate the position of fair-weather wave base (FWWB) which, in the Cardium, lay at about 15 m (Plint and Walker, 1987a).

If the rate of relative sea-level fall exceeds the rate of basin subsidence, a corresponding drop in FWWB will render relatively muddy sediments, formerly deposited *below* FWWB, more susceptible to erosion by both storm and fair-weather wave processes. Increased wave scour of the inner shelf would result in the formation of an erosion surface *in advance of the prograding shoreface sandstones* (Fig. 7b). Wave scour is considered to have been the main agent responsible for the erosion and seaward redistribution of the bulk of the shelf sediments, leaving only a thin lag deposit of siderite-cemented mudstone clasts on the erosion surface. In nearshore areas, these clasts were subsequently incorporated into the base of the prograding shoreface sandstones (Fig. 4), or, in more offshore areas, were blanketed with mudstone following relative sea-level rise (Fig. 6).

The width of the shelf eroded by wave scour would depend on the initial slope of the shelf, the magnitude of the

FIG. 6.—Subtle E2 erosion surface (arrow) formed seaward of shoreface during E2 lowstand. E2 surface bears a veneer of chert granules and a bored and sideritized mudstone intraclast ("I") well 12-14-62-24W5, 1,610 m; core is 7.5 cm wide. Location shown in Figure 7C.

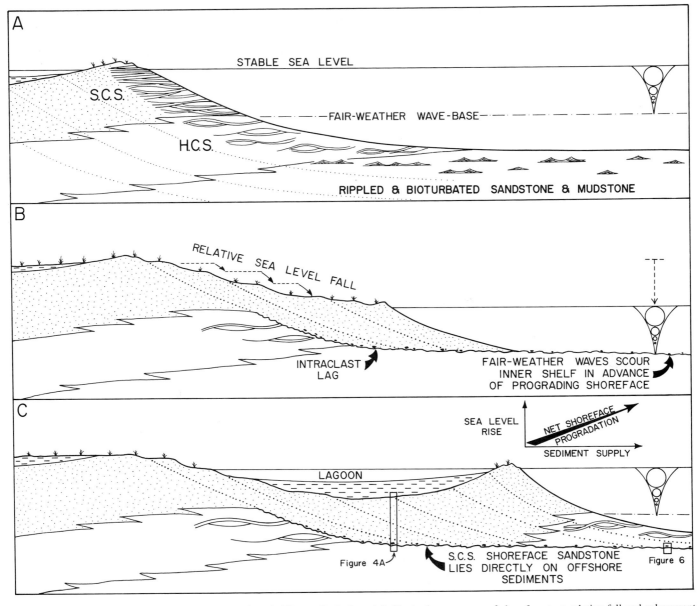

FIG. 7.—Schematic illustration, based on data given in Figures 3, 4, 5, and 6, illustrating response of shoreface to a relative fall and subsequent relative rise of sea level. See text for detailed discussion. SCS = swaley cross-stratification; HCS = hummocky cross-stratification. The relative positions of Figures 4A and 6 are shown in (C).

relative fall in sea level, the rate of subsidence, and the depth of FWWB. In deeper water, beyond the seaward margin of the eroded area, the erosion surface will pass laterally into time-equivalent sediments in which the hiatus may be unrecognizable.

Effects of a relative rise in sea level.—

The rise of the Kakwa Member, relative to the datum between wells 6-8-62-5 and 10-3-62-4 (Fig. 5), and the change from a sharp to a gradationally based sequence are interpreted to result from a relative sea-level rise of about 8 m. The cored wells in this area yield *no* evidence of the landward movement of the shoreline during this rise, suggesting that aggradation of the shoreface kept pace with

sea-level rise as a result of very high rates of sedimentation (Fig. 7c).

The dramatic thickening of the non-marine Musreau Member above the topographically low-lying parts of the Kakwa (N) Member (Figs. 5, 7) suggests that the unusually thick lagoonal deposits (facies association D) accumulated behind the shoreline in response to the T2 transgression (Fig. 2); this situation is directly analogous to that in the Rhone delta plain during the Holocene transgression (Lagaaij and Kopstein, 1964).

The relative rise of sea level at T2 (Fig. 2) initiated a new depositional sequence in which *vertical* aggradation of the relatively muddy Bickerdike Member on the shelf was accompanied by *lateral* progradation of the coeval Kakwa (B) Member at the shoreface.

Rate of shoreface progradation.—

The evidence from the Kakwa Member, as illustrated in Figure 5, suggests that, during coastal progradation between wells 7-10-62-7 and 6-8-62-5, a relative sea-level fall of about 14 m took place. This figure is based on the assumption that the rooted paleosol at the top of each shoreface sequence developed at the same height above mean sea level. During relative sea-level fall, the rate of shoreline progradation (assuming a constant rate of sediment supply) would be expected to increase. This is simply a reflection of the fact that, in shallow water, a given volume of sediment will produce a greater increment of progradation than it would in deeper water. Thus, assuming the rate of subsidence to remain unchanged, falling relative sea level will result in a shoreface sand body that is unusually thin, reflecting both diminished accommodation and the unusually rapid rate of progradation.

EROSION AND DEPOSITION RELATED TO THE E3
AND E4 LOWSTANDS

Stratigraphic Relationships of the Waskahigan and Burnstick Members

Following the T2 transgression, the Kakwa Member prograded northeastward 60 km farther (Figs. 1, 2). Shoreface progradation was interrupted, however, by two more important relative changes of sea level that were responsible for the generation of the E3/T3 and E4/T4 unconformities (Fig. 2). Thus, the more easterly belt of sharp-based shoreface sequences shown in Figure 1 was generated at the beginning of the E3 lowstand, and, similarly, the abrupt eastern limit of the Kakwa shoreface (Fig. 1) marks the position at which the E4 lowstand interrupted progradation. Detailed cross sections showing the correlation of depositional se-

quences and unconformities are given in Plint and others (1986, 1987).

The morphology of the E3/T3 and E4/T4 erosion surfaces seaward of the Kakwa shoreface is summarized in Figure 8. Surface E3 scours into shelf mudstones of the Bickerdike Member and is overlain by a shore-parallel lens of conglomeratic muddy sandstone as thick as 11 m; this is the Waskahigan Member (Fig. 8). Similarly, surface E4 scours into the Hornbeck Member and is overlain by conglomerate of the Burnstick Member. In this cross section, it can be seen that the thickest part of the Burnstick lies seaward of the Waskahigan lens, and that the E4 surface cuts down between wells 6-34-52-17 and 12-33-52-16 to *remove* the Hornbeck and Waskahigan Members completely (Fig. 8).

Facies of the Waskahigan and Burnstick Members

Both the Waskahigan and Burnstick Members show a coarsening-upward sequence that differs markedly from that seen in the underlying Bickerdike and Hornbeck Members (Figs. 2, 8). The lower part of the Waskahigan and Burnstick consists of intensely bioturbated interbeds, millimeters to centimeters thick, of medium- to coarse-grained pebbly sandstone and dark mudstone which rest *abruptly* on bioturbated mudstones and very fine-grained sandstones of the underlying shelf sequence (Fig. 9a). Where bioturbated has not completely homogenized the sediment, sandstones display erosive bases and remnants of wave-ripple lamination (Fig. 9b). Upward, the proportion of coarse material increases, and beds are typically a few centimeters to a few decimeters thick (Figs. 9c, 10c). Thicker sandstone beds may show crude angle-of-repose cross-stratification and frequently contain sideritized mudstone clasts. Occasionally, armored mudballs (Fig. 9e) and gutter casts(?) are seen (Fig. 9f). The uppermost 1 to 2 m of the sequence consists

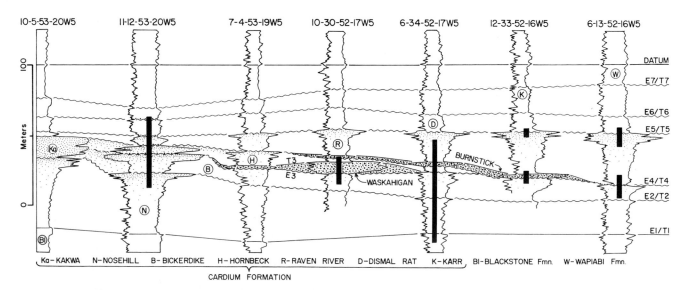

FIG. 8.—Log and core cross section showing relationship of erosion surfaces E3 and E4 to shoreface sandstones of the Kakwa Member and to lowstand shoreface deposits of the Waskahigan and Burnstick members. Note (1) manner in which E3 and E4 erosion surfaces truncate shallowing-upward nearshore sequences in wells 11-12-53-20 and 7-4-53-19; (2) seaward-offlapping relationship of the Waskahigan and Burnstick members; (3) downcutting of the E4 surface between wells 6-34-52-17 and 12-33-52-16 to remove the Hornbeck and Waskahigan members completely. See text for detailed discussion.

of massive to vaguely stratified, relatively poorly sorted pebbly sandstone (Figs. 9d, 10c).

The tops of the Waskahigan and Burnstick members are marked by a gradation, usually over a few decimeters, into dark mudstones deposited during the subsequent transgression (Fig. 10c).

Interpretation of the Waskahigan and Burnstick Members

Sharp-based shoreface sandstones of the Kakwa Member provide evidence which suggests that substantial erosion of the inner shelf occurred through wave scour as a result of the relative fall in sea level at E2 (Figs. 2, 5). The resulting erosion surface extends from the shoreface across the shelf to a depth below which the sea-level fall had no erosive effect.

The fact that unconformities E3 and E4 can also be traced from the shoreface out onto the shelf and *beneath* the conglomeratic deposits of the Waskahigan and Burnstick members strongly suggests a *genetic* link between lowstand erosion of the shoreface and erosion and conglomerate emplacement on the shelf. The linear, seaward-facing erosional scours beneath the Waskahigan and Burnstick members are interpreted to have been cut by wave scour in a new, *erosional* shoreface that developed when the shoreline moved out onto the shelf during relative lowstands of sea level.

On the basis of the *geometry* of the basal-erosion surfaces E3 and E4 and the lithology of the overlying conglomeratic deposits, the Waskahigan and Burnstick members are interpreted as *shoreface* sediments laid down during sea-level lowstands when the shoreface abruptly shifted several tens of kilometers seaward of the highstand shoreline position. The overall coarsening-upward sequence is here interpreted in terms of the progradation of the shoreface. The abundance of mud in the lower part of the sequence suggests deposition below FWWB, and the interbedded sharp-based sandstones were probably introduced during storms. It seems likely that only the relatively clean pebbly sandstone (Fig. 9d) in the uppermost part of the preserved sequence was deposited above FWWB, and that the bulk of the upper shoreface and beach was removed during transgressive erosion. There is the possibility that FWWB in the lowstand shoreface was unusually shallow as a result of wave damping across the relatively shallow, wave-cut platform that must have existed seaward of the erosional shoreface "bevel." This situation may have permitted the accumulation of mud at much shallower depths than in the highstand shoreface.

The presence of a regional unconformity *beneath* each linear conglomerate body *precludes* their interpretation as steadily aggrading and shoaling-upward "offshore bars" or "ridges," gradationally rooted in deep-water shelf mudstones (see discussion in Bergman and Walker, 1986, 1987). The recognition of the erosional "bevels" on the landward side of the conglomerate bodies may provide a solution to the problem, never satisfactorily explained in the literature, of what *localized* the deposition of many ancient examples of "linear shelf ridges."

The contrast in the grain size between fine sandstone of the highstand shoreface (Kakwa Member) and the lowstand shoreface deposits of the Waskahigan, Burnstick, and Carrot Creek members (Fig. 2) is problematic. It seems most likely that regional tilting of the basin margin could have simultaneously effected a relative fall of sea level and increased the gradient of rivers sufficiently to transport a pebbly bedload. An equally perplexing problem concerns the fate of the fine sand that was presumably supplied to the shoreline together with mud and gravel. Lowstand shoreface deposits throughout the Cardium Formation consist mainly of mud and conglomerate with very little fine sand. These problems are the subject of continuing investigation.

EROSION AT THE E5, E6, AND E7 SURFACES

Bergman and Walker (1987) provide a detailed discussion of the geometry and origin of the E5 erosion surface and its associated conglomerate deposits in the Carrot Creek area. In this case, erosion at the lowstand shoreface has cut a "bevel" into the underlying shelf deposits (Raven River Member; Fig. 2) with a relief of about 20 m. As much as 18 m of shoreface conglomerates (Carrot Creek Member) have been deposited against this surface.

Surface E6 (Fig. 2) does not appear to have major relief, but surface E7 shows dramatic erosional relief of about 50 m. The morphology of this surface is essentially similar to that beneath the Waskahigan, Burnstick, and Carrot Creek members (Fig. 2), i.e., a seaward-facing, relatively steeply dipping "bevel" backed by a relatively gently dipping "terrace" (Wadsworth, 1987). The major difference is that, in most areas, the E7 surface bears only a veneer of conglomerate. Representative cross sections for the study area are given in Figures 2 and 3 of Plint and others (1987). Serial log and core cross sections enable the position of the E7 erosional lowstand bevel to be mapped across most of the study area (Fig. 1).

DISCUSSION

Sharp-Based Shoreface Sandstones

Sharp-based shoreface sandstones resting erosively on deeper water mudstones have been predicted to result from falling sea level (Heward, 1981); examples have been described by Roep, and others, (1979), Cant (1984), McCrory and Walker (1986), Eyles and Clark (1986), and Rosenthal and Walker (1987). Cant (1984) interpreted the sharp-based shoreface sandstones in the Lower Cretaceous Spirit River Formation of Alberta to have been deposited at the landward margin of transgressive sequences in water too shallow to permit the accumulation of muds. Cant pointed out that such sharp-based sequences were implicit in many shoreline models, but few ancient examples had yet been documented. The example of simultaneous shoreface progradation and shelf erosion illustrated here from the Cardium Formation may be the first detailed description of the lateral transition from a gradational-based to an erosive-based shoreface in response to a relative fall in sea level.

Although Weller (1960, Fig. 189) illustrated the downward shift of the top of the shoreface in response to falling sea level, the corresponding change to a sharp-based shoreface was not described. Vail and others (1977, fig. 8) re-

9a

9b

9e

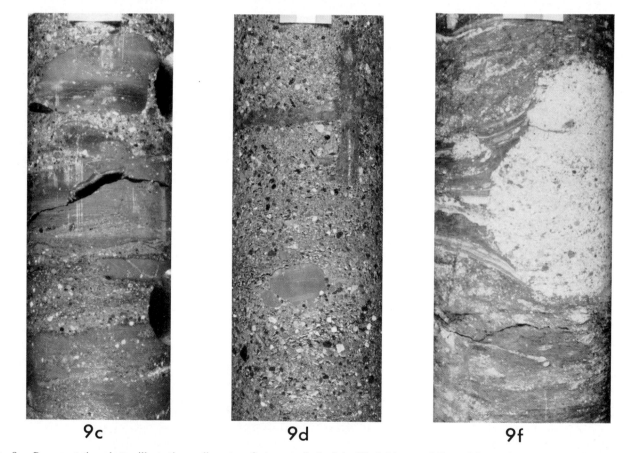

9c

9d

9f

FIG. 9.—Representative photos illustrating sedimentary features typical of the Waskahigan and Burnstick members (scale bar 3 cm). (a) Sharp base of Burnstick Member (arrowed). Note coarse granular sand filling burrows beneath contact. Well 11-30-56-19W5, 2,100 m. (b) Thin, highly bioturbated interbeds of mudstone, very fine sandstone, and coarse granular sandstone typical of lower parts of the members. Well 11-30-56-15W5, 2,199 m. (c) Interbedded and bioturbated mudstone and structureless-to-cross-bedded granular sandstone, typical of middle portions of the members. Note large mudstone clast at top of core and side-filled *Rhizocorallium* burrow immediately beneath. Waskahigan Member, well 12-19-55-18W5, 1,952.3 m. (d) Structureless, poorly sorted pebbly sandstone, typical of upper portions of the members. Well 12-19-55-18W5, 1,951.8 m. (e) Armored mudball in medium-grained sandstone, near top of Burnstick Member. Well 14-7-60-22W5, 1,821.5 m. (f) Gutter cast(?) in Burnstick Member. Well 1-1-57-20W5, 2,173 m.

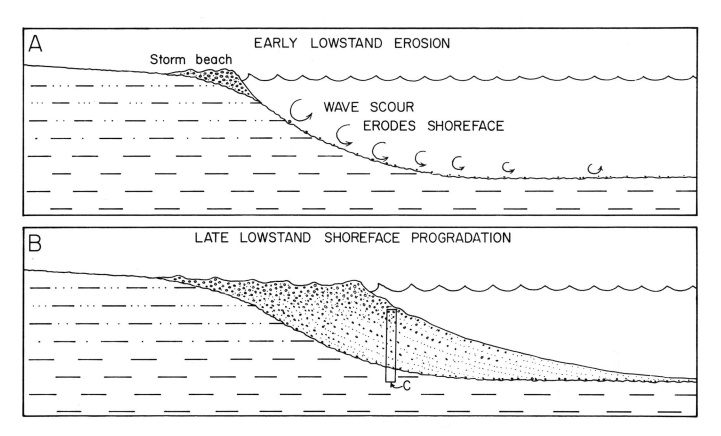

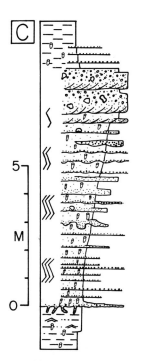

A
EARLY LOWSTAND EROSION
Storm beach
WAVE SCOUR
ERODES SHOREFACE
B
LATE LOWSTAND SHOREFACE PROGRADATION
C
C
5
M
0
Transgressive black mudstone.
Fine-grained chert pebble conglomerate, poorly-developed cross bedding.
Conglomerate often contains large mudstone intraclasts.
Mudstone with sharp-based interbeds of medium-coarse grained granular sandstone.
Occasional gutter casts, armoured mud balls and remnant wave-ripple lamination.
Bioturbation
Slight
Intense
Sharp-based laminae of medium-coarse grained sandstone thoroughly
bioturbated into mudstone.
Erosion surface with granules and pebbles burrowed down into underlying unit.
Bioturbated silty mudstone.

FIG. 10.—Interpretive drawings, based on data in Figures 8 and 9, illustrating origin of the Waskahigan and Burnstick members. (A) Development of erosional shoreface profile during early part of a lowstand. Note scarce shoreface sands and gravels built into relatively narrow storm beach at top of shoreface. (B) Progradation of the shoreface sediment body blankets the erosion surface and effectively halts further erosion of underlying shelf deposits. (C) Schematic lithologic log representative of thicker parts of the Waskahigan and Burnstick members. Deposition was strongly episodic (storm-dominated?), with periodic emplacement of coarse sediment into a dominantly muddy environment. Intense bioturbation has destroyed most sedimentary structures.

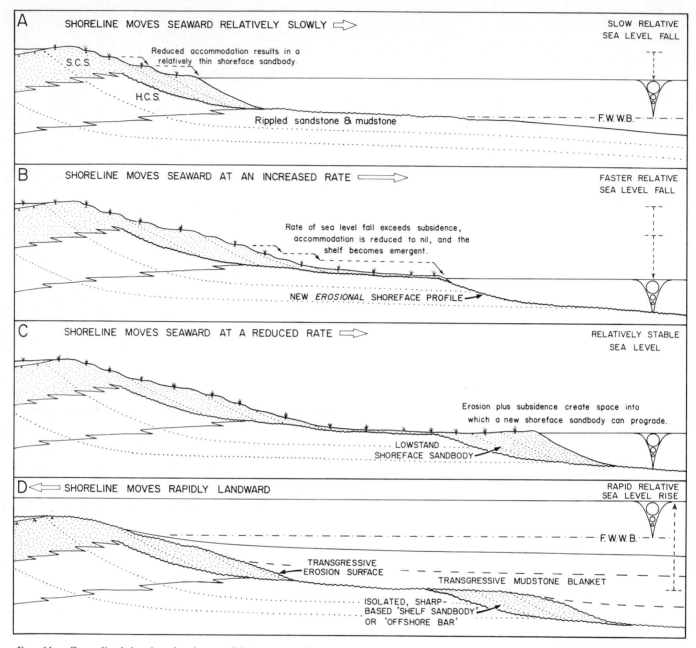

Fig. 11.—Generalized drawing showing possible sequence of events in origin of sharp-based "offshore bars." (A) Well-documented situation shown in Figure 7B. (B) Very low shelf gradient results in rapid basinward movement of shoreline in response to relative fall in sea level. Note insignificant accumulation of shoreface sediments. When shoreline stabilizes, wave scour into underlying shelf deposits will cut a new, energetically stable, *erosional* shoreface profile. (C) Erosion at shoreface creates space into which a *sharp-based* lowstand shoreface sand body can prograde, given adequate sediment supply. (D) Relative rise of sea level results in erosional truncation of lowstand shoreface deposit and landward scouring of coastal plain, removing all evidence of subaerial conditions. Lag gravel veneer is spread over transgressive erosion surface. (D) With continued deepening, relatively deep-water shelf mudstones blanket lowstand shoreface deposits and the eroded top of the underlying, shallowing-upward shelf sequence. The result—a linear, shore-parallel sandstone or conglomerate body, resting *erosively* on offshore mudstones and abruptly overlain by more shelf mudstones.

produced Weller's earlier diagram, which they described in terms of a "downward shift in clinoform pattern," although they stated that this pattern had never been observed on seismic data.

Recent work on Santonian-Campanian shoreface sandstones of the Virgelle Member of the Milk River Formation (McCrory and Walker, 1986) and equivalent Chungo Mem-

ber of the Wapiabi Formation (Rosenthal and Walker, 1987) has shown that, over long distances (>300 km), the base of the SCS shoreface sandstone is sharp, and, in places, shows spectacular loading into the underlying HCS sandstones and mudstones. These features were interpreted as evidence of very rapid progradation of the shoreface in response to a rapid fall in relative sea level, resulting in the abrupt emplacement of thick SCS sandstones directly upon relatively muddy, uncompacted offshore sediments. The sequences in the Virgelle and Chungo sandstones in outcrop are therefore very similar to those in the subsurface, as seen for example in wells 7-10-62-6 and 6-8-62-5 (Fig. 5).

Although the subsurface work reported here cannot always match the lithologic detail possible in outcrop, the use of well logs offers the possibility, through reference to relatively planar upper and lower markers, of *quantifying* the amount of erosion at the base of the shoreface, something difficult or impossible to achieve in most outcrop studies.

Offshore Bars

Current work on Cretaceous shelf sediments in Alberta, including the Viking (Raddysh, 1986; Downing, 1986), Cardium (Plint and others, 1986, 1987; Bergman and Walker, 1986, 1987) and Bad Heart formations (Plint and Walker, 1987b; B. Norris, pers. commun., 1987) has focused attention upon the relationship between shelf erosion during periods of rapid fall in relative sea level and the development of linear sandstone or conglomerate "bars." Although a detailed discussion of results is beyond the scope of this paper, it is clear that the "offshore bars" studied by these workers are all underlain by regional erosion surfaces which are related to relative sea-level lowstands. The linear "bars" were deposited against seaward-facing sigmoidal scours that have as much as 40 m of relief. These scours have been interpreted (Plint and Walker, 1987b; Bergman and Walker, 1987) as the product of wave scour in a lowstand shoreface. Deposition of the "bars" themselves also took place in a lowstand shoreface setting. This interpretation provides a plausible *mechanism* for the localization and orientation of the "bar." Alternative hypotheses (e.g., Swagor, and others, 1976; Campbell, 1979; Rice, 1984; Swift and Rice, 1984) envisage bar deposition to occur in the lee of "pre-existing topography" or a "slope break." Similarly, Campbell (1979) described a series of "transgressive offshore bars" that were underlain by "local unconformities." In this case, "offshore bars" were envisaged to have accreted seaward from the margins of subaerially eroded terraces following marine transgression.

In Figure 11, an alternative mechanism is suggested for the generation of "offshore bars." As already noted, a relative fall in sea level (but at a rate only slightly in excess of the rate of subsidence) results in accelerated shoreface progradation and thinning of the shoreface sand body (Fig. 11A). These conditions prevailed during the E2/T2 erosional episode described in Figures 4–7, when the relative sea-level fall was insufficient to cause shelf emergence, and there was always space available into which an albeit thinned shoreface sand body could prograde. If, however, the rate of relative sea-level fall *exceeded* the rate of subsidence for a period of time, then sediment accommodation would be reduced to nil. Under these conditions, the shoreline would move rapidly seaward, yet would deposit essentially no sediment on the subaerial portion of the shelf (Fig. 11B). When shoreline movement eventually ceased, wave erosion at the shoreline would tend to cut a new shoreface profile (a "bevel") through erosional shoreface retreat (Fig. 11B).

Erosion at the shoreface, together with relatively stable sea level, now provides space into which a new, sharp-based shoreface sand body can prograde (Fig. 11C).

During subsequent marine transgression, erosion would plane off the upper few meters of the lowstand shoreface, together with any subaerial deposits on the emergent shelf, thus removing all evidence of subaerial exposure. As a result, the lowstand shoreface sand body would remain in a seemingly isolated "offshore" position, resting sharply on, and blanketed by, offshore mudstones (Fig. 11D).

This sequence of events appears to provide the best explanation for the origin of the Waskahigan, Burnstick, Carrot Creek and Amundson members of the Cardium Formation (Fig. 2). The broadly comparable geometry and stratigraphic relationships of such well-known "offshore bars" as the "First Mancos sandstone" (Kiteley and Field, 1984) and "Ship Rock sandstone" (Campbell, 1979) may also be explicable in terms of the depositional sequence suggested in Figure 11.

ACKNOWLEDGMENTS

The bulk of this work was conducted during tenure of a post-doctoral fellowship at McMaster University, funded through a Natural Sciences and Engineering Research Council (NSERC) Strategic Grant to R. G. Walker; the manuscript was completed with support from NSERC operating grant No. A1917 to the author. Technical support was provided by Home Oil Ltd., and I am especially grateful to Sid Legget and George Fong for their assistance throughout this study. I thank Jack Whorwood and Ian Craig for preparing the photographs. Ideas have been sharpened through discussions with Michael Covey, Bill Duke, Dale Leckie, David James, and Henry Posamentier. I thank John Van Wagoner and B. E. Bowen for their critical comments, which much improved the final text. I am particularly grateful to Roger Walker for his support, discussion, and encouragement.

REFERENCES

BERGMAN, K. M., AND WALKER, R. G., 1986, Cardium Formation conglomerates at Carrot Creek field: Offshore linear ridges or shoreface deposits?, *in* Moslow, T. F., and Rhodes, E. G., eds., Modern and Ancient Shelf Clastics: A Core Workshop: Society of Economic Paleontologists and Mineralogists, Core Workshop No. 9, p. 217–267.
———, AND ———, 1987, The importance of sea level fluctuations in the formation of linear conglomerate bodies; Carrot Creek Member of Cardium Formation, Cretaceous Western Interior Seaway, Alberta, Canada: Journal of Sedimentary Petrology, v. 57, p. 651–665.
CAMPBELL, C. V., 1979, Model for beach shoreline in Gallup Sandstone (Upper Cretaceous) of northwestern New Mexico: New Mexico Bureau of Mines and Mineral Resources, Circular 164, 32 p.
CANT, D. J., 1984, Development of shoreline-shelf sand bodies in a Cretaceous epeiric sea deposit: Journal of Sedimentary Petrology, v. 54, p. 541–556.
DOWNING, K. P., 1986, The depositional history of the Lower Cretaceous Viking Formation at Joffre, Alberta, Canada: Unpublished M.S. Thesis, McMaster University, Hamilton, Canada, 138 p.
EYLES, N., AND CLARK, B. M., 1986, Significance of hummocky and swaley cross-stratification in Late Pleistocene lacustrine sediments of the Ontario Basin, Canada: Geology, v. 14, p. 679–682.
HEWARD, A. P., 1981, A review of wave-dominated clastic shorelines. Earth Science Reviews, v. 17, p. 223–276.
KITELEY, L. W., AND FIELD, M. E., 1984, Shallow marine depositional environments in the Upper Cretaceous of northern Colorado, *in* Tillman, R. W., and Siemers, C. T., eds., Siliciclastic Shelf Sediments:

370

A. G. PLINT

Society of Economic Paleontologists and Mineralogists Special Publication 34, p. 179–204.

LAGAAIJ, R., AND KOPSTEIN, F. P. H. W., 1964, Typical features of a fluviomarine offlap sequence, in Van Straaten, L. M. J. U., ed., Deltaic and Shallow Marine Deposits: Developments in Sedimentology, v. 1, p. 216–226.

McCRORY, V. L. C., AND WALKER, R. G., 1986, A storm and tidally-influenced prograding shoreline—Upper Cretaceous Milk River Formation of southern Alberta, Canada: Sedimentology, v. 33, p. 47–60.

PLINT, A. G., AND WALKER, R. G., 1987a, Cardium Formation 8. Facies and environments of the Cardium shoreline and coastal plain in the Kakwa field and adjacent areas, northwestern Alberta: Bulletin of Canadian Petroleum Geology, v. 35, p. 48–64.

——, AND ——, 1987b, Morphology and origin of an erosion surface cut into the Bad Heart Formation during major sea level change, Santonian of west-central Alberta, Canada: Journal of Sedimentary Petrology, v. 57, p. 639–650.

——, ——, AND BERGMAN, K. M., 1986, Cardium Formation 6. Stratigraphic framework of the Cardium in subsurface: Bulletin of Canadian Petroleum Geology, v. 34, p. 213–225.

——, ——, AND ——, 1987, Reply to discussion on Cardium Formation 6. Stratigraphic framework of the Cardium in subsurface: Bulletin of Canadian Petroleum Geology, v. 35, p. 365–374.

RADDYSH, H., 1986, Sedimentology of the Viking Formation at Gilby A and B fields, Alberta: Unpublished B.S. Thesis, McMaster University, Hamilton, Canada, 241 p.

RICE, D. D., 1984, Widespread, shallow marine, storm-generated sandstone units in the Upper Cretaceous Mosby Sandstone, central Montana, in Tillman, R. W., and Siemers, C. T., eds., Siliciclastic Shelf Sediments: Society of Economic Paleontologists and Mineralogists Special Publication 34, p. 143–161.

ROEP, T. H. B., BEETS, D. J., DRONKERT, H., AND PAGNIER, H., 1979, A prograding coastal sequence of wave-built structures of Messinian age, Sorbas, Almeria, Spain: Sedimentary Geology, v. 22, p. 135–163.

ROSENTHAL, L. R. P., AND WALKER, R. G., 1987, Lateral and vertical facies sequences in the Upper Cretaceous Chungo Member, Wapiabi Formation, southern Alberta: Canadian Journal of Earth Sciences, v. 24, p. 771–783.

SWAGOR, N. S., OLIVER, T. A., AND JOHNSON, B. A., 1976, Carrot Creek field, central Alberta, in Lerand, M. M., ed., The Sedimentology of Selected Clastic Oil and Gas Reservoirs in Alberta: Canadian Society of Petroleum Geologists, p. 78–95.

SWIFT, D. J. P., AND RICE, D. D., 1984, Sand bodies on muddy shelves: A model for sedimentation in the Western Interior Cretaceous Seaway, North America, Tillman, R. W., and Siemers, C. T., eds., in Siliciclastic Shelf Sediments: Society of Economic Paleontologists and Mineralogists Special Publication 34, p. 43–62.

VAIL, P. R., MITCHUM, R. M., AND THOMPSON, S., 1977, Seismic stratigraphy and global changes of sea level, part 3: Relative changes of sea level from coastal onlap; in Payton, C. E., ed., Seismic Stratigraphy—Application to Hydrocarbon Exploration: American Association of Petroleum Geologists Memoir 26, p. 63–81.

WADSWORTH, J. A., 1987, Geometry and origin of a sequence-bounding unconformity, Cardium Formation, Alberta: Unpublished B.S. Honors Thesis, University of Western Ontario, London, Canada, 54 p.

WALKER, R. G., 1983, Cardium Formation 3. Sedimentology and stratigraphy in the Caroline-Garrington area, Alberta: Bulletin of Canadian Petroleum Geology, v. 31, p. 213–230.

WELLER, J. M., 1960, Stratigraphic principles and practices: Harper and Brothers, New York, 725 p.

CONTROLS ON COAL DISTRIBUTION IN TRANSGRESSIVE-REGRESSIVE CYCLES, UPPER CRETACEOUS, WESTERN INTERIOR, U.S.A.

TIMOTHY A. CROSS

Department of Geology and Geological Engineering, Colorado School of Mines, Golden, Colorado 80401

ABSTRACT: The thickest and most extensive Upper Cretaceous coals of the western interior of the United States occur at the top of, and landward of, shoreface and delta front platforms that are stacked vertically. An explanation for this observation was sought through numerical models derived from the interactions of the three fundamental processes that control stratal geometries and lithofacies distributions. These are eustatic fluctuations, tectonic movement, and quantity of sediment delivered to or produced in a sedimentary basin.

The models show that the fundamental building block of marine-shelf, to coastal-plain stratigraphic sequences is the progradational event, expressed in vertical profile as a shallowing-upward succession of facies. The shallowest facies at the top of one event is capped abruptly by the deepest facies at the base of the subsequent event. This facies asymmetry is modeled by sinusoidal sea-level oscillations superimposed on a constant rate of tectonic subsidence; disharmonic variations in either sea level or tectonic movement are unnecessary to produce this asymmetry.

The models simulate a hierarchical stacking of progradational events that display three geometric patterns: seaward-stepping, landward-stepping, and vertical stacking. The models show that the thickest and most extensive coals accumulate when accommodation space in the lower, potentially coal-bearing portion of the coastal plain is near maximum, and when the rate of sea-level change is balanced by the rate of sediment supplied by progradational events. These factors result in vertical aggradation of coastal-plain facies tracts and vertical stacking of the progradational events.

INTRODUCTION

Upper Cretaceous coals in the Rocky Mountains and Great Plains region of the United States accumulated in swamps occupying coastal positions along the western margin of the western interior seaway. Paleogeomorphic settings that have been ascribed to these coals include alluvial plain, active and abandoned delta plains, and coastal swamps behind barred and non-barred strandlines. Regardless of specific inferred paleogeomorphic settings, there is a more fundamental and overriding commonality of coal occurrence with respect to the position of major coals in the stratigraphic architecture. Numerous studies of the past few decades (Sears and others, 1941; Weimer, 1960; Fassett and Hinds, 1971; Beaumont and others, 1971; Ryer, 1984) have demonstrated that major coals occur at specific positions in the transgressive-regressive sequences that characterize Upper Cretaceous strata of the western interior seaway. In many instances, coals occur at the top of, and landward of, shoreline facies that cap shoaling-upward sequences produced by progradational events. As noted by Beaumont and others (1971) and further documented by Ryer (1984), however, the thickest and most extensive coals occur at the top of, and landward of, shoreline facies in areas where delta-front or shoreface sandstones of successive progradational events are stacked vertically.

This paper proposes an explanation for the preferential distribution of major coals within these vertically stacked progradational units. This explanation is derived from process-response numerical models that simulate sediment deposition within marine-shelf to coastal-plain environments and portray the resulting stratigraphic architecture. The models, similar to those of McGhee and Bayer (1985), Morrow (1986), and Jervey (this volume), are based upon the fundamental, interdependent controls that govern the development of stratigraphic sequences and stratal architecture in sedimentary basins. These are the interactions of eustatic fluctuation, tectonic movement, and quantity of sediment delivered to a sedimentary basin. The models show

that the observed distribution of major coals is related directly to temporal, geographic, and dimensional variations in space in which sediments of different facies tracts may accumulate and be preserved. This space, in turn, is a direct product of the interactions of the fundamental controlling variables. The explanation presented here is an attempt to supply a theoretical basis for the well-documented observation that major coals occur within vertically stacked progradational events.

A detailed discussion of the numerical models is not necessary for the purposes of summarizing the general conclusions derived from those models and proposing an explanation for the distribution of major coals with respect to stratigraphic architecture. A brief summary of the general approach to the numerical simulations is useful, however. For each time step of a model, eustatic change is added to tectonic subsidence, using geologically realistic values for each. Various combinations of eustatic curves having different phases, amplitudes, and periods are superimposed on (convolved with) linear or non-linear (time-varying) subsidence curves. Next, sediment is added to (or removed from) the space made available (or lost) by the resultant relative change in sea level. The amount of sediment that is added is controlled by a non-linear function that incorporates water depth, rate of change in water depth, and distance from shoreline. Corrections for isostatic compensation, compaction of the previously deposited sediment column, and incremental sediment addition are solved iteratively until equilibrium is obtained. This process is repeated for successive time steps over the time span of the model.

CHARACTERISTICS OF TRANSGRESSIVE-REGRESSIVE CYCLES

McGookey (1972), Kauffman (1980) and Weimer (1984), among others, have described five major transgressive-regressive cycles within latest Albian through Maestrichtian strata of the western interior. Individual cycles range in duration from about 3 to 7 million years and, therefore, they apparently are not periodic as the term cyclic normally connotes. Ryer (1984) noted the correspondence in temporal and spatial scales of these cycles to the third-order depositional sequences of Vail and others (1977).

The origin of these third-order transgressive-regressive cycles has been attributed to episodic tectonic events, eustatic fluctuations, episodic variations in sediment volumes delivered to coastal-plain and paralic environments, or to a combination of these. Hancock and Kauffman (1979), for example, argued that Late Cretaceous transgressive and regressive maxima in northern Europe and the western United States were essentially isochronous and inferred a eustatic control. Kauffman (1980) also favored a primary eustatic control based upon the synchroneity of transgressive maxima within one or two faunal zones throughout the western interior. He noted that these synchronous transgressive peaks are recognizable in strata despite variations in thickness or facies that reflect local tectonic or sedimentologic influences.

Others have argued that episodic deformation in the Sevier fold and and thrust belt, which bordered the western margin of the western interior seaway, was a principal or contributing cause of the transgressive-regressive episodes. Jordan (1981) cited the correspondence among mapped paleoshoreline positions, the calculated topographic elevations, and the emplacement of thrust sheets and derived sediment loads. Swift and others (1985) observed that the timing of some thrusting episodes corresponded well with the third-order cycles recognized by Weimer (1984), whereas other episodes did not. Nonetheless, Swift and others (1985) suggested a causal link between the cycles and thrust events. They proposed that thrusting initiated subsidence but, at first, the rate at which sediment was supplied from the orogenic belt exceeded the rate of subsidence and, consequently, regression ensued. As thrusting ceased and topographic relief was reduced by erosion, the rate of sediment supply became less than the rate of subsidence and transgression ensued.

Kauffman (1980) presented yet another explanation. He suggested a correspondence in timing among regional transgressions and maximum subsidence rates within the basin, active thrusting, volcanism, and rapid clastic sedimentation. In this view, episodic uplift caused episodic loading, subsidence, transgression, and increased supply of river-borne detritus into deltaic, coastal-plain, and paralic environments. Conversely, major regressions were associated with times of tectonic and volcanic quiescence, reduced subsidence rates, and decreased sediment supply. Although episodic deformation may have contributed to the development of third-order cycles prior to about 80 Ma, the cessation of Sevier deformation before 75 Ma in Colorado, Utah, and southern Wyoming (Cross, 1986) suggests that tectonism was not the dominant control, either at the time or in the positions that many of the major coals accumulated.

Regardless of the ultimate causes of the third-order transgressive-regressive cycles, their deposits are widespread in the western interior. More important, they share a commonality of stratigraphic geometries and facies distributions within a larger scale stratal architecture, which is discussed below. These shared attributes allow for numerical simulations that fully integrate the interdependent controls of eustatic fluctuation, tectonic movement, and sediment supply. From such simulations a theoretical basis for explaining the distribution of major coals is derived.

Ryer (1984) recounted some of the numerous studies and observations which, over the past few decades, have demonstrated that these Cretaceous third-order cycles are composed of several smaller, asymmetrical, shallowing-upward sequences, each of which represents a discrete progradational event. Following the convention of Vail and others (1977), Ryer termed these progradational units "fourth-order cycles." Similar hierarchies of smaller depositional cycles composing larger cycles have been recognized elsewhere and in other depositional settings. For example, deposits of these discrete progradational events or fourth-order cycles are identical to the genetic sequence and Genetic Increment of Strata of Busch (1959, 1971), the depositional-event of Frazier (1974), and the parasequence of Van Wagoner (1985; Van Wagoner and others, this volume). Ramsbottom (1979) recognized transgressive-regressive sequences of similar scale in the Carboniferous of Europe and termed them mesotherms. They also are similar to, but perhaps of longer duration than, the punctuated aggradational cycles (PACs) of Anderson and Goodwin (1980) and Goodwin and Anderson (1985). Busch and West (1987) provide a useful review of the varieties, scales, and terminologies of hierarchical transgressive-regressive stratigraphic units that have been recognized.

The numerical simulations show that progradational units are arranged systematically in three geometric patterns as a direct consequence of the interactions of eustatic fluctuation, tectonic movement, and sediment supply. The patterns depicted by the numerical simulations are corroborated by numerous stratigraphic studies of Cretaceous deposits of the western interior, as well as strata of other ages elsewhere. The regular, hierarchical pattern formed by the vertical and lateral stacking geometry of these progradational events produces the regional stratigraphic architecture typical of third-order transgressive-regressive cycles. This architecture, discussed below, is similar or identical to the Genetic Sequence of Strata of Busch (1971, 1974), the depositional-episode of Frazier (1974), the PAC sequence of Anderson and Goodwin (1980) and Goodwin and Anderson (1985), the en échelon and stacked cycles of Ryer (1984), and the parasequence set and sequence of Van Wagoner (1985; Van Wagoner and others, this volume) and Posamentier and Vail (this volume).

The internal stratigraphic architecture of a tyipcal third-order transgressive-regressive cycle has the following characteristics, beginning with the most landward position (the transgressive maximum of many authors) attained by a progradational unit (Fig. 1). During progradation, progressively shallower water facies are deposited upon previously deposited deeper water facies, resulting in a shallowing-upward vertical profile at any particular location. The termination of a progradational event is marked abruptly by a temporary landward shift in sites of sediment accumulation and, within the marine realm, a substantial decrease in rate of sediment accumulation and increase in water depth.

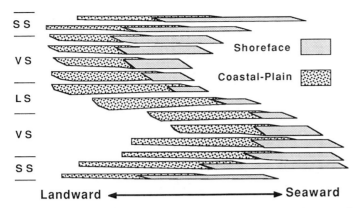

FIG. 1.—Diagrammatic representation of stratigraphic architecture of a typical third-order cycle in Upper Cretaceous strata of the western interior, United States. Fundamental building blocks are progradational events arranged in three geometric patterns of vertical stacking (VS) at the regressive and transgressive maxima, seaward-stepping (SS), and landward-stepping (LS). Only the lower, potentially coal-bearing portion of the coastal plain, including lower alluvial-plain and lower delta-plain environments, is depicted. Relative thickness and lateral extent of coastal plain and shoreface change as a function of the position of the progradational event within the hierarchical stacking pattern.

Consequently, at a constant geographic location, the shallowest water facies that developed at the termination of one progradational event is overlain directly by the deepest water facies of the subsequent progradational event. Relative to the initial and final shoreline positions of the first progradational event, shorelines of the next event are both initiated and terminated in a more seaward position. Because the second event steps in a seaward direction, usually causing a lateral offset of identical facies across the boundary separating the two events, the second event is termed "seaward-stepping." The next few progradational events may also step seaward, that is, each successive event begins and ends in a more seaward position than the preceding one. This overall seaward-stepping geometric pattern of progradational events is equivalent to the progradational parasequence of Van Wagoner (1985) and to the highstand systems tract illustrated in Figure 1 of Van Wagoner and Vail (this volume). The last of these attains a maximum or near-maximum seaward position of all progradational wedges within the third-order transgressive-regressive cycle.

In contrast to seaward-stepping events, the next several progradational events are stacked vertically, such that the initiation and termination of each occur approximately at the same geographic positions. This succession of vertically stacked progradational events occupies the most seaward position within a third-order cycle. It corresponds to the regressive maximum of many authors, to the aggradational parasequences of Van Wagoner (1985; Van Wagoner and others, this volume), and to the lowstand systems tract of Posamentier and Vail (this volume). As before, an individual progradational unit exhibits, in vertical profile, a shallowing-upward facies succession. At any geographic location, the shallowest water facies that developed at the termination of one event is overlain directly by the deepest water facies of the subsequent event, although the differences may be subtle because identical facies usually are not

offset substantially across the event boundaries. The last of these vertically stacked progradational events marks the end of the regressive phase of a third-order cycle.

Deposition of each of the next few progradational events occurs in a progressively more landward position. That is, relative to one progradational event, a subsequent event is both initiated and terminated in a more landward position. This landward-stepping geometric pattern is equivalent to the retrogradational parasequences of Van Wagoner (1985; Van Wagoner and others, this volume), and is the main constituent of the transgressive-systems tract of Posamentier and Vail (this volume).

Termination of the landward-stepping phase of a third-order cycle is marked by a resumption of the vertical stacking of progradational events in the most landward position attained during the cycle. Similar to the vertically stacked events of a regressive maximum, the landward and seaward limits of identical facies tracts within successive progradational events occupy essentially identical geographic positions in this part of a third-order cycle. This phase of vertical stacking corresponds to the transgressive maximum of many authors, the aggradational parasequences of Van Wagoner (1985), and is equivalent to the basal portion of the highstand systems tract of Van Wagoner and others (this volume) and Posamentier and Vail (this volume). The next third-order transgressive-regressive cycle begins with another forward-stepping progradational event.

ORIGIN OF PROGRADATIONAL EVENTS AND STACKING GEOMETRIES

Individual progradational events and their hierarchical geometric arrangements within third-order cycles are simulated by numerical models which incorporate eustatic fluctuation, tectonic movement, and sediment supply (see also McGhee and Bayer, 1985; Morrow, 1986; Jervey, this volume; Posamentier and others, this volume). Consider the example of sinusoidal sea-level oscillations and a constant rate of tectonic subsidence shown in Figure 2. This curve shows the position of the air/sea interface through time at

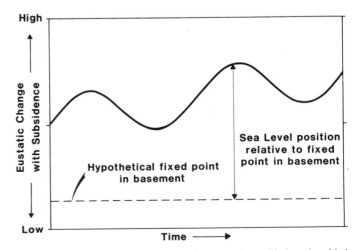

FIG. 2.—Sinusoidal curve representing eustatic oscillations is added to a constant rate of subsidence. Curve shows position of sea/air interface at one geographic location with respect to a fixed point in the basement. Slope of the curve is a function of the rate of subsidence.

one geographic location with respect to a fixed point in the basement. Eustatic fluctuations combine with tectonic subsidence to produce an oscillatory change in sea level superimposed upon a secular increase in water depth. The product of these two variables is a continuously varying space to which sediment may be added or removed.

Jervey (pers. commun., 1982) proposed the term "accommodation" for this space which acts as a potential receptor of sediment. Here, the term "accommodation potential" refers to the cumulative space created or removed by the interaction of eustatic variation and tectonic movement. The accommodation potential at a point in space at a particular time determines the maximum volume of sediment that may accumulate there. This cumulative space is produced by the incremental additions and subtractions of space through time, that is, by changes in accommodation rate through time. Accommodation rate is the rate of creation or removal of space, and accommodation potential is the integral of accommodation rate. Jervey (this volume) and Posamentier and others (this volume) provide extended discussions of accommodation concepts.

As shown in Figure 3, accommodation rate is at a maximum when sea level is rising most rapidly, at the inflection point on the rising eustatic curve. This position represents the maximum incremental addition of new space in which sediment may accumulate at a particular geographic location. Accommodation rate is at a minimum when sea level is falling most rapidly, at the inflection point on the falling eustatic curve. This position represents the maximum incremental loss of potential sediment receptor space. When the rate of sea-level fall equals the rate of subsidence, there is no incremental change in the amount of space made available and the accommodation rate is zero. When the rate of sea-level change is zero, at the peak and trough of the eustatic curve, the accommodation rate is equal to the rate of tectonic subsidence.

Figure 4 illustrates the sedimentary responses to changes in accommodation potential at a single geographic location during two progradational events. In this figure, an extension of the model shown in Figure 2 and one similar to the models of Morrow (1986) and Jervey (this volume), sediment is added to (or removed from) the space made available (or lost) by the combined effects of eustatic change and subsidence. The deepest water occurs at or near the inflection point on the rising sea-level curve, not at the eustatic peak. This corresponds to the time that sea level is rising most rapidly and at the position of maximum accommodation potential (Jervey, this volume). As noted by Pitman (1978), a rapid rise in sea level is commonly, but not necessarily, accompanied by transgression, or landward movement of the strand. During transgression, strand plains founder, rivers transform to estuaries, and sites of sediment accumulation shift landward. Estuaries are efficient sediment traps and, although the flux of sediment to the coastal

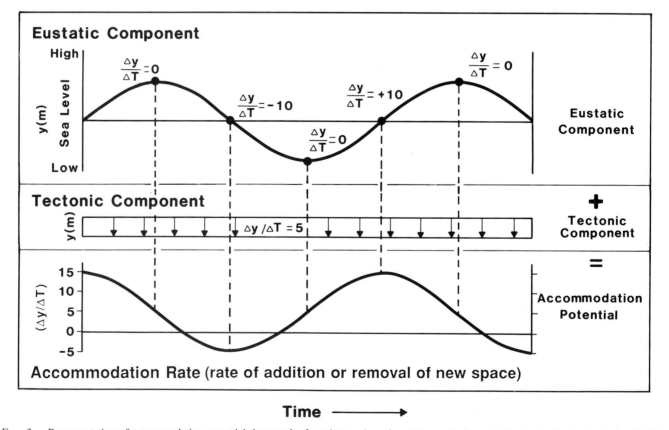

FIG. 3.—Representation of accommodation potential that results from interaction of eustatic oscillations and tectonic subsidence shown in Figure 2. Numerals on curve identify non-dimensional rates of eustatic change. These rates of eustatic change are added to a constant non-dimensional rate of subsidence to construct the curve of accommodation rate.

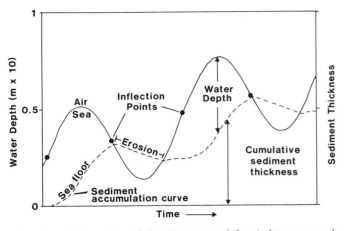

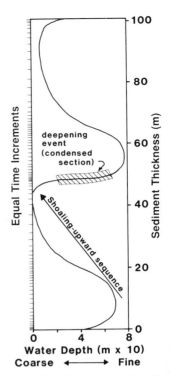

FIG. 4.—Sediment is added to (or removed from) the space made available (or lost) by combined effects of eustatic change and subsidence. Dashed curve shows position of sediment/water interface through time at one geographic location with respect to a fixed point in the basement. Cumulative thickness of sediment through time is measured from horizontal axis to dashed curve. Vertical distance between solid and dashed curves is the measure of water depth (where solid curve is above dashed curve) or times of subaerial exposure and erosion (where dashed curve is above solid curve). Horizontal distance between intersections of the two curves indicates duration of the unconformity.

FIG. 5.—Synthetic stratigraphic column derived from the model shown in Figure 4, but with corrections for sediment compaction and isostatic compensation incorporated. Stratigraphic thickness is shown on vertical axis and water depths are shown on horizontal axis. Equal time increments are represented by tics along vertical axis at left. Where tics are closely spaced, sediment accumulation rates are low. Where tics are widely spaced, sediment accumulation rates are high.

plain may not change, a reduced proportion of the total sediment load is carried into paralic and shelf environments. The sharp contact at the boundary of two progradational events is caused by the rapid rate of sea-level rise that occurs near the inflection point on the curve. As the rate of sea-level rise decreases and the accommodation potential is gradually reduced, sediment is supplied at progressively increasing rates to more seaward positions, gradually filling part of the accommodation space. This reduction in accommodation potential is responsible for the shoaling-upward succession of lithofacies observed in a progradational event. Regression, or seaward movement of the strand, begins near the eustatic peak when the accommodation rate is approximately equal to the rate of tectonic subsidence and continues during falling sea level. Accommodation potential reaches a minimum during the maximum rate of sea-level fall, corresponding to the inflection point of the descending sea-level curve.

These relationships are illustrated in the form of a synthetic stratigraphic column (Fig. 5). This column was derived from the preceding model of two progradational events (Fig. 4), but also includes corrections for sediment compaction and isostatic compensation. The stratigraphic succession is characterized by an initial shallowing-upward sequence of the first progradational event capped abruptly by deeper water sediments at the base of the second shallowing-upward sequence. Rates of sediment accumulation are at a minimum at the base of a progradational event. This occurs near the inflection point of the rising sea-level curve when water depths are greatest. High rates of sediment accumulation occur before the eustatic peak when accommodation potential is high and balanced by the rate of sediment supply. Toward the top of each progradational event, sediment accumulation rates progressively decrease as accommodation potential declines.

It is important to emphasize that this stratigraphic asymmetry, characteristic of the fourth-order cycles in Upper Cretaceous strata of the western interior, was produced by a sinusoidal eustatic variation superimposed on a constant rate of tectonic subsidence. An identical pattern would result if tectonic movement were oscillatory or episodic and sea level were held constant. This discussion is not intended as a demonstration that either tectonic movements or sea-level fluctuations during the Late Cretaceous were sinusoidal. Rather, it is intended to demonstrate that the asymmetrical stratigraphic sequences characteristic of Upper Cretaceous strata do not require disharmonic variations in either sea level or tectonic movement.

A numerical model that illustrates the genesis of the hierarchical stacking geometries of progradational events was generated by using two eustatic curves of different periods and amplitudes superimposed on a constant rate of tectonic subsidence (Fig. 6; see also Jervey, this volume). The short-term eustatic curve is one-half the amplitude and one-tenth the period of the long-term eustatic curve. The short-term curve simulates the progradational events, whereas the long-term curve produces the hierarchical stacking pattern. The results of this model are shown in the more conventional manner of a synthetic stratigraphic column (Fig. 7). Beginning at the base of the column, each successive progradational event is progressively deeper than the preceding one, representing landward-stepping events in stratigraphic cross section. The depths of initiation and cessation of the

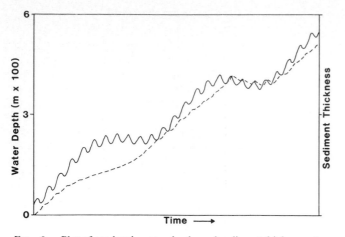

FIG. 6.—Plot of sea level, water depth, and sediment thickness at one point in space produced by convolving two eustatic curves with a constant rate of subsidence. The long-term eustatic curve is twice the amplitude and 10 times the period of the short-term oscillations. Corrections for isostatic compensation and sediment compaction have been made.

next succession of progradational events occur at the same positions, reflecting vertical stacking during the transgressive maximum phase of a third-order cycle. The next series of events is progressively shallower, corresponding to seaward-stepping progradational events in stratigraphic cross section. Finally, the initiation and cessation of the next series of events occur at approximately identical water depths, representing vertical stacking during the regressive maximum phase of a third-order cycle. Both Jervey (this volume) and Posamentier and others (this volume) arrived at similar conclusions.

A more general representation of these relations is shown in Figure 8 in which one short-period, low-amplitude sea-level oscillation is superimposed on another long-period, higher amplitude oscillation. Morrow (1986) illustrated the distortions of short-period sine waves upon which a long-period sine wave or a long-term sea-level rise or fall is superimposed, and discussed the implications for sediment responses to such distortions. In a single long-term cycle, one phase of vertical stacking occurs during the lowest long-term sea level and before the inflection point on the rising long-term sea-level curve. During this period, the rate of sea-level change is sufficiently low that sediment supplied by progradational events keeps pace with the accommodation potential. The rate of long-term sea-level rise and accommodation potential approach a maximum at the inflection point of the long-term sea level curve. During this period, the supply of sediment is insufficient to fill the available space at a single geographic location, and landward-stepping progradation occurs. A second phase of vertical stacking occurs during the highest long-term sea level and before the inflection point on the falling long-term sea-level curve. As before, the rate of sea-level change is sufficiently low that sediment supplied by progradational events keeps pace with the accommodation potential. As the rate of long-term sea-level fall increases, the accommodation potential is reduced, and seaward-stepping progradation results.

The accommodation potential of a point in space at a particular time determines the maximum volume of sediment that may accumulate there. The sediment volume that actually accumulates will be less than the available space if insufficient sediment is transported to that geographic position. Conversely, if more sediment is delivered to that geographic position than can be accommodated, it will be distributed farther seaward and will accumulate there. This

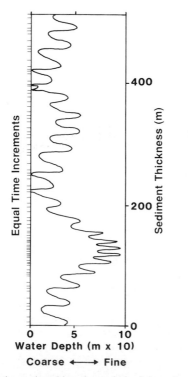

FIG. 7.—Synthetic stratigraphic column derived from the model shown in Figure 6. Labels and symbols as in Figure 5.

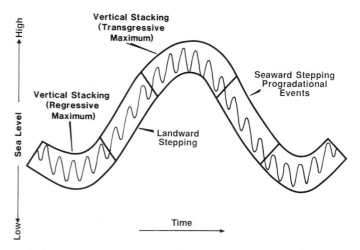

FIG. 8.—Representation of relations among long-term and short-term eustatic fluctuations, accommodation potential, and origin of the hierarchical stacking pattern of fourth-order progradational events. A high-frequency sine wave is convolved with a low-frequency sine wave four times the amplitude and one-sixteenth the frequency. These ratios approximate those commonly observed in Upper Cretaceous strata of the western interior. Positions of seaward-stepping, landward-stepping, and vertically stacked progradational events are shown.

balance between accommodation potential and the sediment volume that actually accumulates in the space available is termed "realized accommodation," where realized accommodation equals the accommodation potential minus the unfilled space. It is the realized accommodation that is geologically significant, as in any marine basin there is always a greater volume of space than may be filled be sediment.

The actual distribution of sediment to various positions within coastal-plain to marine-shelf environments is analogous to a conveyor belt delivering material to a hopper (Fig. 9). Within the time span of a third-order cycle, about 3 to 5 million years, it is likely that climate and corresponding denudational rates will remain essentially constant. Even if denudational rates were instantaneously increased in the source area, for example, by tectonic uplift or emplacement of a thrust sheet, they would subsequently decline little during the time span of a third-order cycle. Thus, during a 3- to 5-million-year period, the sediment load carried by rivers from drainage basins to the coastal plain is essentially constant. By analogy, the conveyor belt delivers material to a hopper at a constant rate. The hopper redistributes its supply of materials to alternative receptors, such that those receptors closest to the hopper will be supplied and filled first, whereas shipments to receptors farther removed from the hopper will be progressively delayed. In the world of depositional systems, the hopper is located upslope of the coastal plain. By analogy, sediments accumulate first in the coastal plain to the limits imposed by the accommodation potential; excess sediment is bypassed to more seaward environments. Thereafter, paralic, shoreface, and marine-shelf environments are supplied with sediment and each is filled progressively to the limit of its respective realized accommodation.

It has been shown that the hierarchical stacking geometry of progradational events within a basin is controlled by geographic variations in accommodation potential through time. Realized accommodation is the sedimentary response to these geographic variations in accommodation potential. The volume of sediment that actually accumulates at any geographic position may be expressed as a gradient in realized accommodation which extends, perpendicular to shoreline, from the coastal plain to the marine shelf. As this gradient changes through time, there will be a differential partitioning of sediment volumes deposited within different facies tracts. The models predict that the realized accommodation of different facies tracts within a single progradational event is determined by the position of that event within the larger scale stacking pattern. In landward-stepping events, when the rate of long-term sea-level rise is at a maximum, realized accommodation gradient is oriented in a landward direction (Fig. 10B). Consequently, in comparison with other events in the hierarchy, proportionally more sediment will accumulate in the coastal plain than in the shoreface. In seaward-stepping events, when the rate of long-term sea-level fall is at a maximum, realized accommodation is oriented in a seaward direction (Fig. 10D). Compared with landward-stepping and vertically stacked events, proportionally more sediment will accumulate in the shoreface than in the coastal plain. In vertically stacked events, the realized accommodation gradient is approximately symmetrical (Fig. 10A, C).

Accompanying these changes in realized accommodation gradient among progradational events are changes in stratal architecture and in the volume, thickness, and lateral distribution of sediment contained within identical facies tracts (Fig. 1). Compared with vertically stacked events, seaward-stepping events are thin and laterally extensive, and landward-stepping events are thick and laterally restricted. In seaward-stepping events and in the transition into vertically stacked events of the regressive maximum, coastal-plain facies tracts are thin and extensive parallel to depositional slope, whereas the shoreface facies tract is thicker and more laterally restricted. By contrast, in the transition from the regressive maximum to landward-stepping events, coastal-plain facies tracts are thick and may be laterally extensive, whereas the shoreface facies tract of approximately

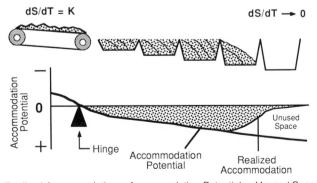

dS/dT = K   **dS/dT → 0**

Realized Accommodation = Accommodation Potential - Unused Space

FIG. 9.—Diagram showing difference between accommodation potential and realized accommodation, using analogy of a conveyor belt delivering sediment at a constant rate to a hopper. Sediment is redistributed from the hopper to additional receptors which, in this example, have increasing accommodation potential in a seaward direction. By contrast, realized accommodation first increases and then decreases in a seaward direction.

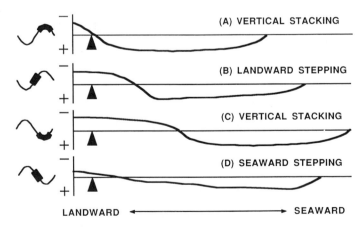

(A) VERTICAL STACKING

(B) LANDWARD STEPPING

(C) VERTICAL STACKING

(D) SEAWARD STEPPING

LANDWARD ◄——————————► **SEAWARD**

FIG. 10.—Model predictions of changes in realized accommodation gradient within progradational events as a function of their position in the hierarchical stacking pattern of third-order cycles. During maximum rate of long-term sea-level rise (B), realized accommodation gradient is shifted landward. During maximum rate of long-term sea-level fall (D), realized accommodation gradient is shifted seaward. Realized accommodation gradient is approximately equal from coastal plain to shoreface in vertically stacked events (A, C).

equal thickness is laterally restricted. Within landward-stepping events and in the transition into vertically stacked events of the transgressive maximum, coastal-plain facies tracts are expanded and the shoreface is thin and laterally restricted. In the transition from the transgressive maximum to seaward-stepping events, coastal-plain facies tracts are thin and the shoreface is expanded.

In summary, the numerical models predict a differential partitioning of sediment volumes into different facies tracts according to the position of individual progradational events within the stacking hierarchy. The volumetric partitioning is a direct consequence of the geographic variations in accommodation potential produced when eustatic oscillations of different periods and amplitudes are convolved with tectonic subsidence. The relationships predicted by the models may be tested by observed distributions and volumes of sediment contained within specific facies tracts. One such test of the model predictions is afforded by the observed distribution of coals within progradational events of Late Cretaceous strata of the western interior. Coals are particularly sensitive indicators of high accommodation potential in the coastal plain. Other factors being equal, such as seasonal stability of a high groundwater table and no flux of detrital sediment into peat-forming mires, the preservation of peat requires a high accommodation potential.

OBSERVED DISTRIBUTION OF COALS

The occurrence of major coals at the top of, and landward of, shoreface platforms within vertically stacked progradational events of third-order cycles is explained by the differences in realized accommodation of specific facies tracts as a function of the position of a progradational event within the hierarchical stacking pattern of fourth-order cycles. Major coals occur within parts of third-order cycles that represent a balance between the maximum potential for vertical aggradation of sediment and availability of sediment. That potential is realized when high accommodation potential is balanced by rate of sediment flux to a particular geographic position.

Cretaceous coals that accumulated along the coastal plain of the western interior seaway occur at the top of, and landward of, shoreline facies within many progradational units, regardless of their hierarchical position within the third-order cycles. The most extensive and thickest coals, however, are restricted to vertically stacked progradational events, which characterize transgressive and regressive maxima.

Ryer (1984) summarized the association of major coals in Utah with respect to their position in the stratigraphic architecture of third-order cycles. Major coals associated with vertically stacked progradational events deposited during transgressive maxima include those in the Kolob-Alton (Dakota Sandstone) and Kaiparowits (Straight Cliffs Formation) coalfields. Major coals associated with vertically stacked progradational events deposited during regressive maxima include those in the Henry Mountains (Ferron Sandstone), Emery (Ferron Sandstone), and Wasatch Plateau Deep (Emery Sandstone) coalfields. These coalfields represent approximately 60 percent of the known Cretaceous coal resources in Utah.

The Ferron Sandstone, studied by Ryer (1981), consists of a series of seaward-stepping, vertically stacked and landward-stepping progradational events deposited during a regressive maximum. The Ferron displays a significant variation in coal development with respect to the position of progradational events within the hierarchical stacking pattern (Fig. 11). Although coals occur within all nine progradational events of the Ferron, they are thinnest, most discontinuous, and often poorest in quality within the seaward- and landward-stepping events. Better quality, thicker, and more extensive coals occur within the six vertically stacked events. Of these, the greatest volume of coal is contained within the upper two events, which represent the end of the vertically stacked phase. These two events occurred when the accommodation potential in the coastal plain was at a maximum, but prior to the maximum rate of long-term sea-level rise that initiated the subsequent landward-stepping progradational events and drowned the coastal plain in this area.

Other examples of major coals within vertically stacked progradational events occur in the southern Rocky Mountains region. Fassett and Hinds (1971) discussed the occurrence of coals within the Fruitland Formation and Pictured Cliffs Sandstone in the San Juan basin. They noted that, although coals occur within most seaward-stepping progradational events, the thickest and most widespread coals are developed within vertically stacked progradational events of the transgressive maximum. These coals are as much as 9 m (30 ft) thick and are elongate parallel to the coast. Beaumont and others (1971) described three major reversals in direction of strandline movements in the Mesaverde Group of the Gallup coalfield. Again, the best coals occur within the vertically stacked progradational events, even though lesser coals occur within most progradational events. These coals accumulated in swamps in the immediate vicinity of, and landward from, the maximum landward position of the marginal marine deposits.

Major coals of the central Rocky Mountains region provide additional examples of coal occurrence with respect to their position in the stratigraphic architecture. Levey (1985) related coals to active progradation of deltas within the Rock Springs Formation in the Green River basin, Wyoming. Inspection of his stratigraphic cross sections (Fig. 12) shows that the thickest and most extensive coals occur upon the tops of progradational events in the lower three-quarters of

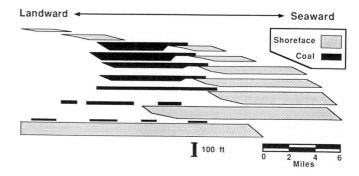

FIG. 11.—Diagrammatic cross section showing positions of coals in the Ferron Sandstone. Adapted from Ryer (1981).

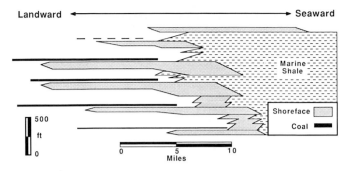

Landward ←→ Seaward

FIG. 12.—Diagrammatic cross section showing positions of coals above delta front sandstones in the Rock Springs Formation. Adapted from Levey (1985).

the formation. Although these events are essentially vertically stacked, they tend to step slightly landward through time. This geometry indicates long-term increase in accommodation potential within the coastal plain. Thinner progradational units and thinner coals occur at the top of the Rock Springs Formation, indicating the initiation of seaward-stepping progradational events and the attendant progressive reduction in accommodation potential within the coastal plain.

Lawrence (1982) described the occurrence of coals within the Adaville Formation in Wyoming. These are developed within a succession of progradational events that begin as slightly seaward-stepping and end with pronounced seaward-stepping patterns (Fig. 13). There is a concomitant decrease in thickness of successive progradational units and the associated coals within each event. This geometry is interpreted as the result of gradually decreasing accommodation potential within the coastal plain that occurs between the maximum sea level and the inflection point of the long-term sea-level fall.

DISCUSSION AND SUMMARY

Studies of the past few decades have demonstrated that the thickest and most extensive coals in Upper Cretaceous strata of the western interior generally are concentrated within vertically stacked progradational events. The numerical models described in this paper provide an explanation for this observation. The models predict that the volumes of sediment deposited during progradational events are partitioned systematically into different facies tracts according to the position of an event within the hierarchical stacking

pattern of a third-order cycle. This partitioning is caused by geographic variations in accommodation potential, or the cumulative potential space in which sediments may accumulate. In shallow-marine through coastal-plain environments, the interactions of eustatic change and tectonic subsidence cause the accommodation potential in the coastal plain to increase and decrease proportionally through time. A gradient perpendicular to the strand is established such that, when the rate of long-term sea-level rise is at a maximum, an increased accommodation potential in the coastal plain allows proportionally more sediment to accumulate there. When the rate of long-term sea-level fall is at a maximum, the accommodation potential in the coastal plain is proportionally decreased and the gradient is oriented seaward. This limits the amount of sediment that may accumulate in the coastal plain and causes proportionally more sediment to accumulate in the shoreface. Major coals occur within parts of third-order cycles which represent a balance between the maximum potential for vertical aggradation of sediment and availability of sediment. That potential is realized when high accommodation potential in the coastal plain is balanced by the rate of sediment flux.

ACKNOWLEDGMENTS

At various stages, Marshall Martin and Mark Baker have been instrumental in software and algorithm development of the numerical simulations. The manuscript was reviewed by Christopher Kendall, Charles Ross, John Van Wagoner, and Cheryl Wilgus. I thank them for their time, attention, and care in the suggestions they made, which helped improve the paper. Mac Jervey, Henry Posamentier, and John Van Wagoner kindly supplied preprints of their articles which appear in this volume; these helped immeasurably in revising the original manuscript. The importance of considering sediment accumulation and stratal architecture from the perspectives of rates of processes and receptor value, or accommodation potential, was first impressed on me by the seminal work of L. L. Sloss (1962) and the companion paper by P. Allen (1964). Subsequent discussions with Mac Jervey solidifed those impressions.

REFERENCES

ALLEN, P., 1964, Sedimentologic models: Journal of Sedimentary Petrology, v. 34, p. 289–293.
ANDERSON, E. J. AND GOODWIN, P. W., 1980, Application of the PAC hypothesis to limestones of the Helderberg Group: Society of Economic Paleontologists and Mineralogists, Eastern Section, Field Trip Guidebook, 32 p.
BEAUMONT, E. C., SHOMAKER, J. W., AND KOTTLOWSKI, F. E., 1971, Stratidynamics of coal deposition in southern Rocky Mountain region, U.S.A., in Shomaker, J. W., Beaumont, E. C., and Kottlowski, F. E., eds., Strippable Low-Sulfur Coal Resources of the San Juan Basin in New Mexico and Colorado: New Mexico Bureau of Mines and Mineral Resources Memoir 25, p. 175–185.
BUSCH, D. A., 1959, Prospecting for stratigraphic traps: American Association of Petroleum Geologists Bulletin, v. 43, p. 2829–2843.
———, 1971, Genetic units in delta prospecting: American Association of Petroleum Geologists Bulletin, v. 55, p. 1137–1154.
———, 1974, Stratigraphic traps in sandstones—Exploration techniques: American Association of Petroleum Geologists Memoir 21, 174 p.
BUSCH, R. M., AND WEST, R. R., 1987, Hierarchal genetic stratigraphy:

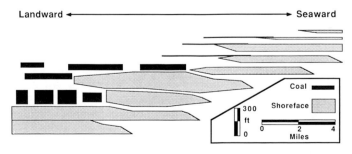

Landward ←→ Seaward

FIG. 13.—Diagrammatic cross section showing positions of coals in the Adaville Formation. Adapted from Lawrence (1982).

A framework for paleoceanography: Paleoceanography, v. 2, p. 141–164.

Cross, T. A., 1986, Tectonic controls of foreland basin subsidence and Laramide style deformation, western United States, *in* Allen, P. A., and Homewood, P., eds., Foreland Basins: International Association of Sedimentologists Special Publication 8: Blackwell Scientific Publications, Oxford, p. 15–39.

Fassett, J. E., and Hinds, J. S., 1971, Geology and fuel resources of the Fruitland Formation and Kirtland Shale of the San Juan Basin, New Mexico and Colorado: U.S. Geological Survey Professional Paper 676, 76 p.

Frazier, D. E., 1974, Depositional-episodes: Their relationship to the Quaternary stratigraphic framework in the northwestern portion of the Gulf Basin: University of Texas at Austin, Bureau of Economic Geology, Geological Circular 74-1, 28 p.

Goodwin, P. W., and Anderson, E. J., 1985, Punctuated aggradational cycles: A general hypothesis of episodic stratigraphic accumulation: Journal of Geology, v. 93, p. 515–533.

Hancock, J. M., and Kauffman, E. G., 1979, The great transgressions of the Late Cretaceous: Journal of the Geological Society of London, v. 136, p. 175–186.

Jordan, T. E., 1981, Thrust loads and foreland basin evolution, Cretaceous, western United States: American Association of Petroleum Geologists Bulletin, v. 65, p. 2506–2520.

Kauffman, E. G., 1980, Major factors influencing the distribution of Cretaceous coal in the Western Interior United States, *in* Carter, L. M., ed., Fourth Symposium on the Geology of Rocky Mountain Coal Proceedings: Colorado Geological Survey, Resource Series 10, p. 1–3.

Lawrence, D. T., 1982, Influence of transgressive-regressive pulses on coal-bearing strata of the Upper Cretaceous Adaville Formation, southwestern Wyoming, *in* Gurgel, K. D., ed., Fifth Symposium on the Geology of Rocky Mountain Coal Proceedings: Utah Geological and Mineral Survey Bulletin 118, p. 32–49.

Levey, R. A., 1985, Depositional model for understanding geometry of Cretaceous coal: Major coal seams, Rock Springs Formation, Green River basin, Wyoming: American Association of Petroleum Geologists Bulletin, v. 69, p. 1359–1380.

McGhee, G. R., and Bayer, U., 1985, The local signature of sea-level changes, *in* Bayer, U., and Seilacher, A., eds., Sedimentary and Evolutionary Cycles: Springer-Verlag, Berlin, p. 98–112.

McGookey, D. P., 1972, Cretaceous System, *in* Mallory, W. W., ed., Geologic Atlas of the Rocky Mountain Region: Rocky Mountain Association of Geologists, p. 190–228.

Morrow, D. W., 1986, The sea-level rise staircase on continental margins and the origin of upward-shoaling carbonate sequences: Bulletin of Canadian Petroleum Geology, v. 34, p. 284–285.

Pitman, W. C., III, 1978, Relationship between eustacy and stratigraphic sequences of passive margins: Geological Society of America Bulletin, v. 89, p. 1389–1403.

Ramsbottom, W. H. C., 1979, Rates of transgression and regression in the Carboniferous of NW Europe: Quarterly Journal of the Geological Society of London, v. 136, p. 147–153.

Ryer, T. A., 1981, Deltaic coals of Ferron Sandstone Member of Mancos Shale: Predictive model for Cretaceous coal-bearing strata of Western Interior: American Association of Petroleum Geologists Bulletin, v. 65, p. 2323–2340.

————, 1984, Transgressive-regressive cycles and the occurrence of coal in some Upper Cretaceous strata of Utah, U.S.A., *in* Rahmani, R. A., and Flores, R. M., eds., Sedimentology of Coal and Coal-Bearing Sequences: International Association of Sedimentologists Special Publication 7: Blackwell Scientific Publications, Oxford, p. 217–227.

Sears, J. D., Hunt, C. B., and Hendricks, T. A., 1941, Transgressive and regressive Cretaceous deposits in southern San Juan basin, New Mexico: U.S. Geological Survey Professional Paper 193-F, p. 110–121.

Sloss, L. L., 1962, Stratigraphic models in exploration: American Association of Petroleum Geologists Bulletin, v. 46, p. 1050–1057.

Swift, D. J. P., Thorne, J. A., and Nummedal, D., 1985, Sequence stratigraphy and petroleum exploration in a foreland basin: Inferences from the Cretaceous Western Interior: 17th Annual Offshore Technology Conference, Proceedings, v. 1, p. 47–54.

Vail, P. R., Mitchum, R. M., and Thompson, S., III, 1977, Seismic stratigraphy and global changes of sea level, part 4: Global cycles of relative changes of sea level, *in* Payton, C. E., ed., Seismic Stratigraphy—Application to Hydrocarbon Exploration: American Association of Petroleum Geologists Memoir 26, p. 83–98.

Van Wagoner, J. C., 1985, Reservoir facies distribution as controlled by sea-level change (abst.): Society of Economic Paleontologists and Mineralogists Mid-Year Meeting, Golden, Colorado, p. 91–92.

Weimer, R. J., 1960, Upper Cretaceous stratigraphy, Rocky Mountain area: American Association of Petroleum Geologists Bulletin, v. 44, p. 1–20.

————, 1984, Relation of unconformities, tectonics, and sea-level changes, Cretaceous of Western Interior, U.S.A., *in* Schlee, J. S., ed., Interregional Unconformities and Hydrocarbon Accumulation: American Association of Petroleum Geologists Memoir 36, p. 7–35.

SEA-LEVEL CHANGES AND TIMING OF TURBIDITY-CURRENT EVENTS IN DEEP-SEA FAN SYSTEMS

V. KOLLA

Elf Aquitaine Petroleum, 1000 Louisiana Avenue, Suite 3800, Houston, Texas 77002;

AND

D. B. MACURDA JR.

The Energists, 10260 Westheimer, Suite 3001, Houston, Texas 77042

ABSTRACT: Although lowstands of sea level greatly favor the development of deep-sea fan systems, the timing and type of turbidite events in these systems may depend not only on the sea-level changes but also on the nature of available sediments, tectonic setting, size, and gradients of the basins. Thus, in basins with steep gradients, located on continental or transitional crust or along active margins close to sedimentary sources that could supply coarse-grained sediments, unchannelled turbidite sand lobes detached from updip channels or valleys might have been deposited during lowstands of sea level. Channel-attached sand lobes and channel levee complexes might have been deposited during sea-level rises in these basins. In contrast, large fan systems, such as the Indus Fan, located off passive margins in oceanic basins with flat gradients, distant sedimentary sources, and a predominantly fine-grained sediment supply have channel-attached lobes and channel levee, overbank complexes that were probably deposited during lowstands and to some extent during highstands of sea level.

INTRODUCTION

It is well known that sea-level changes greatly affected the development of deep-water turbidite and fan systems in the geologic record. We believe, however, that two other factors—type and amount of sediment supplied, and tectonic and geologic setting of the region—also controlled the development of the types of fan systems (Stow and others, 1983–1984). The tectonic and geologic variables include the following: type of continental margin; relief and tectonic setting of hinterland, continental shelf-slope, and basinal areas; distance between source region and shelf-slope areas; width and gradients of shelf-slope areas; and basin size, gradients, and morphology. Secondary factors in the development of turbidite systems include the climatic nature and vegetation of the source region and timing of snowmelt waters or high rainfall during sea-level changes.

We review here the recent turbidite or fan models of Mitchum (1984, 1985), Mutti (1985), and Posamentier and Vail (this volume), and then discuss briefly the internal structure and models of sedimentation of the Indus fan in the Arabian Sea, northwest Indian Ocean, and of a Paleocene-Eocene deep-sea fan in the Porcupine Basin (Seabight), northeast Atlantic. From a comparison of these fan models, we show that the timing and type of turbidite events may be different in different turbidite (fan) systems and that several factors (not just sea level) might have caused these differences.

FAN MODELS OF MUTTI AND MITCHUM

Mutti (1985) synthesized studies of outcrops of the Hecho Group, Spain, and certain sedimentary basins in Italy and distinguished three types of turbidite systems: type I, type II, and type III (Fig. 1). He emphasized sea-level changes in the development of these turbidite systems. In type I systems, during lowstands, large-volume turbidity currents transport the sediments to the lower fan, largely bypassing the upper fan, and deposit them as outer-fan, unchanneled, sand lobes (Fig. 1). The type I system corresponds to the high-efficiency fan type of Mutti (1979). As sea level rises, turbidity-current volumes decrease. If other variables re-

main constant, this decrease results in the deposition of channelled sand bodies in the upper fan and attached lobes in the outer (lower) fan (Fig. 1). These channel-attached lobes (type II systems) correspond to the poorly efficient fans of Mutti (1979). During highstands, small turbidite-current volumes produce channel levee complexes (type III systems) primarily in the upper fan (Fig. 1). Mutti (1985) suggests that the type III system is the ancient equivalent of the modern fan channel levee complexes. Mutti's channel levee complexes are generally small, however, mainly confined to the upper fan. According to Mutti, the slope and upper-fan regions are characterized by large erosional valleys (channels) with coarse-lag deposits formed during the lowstand, corresponding to the type I system. As sea level rises, the erosional valleys are infilled by slump and debris flow deposits; more important, by lenticular sand bodies laid down in small channels without levees, corresponding to the type II system; and finally by small, leveed-channel deposits corresponding to the type III system. In the lower fan, deposits of type II and type III systems show downlap/onlap relation to the type I deposits.

Mitchum (1984, 1985) produced seismic evidence to show that the lower-fan turbidite lobes of certain subsurface fan systems were deposited during lowstands[1] (Fig. 2). Mitchum's model, similar to that of Mutti (1985), implies that during lowstand time, the upper-fan valley is a region of erosion, as sediments largely bypass the region. The seismic-reflection configuration of the lower-fan lobes shows bidirectional downlap with respect to the pre-fan surface. These lobes correspond to the lowstand fan of Posamentier and Vail (this volume) and to the type I turbidite system of Mutti (1985). Furthermore, Mitchum (1984, 1985) documented that, during the initial rise of sea level,[2] channel levee complexes were deposited in the upper fan with on-

---

[1] High rates of eustatic sea-level fall result in maximum sea-level drop for a given basin subsidence (Posamentier and Vail, this volume).

[2] Low rates of eustatic fall or the initial stages of eustatic rise result in a relative rise of sea level for a given basin subsidence (Posamentier and Vail, this volume).

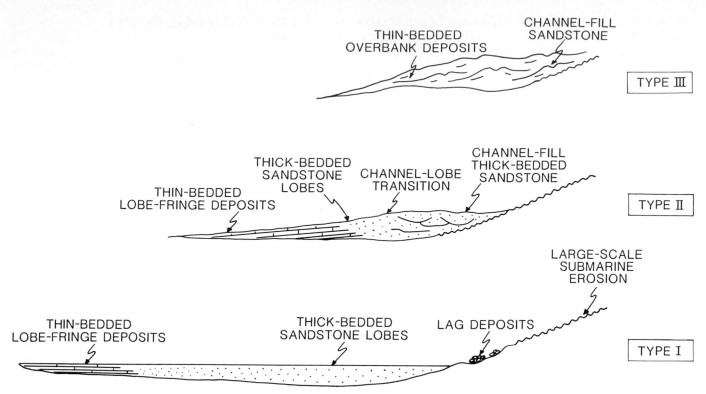

FIG. 1.—Types of turbidite depositional systems. These systems differ from each other mainly in terms of where sand is deposited (from Mutti, 1985).

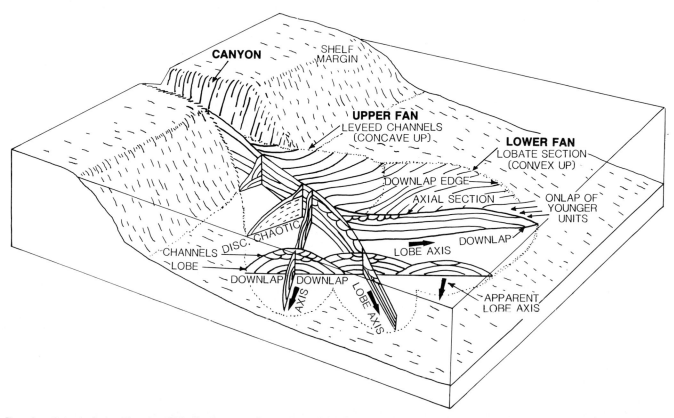

FIG. 2.—Seismic facies diagram of idealized canyon fan system (after Mitchum, 1984, 1985). Lobes deposited in lower fan during lowstands of sea level are mounded (convex-up), with bidirectional downlapping reflection configurations. Overlying the lobes are channels extending updip into leveed channels in the upper fan. Channel levees are characterized by concave-up reflection configurations. Channel levee complexes were deposited mainly during rises or highstands of sea level.

lap/downlap relation to the lower-fan lobes that were deposited during the preceding lowstand. These channel levee complexes may extend to the lower fan and overlie the lower-fan lobes (Fig. 2). The channel levee complexes exhibit concave-up seismic-reflection configurations. According to Posamentier and Vail (this volume), channel levee complexes are part of lowstand wedge deposits, corresponding to type II and III systems of Mutti (1985). As sea level rises further, the canyon that fed the sediments to the fan and that was a zone of erosion during the lowstand is successively back-filled by fine-grained sediments. The fan systems reported by Mitchum (1984) are similar to those of Mutti (1985) in size, or smaller, and the models applicable to these systems are similar, if not identical. It should be noted that the word "fan" of the Indus Fan and other deep-sea fans, as generally used in geologic literature, has a morphologic connotation and includes both the lowstand fan and lowstand wedge deposits of Posamentier and Vail (this volume). Thus, the lowstand fan, according to concepts of Posamentier and Vail, refers only to one stage of the development of deep-sea fans.

The decrease in turbidity-current activity and sediment grain size as sea level rises from lowstands to highstands is generally true for at least the Pleistocene; however, sea level is one of several factors affecting turbidity-current sedimentation (Stow and others, 1983–1984). Precise three-dimensional mapping of turbidite systems in basins of different tectonic settings in relation to coastal onlaps (Posamentier and Vail, this volume), supported by high-resolution biostratigraphic studies, should help in evaluating the relative importance of the different variables. Because of scaling problems and inherent limitations and differences in techniques of studies of modern, subsurface and outcrop fans, characteristics of different fans cannot be strictly compared (Normark and otehrs, 1983–1984). It is instructive, however, to point out that several factors (not just sea level) may affect the timing and size of the turbidity-current events discussed by Mitchum (1985), Mutti (1985), and Posamentier and Vail (this volume).

INDUS FAN

The Indus Fan in the northwest Indian Ocean is bordered on the north and east by the Pakistan-India margin and the Chagos-Laccadive Ridge, on the west by the Owen-Murray ridges, and on the south by the Carlsberg Ridge, a portion of the Mid-Indian Ocean Ridge (Fig. 3). Water depths of the Indus Fan range from about 1,400 m to 4,600 m.

The Indus River system has contributed sediments to form the Indus fan since the Oligocene and Miocene. Based on fairly extensive seismic-data coverage, Kolla and Coumes (1987), Coumes and Kolla (1984), and McHargue and Webb (1986) showed that several canyon complexes existed in the past on the Indus shelf and fed channels on the upper fan that, in turn, transported sediments to the lower fan. One such canyon model is shown in Figure 4. The average width and depth (relief) of the most recent canyon on the Indus shelf-slope are 8 km and 800 m, respectively (Kolla and Coumes, 1987). Similar dimensions have been observed for other canyons on the Indus shelf. Seismic data indicate that

the Indus Canyon sedimentary fill consists of transparent, low-amplitude, discontinuous reflection zones with occasional continuous reflections that are bounded by termination of continuous high-amplitude reflections of the canyon walls. Levees are absent on either side of the canyon walls (Fig. 5).

Canyons may form due to a variety of processes (Shepard, 1981). The depths of the Indus Canyon are too great to be solely attributed to subaerial river cutting during drops in sea level. It is possible that turbidity currents generated at the river mouth during low sea levels could have initiated canyon cutting (Shepard, 1981), but we do not believe that this was the main cause for the formation of the Indus Canyon. We suspect that the canyon originated mainly by mass slumping on the continental slope at a location corresponding to a huge depocenter near the river mouth during a Pleistocene drop in sea level. It grew by subsequent retrograde slumping during the following rise in sea level, as proposed by Coleman and others (1983) for the origin of the Mississippi Canyon. Once the canyon was formed, back cutting during rises in sea level may have been guided by the subaerially cut river valley. Canyon relief may have been subsequently enhanced by sedimentation on the shelf and by processes within the canyon.

Large turbidity currents probably evolved from slumping that initiated the canyon cutting during the lowstand and from other processes both within and at the head of the canyon. Huge loads of sediments released by canyon cutting, as well as sediments delivered at the head of the canyon by the river during the lowstands, were transported by turbidity currents from the shelf-slope region to be deposited in the fan. The canyon in the shelf-slope region was primarily a zone of sediment bypass or degradation (Fig. 5; McHargue and Webb, 1986). During the sea-level rise and highstand, canyon fill was the product of normal shelf processes. The canyon fill consists primarily of mudstone facies with a low-amplitude, discontinuous, transparent seismic-reflection character (facies).

From this description, it is apparent that the Indus Canyon sedimentary characteristics are similar to the canyon (valley) characteristics of slope and upper-fan regions of Mutti (1985) and Mitchum (1984, 1985), which they attributed to changes in sea level. The Indus Canyon sedimentary fill, however, is generally devoid of small leveed channels.

Multichannel and sparker data show numerous large channel levee complexes deposited during different episodes in the upper Indus fan (Fig. 6). Levees may attain 200 to 300 m of relief above the surrounding surface and appear to have developed at the same time that the channels were initiated during lowstands. No extensive erosional zones (Mutti, 1985) or seismically chaotic deposits (Feeley and others, 1985) underlie any of these channel levee complexes. The channel fill consists of basal high-amplitude reflections overlain by low-amplitude, discontinuous, transparent seismic facies (Figs. 5, 6). The channel facies are flanked by wedge-shaped packages with low-amplitude continuous reflections and reflection-free zones adjacent to channels that change distally to moderately high-amplitude continuous reflections characteristic of levee-overbank de-

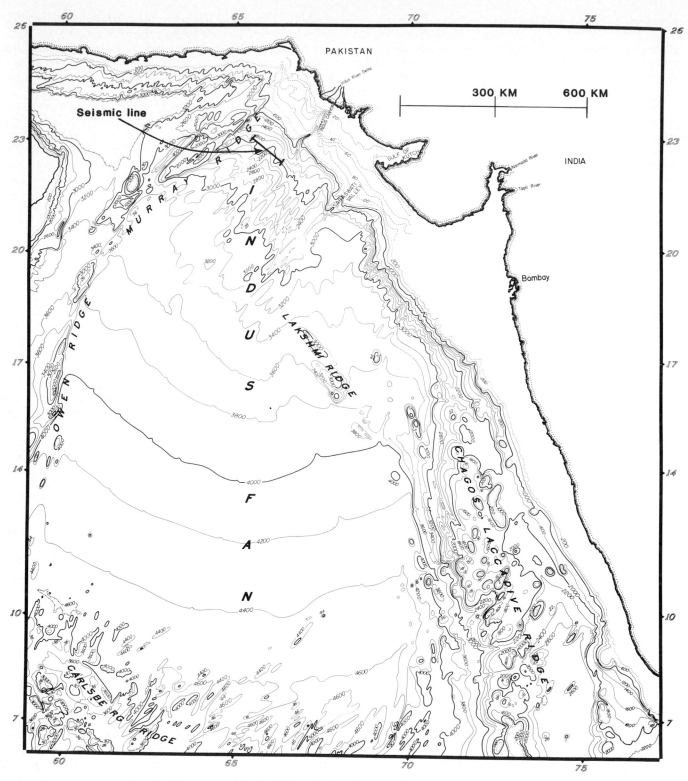

FIG. 3.—Bathymetry of the Indus Fan showing location of multichannel seismic line (Fig. 6).

posits (Figs. 5, 6). After the channel is abandoned, it may be filled by overbank deposits from neighboring channels or by hemipelagic deposits, which produce continuous reflections. Occasionally, very small leveed channels are observed at the sediment/water interface above the large leveed-channel complexes.

Channelled turbidity currents and overbank spilling were the most important mechanisms of sedimentation in the up-

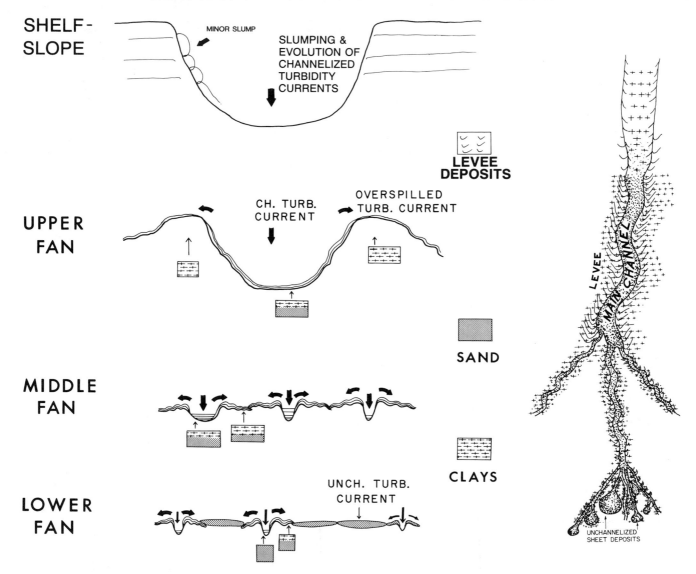

FIG. 4.—A model of Indus Fan sedimentation. Cross sections show sedimentary processes; plan view depicts fan morphology of different regions of the Indus Fan (modified after Kolla and Coumes, 1987).

per Indus Fan (Fig. 4) and resulted in predominantly fine-grained sediments in the upper fan, except in channels where coarse-grained deposits are present (Kolla and Coumes, 1983–1984, 1987), Since the depths and levee heights for the upper-fan channel systems exceed several hundred meters, the turbidity channels in the upper fan must also have been several hundreds of meters thick. As sediments were transported downdip within the confines of levees, channel floor and levee deposits continued to build to increasing heights. Thus, the upper-fan region was a zone of aggradation (Fig. 5; McHargue and Webb, 1986).

The seismic character of the Indus Fan channels indicates deposition from strong to waning turbidity currents, from overbank spilling from nearby channels, and from processes related to changes in sea level. This seismic character, however, does not indicate depositional stages of the type solely attributed to sea-level stands, as postulated by Mutti (1985) for the upper-fan channels (valleys). The ab-

sence of extensive erosional zones below each of the channel levee complexes (Figs. 5, 6) and the initiation of channels and levees at the same time in the upper Indus Fan (Fig. 6) do not conform to Mutti's and Mitchum's models.

The region between the aggradational zone of the upper fan and the degradational zone of the shelf slope is transitional and has leveed channels with erosional bases (Fig. 5). The levees here are lower than those in the upper fan, and appear to have been initiated slightly later than the channels in the transitional zone. The channel fill in the transitional zone consists of abundant low-amplitude, discontinuous seismic facies with common high-amplitude discontinuous reflections (Fig. 5). The high-amplitude facies is less prevalent in the transitional zone than in the aggradational upper fan. The later appearance of levees in the transitional zone may be related to the initial rise of sea level after channel initiation during the lowstand, as postulated by Mutti (1985) and Mitchum (1984, 1985). The

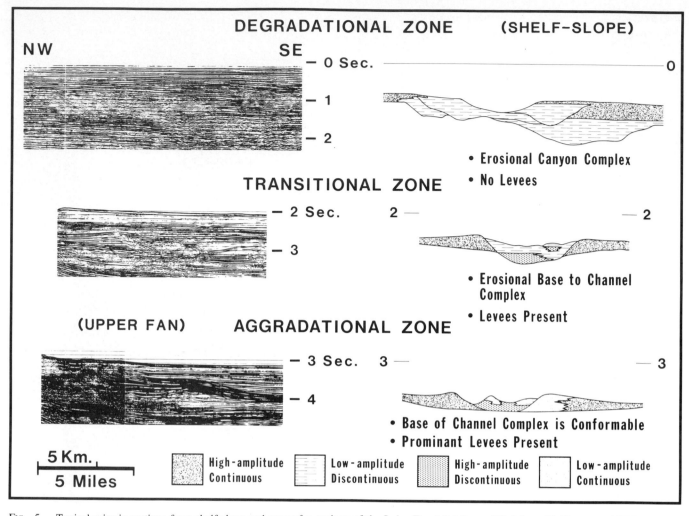

Fig. 5.—Typical seismic sections from shelf slope and upper-fan regions of the Indus Fan (slightly modified from McHargue and Webb, 1986).

slightly later levee formation, however, may also be due to aggradation in the upper fan and subsequent decrease in gradients of the updip sea floor in the transitional zone (McHargue and Webb, 1986). We believe that the upper Indus Fan channels originate in the transitional zone.

By analogy with the Amazon Fan (Damuth and others, 1983), we suggest that channels might have meandered more in the middle fan, where gradients are flatter, than in the upper fan (Fig. 4). Channel abandonment and avulsion may also have been very common in the middle fan. As channels continue downdip, the levees become smaller, corresponding to decreasing availability of fine-grained sediments. There may be less channel meandering but more channel branching in the lower fan as a result of decreasing availability of fine-grained sediments. Some channels continue all the way to the edge of the lower fan, while others terminate at different points along the fan (Fig. 4; Kolla and Coumes, 1987). Channelled turbidity currents and overbank spilling would continue to play significant roles in the channels that endure. Unchannelled turbidity-sheet flow deposition would be the dominant mechanism in front of the terminated channels. Unchannelled deposits are sandier than the levee deposits in the lower fan, however, and pro-

duce a highly opaque echo character on high-resolution seismic records. The unchannelled deposits appear to overlie or underlie the more stratified levee deposits of the adjacent channel. Data from cores suggest that the most recent turbidites in both the lower- and upper-fan regions were deposited during the last glacial lowstand in the Pleistocene (Kolla and Coumes, 1987). The intimate association of large channel levee complexes and inferred unchannelled deposits, apparently attached to channels, correspond to the type III and II systems of Mutti (1985). Mutti's channel levee complexes, however, are small and are primarily confined to the upper fan.

FACTORS CAUSING DIFFERENCES IN FAN MODELS

The depositional stages from lowstands to highstands outlined for the slope and upper-fan regions of Mutti (1985) and Mitchum (1984) are present primarily on the Indus continental slope and to some extent in the transitional region between the upper Indus Fan and the continental slope. Such depositional stages are absent in the upper Indus Fan. Although we cannot relate the turbidite events in the Indus Fan to coastal-onlap curves at this time, it appears from the discussion in the preceding section that the Indus Fan consists of lowstand wedge deposits (Posamentier and Vail, this volume) or type II and III turbidite systems (Mutti, 1985). Type I turbidite system deposits may not be significant in the Indus Fan. The occurrence of type II and III turbidite systems in relation to sea levels is not unique to the Indus Fan but may be applicable to other large fans, e.g., the Bengal and Amazon fans. We now speculate on some possible explanations for the differences between the Indus fan and other models discussed earlier.

The Indus Fan is located mainly in a large oceanic basin off a mature, passive Indian continental margin. The uplifted source region, the Himalayas, which contributes sediments to the Indus Fan through the Indus River system, is far (1,200 km) from the basin (Fig. 7). The gradients of the upper fan, at least during Pleistocene sedimentation, were very low. The turbidite systems reported by Mutti (1985) were deposited in relatively small, elongated, confined troughs or depressions on active margins, with the uplifted source region close to the basins (Fig. 7). The basinal gradients in the upper fan were probably steeper than those of the Indus Fan. The sediments of the Indus Fan were muddier, with less varied grain sizes than those of Mutti's turbidite systems (Nilsen, 1984; Pickering and others, 1986). The subsurface fans reported by Mitchum (1984, 1985) were deposited either in intraslope basins or in small basins at the base of slopes with steep gradients on continental or transitional crust; they probably contain coarser sediments than the large modern fans. We suspect from the steep gradients, coarse but varied sediment grain sizes, and proximity of sediment sources that the depositional sequences in the basins studied by Mutti (1985) and Mitchum (1984) were very sensitive to even small changes in sea level. The upper-fan regions in these basins had better chances to develop extensive erosional zones with no levee formation, while the bulk of the sediment was transported and deposited in the lower fan during low-stands. In terms of extensive erosional areas with no leveed channels, the upper-fan regions of these basins are similar to the continental slope region updip of the upper Indus Fan. During the initial rise in sea level and subsequent highstands, with decrease in volume of turbidity flows and sediment grain size, levee deposition occurred in the turbidite systems studied by Mitchum (1984, 1985) and Mutti (1985). On the other hand, channel levee complexes would more readily develop in the Indus Fan during the lowstand because of the overall dominance of high-mud content and flat gradients. Thus, our seismic data suggest that the channel-attached lobes (type II system) and channel levee complexes (type III system) were deposited in the Indus Fan during the lowstand. Once initiated, the channel levee complexes might have continued to be active during times of rising sea level or highstands, especially on the upper Indus Fan; it is possible that a portion of the upper fan may thus be younger than the lower fan. Nevertheless, it is difficult to see in the Indus Fan, which has been influenced by multiple episodes of sedimentation, the kind of relation between the upper- and lower-fan deposits documented by Mitchum (1984, 1985; Kolla and Coumes, 1987). We believe, however, that turbidity-current transport for hundreds of kilometers to the lower-fan regions (with flat gradients) in large oceanic basins such as the In-

dus Fan could occur only through leveed channels and that these channel levee complexes in the Indus Fan were mainly initiated during sea-level lowstands.

Some other factors that may influence the timing and type of turbidite events, such as timing of major snowmelt and high rainfall during changes in sea level, types of clay mineralogy in the sediments, and local tectonism (uplift or subsidence) may also influence fan sedimentation.

A DEEP-SEA FAN IN THE PORCUPINE BASIN, NORTHEAST ATLANTIC

Porcupine Basin (Seabright), located southwest of Ireland in the Northeast Atlantic, is bordered by the Porcupine Bank to the north and west and by the Goban Spur to the south (Fig. 8). Several Paleocene-Eocene deep-sea fan complexes can be distinguished in the Porcupine Basin from the interpretation of multichannel seismic data. One of these fan complexes, located in the northwestern part of the basin (Fig. 8), is elongate in outline (32–39 km long and 26 km wide) and trends northwest-southeast. The anatomy of this Porcupine fan is shown on a strike multichannel seismic line in Figure 9.

From bottom to top, four sequences (I, II, III, and IV) can be distinguished in the Porcupine fan (Fig. 9). Sequence I is the lowest unit of the fan and can be traced completely across the Porcupine Basin. Its external geometry is essentially a sheet grading into a broad mound shape toward the top, and its reflection character is uniform. These characteristics probably indicate sheet flow deposition in the distal fan. Sequence II is also a sheet, more local in distribution and more mounded than sequence I. Locally, it has shingled clinoforms and also indicates deposition largely from unchannelled sheet flows. Presence of channels is inferred in the sequences on the seismic line from hummocky and mounded topography. Sequence III is the first unit that shows a channelled character (e.g., at A in sequence III). Sequence IV is highly channelled; different channel configurations, active during different periods, can be identified within this sequence. For example, channels at B were active earlier than those at C. The eastern part of sequence IV shows complex channel phenomena. The channels of sequences III and IV can be extended to both the updip and downdip areas of the fan complex. The channels become fewer, larger, and leveed in the updip areas and smaller in the downdip areas. Both sequences III and IV are individualized and mounded; their boundaries cannot be traced from one fan complex to another in the basin, unlike the boundaries of sequences I and II. Sequences III and IV appear to have resulted primarily from channelled flows. The combined thickness of sequences III and IV is significantly more than that of sequences I and II. The deposits on the top of sequence IV probably represent a passive-fill facies that progressively blanketed the mounded fan.

The occurrence of channelled sequences (III and IV) overlying unchannelled sequences (I and II) may reflect increased turbidity-current activity or progradation of more proximal fan facies onto more distal fan deposits in the course of overall fan development. It may reflect the timing of sea-level fall and rise during a single eustatic cycle (Posamentier and Vail, this volume). At present, we are not able to relate the Porcupine fan sequences to coastal-onlap curves,

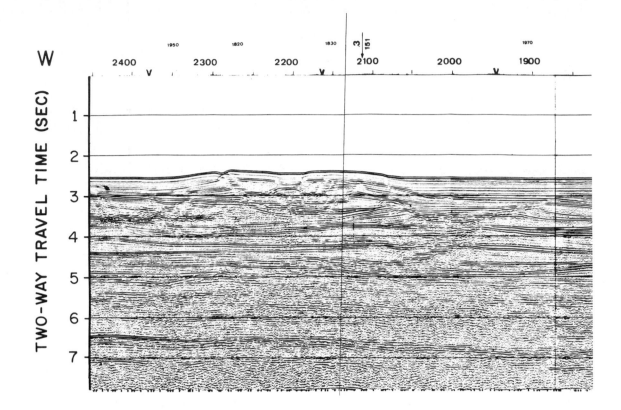

MCS

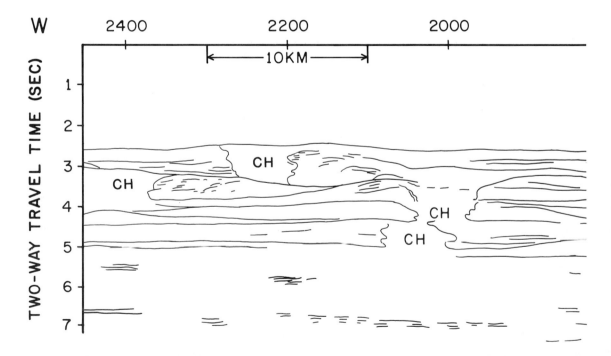

FIG. 6.—Multichannel seismic strike line showing several layers of vertically stacked channel levee complexes. CH =

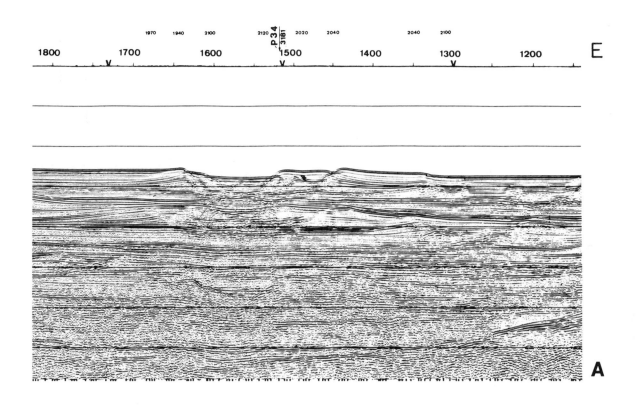

A

LINE 12

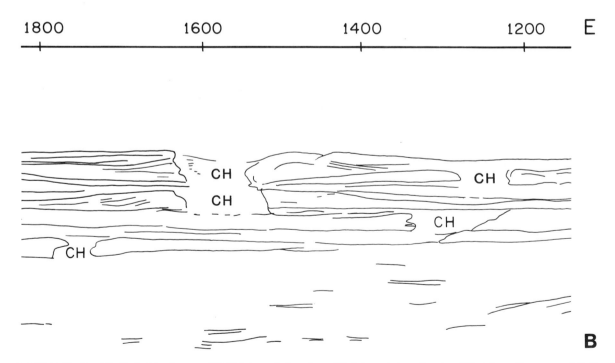

B

locations of channels flanked by levees. (A) Uninterpreted section. (B) Interpreted section (modified from Coumes and Kolla, 1984).

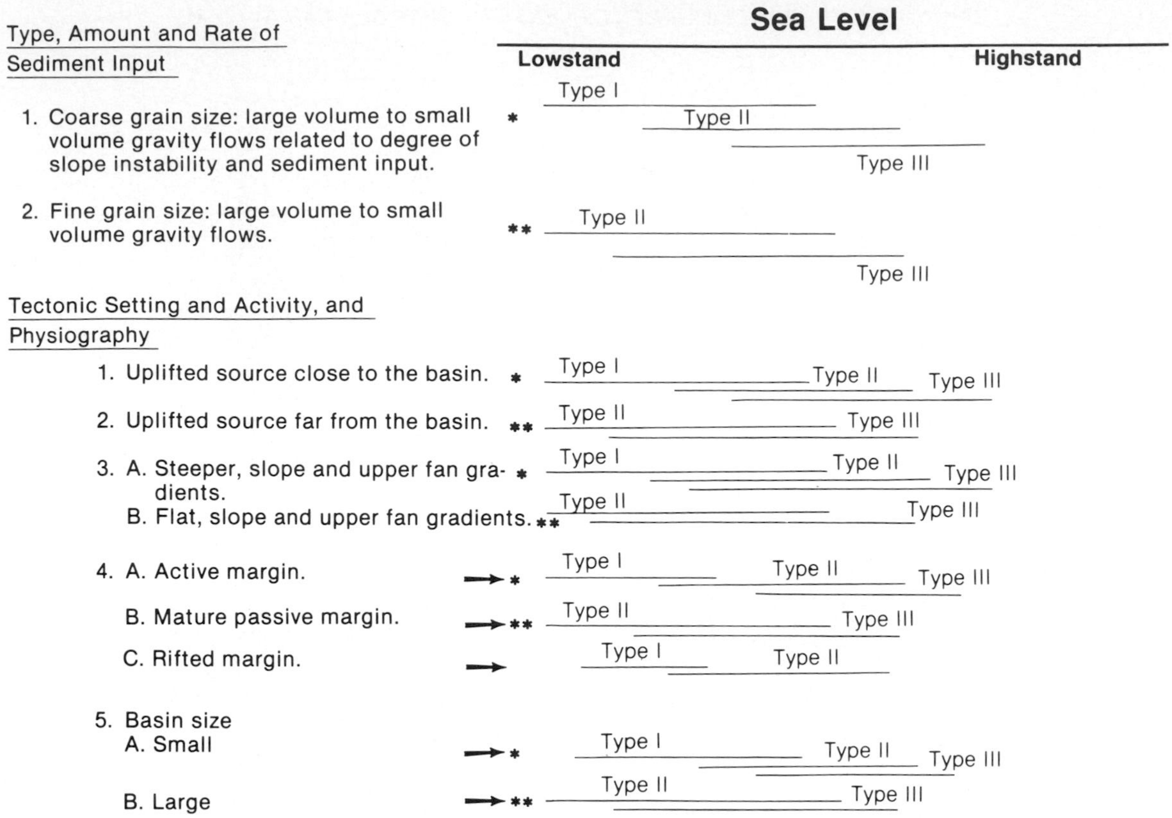

Sea Level

Lowstand   Highstand

Type, Amount and Rate of Sediment Input

1. Coarse grain size: large volume to small volume gravity flows related to degree of slope instability and sediment input.
   * Type I
   Type II
   Type III

2. Fine grain size: large volume to small volume gravity flows.
   ** Type II
   Type III

Tectonic Setting and Activity, and Physiography

1. Uplifted source close to the basin.   *   Type I   Type II   Type III

2. Uplifted source far from the basin.   **   Type II   Type III

3. A. Steeper, slope and upper fan gradients.   *   Type I   Type II   Type III
   B. Flat, slope and upper fan gradients.   **   Type II   Type III

4. A. Active margin.   → *   Type I   Type II   Type III
   B. Mature passive margin.   → **   Type II   Type III
   C. Rifted margin.   →   Type I   Type II

5. Basin size
   A. Small   → *   Type I   Type II   Type III
   B. Large   → **   Type II   Type III

FIG. 7.—Sea-level stands, type and amount of sediment input, tectonic settings, timing and type of turbidite events. See text for explanation of types of turbidite systems. *Systems with timings similar to those of Mutti (1985). **Systems with timings similar to those of the Indus Fan.

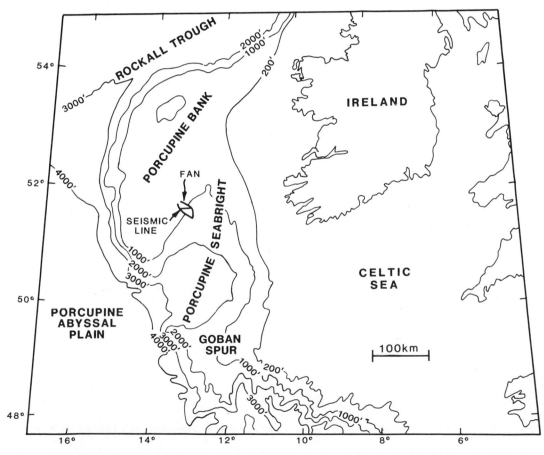

FIG. 8.—Porcupine Seabight (Basin) in North Atlantic showing locations of Porcupine fan and multichannel seismic line (Fig. 9).

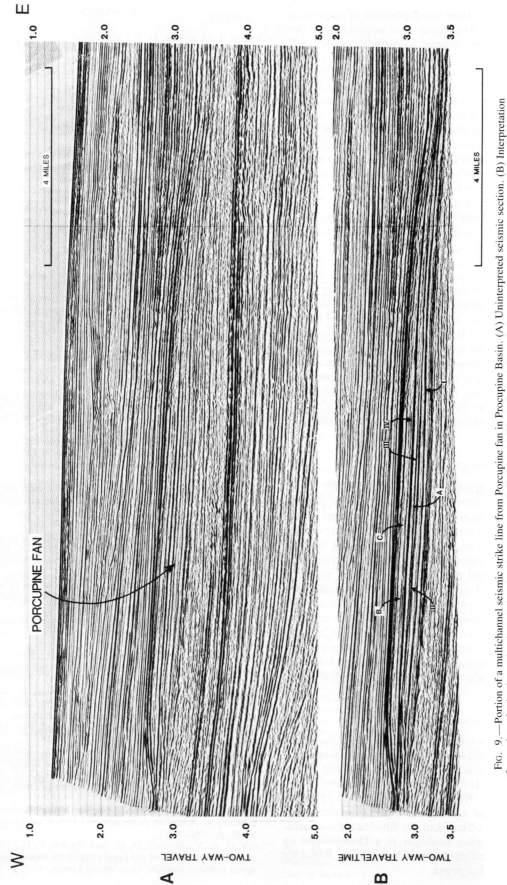

Fig. 9.—Portion of a multichannel seismic strike line from Porcupine fan in Porcupine Basin. (A) Uninterpreted seismic section. (B) Interpretation of a portion of seismic section in (A) showing fan sequences I, II, III, and IV. A, B, and C on (B) locate examples of channel configurations within sequences III and IV.